W9-AFD-402

Operator inequalities

This is Volume 147 in
MATHEMATICS IN SCIENCE AND ENGINEERING
A Series of Monographs and Textbooks
Edited by RICHARD BELLMAN, *University of Southern California*

The complete listing of books in this series is available from the Publisher upon request.

OPERATOR INEQUALITIES

JOHANN SCHRÖDER

Mathematical Institute
University of Cologne
Cologne, Federal Republic of Germany

ACADEMIC PRESS 1980

A Subsidiary of Harcourt Brace Jovanovich, Publishers

New York London Toronto Sydney San Francisco

ACADEMIC PRESS, INC.
111 Fifth Avenue, New York, New York 10003

United Kingdom Edition published by
ACADEMIC PRESS, INC. (LONDON) LTD.
24/28 Oval Road, London NW1 7DX

Library of Congress Cataloging in Publication Data

Schröder, Johann, Date.
Operator inequalities.

(Mathematics in science and engineering)
Includes bibliographical references and index.
1. Operator theory. 2. Inequalities (Mathematics)
I. Title. II. Series.
QA329.S37 515.7'24 79-28754
ISBN 0-12-629750-9

PRINTED IN THE UNITED STATES OF AMERICA

80 81 82 83 9 8 7 6 5 4 3 2 1

Contents

Preface ix
Acknowledgments xiii
Some remarks concerning notation xv

Chapter I Some results on functional analysis

1 Basic concepts of order theory 1
  1.1 Ordered spaces 2
  1.2 Ordered linear spaces 2
  1.3 Positive linear functionals and operators 9
  1.4 Convergence and continuity 11
2 Convex sets 14
  2.1 Basic concepts 14
  2.2 Description by functionals 17
  2.3 The special case of cones 19
3 Fixed-point theorems and other means of functional analysis 22
  3.1 Compact operators in a Banach space 22
  3.2 Iterative procedures in ordered spaces 24
  Notes 26

Chapter II Inverse-positive linear operators

1 An abstract theory 29
1.1 The concept of an inverse-positive operator, general assumptions 30
1.2 Basic results 31
1.3 A more detailed analysis 33
1.4 Connected sets of operators 34
1.5 Noninvertible operators 36
2 M-matrices 37
2.1 Definitions and notation 38
2.2 General results for square matrices 39
2.3 Off-diagonally negative matrices 42
2.4 Eigenvalue problems 47
2.5 Iterative procedures 54
3 Differential operators of the second order, pointwise inequalities 61
3.1 Assumptions and notation 62
3.2 Basic results on inverse-positivity 64
3.3 Construction of majorizing functions 68
3.4 Sets of inverse-positive regular differential operators 71
3.5 Inverse-positivity under constraints 76
3.6 Stronger results on positivity in the interior 78
3.7 Strictly inverse-positive operators 81
3.8 Equivalence statements 85
3.9 Oscillation and separation of zeros 86
3.10 Eigenvalue problems 88
3.11 Differential operators on noncompact intervals 90
4 Abstract theory (continuation) 94
4.1 The method of reduction 95
4.2 Z-operators and M-operators 104
4.3 Operators $\lambda I - T$ with positive T 112
4.4 Inverse-positivity for the case $K^c = \varnothing$ 118
5 Numerical applications 121
5.1 A convergence theory for difference methods 122
5.2 A posteriori error estimates 132
6 Further results 144
6.1 Differential operators of the fourth order 144
6.2 Duality 148
6.3 A matrix of Lyapunov 149
Notes 150

Chapter III Two-sided bounds for second order differential equations

1 The concept of an inverse-monotone operator 160
2 Inverse-monotone differential operators 162
2.1 Simple sufficient conditions 162
2.2 More general results 166
2.3 Miscellaneous results 174
3 Inverse-monotonicity with weak differential inequalities 180
3.1 Description of the problem 181
3.2 Monotone definite operators 185

3.3 Further constructive sufficient conditions 188
3.4 L_1-differential inequalities 193
3.5 Stronger positivity statements for linear problems 195
4 Existence and inclusion for two-point boundary value problems 198
4.1 The basic theorem 200
4.2 Bounds for the derivative of the solution 203
4.3 Quadratic growth restrictions, Nagumo conditions 207
4.4 Singular problems 212
4.5 Different approaches 216
5 Existence and inclusion by weak comparison functions 220
5.1 Application of Schauder's fixed-point theorem 220
5.2 Existence by the theory on monotone definite operators 222
6 Error estimates 223
6.1 Construction of error bounds by linear approximation 223
6.2 Initial value problems 224
6.3 Boundary value problems 226
Notes 228

Chapter IV An estimation theory for linear and nonlinear operators, range–domain implications

1 Sufficient conditions, general theory 234
1.1 Notation and introductory remarks 234
1.2 A continuity principle 235
2 Inverse-monotone operators 237
2.1 Abstract results using majorizing elements 238
2.2 Nonlinear functions in finite-dimensional spaces 245
2.3 Functional-differential operators of the second order 248
2.4 A different approach for initial value problems 251
3 Inclusion by upper and lower bounds 259
4 Range–domain implications for linear and concave operators 261
4.1 Abstract theory 262
4.2 Applications to non-inverse-positive differential operators 264
Notes 272

Chapter V Estimation and existence theory for vector-valued differential operators

1 Description of the problem, assumptions, notation 275
2 Two-sided bounds by range–domain implications 279
3 Existence and inclusion by two-sided bounds 287
3.1 The basic theorem 288
3.2 Simultaneous estimation of the solution and its derivative 290
3.3 Growth restrictions concerning u' 293
3.4 Application of the methods to other problems 295
4 Pointwise norm bounds by range–domain implications 299
4.1 Notation and auxiliary means 299
4.2 General estimation theorems for Dirichlet boundary conditions 301

4.3 Special cases and examples 305
4.4 More general operators and other boundary conditions 312
4.5 First order initial value problems 316
5 Existence and estimation by norm bounds 323
5.1 The basic theorem 324
5.2 Simultaneous estimation of u^* and $(u^*)'$ 326
5.3 Quadratic growth restrictions 329
6 More general estimates 332
6.1 Estimation by two-sided bounds and norm bounds 333
6.2 Estimates by Lyapunov functions 335
Notes 340

References 343

Notation index 359
Subject index 361

Preface

This book is concerned with inequalities that are described by operators or, briefly, with *operator inequalities.* These operators may be matrices, differential operators, integral operators, etc. Special emphasis is given to differential operators related to boundary value problems.

The main interest is in results that allow one to derive properties of an unknown element (vector, function, ...) from operator inequalities that involve this element. These properties usually are formulated as estimates in terms of order relations, norms, or other means. The estimation methods can, for instance, be employed to investigate *operator equations* (boundary value problems, linear algebraic systems of equations, etc.). In particular, these methods can be used to derive statements on the qualitative and quantitative behavior of solutions and to prove uniqueness statements. Moreover, the methods are helpful as a tool in proving the existence of solutions that satisfy certain estimates.

Major subjects of this book are the theory of *inverse-positive linear operators* with a series of applications to M-matrices, the boundary maximum principle, oscillation theory, eigenvalue theory, convergence proofs, etc., and a corresponding theory on *inverse-monotone nonlinear operators.* Another topic, in which estimates are combined with existence statements, is the theory of *comparison functions* for (scalar-valued and vector-valued) differential operators of the second order. This theory has proved to be very useful, for instance, in the investigation of nonlinear eigenvalue problems and singular perturbation problems and also for numerical error estimates.

Described in a more general way, the basic results have the form of *input–output statements,* where properties (estimates) of an unknown element (vector, function, . . .) are derived from known properties (estimates) of the image Mv of this element v under an operator M. Since here properties of elements in the domain of the operator are derived from properties of elements in its range, we speak also of *range–domain statements* and *range–domain implications.* Such implications can formally be written as $Mv \in C \Rightarrow v \in K$. The implication $Mv \leq Mw \Rightarrow v \leq w$, which describes inverse-monotone operators, is a special case. By properly applying such implications, together with Schauder's fixed-point theorem or other results of existence theory to a given equation, one can often prove the existence of a solution that lies in a certain set K.

The theory of inverse-monotone operators and the theory of comparison functions yield *two-sided estimates* $\varphi \leq v \leq \psi$. In addition, for vector-valued functions v, *pointwise norm estimates* $\| v(x) \| \leq \psi(x)$ are investigated. The use of such estimates instead of two-sided estimates can have certain specific advantages (for example, in stability theory). Estimation by more general Lyapunov-type functions is outlined briefly.

As the title of the book suggests, we employ abstract terms in developing our theory and methods. Abstract results, however, are not considered here as ends in themselves, but as means to obtain results for concrete problems. The degree to which abstract means are used is accommodated to the nature of the various topics considered. For example, to treat inverse-positive linear operators, first an abstract theory is developed and then this theory is applied to matrices, differential operators, etc. A theory on abstract inverse-monotone linear operators is also formulated, but most results on concrete problems are derived directly. More generally, the variety of possible estimates and the variety of nonlinear operators which may be of practical interest is so immense that we do not consider it feasible for our purpose to derive general abstract results on estimation. Rather, on one hand, we investigate some problems by a simpler direct approach in detail and, on the other hand, describe certain *methods* in abstract terms (the continuity principle, the method of absorption, the method of modification, etc.), thus providing the tools for obtaining results on problems not considered here in an analogous way.

Except for a section on M-matrices and a few other results, we concentrate on (scalar-valued and vector-valued) ordinary differential operators and corresponding differential equations. Here the emphasis is on operators of the second order and boundary value problems, although first order initial value problems and differential operators of the fourth order are also considered. With the methods explained, it is easy to generalize the results to certain functional-differential operators, such as integro-differential operators. This is carried out for some estimates; for others, the generalization is left to the reader. Similarly, some of the results derived here can easily be carried over to abstract differential operators, in particular, to differential operators in Banach spaces. Here again, this is performed explicitly in some cases, while others are left as problems, etc.

Because of size limitations on the book, we essentially had to omit in this volume partial differential operators and partial differential equations—with some regret. We point out, however, that most of the results which have the form of range–domain implications can be carried over without any essential difficulty to elliptic-parabolic operators of the second order. This applies to the theory of inverse-monotone operators as well as to pointwise norm estimates. Thus we hope that the reader will find this presentation helpful also for treating partial differential equations. Of course, results for partial differential equations, including existence statements, are, in general, more difficult to achieve.

The topics of this monograph and their presentation correspond to the scientific interest of the author. The topics presented intersect with some well-established theories and touch others. The notes which follow each chapter are not intended to give surveys of all these various theories, but rather to help the reader in further studying the subjects treated here, and to indicate some further results related to these subjects.

Acknowledgments

I owe sincere thanks to many co-workers. Christa-Maria Gruner typed the manuscript perfectly, never tiring in making subsequent changes and additions. Ute Gärtel read the manuscript and the galleys and checked proofs and examples. William M. Snyder, Jr. corrected my use of the English language. Gerd Hübner programmed algorithms and calculated numerical examples.

Part of the research work contained in this book was carried out—long ago—in the Boeing Mathematics Research Laboratory, while I was on leave from the University of Cologne. I would like to express my appreciation to Burton H. Colvin for inviting me there, and I particularly thank my then office mate Roger J. B. Wets for many fruitful discussions. Thanks are also due to my former collaborators Ulrich Trottenberg, Klaus Stüben, and Kristian Witsch for pleasant cooperation and discussions.

I acknowledge with gratitude the support given by the European Research Office of the United States Army under Grant No. DAERO 78-G-013. The Royal Society of Edinburgh, SIAM, and Springer-Verlag did not hesitate in agreeing that parts of my papers [1977b, 1978, and 1961b, 1977a], for which they have the respective copyrights, be incorporated in the book. Finally, I sincerely thank Academic Press, not only for the careful printing and rapid publication, but also for responding very nicely to a series of special wishes concerning mathematical symbols, arrangements of the text, etc.

Some remarks concerning notation

The book is divided into five chapters, I–V. When in some chapter we refer to a formula of another chapter, the number of the latter chapter is adjoined to the formula. (For example, formula (I.3.4) is formula (3.4) in Chapter I.) An analogous notation will be used for theorems, examples, etc.

Most of our mathematical notation will be explained in the text of Chapter I. In the following, we clarify our use of some logical symbols and, in addition, provide a list of function spaces. First of all, we emphasize that *all vectors and functions which will occur in the text are assumed to be real-valued*, unless stated differently.

Logical implications will often be described by the sign $\Rightarrow$. Our use of this symbol may be explained by considering the following simple statement (i) as an example.

(i) *If both u and v are elements of the set R and if $u \geqslant o$ and $v \geqslant o$, then $u + v \geqslant o$.*

This statement may also be written in one of the following three forms, depending on the context and on the emphasis we want to give the statement.

(ii) *For all $u, v \in R$*

$$\left.\begin{array}{l} u \geqslant o \\ v \geqslant o \end{array}\right\} \quad \Rightarrow \quad u + v \geqslant o.$$

(iii) *For all $u, v \in R$*

$$u \geqslant o, \quad v \geqslant o \quad \Rightarrow \quad u + v \geqslant o.$$

(iv) $u \in R, v \in R, u \geqslant o, v \geqslant o \Rightarrow u + v \geqslant o.$

Thus, commas such as those in (iv) represent a logical connection stronger than the sign $\Rightarrow$. If we want to separate two implications of this type, we shall use a semicolon (see formula (I.1.1), for example).

The sign $\Leftrightarrow$, which replaces the term "if and only if," will be used analogously. Terms like $(n = 1, 2, \ldots, p)$, $(0 \leqslant x \leqslant 1)$, and $(\ell \in \mathscr{L})$ indicate that the relations or statements preceding these terms are assumed to hold for all values of the respective parameters n, x, and ℓ which are described in parentheses.

Now we explain some symbols for function spaces by listing the properties of their elements. Here, we restrict ourselves to functions on $[0, 1]$. An analogous notation will be used for functions on other intervals, such as $[0, \infty)$, and also for functions on more general sets Ω.

$\mathbb{R}[0, 1]$: real-valued functions on $[0, 1]$;
$\mathbb{R}_b[0, 1]$: bounded real-valued functions on $[0, 1]$;
$D_m[0, 1]$: functions $u \in \mathbb{R}[0, 1]$ which are m-times differentiable on $[0, 1]$;
$C_m[0, 1]$: functions $u \in \mathbb{R}[0, 1]$ which have continuous derivatives up to the order m on $[0, 1]$;
$C_{j,m}[x_0, x_1, \ldots, x_k, x_{k+1}]$ (with $0 = x_0 < x_1 < \cdots < x_k < x_{k+1} = 1$; $0 \leqslant j < m$): functions $u \in C_j[0, 1]$ such that $u(x) = u_i(x)$ $(x_i < x < x_{i+1})$ for some $u_i \in C_m[x_i, x_{i+1}]$ $(i = 0, 1, \ldots, k)$;
$C'_{j,m}[0, 1]$: functions which belong to some space $C_{j,m}[x_0, x_1, \ldots, x_k, x_{k+1}]$;
$C_m[x_0, x_1, \ldots, x_k, x_{k+1}]$: functions $u \in \mathbb{R}[0, 1]$ such that $u(x) = u_i(x)$ $(x_i < x < x_{i+1})$ for some $u_i \in C_m[x_i, x_{i+1}]$ $(i = 0, 1, \ldots, k)$, $u(0) = u_1(0)$, $u(1) = u_k(1)$, and for each $i \in \{1, 2, \ldots, k\}$ either $u(x_i) = u_{i-1}(x_i)$ or $u(x_i) = u_i(x_i)$;
$C'_m[0, 1]$: functions which belong to some space $C_m[x_0, x_1, \ldots, x_k, x_{k+1}]$.

Derivatives at the endpoints of the interval indicated in the symbol are interpreted to be one-sided derivatives. The functions considered may also be defined on an interval larger than the interval indicated. For example, more precisely, $C_2(0, 1]$ is the set of all functions u which are defined on an interval $\Omega \supset (0, 1]$ such that the restrictions to $(0, 1]$ have continuous second derivatives. With this interpretation terms such as $C_0[0, 1] \cap C_2(0, 1]$ have an obvious meaning.

Spaces of vector-valued functions $u = (u_i)$ with n components u_i $(i = 1, 2, \ldots, n)$ are denoted by an upper index n, analogous to the notation $\mathbb{R}^n$ for the set of real n-vectors. For example,

$$u \in \mathbb{R}^n[0, 1] \quad \Leftrightarrow \quad u = (u_i) \quad \text{with} \quad u_i \in \mathbb{R}[0, 1] \quad (i = 1, 2, \ldots, n).$$

$$u \in C_2^n[0, 1] \quad \Leftrightarrow \quad u = (u_i) \quad \text{with} \quad u_i \in C_2[0, 1] \quad (i = 1, 2, \ldots, n).$$

For $u, v \in \mathbb{R}^n[0, 1]$

$[u, v]_m$ is the set of all $w \in C_m^n[0, 1]$ such that

$$u_i(x) \leqslant w_i(x) \quad \text{and} \quad w_i(x) \leqslant v_i(x) \quad (0 \leqslant x \leqslant 1; i = 1, 2, \ldots, n);$$

$$\|u\|_\infty = \sup\{|u_i(x)| : 0 \leqslant x \leqslant 1; \quad i = 1, 2, \ldots, n\}.$$

Moreover, $e \in \mathbb{R}^n[0, 1]$ is defined by

$$e = (e_i), \quad e_i(x) = 1 \quad (0 \leqslant x \leqslant 1; \quad i = 1, 2, \ldots, n).$$

Chapter I

Some results on functional analysis

In this first chapter functional analytical terms and results are presented which are needed in the succeeding chapters. Section 1 is concerned with simple order theoretical concepts. These will be used, for instance, in the theory of inverse-positive linear operators and its applications in Chapter II. In Section 2 the order theory of Section 1 is generalized by considering arbitrary convex sets instead of (order) cones. Since here the approach is different, Section 2 also provides a new understanding of the order theory. The results of Section 2 will be used mainly in the theory of linear range–domain implications in Section IV.4. In Section 3 we collect, without proof, a series of results in functional analysis which are used in existence theories. For instance, we formulate Schauder's fixed-point theorem, which will be applied in Sections III.4, III.6, V.3, and V.5.

1 BASIC CONCEPTS OF ORDER THEORY

In this section mainly those simple order theoretical concepts are described which will be used in the theory on inverse-positive operators. By studying Section II.1, the reader may simply check which of the terms and results below are needed.

1.1 Ordered spaces

A set R is called *ordered* if for some elements u, v in R a relation $u \leqslant v$ is defined such that

$$
\begin{array}{llll}
u \leqslant v, & v \leqslant w \quad \Rightarrow & u \leqslant w & (\textit{transitivity}), \\
u \leqslant u & & & (\textit{reflexivity}), \\
u \leqslant v, & v \leqslant u \quad \Rightarrow & u = v & (\textit{antisymmetry})
\end{array}
$$

whenever $u, v, w \in R$. The relation $\leqslant$ is then called an *order* or *order relation.* We shall also use the term *ordered space*, and write $(R, \leqslant)$ to indicate which symbol is used for the order relation. Moreover, we define

$$
u < v \quad \Leftrightarrow \quad u \leqslant v, \quad u \neq v; \qquad v \geqslant u \quad \Leftrightarrow \quad u \leqslant v;
$$
$$
v > u \quad \Leftrightarrow \quad u < v.
$$

An essential part of our theory in the succeeding chapters is the simultaneous use of several different order relations, for which we then shall have to use different symbols. The following concept allows us to compare such order relations. Let both $\leqslant$ and $\overset{\prime}{\leqslant}$ be orders in R. Then the order relation $\overset{\prime}{\leqslant}$ is said to be *stronger* than $\leqslant$, and the relation $\leqslant$ is said to be *weaker* than $\overset{\prime}{\leqslant}$, if $u \overset{\prime}{<} v$ for at least one pair of elements in R and if the following implications hold for all $u, v, w \in R$:

$$
u \overset{\prime}{\leqslant} v \quad \Rightarrow \quad u \leqslant v; \qquad u \leqslant v, \quad v \overset{\prime}{<} w \quad \Rightarrow \quad u \overset{\prime}{<} w;
$$
$$
u \overset{\prime}{<} v, \quad v \leqslant w \quad \Rightarrow \quad u \overset{\prime}{<} w.
$$

On the other hand, we shall often use the same symbol $\leqslant$ for order relations in different spaces $(R, \leqslant)$ and $(S, \leqslant)$ occurring in the theory, even when the sets R and S have the same elements. In each case the meaning of the inequalities will be clear from the context. Of course, R *may* be a subset of S, and for $u, v \in R$ the relation $u \leqslant v$ *may* have the same meaning whether $\leqslant$ denotes the order in R or the order in S. In this case we write $(R, \leqslant) \subset (S, \leqslant)$ and say that the order in R is *induced* by the order in S.

Suppose that M denotes an operator which maps an ordered space $(R, \leqslant)$ into an ordered space $(S, \leqslant)$. This operator is called *monotone* if for all $u, v \in R$ the inequality $u \leqslant v$ implies $Mu \leqslant Mv$. If $u \leqslant v$ implies $Mu \geqslant Mv$, then M is said to be *antitone*. In particular, this notation is applied to functionals on R, that is, operators mapping R into $\mathbb{R}$.

1.2 Ordered linear spaces

Ordered spaces which occur in applications usually have also an algebraic structure such that certain compatibility conditions are satisfied. First, let us introduce such spaces without using the definitions of the preceding section.

A set R is called an *ordered linear space* provided

(1) R is a (real) linear space (with zero element o),
(2) for certain $u \in R$ a relation $u \geqslant o$ is defined such that $o \geqslant o$ and for all $u, v \in R$ and all $\lambda \in \mathbb{R}$

$$u \geqslant o, \quad v \geqslant o \quad \Rightarrow \quad u + v \geqslant o; \qquad u \geqslant o, \quad \lambda \geqslant 0 \quad \Rightarrow \quad \lambda u \geqslant o; \quad (1.1)$$

$$u \geqslant o, \quad -u \geqslant o \quad \Rightarrow \quad u = o. \quad (1.2)$$

The elements $u \geqslant o$ are called *positive*. We shall write $u > o$ ($o < u$) if $u \geqslant o$ and $u \neq o$.

In such a space R an order relation $u \leqslant v$ is defined by $v - u \geqslant o$, and this relation satisfies the *compatibility conditions*:

$$u \leqslant v \quad \Rightarrow \quad u + w \leqslant v + w; \qquad u \leqslant v, \quad \lambda \geqslant 0 \quad \Rightarrow \quad \lambda u \leqslant \lambda v,$$

for all $u, v, w \in R$ and $\lambda \in \mathbb{R}$. On the other hand, an ordered linear space can, equivalently, be defined to be a linear space which is ordered with respect to $\leqslant$ such that this relation satisfies the above compatibility conditions. An order relation with this property is called a *linear order*.

From the above axioms, the "usual" rules of calculation can be deduced, for example: $u \leqslant v, \lambda < 0 \Rightarrow \lambda u \geqslant \lambda v$. We shall not list these rules here, but use them implicitly in the following.

Example 1.1. In the space $R = \mathbb{R}^n$ with elements $u = (u_k)$, a linear order $\leqslant$ may be induced by defining the inequality $u \geqslant o$ by one of the following sets of relations.

(i) *natural order*

$$u_k \geqslant 0 \qquad (k = 1, 2, \ldots, n);$$

(ii) *natural strict order*

$$u_k > 0 \quad (k = 1, 2, \ldots, n), \qquad \text{or} \qquad u_k = 0 \quad (k = 1, 2, \ldots, n);$$

(iii) *lexicographic order*

$$u_1 \geqslant 0, \qquad u_{k+1} \geqslant 0 \qquad \textit{if} \quad u_1 = u_2 = \cdots = u_k = 0$$
$$(k = 1, 2, \ldots, n - 1).$$

The natural order is the one most commonly used. Observe that for this order the inequality $u > o$ means that $u \geqslant o$ and $u_j > 0$ for at least *one* component. When using the natural order relation, however, we shall also be interested in vectors u with *all* components $u_k > 0$. Such vectors then will be described by a different symbol $\succ$ explained below.

Of course the meaning of the relation $u > o$ depends on the definition of $u \geqslant o$. For example, if $\leqslant$ now denotes the order in (ii), then $u > o$ if and only if $u_k > 0$ for *all* components. □

Example 1.2. Let R denote a linear subspace of $\mathbb{R}[0, 1]$, the space of all real-valued functions on $[0, 1]$. Then the *natural order* in R is defined by $u \geqslant o \Leftrightarrow u(x) \geqslant 0$ $(0 \leqslant x \leqslant 1)$. With respect to this order, R is an ordered linear space. □

Example 1.3. For any set $\Omega \subset \mathbb{R}$ define $\mathbb{R}(\Omega)$ to be the set of all real-valued functions on Ω and $\mathbb{R}^n(\Omega)$ to be the set of all n-vectors $u = (u_k)$ with components $u_k \in \mathbb{R}(\Omega)$ $(k = 1, 2, \ldots, n)$. Each linear subspace $R \subset \mathbb{R}^n(\Omega)$ is an ordered linear space with respect to the *natural order* defined by

$$u \geqslant o \quad \Leftrightarrow \quad u_k(x) \geqslant 0 \qquad (x \in \Omega; \quad k = 1, 2, \ldots, n). \quad \square$$

Most of our applications in the succeeding chapters will be concerned with spaces of n-vectors, scalar-valued functions, or vector-valued functions. For convenience, let us therefore agree to the following convention: In this book, unless stated differently, the relation $\leqslant$ between elements of $\mathbb{R}^n(\Omega)$ (or $\mathbb{R}^n$) always denotes the natural order, and the signs $\geqslant$, $<$, $>$ are interpreted correspondingly.

Ordered linear spaces can also be introduced by characterizing the *set* of all positive elements. A subset K of a real linear space is called a *cone* if

$$\begin{gathered} o \in K, \qquad K + K \subset K, \qquad \lambda K \subset K \qquad \text{for} \quad \lambda > 0, \\ K \cap (-K) = \{o\}. \end{gathered} \tag{1.3}$$

The set K of elements $u \geqslant o$ in an ordered linear space $(R, \leqslant)$ constitutes a cone, the corresponding *positive cone* (*order cone*). If, on the other hand, a cone K is defined in a linear space R, this cone *induces a linear order* $\leqslant$ given by $u \geqslant o \Leftrightarrow u \in K$. The corresponding positive cone is K itself.

We therefore have two possibilities to describe properties of ordered linear spaces. We may use either order relations or cones. Both possibilities have advantages and we shall, therefore, use both of them. For example, we shall also write (R, K) instead of $(R, \leqslant)$, or $(R, \leqslant, K)$.

Example 1.4. In $\mathbb{R}^3$ the set K of all $u = (u_k)$ with $u_3 \geqslant 0$, $u_3^2 \geqslant u_1^2 + u_2^2$ constitutes a cone (which is sometimes called the "ice cream cone"). □

To introduce a linear order in a given real linear space R, it is sometimes more convenient to first define the relation $u > o$ and then define $u \geqslant o$ by

$$u \geqslant o \quad \Leftrightarrow \quad u > o \quad \text{or} \quad u = o. \tag{1.4}$$

If $(R, \leqslant)$ is an ordered linear space, the relation $>$ satisfies

$$\begin{gathered} u > o \quad \Rightarrow \quad u \neq o; \qquad u > o, \quad v > o \quad \Rightarrow \quad u + v > o; \\ u > o, \quad \lambda > 0 \quad \Rightarrow \quad \lambda u > o. \end{gathered} \tag{1.5}$$

If, on the other hand, in a linear space R a relation $u > o$ is defined such that the above conditions are satisfied, then a linear order is introduced by (1.4).

In an ordered linear space $(R, \leqslant)$ we define the relation $u \succ o$ by the property that for each $v \in R$ a number n with $nu + v \geqslant o$ exists. An element $u \succ o$ is called *strictly positive*. It can be shown that condition (1.5) holds for $\succ$ in place of $>$. Thus, by defining $u \succcurlyeq o \Leftrightarrow u \succ o$ or $u = o$, a linear order $\preccurlyeq$ is given, which will be called the *strict order induced by* $\leqslant$ (or the *strict order corresponding to* $\leqslant$). If K is the positive cone of the order $\leqslant$, the cone corresponding to $\preccurlyeq$ will be denoted by K_s.

Proposition 1.1. *If $(R, \leqslant)$ is an ordered linear space, then $(R, \preccurlyeq)$ is also an ordered linear space.*

Example 1.1 (continued). For the three relations (i)–(iii) in $\mathbb{R}^n$ defined above, we have $u \succ o$ if and only if u is an interior point of the corresponding order cone (in the usual topology). Thus, (ii) is the strict order induced by (i). Moreover, the strict order corresponding to (ii) is (ii) itself. □

As the definition of the strict order and the above example show, the strict order depends on the relation $\leqslant$ defined in R. The next two examples will illustrate the dependence of the strict order on the set R.

Example 1.5. Suppose that R is a linear subspace of $\mathbb{R}[0, 1]$ (endowed with the natural order relation). We shall describe the strictly positive elements for various R.

(i) In $R = \mathbb{R}[0, 1]$ no strictly positive element exists; therefore,

$$u \succcurlyeq o \quad \Leftrightarrow \quad u = o.$$

(ii) For $R = \mathbb{R}_b[0, 1]$, the set of all bounded functions in $\mathbb{R}[0, 1]$,

$$u \succ o \quad \Leftrightarrow \quad \inf\{u(x) : x \in [0, 1]\} > 0.$$

(iii) For $R = C_0[0, 1]$,

$$u \succ o \quad \Leftrightarrow \quad u(x) > 0 \qquad (0 \leqslant x \leqslant 1).$$

(iv) For $R = \{u \in C_4[0, 1] : u(0) = u'(0) = u''(0) = u'''(0) = 0\}$,

$$u \succ o \quad \Leftrightarrow \quad u(x) > 0 \quad (0 < x \leqslant 1), \quad u^{\mathrm{iv}}(0) > 0.$$

(v) For $R = \{u \in C_2[0, 1] : u(0) = u(1) = 0\}$,

$$u \succ o \quad \Leftrightarrow \quad u(x) > 0 \quad (0 < x < 1), \quad u'(0) > 0, \quad u'(1) < 0.$$

□

Example 1.6. Let R be a given linear subspace of $\mathbb{R}[0, \infty)$ (endowed with the natural order relation).

(i) For $R = C_0[0, \infty)$, we have $u \succcurlyeq o \Leftrightarrow u = o$.

(ii) For $R = \mathbb{R}_b[0, \infty)$, the set of all bounded functions on $[0, \infty)$,

$$u \succ o \quad \Leftrightarrow \quad \inf\{u(x) : x \in [0, \infty)\} > 0.$$

□

In all cases of Example 1.1, the strictly positive elements are the interior points of the respective positive cone. The same statement can be made for Examples 1.5 and 1.6, if, in each case considered, a suitable topology is introduced (cf. Section 2.3). However, by using the very simple term of a strictly positive element, we need not introduce topological concepts in most of the theory below. In particular, we need not define open sets in each case.

On the other hand, we shall also need an order theoretic term which can replace the concept of a (topologically) closed set in many circumstances. An ordered linear space $(R, \leqslant)$ is called *Archimedian* and the relation $\leqslant$ is also called Archimedian, provided the following property holds for each $u \in R$: If $nu + v \geqslant o$ for some $v \in R$ and all $n \in \mathbb{N}$, then $u \geqslant o$.

Proposition 1.2. *If $(R, \leqslant) = (R, K)$ is an ordered linear space, the following three statements are equivalent:*

(i) *$(R, \leqslant)$ is Archimedian.*

(ii) *For arbitrary $u, v \in R$,*

$$u + \lambda_0 v \geqslant o \qquad \text{for} \quad \lambda_0 = \inf\{\lambda : u + \lambda v \geqslant o\}, \qquad \text{if } \lambda_0 \text{ is finite.}$$

(iii) *The intersection of K with any line $\{u = x + \lambda y : -\infty < \lambda < \infty\}$ is closed (with respect to the line topology).*

Proof. (i) *implies* (ii). Suppose that λ_0 is finite and $u + \lambda_1 v \in K$ for some $\lambda_1 > \lambda_0$. Then, since the intersection of K and the line $g = \{w : w = u + \lambda v, -\infty < \lambda < \infty\}$ is convex, $u + \lambda v \in K$ for $\lambda_0 < \lambda \leqslant \lambda_1$. Thus, for some $N \in \mathbb{N}$ and all $n \in \mathbb{N}$, $u + (\lambda_0 + (N + n)^{-1})v \geqslant o$, i.e., $n(u + \lambda_0 v) + v + N(u + \lambda_0 v) \geqslant o$. The last relation implies $u + \lambda_0 v \geqslant o$ for Archimedian R.

(ii) *implies* (i). Let $u, v \in R$ satisfy $nu + v \geqslant o$ $(n = 1, 2, \ldots)$. Then $u + \lambda v \geqslant o$ for $\lambda = n^{-1}$ $(n = 1, 2, \ldots)$. It follows that $u + \lambda_1 v \geqslant o$ for some $\lambda_1 \leqslant 0$. This is obvious if $\lambda_0 = -\infty$; otherwise it follows from (ii). Since in particular $u + v \geqslant o$, we obtain $u = \alpha(u + v) + (1 - \alpha)(u + \lambda_0 v) \geqslant o$ for $\alpha = -\lambda_0(1 - \lambda_0)^{-1}$. Thus R is Archimedian.

The equivalence of (ii) and (iii) is easily seen. □

Example 1.1 (continued). The order cone which corresponds to the natural order (i) obviously has the third property in the above proposition. Thus this order relation is Archimedian. On the other hand, the natural strict order and the lexicographic order are not Archimedian. □

Example 1.7. We denote by $\mathbb{R}^{n,m}$ the space of all (real) $n \times m$ matrices $A = (a_{ik})$ and by O the null matrix, i.e., the null element of this linear space. The natural order in $\mathbb{R}^{n,m}$ is defined elementwise, $A \geqslant O \Leftrightarrow a_{ik} \geqslant 0$ for all elements of A. This order is Archimedian; moreover, $A \succ O$ if and only if $a_{ik} > 0$ for all elements of A. □

The next proposition shows that a strict order relation $\preccurlyeq$ never is Archimedian.

Proposition 1.3. *Suppose that* $(R, \leqslant)$ *is an ordered linear space and that for some* $u, v \in R$ *the set* $\Lambda = \{\lambda : u + \lambda v \geqslant o\}$ *is not empty and bounded below. Then* $u + \lambda_0 v \not\succ o$ *for* $\lambda_0 = \inf \Lambda$.

Proof. The inequality $u + \lambda_0 v \succ o$ implies $-v + n(u + \lambda_0 v) \geqslant o$ for some $n \in \mathbb{N}$, that is, $u + (\lambda_0 - n^{-1})v \geqslant o$. This inequality contradicts the definition of λ_0. □

For certain function spaces such as $\mathbb{R}[0, 1]$ we do not only want to use the natural (pointwise) order, but, simultaneously, also other order relations, which will often be denoted by $\leqslant'$. If two linear orders $\leqslant$ and $\leqslant'$ are defined on R, then $\leqslant'$ is said to be *stronger* than $\leqslant$ (and $\leqslant$ is said to be *weaker* than $\leqslant'$), if at least one element $u >' o$ exists and the following implications hold for all $u, v \in R$:

$$u \geqslant' o \quad \Rightarrow \quad u \geqslant o; \qquad u >' o, \quad v \geqslant o \quad \Rightarrow \quad u + v >' o. \tag{1.6}$$

This definition is consistent with the corresponding definition in Section 1.1. Under the above conditions we shall also say that $\leqslant'$ is a *strong order in* $(R, \leqslant)$.

Example 1.8. Order relations in $\mathbb{R}^n$ which are stronger than the natural order, are given by (a) and (b):

(a) $u >' o \quad \Leftrightarrow \quad u > o, \quad u_k > 0 \quad$ for all $\quad k \in J$,

where J denotes a given subset of $\{1, 2, \ldots, n\}$, which may be empty.

(b) $u >' o \quad \Leftrightarrow \quad u \geqslant o, \quad u_{k_i} > 0 \quad$ for some $\quad k_i \in J_i \quad (i = 1, 2, \ldots, p)$,

where the J_i are nonempty pairwise disjoint sets with $J_1 \cup \cdots \cup J_p = \{1, 2, \ldots, n\}$. □

Example 1.9. For each subspace $R \subset \mathbb{R}[0, 1]$ (including $R = \mathbb{R}[0, 1]$) a linear order $\leqslant'$ which is stronger than the natural order is given by

$$u >' o \quad \Leftrightarrow \quad u(x) > 0 \qquad (0 \leqslant x \leqslant 1).$$

For $R = C_0[0, 1]$, the following order is also of interest:

$$u >' o \quad \Leftrightarrow \quad u \geqslant o, \quad u(0) > 0, \quad u(1) > 0. \quad \square$$

Proposition 1.4. *If* $(R, \leqslant)$ *is an ordered linear space which contains a strictly positive element, then the strict order* $\preccurlyeq$ *is stronger than* $\leqslant$. *Moreover the strict order is stronger than any other strong order in* $(R, \leqslant)$.

For elements in an ordered linear space $(R, \leqslant)$ we have to distinguish between order intervals and segments.

In an ordered space $(R, \leqslant)$ the notion of an *order interval* $[u, v]$ and corresponding symbols $(u, v]$, $[u, v)$, and (u, v) are defined by

$$[u, v] = \{x \in R : u \leqslant x \leqslant v\}, \qquad (u, v] = \{x \in R : u < x \leqslant v\}$$

and analogous formulas. Here, it is assumed that $u, v \in R$ and $u \leqslant v$ or $u < v$, respectively. For clearness, we shall sometimes write $[u, v]_R$ instead of $[u, v]$. This notion will also be used, if u, v do not necessarily belong to R, but to a space $(S, \leqslant) \supset (R, \leqslant)$. Moreover, whenever an order interval $[u, v]$ without such a lower index occurs, it is assumed that $[u, v] = [u, v]_R$ where $(R, \leqslant)$ denotes the space defined in the context.

In a linear space R the *segment* $[u : v]$ and corresponding sets $(u : v]$, $[u : v)$, $(u : v)$ for $u, v \in R$ are defined by

$$[u : v] = \{x = (1 - t)u + tv : 0 \leqslant t \leqslant 1\},$$

$$(u : v] = \{x = (1 - t)u + tv : 0 < t \leqslant 1\}$$

and analogous formulas. Observe that, for example, $[u : u] = (u : u] = \{u\}$. Intervals in $\mathbb{R}$ will, in general, be denoted by $[u, v]$, $(u, v]$, etc.

The order interval will also be used in an *interval arithmetic*, which may be best explained by the following example. If F is a function on R with values in $\mathbb{R}$, then $F([u, v])$ denotes the *set* of all $F(x)$ with $x \in [u, v]$. Moreover, the condition $0 \leqslant F([u, v])$ means that $0 \leqslant F(x)$ holds for all $x \in [u, v]$. Other similar notations are to be interpreted in a corresponding way.

Problems

1.1 A transitive and reflexive relation $\leqslant$ in R, which is not necessarily antisymmetric, is called a *preorder*. A *preordered linear space* is defined in the same way as an ordered linear space except that condition (1.2) is omitted. A set K in a linear space which has the first three properties in (1.3) is sometimes called a *wedge*. Most of the theory of this section on ordered linear spaces (and cones) can be generalized to preordered linear spaces (and wedges). Carry out this generalization. In particular, define $u \succ o$ for preordered linear spaces $(R, \leqslant, K)$ with $K \neq R$. For what reason should $K \neq R$ be required? Prove that a linear preorder $\leqslant$ is stronger than itself if and only if it is an order and $u > o$ for some u. Also show that the strict order relation $\preccurlyeq$ in a preordered linear space $(R, \leqslant, K)$ with $K \neq R$ is a linear order (provided $u \succ o$ for some u).

1.2 Prove Proposition 1.4.

1.3 Let $\leqslant$ denote the natural order in the space l_2 of all sequences $\{u_i\}$ with $\sum_{i=1}^{\infty} (u_i)^2 < \infty$, that is, $u \geqslant o \Leftrightarrow u_i \geqslant 0$ $(i = 1, 2, \ldots)$. Prove that $(l_2, \leqslant)$ does not contain a strictly positive element and that a strong order is defined by $u \succ o \Leftrightarrow u_i > 0$ $(i = 1, 2, \ldots)$.

1.4 Let R denote the set of all hermitian complex-valued $n \times n$-matrices A, B, ... This set becomes a real linear space, when only the multiplication λA with $\lambda \in \mathbb{R}$ is considered. Prove that $(R, \leqslant)$ is an ordered linear space, if $A \geqslant O$ means that A is positive semidefinite. Characterize the matrices $A \succ O$. (Here O denotes the null matrix.)

1.3 Positive linear functionals and operators

Suppose that $(R, \leqslant) = (R, K)$ is an ordered linear space. Denote by R^* the linear space of all (real-valued) linear functionals ℓ on R, i.e., the space of all linear mappings $\ell : R \to \mathbb{R}$, with the null element (null functional) o defined by $ou = 0$ $(u \in R)$.

A functional $\ell \in R^*$ is said to be positive, $\ell \geqslant o$, if for all $u \in R$, $u \geqslant o \Rightarrow \ell u \geqslant 0$. In this way a linear preorder is introduced in R^* (see Problem 1.1). In the theory below it is often convenient to characterize the cone K or certain subsets of K by positive linear functionals, as in the following definition.

A set $\mathscr{L} \subset R$ is said to *describe the positive cone* K in R (or to *generate* K) if $\ell > o$ (i.e., $\ell \geqslant o$ and $\ell \neq o$) for all $\ell \in \mathscr{L}$ and if for all $u \in R$

$$u \geqslant o \quad \Leftrightarrow \quad \ell u \geqslant 0 \quad \text{for all} \quad \ell \in \mathscr{L}. \tag{1.7}$$

Moreover, in case R has a strictly positive element, $\mathscr{L}$ is said to *describe* K *completely*, if, in addition, for all $u \in R$

$$u \succ o \quad \Leftrightarrow \quad \ell u > 0 \quad \text{for all} \quad \ell \in \mathscr{L}. \tag{1.8}$$

Obviously, if (a $u \succ o$ exists and) $\mathscr{L}$ describes K completely, then for all $u \in R$

$$u \geqslant o, \quad u \nsucc o \quad \Leftrightarrow \quad u \geqslant o, \quad \ell u = 0 \quad \text{for some} \quad \ell \in \mathscr{L}. \tag{1.9}$$

Example 1.10. The positive cone in $C_0[0, 1]$ (with respect to the natural order relation) is completely described by the set $\mathscr{L} = \{\ell_x : 0 \leqslant x \leqslant 1\}$ of all *point functionals* defined by $\ell_x u = u(x)$. □

Example 1.11 (compare Example 1.5(v)). The positive cone K in $R = \{u \in C_2[0, 1] : u(0) = u(1) = 0\}$ is described, but not completely described by the set $\mathscr{L}$ of all point functionals ℓ_x $(0 \leqslant x \leqslant 1)$. However, K is completely described by the set $\mathscr{L} = \{\ell_{(x)} : 0 \leqslant x \leqslant 1\}$, where $\ell_{(x)} = \ell_x$ for $0 < x < 1$, $\ell_{(0)}u = u'(0)$, $\ell_{(1)}u = -u'(1)$. □

Concerning the existence of sets $\mathscr{L}$ with properties (1.7), (1.8) we only state here the following proposition without proof. (See, however, Theorem 2.9.)

Proposition 1.5. *Suppose that* $(R, \leqslant, K)$ *is Archimedean and has a strictly positive element. Then the set* $\mathscr{L}$ *of all linear functionals* $\ell > o$ *on* R *describes* K *completely.*

The knowledge of this result is neither sufficient nor necessary for applying most of the theory below; we need *construct* such a set $\mathscr{L}$ as in the above example. Also it is often inconvenient to work with the set of *all* $\ell > o$. In some cases it is even more convenient to generalize the above concept by allowing $\mathscr{L}$ to contain also nonlinear functionals (see Example 2.4).

Now assume that $(S, \leqslant, C)$ denotes another ordered linear space. A linear operator $A : R \to S$ is said to be *positive*, $A \geqslant O$, if for all $u \in R$, $u \geqslant o \Rightarrow Au \geqslant o$. Here O denotes the null operator with $Ou = o\ (u \in R)$. Moreover, we call A *pointwise strictly positive*, if for all $u \in R$, $u > o \Rightarrow Au \succ o$. Clearly, A can have the latter property only if S contains a strictly positive element. Finally, A is said to be *(linearly) monotonically decomposable* if $A = A_1 - A_2$ with positive linear operators mapping R into S.

Example 1.12. Let $R = \mathbb{R}^m$ and $S = \mathbb{R}^n$, both spaces endowed with the natural order relation. Each linear operator $A : \mathbb{R}^m \to \mathbb{R}^n$ is identified with the corresponding $n \times m$ matrix $A = (a_{ik})$. One easily verifies that $A \geqslant O$ if and only if $a_{ik} \geqslant 0$ for all elements a_{ik}, and that A is pointwise strictly positive if and only if $a_{ik} > 0$ for all elements. Moreover, each such A is monotonically decomposable, with A_1 having the elements $\frac{1}{2}(|a_{ik}| + a_{ik})$. □

Comparing the preceding example with Example 1.7, we see that a matrix A is strictly positive if and only if it is pointwise strictly positive. Such a statement does not generally hold for linear operators. By the above definition of a positive operator a linear preorder is introduced in the linear space $L[R, S]$ of all linear operators $A : R \to S$. Thus, the term $A \succ O$ is defined as in Section 1.1, but it will play no essential role in the theory below. (Here we write O for the null element, instead of o.)

Problems

1.5 For each of the following spaces R find a set $\mathscr{L}$ which describes the positive cone (of the natural order relation) completely,

(a) $R = \{u \in C_2[0, 1] : u(0) = u'(0) = 0\}$.
(b) $R = \{u \in C_4[0, 1] : u(0) = u'''(0) = u'(1) = u''(1) = 0\}$.
(c) $R = \mathbb{R}_b[0, 1]$.
(d) $R = l_\infty$, the set of all sequences $u = (u_i)$ with $\sup_i |u_i| < \infty$.

(Here, $u \geqslant o \Leftrightarrow u_i \geqslant 0\ (i = 1, 2, \ldots)$.)

1.6 The set of all positive linear functionals $\ell \geqslant o$ on an ordered (or preordered) linear space $(R, \leqslant, K)$ is called the *dual wedge* K^* (or the *dual cone*, provided $K^* \cap (-K^*) = \{o\}$). The definition of a strictly positive element can also be applied to the preordered linear space (R^*, K^*). Prove that $K \neq \{o\}$ implies $K^* \neq R^*$. Moreover, show that $\ell \in R^*$ is pointwise strictly positive, i.e., $u \in R$, $u > o \Rightarrow \ell u > 0$, if $\ell \succ o$. (In general, the converse is not true!)

1.7 Let $A = (a_{ik}) \in \mathbb{R}^{n,n}$ satisfy $|a_{ik}| = (-1)^{i+k} a_{ik}$. Define a linear order in $\mathbb{R}^n$ such that $A : \mathbb{R}^n \to \mathbb{R}^n$ is positive with respect to this order.

1.4 Convergence and continuity

Let $(R, \leqslant)$ denote an ordered linear space. First we introduce a simple notion of convergence for sequences $\{u_n\}$ and a corresponding notion of continuity in order theoretic terms and prove some corresponding results, which will be used in Section II.1.4, for example. Unless stated differently, the indices of a sequence shall always have the values 1, 2, ... An *antitone zero sequence* $\{\lambda_n\} \subset \mathbb{R}$ is a sequence such that $0 \leqslant \lambda_{n+1} \leqslant \lambda_n$, $\lim \lambda_n = 0$.

For given $\zeta \geqslant o$ in R, a *ζ-limit of a sequence* $\{u_n\} \subset R$ is an element $u = \zeta\text{-lim } u_n \in R$, such that there exists an antitone zero sequence $\{\lambda_n\}$ with $-\lambda_n \zeta \leqslant u - u_n \leqslant \lambda_n \zeta$ for all sufficiently large n. Obviously, $u = \zeta\text{-lim } u_n$ if and only if for each $\varepsilon > 0$ a number $N(\varepsilon)$ exists such that $u - u_n \in \varepsilon[-\zeta, \zeta]$ for $n \geqslant N(\varepsilon)$.

Proposition 1.6. *Suppose that $(R, \leqslant)$ is Archimedian, $\{u_n\}$ denotes a sequence in R, and $\zeta \in R$ a positive element.*

(i) *The sequence $\{u_n\}$ has at most one ζ-limit.*
(ii) *If $u = \zeta\text{-lim } u_n$ and $u \succ o$, then $u_n \succ o$ for all sufficiently large n.*
(iii) *If $u = \zeta\text{-lim } u_n$ and $u_n \geqslant o$ $(n = 1, 2, \ldots,)$, then $u \geqslant o$.*

These statements follow easily from the definitions which are involved.

Suppose now that $\{u(t) : 0 \leqslant t \leqslant 1\}$ is a family of elements $u(t) \in R$. For given $\zeta \geqslant o$ in R, we shall say that $u(t)$ is *ζ-continuous* at $t = t_0 \in [0, 1]$, if $u(t_0) = \zeta\text{-lim } u(t_n)$ for each sequence $\{t_n\} \subset [0, 1]$ with $\lim t_n = t_0$. In other words, for each $\varepsilon > 0$ there exists a $\delta(\varepsilon) > 0$ such that $u(t) - u(t_0) \in \varepsilon[-\zeta, \zeta]$ for all $t \in [0, 1]$ with $|t - t_0| < \delta(\varepsilon)$.

Often the above notion of convergence and continuity can also be defined by a seminorm. If $z \in R$ is strictly positive, then the functional

$$\|u\|_z = \inf\{\alpha \geqslant 0 : -\alpha z \leqslant u \leqslant \alpha z\}$$

is a seminorm, that is, $\|u\|_z \geqslant 0$, $\|\lambda u\|_z = |\lambda|\, \|u\|_z$, $\|u + v\|_z \leqslant \|u\|_z + \|v\|_z$ for $u, v \in R$, $\lambda \in \mathbb{R}$. Obviously, $u = z\text{-lim } u_n$ if and only if $\lim \|u - u_n\|_z = 0$. Moreover, $u = z\text{-lim } u_n$, if $u = \zeta\text{-lim } u_n$ for some $\zeta \geqslant o$.

Proposition 1.7. *If the ordered linear space $(R, \leqslant)$ is Archimedian and z is a strictly positive element in R, then $\| \ \|_z$ is a norm, that is, $\|u\|_z > 0$ for each $u \neq o$ in R.*

Example 1.13. Let $R = C_0[0, 1]$ be furnished with the natural order relation. For $z \in R$ with $z(x) > 0$ $(0 \leqslant x \leqslant 1)$, $\|u\|_z$ is the weighted maximum norm $\|u\|_z = \max\{|u(x)/z(x)| : 0 \leqslant x \leqslant 1\}$. □

If $z \succ o$ and $\zeta \succ o$ are different elements in R, then the corresponding seminorms are equivalent, i.e., there exists a $\gamma > 0$ such that $\gamma^{-1}\|u\|_z \leqslant \|u\|_\zeta \leqslant \gamma\|u\|_z$ for all $u \in R$. This follows from the fact that $\gamma^{-1}z \leqslant \zeta \leqslant \gamma z$ for some $\gamma > 0$. Consequently, the convergence of a sequence with respect to the seminorm $\| \ \|_z$ does not depend on the choice of $z \succ o$.

If the ordered linear space $(R, \leqslant)$ contains a strictly positive element, we shall say that a sequence $\{u_n\}$ *order converges* (o-converges) toward an *order limit* $u =$ o-lim u_n if $\lim \|u - u_n\|_z = 0$ for some $z \succ o$. Correspondingly, a family $\{u(t) : 0 \leqslant t \leqslant 1\} \subset R$ is o-*continuous* at $t_0 \in [0, 1]$ if $u(t_0) =$ o-lim $u(t_n)$, for each sequence $\{t_n\} \subset [0, 1]$ with $\lim t_n = t_0$. Moreover, the family is simply called o-continuous if it is o-continuous at each $t_0 \in [0, 1]$.

The above remarks result in the following statement.

Proposition 1.8. *Let the ordered linear space $(R, \leqslant)$ contain a strictly positive element. Then $u = \zeta$-lim u_n for some $\zeta \geqslant o$ if and only if $u =$ o-lim u_n. Moreover, a family $\{u(t) : 0 \leqslant t \leqslant 1\} \subset R$ is ζ-continuous at $t_0 \in [0, 1]$ for some $\zeta \geqslant o$ if and only if it is* o-*continuous at $t_0 \in [0, 1]$.*

An immediate consequence of Propositions 1.6 and 1.8 is the following result.

Proposition 1.9. *Suppose that $(R, \leqslant)$ is Archimedian and contains a strictly positive element, and let $\{u(t) : 0 \leqslant t \leqslant 1\}$ denote an* o-*continuous family in R.*

Then $\{t \in [0, 1] : u(t) \geqslant o\}$ is a closed subset of $[0, 1]$, and $\{t \in [0, 1] : u(t) \succ o\}$ is a (relatively) open subset of $[0, 1]$.

Now we add some results on the continuity or, equivalently, boundedness of positive operators which will not be needed in our abstract theory in Section II.1, but are of interest for certain applications.

Suppose that $(R, \leqslant)$ and $(S, \leqslant)$ are ordered linear spaces and that these spaces contain strictly positive elements $z \succ o$ and $\zeta \succ o$, respectively. For simplicity of presentation, we assume that the corresponding seminorms are definite, that means they are norms. All topological terms shall here be interpreted with respect to these norms. For example, $\alpha \geqslant 0$ is a bound of $A \in L[R, S]$ if

$$\|Au\|_\zeta \leqslant \alpha\|u\|_z \qquad \text{for all} \quad u \in R. \tag{1.10}$$

Under these assumptions we obtain

Theorem 1.10. *If $(S, \leqslant)$ is Archimedian, then the following statements hold:*

(i) *$A \in L[R, S]$ has the bound $\alpha \geqslant 0$ if and only if*

$$-z \leqslant u \leqslant z \quad \Rightarrow \quad -\alpha\zeta \leqslant Au \leqslant \alpha\zeta \qquad \text{for all} \quad u \in R. \tag{1.11}$$

(ii) *Each positive operator $A \in L[R, S]$ is bounded, and $\alpha \geqslant 0$ is a bound of A if and only if $Az \leqslant \alpha\zeta$.*

(iii) *Each monotonically decomposable operator $A \in L[R, S]$ is bounded. If $A = A_1 - A_2$ with $A_1 \geqslant O$, $A_2 \geqslant O$, and $(A_1 + A_2)_z \leqslant \alpha\zeta$ for some $\alpha \geqslant 0$, then α is a bound of A.*

Proof. To show the equivalence of (1.10) and (1.11), stated in (i), we first suppose that (1.11) holds. Since $\| \ \|_z$ is a norm, we may assume that $\|u\|_z = 1$. Then, $-(1 + n^{-1})z \leqslant u \leqslant (1 + n^{-1})z$ and, consequently, $-\alpha(1 + n^{-1})\zeta \leqslant Au \leqslant \alpha(1 + n^{-1})\zeta$, for all $n \in \mathbb{N}$. It follows $\|Au\|_\zeta \leqslant \alpha\|u\|_z$, by the definition of the norms.

Now, let (1.10) be true. Then, $-z \leqslant u \leqslant z$ implies $\|u\|_z \leqslant 1$. Consequently, $\|Au\|_\zeta \leqslant \alpha$, so that $-\alpha(1 + n^{-1})\zeta \leqslant Au \leqslant \alpha(1 + n^{-1})\zeta$, for all $n \in \mathbb{N}$. Since S is Archimedian, we conclude that also $-\alpha\zeta \leqslant Au \leqslant \alpha\zeta$.

To prove (iii), let $-z \leqslant u \leqslant z$. Then, $Au = A_1u + A_2(-u) \leqslant (A_1 + A_2)z$, and $Au \geqslant -(A_1 + A_2)z$. Since $\zeta \succ o$, there exists some $\alpha \geqslant 0$ with $(A_1 + A_2)z \leqslant \alpha\zeta$. Thus, for any α with this property, the inequality (1.10) follows with statement (i).

Since each $A \geqslant O$ is monotonically decomposable, we already have proved one part of statement (ii). It remains to be shown that $Az \leqslant \alpha\zeta$, if (1.10) holds. This follows from (i) applied to $u = z$. □

Of course, these results apply also to linear functionals $\ell: R \to \mathbb{R}$.

The special case where $(R, \leqslant) = (S, \leqslant)$ and $z = \zeta$ is of particular interest. The corresponding operator norm is denoted by

$$\|A\|_z := \inf\{\alpha \geqslant 0 : \|Au\|_z \leqslant \alpha\|u\|_z, u \in R\}.$$

Corollary 1.10a. *Suppose that $(R, \leqslant)$ is an Archimedian ordered linear space and that $A \in L[R, R]$ is positive. Then for any $z \succ o$ in R, $\|A\|_z \leqslant \alpha \Leftrightarrow Az \leqslant \alpha z$, and $\|A\|_z < 1 \Leftrightarrow Az \prec z$.*

Proof. The first statement is contained in part (ii) of Theorem 1.10. The second follows from the fact that $Az \prec z$ if and only if $Az \leqslant \alpha z$ for some $\alpha \in [0, 1)$. □

For example, this result can be applied when a fixed-point equation $u = Au + r$ with $r \in R$ is given. If $\|A\|_z < 1$ for some $z \succ o$, existence and uniqueness statements as well as error estimates follow from the fixed-point theorem of Banach.

Problem

1.8 The preceding theory can be generalized to preordered linear spaces. Carry out this generalization.

2 CONVEX SETS

A series of concepts and results for cones discussed in the preceding section will here be carried over—in a generalized form—to arbitrary convex sets. The theory will be used in Section IV.4 in order to prove range–domain implications. In addition, the theory below also yields an extension and a new understanding of several order theoretical terms and results of Section 1 (see Section 2.3).

2.1 Basic concepts

Let R be a real linear space, X a nonempty convex subset of R, and $A \subset X$. Of the terms defined below, the following three will be used directly in Section IV.4: The *core* A_X^c *of* A *relative to* X, the (*algebraic*) *boundary* $\partial_X A$ *of* A *relative to* X, and the term of a *segmentally* X*-closed set*. For convex sets A, in which we are mainly interested, these concepts will be characterized in Proposition 2.2 in a particularly simple way. The equivalence statements in this proposition can also be used as a definition, so that for many purposes it is not necessary to study the more general approach preceding Proposition 2.2.

In most applications X will equal R. In this case, the set X will not be indicated in the definitions; we then simply speak of the *core* A^c *of* A, the (*algebraic*) *boundary* ∂A *of* A and a *segmentally closed set* A. When using these simple terms (for instance, in Section IV.4), we need not introduce a topology in R.

The *core* (or *algebraic interior* or *internal set*) *of* A *relative to* X is defined to be the set A_X^c of all elements $u \in A$ with the following property: for each $v \in X$ with $v \neq u$, there exists an element $w \in (u : v]$ such that $[u : w] \subset A$. Moreover, we write $A_X^a = (X - A)_X^c$ and call $\partial_X A = X \sim (A_X^c \cup A_X^a)$ the *algebraic boundary* (or *bounding set*) *of* A *relative to* X. Finally, A is called *algebraically* X*-closed* if $\partial_X A \subset A$; the *algebraic* X*-closure* of A is $\overline{A}_X = A \cup \partial_X A$.

As mentioned before, we shall omit the letter X in the case $X = R$, so that, for example, $\overline{A} = A \cup \partial A$ is the algebraic closure of A. If the space R is endowed with a topology, the topological interior, closure, and boundary of A will be denoted by A^i, $\bar{A}$, and δA, respectively.

Example 2.1. Let $R = \mathbb{R}^2$, $X = \{u \in R : u_2 \geqslant 0\}$, $A = \{u \in R : u_1 \geqslant 0, u_2 > 0\}$. Then $A_X^c = A^i$, $A_X^a = X \sim \bar{A}$, $\partial_X A = \delta A$, and $\overline{A}_X = \bar{A}$. □

By definition of the above terms, we have $X = A_X^c \cup A_X^a \cup \partial_X A$ with the three sets on the right being pairwise disjoint. From this equation, a series of other relations follows, for example, $\overline{A}_X = A_X^c \cup \partial_X A$, $X = \overline{A}_X \cup A_X^a$.

For $u \in X$ the following characterizations by sequences are derived from the definitions: *$u \in \overline{A}_X$ if and only if there exist a $v \in X$ and an antitone zero sequence $\{\mu_n\}$ such that $u + \mu_n(v - u) \in A$ $(n = 1, 2, \ldots)$; $u \in \partial_X A$ if and only if $u \notin A_X^c$ and, moreover, v and $\{\mu_n\}$ as specified above exist.* Finally, *A is algebraically X-closed if and only if $u \in A$ whenever $u + \mu_n(v - u) \in A$ $(n = 1, 2, \ldots)$ for some $v \in X$ and an antitone zero sequence $\{\mu_n\}$.*

This characterization of algebraically X-closed sets leads to the following concept and result. The set A is called *segmentally X-closed* if and only if for each segment $[u : v] \subset X$ the set of all $t \in [0, 1]$ with $(1 - t)u + tv \in A$ is closed; in other words: each segment in X meets A in a closed subset of the segment. Moreover, a *segmentally closed set* is defined to be a segmentally R-closed set.

Proposition 2.1. *The set A is segmentally X-closed if and only if it is algebraically X-closed.*

If A is convex, then a *fixed* sequence $\{u + \mu_n(v - u)\}$ as specified above belongs to A if and only if $u + \mu(v - u) \in A$ for $0 < \mu \leqslant \mu_1$, that is, $u + \lambda(w - u) \in A$ for $0 < \lambda \leqslant 1$ and $w = u + \mu_1(v - u)$. Consequently, for convex A all terms introduced above can also be characterized using intervals $(u : w]$ or using *arbitrary* sequences of the above type. In particular we obtain the following result.

Proposition 2.2. *For convex A the following statements hold.*

(i) *$u \in R$ belongs to A_X^c, the core of A relative to X, if and only if $u \in A$ and to each $v \in X$ an $n \in \mathbb{N}$ exists such that $(n + 1)^{-1}(nu + v) \in A$.*

(ii) *$u \in R$ belongs to $\partial_X A$, the algebraic boundary of A relative to X, if and only if $u \notin A_X^c$ and a $v \in X$ exists such that $(n + 1)^{-1}(nu + v) \in A$ for all $n \in \mathbb{N}$.*

(iii) *A is segmentally X-closed if and only if for each fixed $u, v \in X$ the relations $(n + 1)^{-1}(nu + v) \in A$ $(n = 1, 2, \ldots)$ can only hold if $u \in A$.*

For determining $\partial_X A$ observe also that $\partial_X A = A \sim A_X^c$ for segmentally X-closed A.

Proposition 2.3. *For convex A*

(i) *$u \in \overline{A}_X$ if and only if $u \in X$ and $(u : v] \subset A$ for some $v \in A$.*

(ii) *$[u : v) \subset A_X^c$ whenever $u \in A_X^c$, $v \in A$.*

Proof. (i) follows from the remarks above.

(ii) We assume $v = o$, without loss of generality, and choose any $x \in X$. Since $u \in A_X^c$, there exists a $\lambda \in (0, 1)$ such that $u_\lambda := (1 - \lambda)u + \lambda x \in A$ and hence $su_\lambda \in A$ $(0 \leqslant s \leqslant 1)$. Given any $\mu \in (0, 1)$, one can find $\sigma \in (0, 1)$ and $\tau \in (0, 1)$ satisfying $(1 - \tau)w + \tau x = \sigma u_\lambda \in A$ for $w = \mu u$; consequently, $(1 - t)w + tx \in A$ for $0 \leqslant t \leqslant \tau$. Thus, each $w \in [u : v)$ belongs to A_X^c. □

Now we assume that A is convex. Then a set $E \subset A$ is called an *extremal subset* of A if $E \neq \emptyset$ and the following property holds:

$$\left.\begin{array}{ll} u \in A, \quad v \in A, & \\ (1 - \lambda)u + \lambda v \in E & \text{for some} \quad \lambda \in (0, 1) \end{array}\right\} \Rightarrow [u : v] \subset E. \tag{2.1}$$

An extremal subset may consist of only one element, which then is called an *extremal element* or *extremal point* of A.

Proposition 2.4. *A nonempty set $E \subset A$ is an extremal subset of the convex set A if and only if*

$$u \in A \sim E, \quad v \in A \quad \Rightarrow \quad [u : v) \subset A \sim E. \tag{2.2}$$

Proof. Suppose that E is an extremal subset, $u \in A \sim E$, and $v \in A$, but $x \in E$ for some $x \in [u : v)$. Then (2.1) implies $[u : v] \subset E$, which contradicts our assumptions.

Now suppose that the relations in the premise of (2.1) hold, but $y \in A \sim E$ for some $y \in [u : v]$. Then $x := (1 - \lambda)u + \lambda v$ belongs to $[u : y)$ or to $(y : v]$. In the latter case, (2.2) applied to y and v yields $[y : v) \subset A \sim E$, which contradicts $x \in E$. The first case is treated analogously. □

Proposition 2.5. (i) *$E = A \cap \partial A$ is an extremal subset of A if $E \neq \emptyset$.*
(ii) *If E is an extremal subset of A, then either $E = A$ or $E \subset \partial A$.*

Proof. (i) One verifies (2.2) for $A \sim E = A^c$ using Proposition 2.3(ii).
(ii) Suppose that $E \not\subset \partial A$, so that $A^c \cap E$ contains an element u. For any given $v \in A$, $v \neq u$, there exists an $\varepsilon > 0$ such that $u_\varepsilon := (1 - \varepsilon)u + \varepsilon(-v) \in A$ and hence $[u_\varepsilon : v] \subset E$, according to (2.1). Thus $A \subset E$. □

Finally, we briefly discuss certain relations between some of the terms introduced above and corresponding topological concepts. For simplicity, we restrict ourselves to considering a seminormed linear space R. It is quite obvious that $A^i \subset A^c$ for any $A \subset R$ and hence $\bar{A} \supset \overline{A}$. We are particularly interested in cases such that equality holds in these relations.

Theorem 2.6. *Let A denote a convex subset of a seminormed linear space R such that $A^i \neq \emptyset$.*

Then $A^i = A^c$, $\bar{A} = \overline{A}$, and $\delta A = \partial A$. In particular, A is topologically closed if and only if A is segmentally closed.

Proof. Let $z \in A^i$. To prove $\bar{A} \subset \overline{A}$ one shows that for each $u \in \delta A$ the set $[z : u)$ belongs to A, so that $u \in \overline{A}$. To prove $A^c \subset A^i$, let $u \in A^c$ and $w \in \partial A$ such that $u \in [z:w)$. There exists a sphere with center z such that the convex hull of this sphere and w belongs to A, except possibly the point w itself. This subset of A contains a sphere with center u; consequently $u \in A^{\mathrm{i}}$. □

Problem

2.1 Let $A = S_1 \cup S_2 \cup T$ be a subset in $\mathbb{R}^2$ where S_1, S_2 are two closed spheres which have exactly one point in common and T denotes the tangent of the spheres in this point. Determine A^{c}, ∂A, $(A^{\mathrm{c}})^{\mathrm{c}}$, $\partial(A^{\mathrm{c}})$.

2.2 Description by functionals

Let A denote a convex subset of a (real) linear space R. This set and related sets such as A_X^{c} and $\partial_X A$ can often be described by functionals or extended functionals. An *extended functional* ℓ on R is a (linear or nonlinear) function on R which may assume not only real values but also the values $+\infty$, $-\infty$, that means $\ell: R \to \{\mathbb{R}, -\infty, +\infty\}$. Often an extended functional will simply be called a functional.

A set $\mathscr{L}$ of extended functionals $\ell \neq o$ on R is said to *describe* A if to each $\ell \in \mathscr{L}$ a real number μ_ℓ exists such that for all $u \in R$

$$u \in A \quad \Leftrightarrow \quad \ell u \geqslant \mu_\ell \quad \text{for all} \quad \ell \in \mathscr{L}. \tag{2.3}$$

Moreover, in case $A^{\mathrm{c}} \neq \emptyset$, $\mathscr{L}$ is said to *describe A completely* if, in addition,

$$u \in A^{\mathrm{c}} \quad \Leftrightarrow \quad \ell u > \mu_\ell \quad \text{for all} \quad \ell \in \mathscr{L}.$$

In some of the later applications A will be described by a single functional $\Phi: R \to \{\mathbb{R}, -\infty\}$ which is *concave*, i.e.,

$$\Phi((1-t)u + tv) \geqslant (1-t)\Phi u + t\Phi v \qquad \text{for} \quad 0 \leqslant t \leqslant 1$$

and all $u, v \in R$ with finite Φu, Φv. For this case we formulate

Proposition 2.7. *Suppose that* $A \subset R$ *is a segmentally closed set and* $\Phi: R \to \{\mathbb{R}, -\infty\}$ *a concave functional such that* (i) $u \in A$ *if and only if* $u \in R$ *and* $\Phi u \geqslant 0$, (ii) $\Phi z > 0$ *for some* $z \in R$. *Define* X *to be the set of all* $u \in R$ *with finite* Φu.

Then both A *and* X *are convex, and for* $u \in R$

$$u \in A_X^{\mathrm{c}} \quad \Leftrightarrow \quad \Phi u > 0; \qquad u \in \partial_X A \quad \Leftrightarrow \quad \Phi u = 0.$$

Proof. The convexity of the sets is obvious. If $\Phi u > 0$ and $w \in X$, then a $\lambda \in (0, 1]$ exists such that $\Phi((1-\lambda)u + \lambda w) \geqslant (1-\lambda)\Phi u + \lambda \Phi w > 0$; consequently $u \in A_X^{\mathrm{c}}$. On the other hand, if $u \in A_X^{\mathrm{c}}$, then $[u - \lambda(z-u) : z] \subset A$ for small enough $\lambda > 0$, and therefore,

$$\begin{aligned}\Phi u &= \Phi[(1+\lambda)^{-1}(u - \lambda(z-u)) + \lambda(1+\lambda)^{-1}z] \\ &\geqslant (1+\lambda)^{-1}\Phi(u - \lambda(z-u)) + \lambda(1+\lambda)^{-1}\Phi z \geqslant \lambda(1+\lambda)^{-1}\Phi z > 0.\end{aligned}$$

Thus, the first equivalence statement is verified; the second is an easy consequence. □

Another use of nonlinear functionals will be described in Example 2.4.

The existence of a set $\mathscr{L}$ which describes a convex set completely can be deduced from the following theorem, which we state without proof.

Theorem 2.8. *Suppose that A and B are nonempty convex subsets of R with $A^c \neq \varnothing$ and $A^c \cap B = \varnothing$. Then a linear functional $\ell \in R^*$ exists such that*

$$\ell u \geqslant \ell v \quad \text{for all} \quad u \in A, \quad v \in B;$$
$$\ell u > \ell v \quad \text{for all} \quad u \in A^c, \quad v \in B.$$

In particular this result can be applied to a set B consisting of a single point $v \in \partial A$, so that we obtain

Corollary 2.8a. *To each point $v \in \partial A$ of a convex set A with $A^c \neq \varnothing$ there exists a functional $\ell \in R^*$ such that $\ell v = \mu_\ell$, and $\ell u > \mu_\ell$ for all $u \in A^c$, where $\mu_\ell = \inf_{w \in A} \ell w$.*

A functional $\ell \neq o$ in R^* such that $\ell v = \mu_\ell$ for some $v \in A$ (with μ_ℓ as in the above corollary) is said to *support A at v*. Moreover, a point (an element) $v \in A$ to which a functional $\ell \in R^*$ exists which supports A at v is called a *support point of A*. A *nonsupport point of A* hence is an element $u \in A$ such that $\ell u > \mu_\ell$ for all $\ell \neq o$ in R^*.

Theorem 2.9. *Suppose that $A \subset R$ is a segmentally closed convex set with $A^c \neq \varnothing$ and denote by $\mathscr{L}$ the set of all functionals $\ell \neq o$ in R^* which support A at some $v \in \partial A$. Then $\mathscr{L}$ describes A completely.*

Proof. It suffices to verify (2.3) for $\mu_\ell = \inf_{w \in A} \ell w$. We only show that $u \in A$, if $\ell u \geqslant \mu_\ell$ for all $\ell \in \mathscr{L}$, leaving the remaining proof to the reader.

Suppose the above inequalities hold for all $\ell \in \mathscr{L}$, but $u \notin A$. Due to our assumptions there exist a $z \in A^c$ and a $t \in (0, 1)$ such that $v := (1 - t)u + tz \in \partial A$. For the functional ℓ which supports A at v, we have $\ell u \geqslant \mu_\ell$ and $\ell z > \mu_\ell$ (cf. Corollary 2.8a). Consequently $\ell v > \mu_\ell$, which contradicts the definition of ℓ. □

Example 2.2. The set $A = \{u : u_1 u_2 \geqslant 1,\ u_1 > 0\} \subset \mathbb{R}^2$ is completely described by the set $\mathscr{L}$ of all functionals $\ell u = \alpha_1 u_1 + \alpha_2 u_2$ with $\alpha_1 > 0$, $\alpha_2 > 0$. Observe that the equation $\ell u = \mu_\ell := \inf_{w \in A} \ell w$ describes a tangent at ∂A and that, on the other hand, each tangent can be written in this way. We remark that the functionals $\ell_1 u = u_1$ and $\ell_2 u = u_2$, in a more general terminology, also are said to support A. However, the corresponding lines $\ell u = \mu_\ell$ have empty intersections with A. Of course, if these functionals are added to $\mathscr{L}$ the resulting set also describes A completely. □

Corollary 2.9a. *Under the assumptions of Theorem 2.9 an element $u \in R$ is a support point of A if and only if $u \in \partial A$, and u is a nonsupport point of A if and only if $u \in A^c$.*

Problem

2.2 For a convex set A the (generalized) *Minkowski functional* corresponding to $w \in A$ is defined by $\varphi u = \inf\{\alpha > 0 : u \in w + \alpha(A - w)\}$. Apply Proposition 2.7 to $\Phi u = 1 - \varphi u$.

2.3 The special case of cones

Now we apply the concepts and results of the previous sections to the case in which $(R, \leqslant)$ is an ordered linear space and A equals the positive cone K. Here the statements in Propositions 2.2 and 2.3 can be modified using the fact that $u \in K$ if and only if $\lambda u \in K$ for some $\lambda > 0$. In this way we obtain the next result from Proposition 2.2(i), (iii).

Proposition 2.10. (i) *An element $u \in R$ is strictly positive ($u \succ o$) if and only if $u \in K^c$.*

(ii) *The order relation $\leqslant$ is Archimedian if and only if the positive cone is segmentally closed.*

Proposition 2.11. (i) *$u \in \overline{K}$ if and only if $u + \lambda w \geqslant o$ for some $w \geqslant o$ and all $\lambda > 0$.*

(ii) *$\overline{K}$ is a wedge.*

(iii) *$\overline{K}$ is a cone if and only if for arbitrary $u, v \in R$ the following implication holds:*

$$-n^{-1}v \leqslant u \leqslant n^{-1}v \qquad (n = 1, 2, \ldots) \quad \Rightarrow \quad u = o. \tag{2.4}$$

(iv) *Suppose $(R, \leqslant)$ contains an element $z \succ o$. Then the seminorm $\| \ \|_z$ is definite if and only if $\overline{K}$ is a cone.*

Proof. (i) follows from Proposition 2.3(i).

(ii) Suppose that $u, v \in \overline{K}$. Then there exist $x \geqslant o$, $y \geqslant o$ such that $u + \lambda x \geqslant o$ and $v + \lambda y \geqslant o$ for all $\lambda > 0$; consequently, $\alpha u + \beta v + \lambda(\alpha x + \beta y) \geqslant o$ for all $\lambda > 0$, $\alpha \geqslant 0$, $\beta \geqslant 0$. Since $\alpha x + \beta y \geqslant o$, we conclude that $\alpha u + \beta v \in \overline{K}$ for all $\alpha \geqslant 0$, $\beta \geqslant 0$. Therefore, $\overline{K}$ is a wedge.

(iii) First suppose that $\overline{K}$ is a cone and the premise of (2.4) holds for some $u, v \in R$. The required inequalities imply $u + \lambda v \geqslant o$ and $-u + \lambda v \geqslant o$ for all $\lambda > 0$, so that $u \in \overline{K}$, $-u \in \overline{K}$ and hence $u = o$.

Now we assume that (2.4) holds and that $u \in \overline{K}$, $-u \in \overline{K}$ for some $u \in R$. Then there exist $x \geqslant o$, $y \geqslant o$ such that $u + \lambda x \geqslant o$, $-u + \mu y \geqslant o$ for all $\lambda > 0, \mu > 0$. For $\lambda = n^{-1}$ and $\mu = 2\lambda$, a suitable linear combination of these inequalities yields $u + n^{-1}v \geqslant o$ with $v = 2(x + y)$. Similarly, $-u + n^{-1}v \geqslant o$ is obtained. Since these inequalities hold for all $n \in \mathbb{N}$, we have $u = o$, according to (2.4).

(iv) follows from (iii). □

Theorem 2.6 immediately applies to the cone $A = K$, so that, due to Proposition 2.10, strictly positive elements and Archimedian orders can also be characterized by topological terms, provided the topological interior K^{i} (with respect to a seminorm topology) is not empty. In particular we obtain

Proposition 2.12. *Suppose that the ordered linear space $(R, \leqslant)$ contains a strictly positive element. Then for $u \in R$,*

$$u \in K^{\mathrm{i}} \quad \Leftrightarrow \quad u \in K^{\mathrm{c}} \quad \Leftrightarrow \quad u \succ o;$$

and

$$K = \bar{K} \quad \Leftrightarrow \quad K = \overline{K} \quad \Leftrightarrow \quad (R, \leqslant) \text{ is Archimedian,}$$

where K^{i} and $\bar{K}$ are defined with respect to the order topology.

The next results are concerned with the relation between strong order relations and extremal sets. For an extremal subset E of K define $K_E = (K \sim E) \cup \{o\}$.

Proposition 2.13. *If E is an extremal subset of the cone K, then*

(i) $u \in E \Rightarrow \lambda u \in E$ *for all* $\lambda \geqslant 0$;
(ii) $u \in K \sim E,\ v \in K \Rightarrow u + v \in K \sim E$;
(iii) K_E *is a cone.*

Proof. (i) If $u \in E \subset K$, then $w := \mu u \in K$ for each (fixed) $\mu > 1$. Since $u = (1 - \lambda)o + \lambda w \in E$ for $\lambda = \mu^{-1}$, property (2.1) yields $[o : w] \subset E$.

(ii) If $u \in K \sim E$, $v \in K$, but $u + v \in E$, then also $\frac{1}{2}(u + v) \in E$ according to (i). In this case (2.1) yields $[u : v] \subset E$, contradicting the assumption. (iii) is an easy consequence of (i) and (ii). □

Since, for any extremal subset $E \neq K$ of K, K_E is a cone, a linear order $\stackrel{\scriptscriptstyle E}{\leqslant}$ is induced in $(R, \leqslant)$ by

$$u \stackrel{\scriptscriptstyle E}{\geqslant} o \quad \Leftrightarrow \quad u \in K_E, \qquad \text{or equivalently,} \qquad u \stackrel{\scriptscriptstyle E}{>} o \quad \Leftrightarrow \quad u \in K \sim E. \tag{2.5}$$

Proposition 2.14. (i) *For a given extremal subset $E \neq K$ of the positive cone K in $(R, \leqslant)$, the order relation $\stackrel{\scriptscriptstyle E}{\leqslant}$ defined by* (2.5) *is stronger than $\leqslant$.*

(ii) *For a given linear order $\stackrel{\scriptscriptstyle E}{\leqslant}$ in $(R, \leqslant)$ which is stronger than $\leqslant$, the set $E \subset K$ defined by* (2.5) *is an extremal subset $E \neq K$ of K.*

Proof. (i) follows immediately from statement (ii) in Proposition 2.13.

(ii) For $u \in K \sim E$ and $v \in K$, that is, $u \succ o$ and $v \geqslant o$, we have

$$(1 - \lambda)u \succ o \qquad \text{and} \qquad v \geqslant o \qquad (0 \leqslant \lambda < 1),$$

hence $(1 - \lambda)u + \lambda v \succ o$, so that $[u: v) \subset K \sim E$. Thus E is an extremal set, according to Proposition 2.4. □

As this proposition shows, there is a one-to-one correspondence between strong orders and extremal sets $E \neq K$.

Example 2.3. (a) Let $K^c \neq \{o\}$. Then $E = \partial K \cap K$ is the extremal subset corresponding to the strict order $\lesssim$ in R.

(b) Let $K \neq \{o\}$. Then $E = \{o\}$ is the extremal subset corresponding to the order $\leqslant$. □

It is essential for our theory on operator inequalities in Archimedian ordered linear spaces that we distinguish a subset of the positive cone K by certain properties from the complementary part of K. If $K^c \neq \varnothing$, we shall, in general, choose ∂K to be this subset. However, if $K^c = \varnothing$, one has $K = \partial K$, so that the complement is empty. In such a case one may choose $\partial_X K$ for a suitable set X.

We choose an element $p > o$ in $(R, \leqslant)$ and define *p-strictly positive elements* by

$$u \succ_p o \quad \Leftrightarrow \quad u \geqslant \alpha p \qquad \text{for some} \quad \alpha > 0. \tag{2.6}$$

Moreover, we denote by B_p the set of all $u \in R$ which are *p-bounded below*, that is, $u \geqslant -\gamma p$ for some $\gamma \in \mathbb{R}$. Clearly, if $p \succ o$, then $u \succ_p o \Leftrightarrow u \succ o$, and $B_p = R$.

Defining $\Phi u = \sup\{\lambda \in \mathbb{R} : u \geqslant \lambda p\}$ one derives the following statements from Proposition 2.7.

Proposition 2.15. *Suppose that $(R, \leqslant)$ is Archimedian and define $X = B_p$ for some $p > o$ in R. Then for all $u \in R$,*

$$u \succ_p o \quad \Leftrightarrow \quad u \in K_X^c; \qquad u \in \partial_X K \quad \Leftrightarrow \quad u \not\succ_p o, \quad u \geqslant o.$$

Suppose that $\mathscr{L} \subset R^*$ describes K and that $p > o$ is a nonsupport point of K with respect to $\mathscr{L}$, that means no $\ell \in \mathscr{L}$ supports K at p. Then $\Phi u = \inf\{\ell u(\ell p)^{-1} : \ell \in \mathscr{L}\}$ and, in particular, $u \succ_p o \Leftrightarrow \inf\{\ell u(\ell p)^{-1} : \ell \in \mathscr{L}\} > 0$.

Example 2.4. Let $R = l_\alpha$ with $1 \leqslant \alpha \leqslant \infty$ and define $u = \{u_i\} \geqslant o$ by $u_i \geqslant 0$ $(i = 1, 2, \ldots)$. Clearly, the set $\mathscr{L}_0$ of all functionals ℓ_i $(i = 1, 2, \ldots)$ defined by $\ell_i u = u_i$ describes the cone K, and $p \in R$ is a nonsupport point of K if and only if $p_i > 0$ $(i = 1, 2, \ldots)$. If p has this property then, for example, $u \succ_p o \Leftrightarrow \inf_i u_i/p_i > 0$.

However, we may also describe the points $u \succ_p o$ in a different way. Define $\mathscr{L}$ to be the union of $\mathscr{L}_0$ and the (nonlinear) functional $\ell_\infty u = \liminf_i u_i$.

Obviously, if $p_i > 0$ $(i = 1, 2, \ldots)$, then

$$\begin{aligned} u \geqslant o \quad &\Leftrightarrow \quad \ell u \geqslant 0 \quad \text{for all} \quad \ell \in \mathscr{L}; \\ u \succ_p o \quad &\Leftrightarrow \quad \ell u > 0 \quad \text{for all} \quad \ell \in \mathscr{L}. \end{aligned} \tag{2.7}$$

Thus, $\mathscr{L}$ has properties analogous to (1.7) and (1.8).

For the space $R = l_\infty$ we have $K^c \neq \varnothing$ and hence $u \succ_p o \Leftrightarrow u \succ o$, for $p \in K^c$, so that the above set $\mathscr{L}$ describes K completely. According to Proposition 1.5, the cone K can also be described completely by a set of linear functionals. However, it is sometimes more convenient to use the above set $\mathscr{L}$, which contains the nonlinear functional ℓ_∞. □

Problems

2.3 Derive Proposition 1.5 from Theorem 2.9.

2.4 Let $R = C_0[0, \infty)$ and $p \in R$ with $p(x) > 0$ $(0 \leqslant x < \infty)$. Find a set $\mathscr{L}$ of functionals on R such that (2.7) holds.

2.5 Consider an order interval $A = [\varphi, \psi]$ in an Archimedian ordered linear space $(R, \leqslant)$. Show that A is segmentally closed and $A^c = \{u \in R : \varphi \prec u, u \prec \psi\}$.

2.6 Let $(R, \leqslant)$ and A be as in the preceding problem and suppose that $p, q \in R$ satisfy $p > o$, $q > o$. Denote by X the set of all $u \in R$ such that $u \in [\varphi - \gamma p, \psi + \gamma q]$ for some $\gamma \in \mathbb{R}$. Show that $u \in A^c_X$, if and only if $\varphi + \alpha p \leqslant u \leqslant \psi - \alpha q$ for some $\alpha > 0$.

2.7 Let $(R, \leqslant)$ and A be as in Problem 2.5 and suppose that $\overset{s}{\leqslant}$ is a strong order in R. Show that the set $E \subset A$ defined by $A \sim E = \{u \in R : \varphi \overset{s}{<} u \overset{s}{<} \psi\}$ with $\varphi \overset{s}{<} \psi$ is an extremal set of A.

3 FIXED-POINT THEOREMS AND OTHER MEANS OF FUNCTIONAL ANALYSIS

In this section we collect some results on functional analysis which can be applied to prove the existence of solutions of operator equations. The proofs will be omitted or sketched only.

3.1 Compact operators in a Banach space

Suppose that R is a (real or complex) Banach space with norm $\| \ \|$. A set $D \subset R$ is called *relatively compact* if each sequence $\{u_\nu\}$ in D contains a convergent subsequence (with limit in R). An operator T which maps its domain $D \subset R$ into R is said to be *compact* if T maps each bounded subset of D into a relatively compact set.

Theorem 3.1 (*Fixed-point theorem of Schauder*). *Suppose that T is a continuous operator which maps a closed convex subset $D \neq \varnothing$ of R into a relatively compact set $TD \subset D$.*

Then there exists a fixed point $u^* = Tu^* \in D$.

The next theorem can be deduced from the theory of Leray–Schauder on the degree of mapping.

Theorem 3.2. *Suppose that D is a closed bounded convex subset of R with $o \in D^{\mathrm{i}}$ and $T: D \to R$ a continuous compact operator such that for each $\alpha \in (0, 1)$*

$$\left.\begin{array}{l} u \in D \\ u - \alpha T u = o \end{array}\right\} \quad \Rightarrow \quad u \in D^{\mathrm{i}}. \tag{3.1}$$

Then there exists a fixed point $u^* = Tu^* \in D$.

Now suppose that $T: R \to R$ is a bounded linear operator, and define $\mathbb{K} = \mathbb{R}$ or $\mathbb{K} = \mathbb{C}$, if R is a real or complex Banach space, respectively. We say that T is *invertible* (or that the inverse operator T^{-1} exists) if the equation $Tu = o$ has the only solution $u = o$ in R. Then the *inverse operator* T^{-1} is defined on TR by $T^{-1}v = u$ with $v = Tu$. A number $\lambda \in \mathbb{K}$ belongs to the *residual set* of T if $\lambda I - T$ is invertible and $(\lambda I - T)^{-1}$ is a bounded operator defined on all of R. The *spectrum* of T consists of all $\lambda \in \mathbb{K}$ which do not belong to the residual set. The spectrum of T is a closed bounded subset of $\mathbb{K}$.

If R is a complex Banach space ($\mathbb{K} = \mathbb{C}$), then the *spectral radius* $\rho(T)$ of T is the maximum of all values $|\lambda|$ with λ in the spectrum of T. For the case $\mathbb{K} = \mathbb{R}$ the spectral radius $\rho(T)$ is defined by $\rho(T) = \rho(\hat{T})$ with $\hat{T}: \hat{R} \to \hat{R}$ denoting the *complexification* of T. Here $\hat{R}$ is the complex Banach space of all $\hat{u} = u_1 + iu_2$ ($u_1, u_2 \in R$, $i = \sqrt{-1}$) with the norm $\|\hat{u}\| = \sup\{\|u_1 \cos \vartheta + u_2 \sin \vartheta\| : 0 \leqslant \vartheta \leqslant 2\pi\}$ and $\hat{T}\hat{u} = Tu_1 + iTu_2$. In most applications one need not "construct" $\hat{T}$ in this manner; instead, T occurs as the "natural" restriction of $\hat{T}$. For example, if $R = \mathbb{R}^n$ and T is given by a real $n \times n$ matrix, then $\hat{T}: \mathbb{C}^n \to \mathbb{C}^n$ is an operator defined by the same matrix. Thus $\rho(T)$ is the maximum of all $|\lambda|$ with λ being a complex eigenvalue of the matrix.

An *ordered Banach space* is defined to be a Banach space endowed with a linear order which is compatible with the norm; that means each convergent sequence $\{u_n\}$ of positive elements in R has a positive limit. An *ordered normed linear space* is defined in a corresponding way.

Theorem 3.3. *Suppose that T is a positive compact linear operator which maps an ordered Banach space into itself, and that $\rho(T) > 0$.*

Then there exists a $\varphi > o$ in R such that $T\varphi = \lambda\varphi$ for $\lambda = \rho(T)$.

3.2 Iterative procedures in ordered spaces

Now suppose that $(R, \leqslant)$ is an ordered space such that in R the notion of a convergent sequence $\{u_\nu\}$ and the notion of the limit $u = \lim u_\nu$ are defined (we shall also write $u_\nu \to u$). These terms may be induced by a norm, or by other means. We assume only that (i) each sequence has at most one limit; (ii) if $u_\nu = u \in R$ $(\nu = 1, 2, \ldots)$, then $u_\nu \to u$; (iii) if $u_\nu \to u$, then $u_{k_\nu} \to u$ for each monotone subsequence $\{k_\nu\} \subset \{\nu\}$; (iv) if $u_\nu \to u$ and $u_\nu \leqslant v$ $(\nu = 1, 2, \ldots)$, then $u \leqslant v$. Under these conditions R is called an *ordered limit space* (*ordered L-space*).

By T we denote an operator which maps at set $D \subset R$ into R. This set is assumed to be *order convex*, i.e., $[u, v] \subset D$ for $u, v \in D$. The operator T is called *L-continuous* if $\lim Tu_\nu = T(\lim u_\nu)$ for each convergent sequence $\{u_\nu\}$ in D with limit in D.

Theorem 3.4. *Suppose that T is monotone and x_0, y_0 denote elements in D which satisfy*

$$x_0 \leqslant Tx_0, \qquad x_0 \leqslant y_0, \qquad Ty_0 \leqslant y_0. \tag{3.2}$$

Then the following statements hold.

(i) *The formulas*

$$x_{\nu+1} = Tx_\nu, \qquad y_{\nu+1} = Ty_\nu \qquad (\nu = 0, 1, 2, \ldots)$$

define sequences $\{x_\nu\}$, $\{y_\nu\}$ such that for all integers $\nu \geqslant 0$

$$x_\nu \leqslant x_{\nu+1} \leqslant y_{\nu+1} \leqslant y_\nu \tag{3.3}$$

and

$$T[x_\nu, y_\nu] \subset [x_\nu, y_\nu]. \tag{3.4}$$

(ii) *If T has a fixed point $u^* = Tu^* \in [x_0, y_0]$, then*

$$x_\nu \leqslant u^* \leqslant y_\nu \qquad (\nu = 0, 1, 2, \ldots). \tag{3.5}$$

(iii) *If T is L-continuous and if the sequences $\{x_\nu\}$, $\{y_\nu\}$ converge toward limits x^*, y^*, respectively, then these limits are fixed points of T and each fixed point $u^* \in [x_0, y_0]$ of T satisfies $x^* \leqslant u^* \leqslant y^*$.*

Proof. (i) Applying the monotone operator T to the second inequality in (3.2) we obtain $Tx_0 \leqslant Ty_0$, so that (3.3) holds for $\nu = 0$. The general statement (3.3) is proved by induction. If $x_\nu \leqslant x \leqslant y_\nu$ for some ν, then $x_\nu \leqslant x_{\nu+1} = Tx_\nu \leqslant Tx \leqslant Ty_\nu = y_{\nu+1} \leqslant y_\nu$, so that (3.4) holds. (ii) and (iii) are verified by similar means. □

When we want to establish the existence of a fixed point of T under the general assumptions of the above theorem, we may, for example, exploit property (3.4) or try to verify the additional assumptions in (iii). In the following remarks we shall sketch some ways to proceed.

Remarks. 1. If $(R, \leqslant)$ is an ordered Banach space, then $[x_0, y_0]$ is a closed convex subset. Thus, if $T[x_0, y_0]$ is relatively compact, Schauder's fixed-point theorem yields the existence of a fixed point $u^* \in [x_0, y_0]$, for which (3.5) holds also.

2. Suppose that R is an ordered Banach space which is *normal*, i.e., there exists a constant γ such that $o \leqslant u \leqslant v$ implies $\|u\| \leqslant \gamma\|v\|$ (for $u, v \in R$). Then the sequence $\{x_\nu\}$ is bounded. Thus, if T is compact, the image $\{Tx_\nu\}$ contains a convergent subsequence with limit x^*. Since the sequence $\{x_\nu\}$ is monotone, this sequence, too, converges toward x^*. Analogous arguments hold for the sequence $\{y_\nu\}$. Assuming, in addition, that T is continuous, one can apply statement (iii) of the theorem.

3. A subset $A \subset R$ is called *order-bounded* if elements $x, y \in R$ exist such that $x \leqslant u \leqslant y$ for all $u \in A$. If the ordered L-space R has the additional property that each order-bounded monotone (or antitone) sequence $\{u_\nu\}$ converges toward a limit $u^* \in R$ and if $T(\lim u_\nu) = \lim Tu_\nu$ for each such sequence, then the existence of fixed points x^* and y^* as in (iii) follows.

4. Sometimes it is helpful to extend a given equation $u = Tu$ in R to an equation $\hat{u} = \hat{T}\hat{u}$ in a larger space $\hat{R}$, then prove the existence of a fixed point of $\hat{T}$ and, finally, verify that this fixed point belongs to R.

Suppose, for instance, that the properties used in the third remark do not hold for R and T, but for an ordered L-space $(\hat{R}, \leqslant) \supset (R, \leqslant)$ and an extended operator $\hat{T}: [x_0, y_0]_{\hat{R}} \to \hat{R}$. Then the existence of a fixed point of $\hat{T}$ follows as in Remark 3. If, in addition, $\hat{T}\hat{u} \in R$ for $\hat{u} \in \hat{R}$, then this fixed point belongs to R and thus is a fixed point of T. (Similar arguments hold if $\hat{T}^m\hat{u} \in R$ for some $m \in \mathbb{N}$.)

Instead of requiring that T be monotone we assume now that Tu can be estimated by

$$H_1[u, u] \leqslant Tu \leqslant H_2[u, u] \qquad \text{for} \quad u \in D,$$

where $H_i[u, v] \in R$ $(i = 1, 2)$ are defined for $u, v \in D$ and satisfy

$$H_i[u_1, v_1] \leqslant H_i[u_2, v_2] \qquad \text{for} \quad u_1 \leqslant u_2, \quad v_1 \geqslant v_2.$$

If $H_1[u, v] = H_2[u, v] =: H[u, v]$ for $u, v \in D$ and hence

$$Tu = H[u, u] \qquad \text{for} \quad u \in D,$$

we call T *monotonically decomposable.* An important special case is described by

$$H[u, v] = T_1 u - T_2 v \quad \text{with monotone operators} \quad T_i : D \to R \quad (i = 1, 2).$$

In particular, we have a corresponding presentation of T, if T is monotone ($T = T_1$) or antitone ($T = -T_2$).

Theorem 3.5. *Suppose that mappings H_1, H_2 as described above exist and that x_0, y_0 denote elements in D which satisfy*

$$x_0 \leqslant H_1[x_0, y_0], \qquad x_0 \leqslant y_0, \qquad H_2[y_0, x_0] \leqslant y_0.$$

Then the formulas

$$x_{\nu+1} = H_1[x_\nu, y_\nu], \qquad y_{\nu+1} = H_2[y_\nu, x_\nu] \qquad (\nu = 0, 1, 2, \ldots),$$

define sequences $\{x_\nu\}$, $\{y_\nu\}$ in D such that (3.3) *and* (3.4) *hold. Moreover, statement* (ii) *of Theorem* 3.4 *is true for this case also.*

To derive the existence of a fixed point of T with this theorem one may apply Schauder's fixed point theorem exploiting property (3.5) (see Remark 1). Moreover, under suitable conditions one can also prove that the sequences $\{x_\nu\}$, $\{y_\nu\}$ converge. Observe, however, that (for L-continuous H_1, H_2) the limits x^*, y^* satisfy $x^* = H_1[x^*, y^*]$, $y^* = H_2[y^*, x^*]$, and thus, in general, need not be fixed points of T.

For linear operators, additional results can be proved. In Section II.2.4 such results will be formulated for matrices. The statements derived there can be carried over to more general linear operators.

NOTES

Sections 1 and 2

Most of the material contained in Sections 1 and 2 can be found in any of the available books on ordered linear spaces. For example, an approach similar to ours was used by Jameson [1970] (see also the lecture notes of Bauer [1974]). Krein and Rutman [1948] presented much of the theory on ordered Banach spaces which contain a strictly positive element, denoting such an element by $u \gg o$. (The notation $u \succ o$ is more convenient for our purposes, in particular, since we have to combine this order relation with an equality sign, as in $u \preccurlyeq v$.)

In the theory of convex sets the terms *algebraically closed* (algebraisch abgeschlossen), *algebraic boundary* (algebraischer Rand), etc. were used by Köthe [1966]. The more general concepts, such as the *core relative to* X were introduced by Klee [1969]. A proof of Theorem 2.8 can, for example, be found in Krein and Rutman [1948] and Kelley and Namoika [1963]. Proposition 2.14 was taken from Schröder [1970a].

Section 3

For a proof of Schauder's fixed-point theorem see, for example, Edwards [1965, Corollary 3.6.2]. A simple proof of Theorem 3.2 which does not use the degree theory was given by Schaefer [1955]. Concerning a proof of Theorem 3.3, see Krein and Rutman [1948, Theorem 6.1] or Schaefer [1966, p. 265].

Iterative procedures with (abstract) monotone operators were first investigated by Kantorovič [1939] (in countably complete lattices). The case of (abstract) antitone operators was considered by Schröder [1956a] and Collatz and Schröder [1959]. Monotonically decomposable operators were introduced by Schröder [1959].

The results on iterative procedures with monotone operators can be applied to investigate iterative procedures in a generalized metric space, using a majorant principle. This was first undertaken by Kantorovič [1939] who used elements of a lattice as generalized norms. Schröder [1956b, 1961c] defined distances $\delta(u, v)$ as elements of a more general ordered limit space. In the second paper mentioned, the following four properties were required of $\delta(u, v)$. (i) $\delta(u, v) = o \Leftrightarrow u = v$; (ii) $\delta(u_\nu, v_\nu) \leqslant \xi_\nu, \xi_\nu \to o \Rightarrow \delta(u_\nu, v_\nu) \to o$ (*positivity in limit*); (iii) $\delta(u_\nu, v_\nu) \to o \Rightarrow \delta(v_\nu, u_\nu) \to o$ (*symmetry in limit*); (iv) $\delta(u, v) \leqslant \delta(u, v) + \delta(w, v)$. Concerning further literature on this subject see Bohl [1974], Collatz [1964], Ortega and Rheinbold [1970].

Chapter II

Inverse-positive linear operators

This chapter is devoted to theory and applications of inverse-positive linear operators M, which are described by the property that $Mu \geqslant o$ implies $u \geqslant o$. These are operators M which have a positive inverse M^{-1}. For an inverse-positive operator M one can derive estimates of $M^{-1}r$ from properties of r, without knowing the inverse M^{-1} explicitly. Obviously, this property can be used to derive a priori estimates for solutions of equations $Mu = r$. However, there are important further applications, a series of which will be discussed either for abstract operators or for special operator classes. For example, if M is inverse-positive, an equation $Mu = Bu$ with a nonlinear operator B may be transformed into a fixed-point equation $u = M^{-1}Bu$, to which then a monotone iteration method or other methods may be applied, if B has suitable properties. Moreover, for inverse-positive M the eigenvalue problem $M\varphi = \lambda\varphi$ is equivalent to $T\varphi = \kappa\varphi$ with $\kappa = \lambda^{-1}$ and $T = M^{-1} \geqslant O$. Thus the theory on inverse-positive operators is closely related to the eigenvalue theory of positive operators, in particular to the Perron–Frobenius theory on positive eigenelements.

Section 1 mainly provides sufficient conditions and necessary conditions for inverse-positivity. Here, the *monotonicity theorem* is the basic result on which most of the further theory depends. To certain classes of operators this theory can be applied in a comparably easy way, yielding at the same time a rich theory on these operators. In particular, this is true for operators in finite-dimensional spaces defined by square matrices $M = (m_{ik})$ with $m_{ik} \leqslant 0$ for $i \neq k$ (Z-matrices), and for second order differential operators related to boundary value problems. Such operators are treated in Sections 2 and 3, respectively.

On the other hand, for other classes of operators, such as differential operators of higher order, the theory of inverse-positivity is rather complicated. The crucial assumption in the monotonicity theorem is the requirement that the operator M be *pre-inverse-positive*. In all our applications to special operators it turns out that verifying pre-inverse-positivity is either "trivial" or very complicated. Moreover, operators for which pre-inverse-positivity is easily verified usually have the much stronger property that they are Z-operators defined by the property that all operators $M_\lambda u = Mu + \lambda u$ ($\lambda \in \mathbb{R}$) are also pre-inverse-positive.

Section 4 deals with this situation, continuing the abstract theory. On one hand, more sophisticated methods for proving inverse-positivity are discussed, which essentially means, methods for proving pre-inverse-positivity. On the other hand, a theory of Z-operators M and Z-pairs (M, N) of operators is developed. Operators $M = I - T$ with positive T constitute a special case of such Z-operators.

Section 5 is concerned with numerical applications. Some further results are described in Section 6.

1 AN ABSTRACT THEORY

Simple sufficient conditions and necessary conditions for a linear operator M being inverse-positive are provided in Section 1.2, where the *monotonicity theorem* is the main result. In Section 1.3 it is shown that stronger and more detailed statements on the "positivity" of u can often be derived from $Mu \geqslant o$. This statement will, for instance, be used in Section 3 to prove the strong maximum principle and oscillation properties for differential equations. For certain types of operators, such as those treated in Sections 2 and 3, the theory of sets of inverse-positive operators in Section 1.4 leads to interesting "natural" characterizations of inverse-positive operators, which are different from those directly derived with the theory in Section 1.2. While inverse-positive linear operators in ordered linear spaces always are invertible, certain noninvertible operators with similar properties, as considered in Section 1.5, are also of interest, in particular, in the theory of eigenvalue problems.

1.1 The concept of an inverse-positive operator, general assumptions

Suppose that $(R, \leqslant)$ and $(S, \leqslant)$ are ordered linear spaces with positive cones K and C, respectively, and that M is a linear operator which maps R into S. (The order relations in R and S are denoted by the same symbol $\leqslant$, although they may be different, even if the sets R and S have the same elements.)

The operator M is called *inverse-positive* (with respect to the given order relations) if and only if for all $u \in R$

$$Mu \geqslant o \quad \Rightarrow \quad u \geqslant o, \tag{1.1}$$

or, equivalently, $Mu \in C \Rightarrow u \in K$. When we want to indicate the cones, we may use the term (C, K)*-inverse-positive* or, more briefly, (C, K)*-inverse.*

This notation has been introduced because of the following fact.

Proposition 1.1. *The operator M is inverse-positive if and only if the inverse M^{-1} exists and is positive.*

Proof. Let M be inverse-positive and $Mu = o$. Then $Mu \geqslant o$ and $M(-u) \geqslant o$, and consequently, $u \geqslant o$ and $-u \geqslant o$, so that $u = o$. Thus, M^{-1} exists. If M^{-1} exists, however, then statement (1.1) is equivalent to $U \geqslant o \Rightarrow M^{-1}U \geqslant o$, for $U = Mu$. □

We shall also consider stronger properties than (1.1), such as the following, which involves the strict order $\prec$ in R. The operator M is called *strictly inverse-positive* if for all $u \in R$

$$Mu \geqslant o \quad \Rightarrow \quad u \succ o.$$

In the theory of this section we shall assume that $M : R \to S$ is a given linear operator and that the spaces R and S satisfy the following two assumptions (A_1) and (A_2), unless otherwise indicated. These assumptions will also be referred to in later sections.

(A_1) *Let $(R, \leqslant)$ and $(S, \leqslant)$ be ordered linear spaces with order cones K and C, respectively, such that $(R, \leqslant)$ is Archimedian and contains a strictly positive element.*

(A_2) *Suppose that in the space S a further linear order $\preccurlyeq$ with order cone W is given which is stronger than the relation $\leqslant$ in S.* (This relation $\preccurlyeq$ will simply be called the *strong order* in S.)

Terms such as positive and inverse-positive are always interpreted with respect to the order relations denoted by $\leqslant$. Many of the following results hold also under weaker assumptions. For example, one may consider pre-ordered spaces R and S. Then, however, an inverse-positive linear operator need not be invertible.

1.2 Basic results

We are interested in sufficient conditions and necessary conditions such that a given linear operator $M : R \to S$ is inverse-positive. From the following basic theorem many other results will be derived in this section and other following sections.

Theorem 1.2 (*Monotonicity theorem for linear operators*). *Let the following two conditions be satisfied.*

I. *For all* $u \in R$

$$\left.\begin{array}{r} u \geqslant o \\ Mu \mathrel{\dot{>}} o \end{array}\right\} \quad \Rightarrow \quad u \succ o.$$

II. *There exists an element* $z \in R$ *such that*

$$z \geqslant o, \qquad Mz \mathrel{\dot{>}} o.$$

Then the operator M *is inverse-positive.*

Proof. Because of condition I, the element z is strictly positive. Now suppose that $Mu \geqslant o$, but $u \not\geqslant o$ for some $u \in R$. Define $v(\lambda) = (1 - \lambda)u + \lambda z = u + \lambda(z - u)$, so that $v(0) = u \not\geqslant o$ and $v(1) = z \succ o$. According to Propositions I.1.2 and I.1.3, there exists a smallest $\lambda_0 \in (0, 1)$ with $v(\lambda_0) \geqslant o$, $v(\lambda_0) \nsucc o$. Because of condition II, we have $Mv(\lambda_0) = (1 - \lambda_0)Mu + \lambda_0 Mz \mathrel{\dot{>}} o$. These inequalities together contradict condition I applied to $v(\lambda_0)$. Thus, $Mu \geqslant o \Rightarrow u \geqslant o$. □

Corollary 1.2a. *If* M *is inverse-positive and satisfies condition* I, *then for all* $u \in R$

$$Mu \mathrel{\dot{>}} o \quad \Rightarrow \quad u \succ o; \qquad Mu \mathrel{\dot{\geqslant}} o \quad \Rightarrow \quad u \succcurlyeq o. \tag{1.2}$$

Proof. Since $Mu \mathrel{\dot{\geqslant}} o \Rightarrow Mu \geqslant o \Rightarrow u \geqslant o$, the statement follows from condition I. □

Notice that (1.2) means strict inverse-positivity of M, if $U \mathrel{\dot{>}} o$ is equivalent to $U > o$.

An operator M which satisfies condition I of the above theorem is said to be *pre-inverse-positive*. An element $z \in R$ such that $z \succ o$, $Mz \mathrel{\dot{>}} o$ is called a *majorizing element for* M. Thus, the monotonicity theorem can be reformulated as follows.

If M *is pre-inverse-positive and if there exists a majorizing element for* M, *then* M *is inverse-positive.*

Of course, the above notation depends on the given order relations. If we consider both relations $\leqslant$ (in R and S) as fixed, the notation still depends on the strong order $\overset{\prime}{\leqslant}$ which has been chosen. When we want to indicate this dependence, we shall use the terms *pre-inverse-positive with respect to* $\overset{\prime}{\leqslant}$ or *W-pre-inverse-positive*. The weaker the relation $\overset{\prime}{\leqslant}$ is, the weaker condition II is and the stronger condition I is. We shall say that M is *weakly pre-inverse-positive*, if $u \geqslant o$ and $Mu \succ o$ together imply $u \succ o$; and we shall say that M is *strictly pre-inverse-positive*, if $u \geqslant o$ and $Mu > o$ together imply $u \succ o$. Moreover, $z \in R$ is called a *strictly majorizing element* or a *weakly majorizing element*, if $z \succ o$ and $Mz \succ o$ or $Mz > o$, respectively.

The following propositions provide simple tools which can be helpful for proving pre-inverse-positivity (for a given strong order $\overset{\prime}{\leqslant}$ in S). The proofs are left to the reader.

Proposition 1.3. *The operator M is pre-inverse-positive if for each $u \in R$ with $u \geqslant o, u \nsucc o$ there exists a linear functional ℓ on S such that $\ell Mu \leqslant 0$ and $\ell U > 0$ for all $U \gtrdot o$ in S.*

Proposition 1.4. *Let $M = A - B$ with linear operators $A, B : R \to S$ such that A is pre-inverse-positive and B is positive. Then M is pre-inverse-positive.*

When the element Mu has stronger positivity properties, the conditions on z can be weakened.

Corollary 1.2b. *If in Theorem* 1.2 *the inequalities $z \succ o$ and $Mz \geqslant o$ are required, instead of $z \geqslant o$ and $Mz \gtrdot o$, then for all $u \in R$ the inequalities $Mu \geqslant o$ and $Mu + Mz \gtrdot o$ imply $u \geqslant o$.*

The proof is similar to the proof of the theorem except that, for $\lambda_0 > 0$, the inequality $Mv(\lambda_0) \gtrdot o$ is derived from $Mu \geqslant o$, $Mz \geqslant o$, and $Mu + Mz \gtrdot o$. For example, this corollary yields the following result.

Theorem 1.5. *Let M be pre-inverse-positive and suppose that there exists an element $z \in R$ with $z \succ o$, $Mz \geqslant o$. Then for all $u \in R$, $Mu \gtrdot o \Rightarrow u \succ o$.*

Proof. Let $Mu \gtrdot o$. Then we have $Mu \geqslant o$ and $Mu + Mz \gtrdot o$, so that $u \geqslant o$ by the above corollary. Moreover, $u \succ o$ follows, because M is pre-inverse-positive. □

Essentially, the assumptions of the above theorems are also necessary, as formulated in the following theorem.

Theorem 1.6. (i) *If M is inverse-positive, then M is weakly pre-inverse-positive and each $z \in R$ with $Mz \succ o$ is a majorizing element.*

(ii) *If* (1.2) *holds, then M is pre-inverse-positive* (*with respect to* $\overset{\prime}{\leqslant}$). (iii) *Each strictly inverse-positive M is strictly pre-inverse-positive.*

Proof. (i) Let $u \geqslant o$ and $Mu \succ o$. Then, for each $v \in R$, there exists an $n \in \mathbb{N}$ such that $M(nu + v) = nMu + Mv \geqslant o$. This inequality implies

$nu + v \geqslant o$. Thus, $u \succ o$, according to the definition of the strict order relation. The remaining statement in (i) and the statements (ii) and (iii) are obvious. □

Problems

In all these problems, assumptions (A_1) and (A_2) are supposed to be satisfied.

1.1 Let $B: R \to R$ denote a positive linear operator. Prove that (a) $I - B$ is inverse-positive, if $z \geqslant o$ and $z \succ Bz$ for some $z \in R$; (b) $I - tB$ is inverse-positive for small enough $t > 0$.
1.2 Suppose that $A, M \in L[R, S]$ and that A satisfies both conditions I and II of the monotonicity theorem. Prove that (a) M is inverse-positive if $A \leqslant M$ and M satisfies I; (b) M is inverse-positive if $A \geqslant M$ and M satisfies II.
1.3 Derive relations between the following four properties (a)–(d), which are understood to hold for all $u \in R$:

$$\text{(a)} \quad Mu \geqslant o \quad \Rightarrow \quad u \succcurlyeq o, \qquad \text{(b)} \quad Mu \succcurlyeq o \quad \Rightarrow \quad u \succcurlyeq o,$$
$$\text{(c)} \quad Mu \geqslant o \quad \Rightarrow \quad u \geqslant o, \qquad \text{(d)} \quad Mu \succcurlyeq o \quad \Rightarrow \quad u \geqslant o.$$

In particular, prove that (d) implies (c) if there exists a $z \in R$ with $Mz \succ o$, and that (d) implies (b).
1.4 Prove that $M = A - B$ with $A, B \in L[R, S]$ is pre-inverse-positive, if the following three conditions hold: (i) A satisfies both conditions I, II of the monotonicity theorem (so that A^{-1} exists), (ii) $AR = S$, (iii) $A^{-1}B \geqslant O$ as an operator in $L[R, R]$.

1.3 A more detailed analysis

If the pre-inverse-positivity of M can be derived in a certain way, then additional and more detailed statements can be made about the inequalities which follow from $Mu \geqslant o$. Suppose that $\mathscr{L}$, $\mathscr{L}_1$, and $\mathscr{L}_2$ are given sets of positive linear functionals on R such that $\mathscr{L}$ describes K completely, $\mathscr{L}_1 \neq \varnothing$, and $\mathscr{L}_1 \cup \mathscr{L}_2 = \mathscr{L}$. (For a definition of these terms see Section I.1.3. The sets $\mathscr{L}_1$ and $\mathscr{L}_2$ need not be disjoint.) Now we do not assume that a strong order $\stackrel{\centerdot}{\leqslant}$ is given in S (as required in assumption (A_2)), but rather define such an order relation in the proof of the following theorem.

Theorem 1.7. (A) *The operator M is inverse-positive if the following two conditions are satisfied:*

I. (a) *For each $\ell \in \mathscr{L}_1$ there exists a positive functional $f = f_\ell \in S^*$ such that for all $u \in R$*

$$u \geqslant o, \quad \ell u = 0 \quad \Rightarrow \quad f Mu \leqslant 0. \tag{1.3}$$

(b) *For each* $\ell \in \mathscr{L}_2$ *and all* $u \in R$

$$u > o, \quad \ell u = 0 \quad \Rightarrow \quad Mu \not\geqslant o. \tag{1.4}$$

II. *There exists an element* $z \in R$ *such that*

$$z \geqslant o, \qquad Mz \geqslant o, \qquad f_\ell Mz > 0 \qquad \textit{for all} \quad \ell \in \mathscr{L}_1.$$

(B) *If* M *is inverse-positive and condition* Ib *holds, then for all* $u \in R$

$$Mu \geqslant o \quad \Rightarrow \quad \begin{cases} u \geqslant o \\ \ell u > 0 \ \ \textit{for all} \quad \ell \in \mathscr{L}_2, \quad \textit{in case} \quad u \neq o. \end{cases} \tag{1.5}$$

Proof. Suppose that conditions I and II hold. For $U \in S$, let $U \succ o$ if and only if $U \geqslant o$ and $f_\ell U > 0$ for all $\ell \in \mathscr{L}_1$. Then the assumptions of Theorem 1.2 are satisfied. We need only prove that M is pre-inverse-positive with respect to the corresponding relation $\lessdot$.

Let $u \geqslant o$ and $Mu \succ o$, but $u \not\succ o$, so that $\ell u = 0$ for some $\ell \in \mathscr{L}$. Because of condition Ia, this relation $\ell u = 0$ cannot be true for any $\ell \in \mathscr{L}_1$. Since $\mathscr{L}_1 \neq \varnothing$, we have $u > o$. Therefore, condition Ib implies that $\ell u = 0$ cannot hold for any $\ell \in \mathscr{L}_2$ either. Consequently, $Mu \geqslant o$ implies $u \geqslant o$. Moreover, in case $u > o$, condition Ib yields $\ell u > 0$ for all $\ell \in \mathscr{L}_2$. □

Corollary 1.7a. *Let conditions* I, II *in Theorem* 1.7 *be satisfied. If for given* $u \in R$ *and some* $\ell_0 \in \mathscr{L}_1$ *the inequalities* $Mu \geqslant o$ *and* $f_{\ell_0} Mu > 0$ *hold, then* $u \geqslant o$, $\ell u > 0$ *for all* $\ell \in \mathscr{L}_2$, *and* $\ell_0 u > 0$.

This statement remains true if z *satisfies the following inequalities, instead of those in condition* II:

$$z \succ o, \qquad Mz \geqslant o, \qquad f_\ell Mz \begin{cases} \geqslant o & \textit{for all} \quad \ell \in \mathscr{L}_1 \quad \textit{with} \quad f_\ell Mu > 0 \\ > o & \textit{for all} \quad \ell \in \mathscr{L}_1 \quad \textit{with} \quad f_\ell Mu = 0. \end{cases}$$

Proof. This corollary follows from Corollary 1.2b in essentially the same way as Theorem 1.7 was derived from Theorem 1.2. □

1.4 Connected sets of operators

When the monotonicity theorem is applied, one has to prove the existence of a majorizing element z. The actual construction of such an element can be avoided by means of the following results. The idea is to *connect* the given operator M with a simpler operator M_0 for which inverse-positivity is easily proved, and then to conclude by continuity arguments that M is also inverse-positive. In the following, pre-inverse-positivity is interpreted with respect to the given strong order $\lessdot$ in $(S, \leqslant)$. The term o-continuous is defined as in Section I.1.4.

Theorem 1.8 (*Continuation theorem*). *Let $\mathcal{M}$ denote a set of linear operators $M: R \to S$ with the following properties*:

(i) *Each $M \in \mathcal{M}$ is pre-inverse-positive.*

(ii) *Each $M \in \mathcal{M}$ is one-to-one.*

(iii) *The ranges of all $M \in \mathcal{M}$ have at least one element $r \succ o$ in common.*

(iv) *For each pair of operators $M_0, M_1 \in \mathcal{M}$, there exists a family $\{M(t): 0 \leqslant t \leqslant 1\} \subset \mathcal{M}$ with $M(0) = M_0, M(1) = M_1$ such that $z(t) = M^{-1}(t)r$ is* o-*continuous on* $[0, 1]$.

Then, if $\mathcal{M}$ contains at least one inverse-positive operator, all operators $M \in \mathcal{M}$ are inverse-positive.

Corollary 1.8a. *The theorem remains true if assumptions* (ii)–(iv) *are replaced by the following one.*

For each pair $M_0, M_1 \in \mathcal{M}$, there exists a family $\{M(t): 0 \leqslant t \leqslant 1\} \subset \mathcal{M}$ with $M(0) = M_0$, $M(1) = M_1$ and an o-*continuous family $\{z(t): 0 \leqslant t \leqslant 1\} \subset R$ such that*

$$M(t)z(t) \succ o \qquad (0 \leqslant t \leqslant 1). \tag{1.6}$$

Proof. Under the assumptions of the theorem, the elements $z(t) = M^{-1}(t)r$ have the properties required in the corollary. Therefore, we assume only that the assumptions of the corollary are satisfied.

Suppose that $M_0 \in \mathcal{M}$ is inverse-positive, that M_1 is any other operator in $\mathcal{M}$, and that $\{M(t): 0 \leqslant t \leqslant 1\}$ is a connecting family as in the assumptions. Define T to be the set of all $t \in [0, 1]$, such that $M(t)$ is inverse-positive.

We derive from (1.6) that $z(t) \geqslant o$ for all $t \in T$. On the other hand, if $z(t) \geqslant o$ for some $t \in [0, 1]$, then $M(t)$ is inverse-positive according to the monotonicity theorem. Thus $T = \{t \in [0, 1]: z(t) \geqslant o\}$. Since (1.6) holds and all operators $M(t)$ are pre-inverse-positive, we even have $T = \{t \in [0, 1]: z(t) \succ o\}$.

These two formulas for T show that this set, at the same time, is a closed subset and a (relatively) open subset of $[0, 1]$ (cf. Proposition I.1.9). Moreover, T is not empty because $0 \in T$. Consequently, T is the whole interval $[0, 1]$, so that M_1 is inverse-positive. □

Let us briefly discuss the assumptions of the above theorem. For many operator classes, such as certain differential operators, assumptions (iii) and (iv) are easily verified, if (ii) holds and $\mathcal{M}$ is "connected" in a certain sense. One needs here a suitable theory on the existence of solutions and their continuous dependence on parameters. Now suppose that all operators in such an operator class are pre-inverse-positive. Then *property* (ii) *essentially determines which* (connected) *sets $\mathcal{M}$ can be used.*

For inverse-positive operators additional monotonicity properties such as (1.2) or (1.5) can often be derived by verifying additional assumptions. (Compare Corollary 1.2a and Theorem 1.7, B, for example.) Obviously, the continuation theorem is also helpful in proving such additional properties.

1.5 Noninvertible operators

The monotonicity theorem can be generalized such that it can be applied to certain noninvertible operators. The corresponding results will be used, for example, to study eigenvalue problems.

Theorem 1.9. *Let the following two conditions be satisfied.*

I. *For all* $u \in R$

$$\left.\begin{array}{r} u > o \\ Mu \geqslant o \end{array}\right\} \Rightarrow u \succ o. \tag{1.7}$$

II. *There exists an element* $z \in R$ *such that*

$$z > o, \qquad Mz \geqslant o. \tag{1.8}$$

Then the following three properties (i)–(iii) *are equivalent, and properties* (i′)–(iii′) *are also equivalent. Moreover, the operator* M *has either all properties* (i)–(iii) *or all properties* (i′)–(iii′).

(i) $z \succ o$, $Mz > o$.
(ii) $u \in R$, $Mu \geqslant o \Rightarrow u \succcurlyeq o$.
(iii) $u \in R$, $Mu = o \Rightarrow u = o$.
(i′) $z \succ o$, $Mz = o$.
(ii′) $u \in R$, $Mu \geqslant o \Rightarrow Mu = o$, $u = \alpha z$ *for some* α.
(iii′) *There exists an element* $u \in R$ *with* $u \neq o$, $Mu = o$.

The following lemma is the main tool of the proof.

Lemma 1.10. *Under the assumptions of the theorem,* $Mu \geqslant o$ *implies that either* $u \geqslant o$ *or*

$$u = \alpha z \quad \text{for some} \quad \alpha < 0, \qquad \text{and} \qquad Mu = Mz = o. \tag{1.9}$$

Proof of the Lemma. We proceed as in the first part of the proof of Theorem 1.2 with two exceptions. Now we obtain $Mv \geqslant o$ for $v = v(\lambda_0)$, and we reach a contradiction only if $v \neq o$. If $v = o$, the relations (1.9) follow from $o = (1 - \lambda_0)u + \lambda_0 z$, $o = Mv = (1 - \lambda_0)Mu + \lambda_0 Mz$, $0 < \lambda_0 < 1$, and $Mu \geqslant o$, $Mz \geqslant o$. □

Proof of the Theorem. Because of condition I, the element z satisfies $z \succ o$, $Mz \geqslant o$. Therefore, either (i) or (i′) holds, so that we only have to prove the equivalence in the statements of the theorem.

If (i) holds, then (1.9) cannot occur, so that (ii) follows from the above lemma. Property (ii) implies (iii), according to Proposition 1.1. Finally, (iii) prevents $Mz = o$, so that (iii) implies (i).

Properties (i′), (iii′) are equivalent, since properties (i), (iii) are. Let (i′) hold and suppose that $u \in R$ satisfies $Mu \geqslant o$. Then $M(u + \lambda z) = Mu \geqslant o$ for all $\lambda \in (-\infty, \infty)$. Thus, according to Lemma 1.10, for each λ three cases may occur: $u + \lambda z \succ o$, or $u + \lambda z = o$, or $u + \lambda z = \beta z$ for some $\beta < 0$ and $M(u + \lambda z) = Mz = o$. Since the cone K does not contain a line, the first case cannot occur for all λ. However, if the second or third case occurs for some λ. then $Mu = M(u + \lambda z) - \lambda Mz = o$ and there exists some α with $u = \alpha z$. This shows that (i′) implies (ii′). If (ii′) holds, (i) cannot be true, so that (i′) is satisfied. □

Corollary 1.9a. *Suppose that $\mathscr{L} \subset R^*$ describes the cone K completely. Then condition* I *of Theorem* 1.9 *is satisfied if for each $\ell \in \mathscr{L}$ and all $u \in R$ the relations $u > o$, $\ell u = 0$ imply $Mu \not\geqslant o$.*

The class of all operators satisfying the above assumptions I, II can also be characterized as follows. (The proof is left to the reader.)

Corollary 1.9b. (a) *An invertible linear operator $M: R \to S$ satisfies both assumptions* I, II *of Theorem* 1.9 *if and only if M is strictly inverse-positive and its range contains an element $> o$.*

(b) *A noninvertible linear operator $M: R \to S$ satisfies both assumptions* I, II *of Theorem* 1.9 *if and only if the range of M does not contain an element $> o$ and the null space of M is one-dimensional and contains an element $\succ o$.*

2 M-MATRICES

As a first example for the application of the theory in Section 1 we consider here finite square matrices M which have an inverse $M^{-1} = (\mu_{ik})$ such that $\mu_{ik} \geqslant 0$ for all indices i, k. In other words, we consider inverse-positive linear operators $M: \mathbb{R}^n \to \mathbb{R}^n$ where $\mathbb{R}^n$ is endowed with the natural (componentwise) order relation. Such operators will always be identified with the corresponding matrices, so that we shall speak of inverse-positive matrices, etc.

In Section 2.2 results on general square matrices are derived from the abstract theory above. The sections then following are mainly concerned with inverse-positive off-diagonally negative matrices (M-matrices). In the sense of our theory off-diagonally negative matrices M (Z-matrices) are distinguished by the fact that all matrices $M + \lambda I$ $(-\infty < \lambda < \infty)$ are weakly pre-inverse-positive. Moreover, for such matrices M one can formulate rather simple sufficient conditions such that all matrices $M + \lambda I$ $(-\infty < \lambda < \infty)$ are pre-inverse-positive with respect to certain strong order relations $\leqslant^{\prime}$, different from the strict natural order $\prec$. These properties not only make it comparably easy to apply the abstract theory to Z-matrices, but enable us also to develop a theory of certain eigenvalue problems $B\varphi = \lambda A\varphi$, in particular, eigenvalue problems $-M\varphi = \lambda\varphi$ with a Z-matrix M (Section 2.4). Moreover, for Z-matrices M, the inverse-positivity is very closely related and often equivalent to the convergence of certain iterative procedures, in particular, procedures based on a regular splitting $M = A - B$ (Section 2.5).

M-matrices will further be applied in Section 5, where the convergence of certain difference methods for differential equations is proved.

The theory in Section 2.2 on general square matrices can immediately be carried over to the case in which $\mathbb{R}^n$ is endowed with an arbitrary linear order $\leqslant$, or even to the case in which the order relations $\leqslant$ in the domain and range of $M: \mathbb{R}^n \to \mathbb{R}^n$ are different. For a generalization of the theory of Z-matrices and M-matrices, see Section 4.2.

2.1 Definitions and notation

We propose to apply the abstract theory in Section 1 to the case in which $M = (m_{ik})$ denotes an $n \times n$ matrix, which is identified with the corresponding linear operator $M: \mathbb{R}^n \to \mathbb{R}^n$, and $\leqslant$ denotes the natural order relation in $\mathbb{R}^n$. The results of Section 1 can immediately be reformulated for this case, replacing the term operator by the term matrix, etc. (Observe that assumption (A_1) in Section 1.1 is here satisfied.)

All order relations for matrices can now be interpreted elementwise. For convenience, we briefly collect some definitions. A matrix $A = (a_{ik}) \in \mathbb{R}^{n,p}$, that is, an $n \times p$ matrix A is called *positive* ($A \geqslant O$) when $a_{ik} \geqslant 0$ for all its elements, and A is *strictly positive* ($A \succ O$) when $a_{ik} > 0$ for all its elements. $A > O$ means that $A \geqslant O$ but $A \neq O$; and $A \succcurlyeq O$ means that $A \succ O$ or $A = O$. Here O denotes the $n \times p$ null matrix (the dimension will always be clear). Moreover, $|A|$ is the matrix $(|a_{ik}|)$, $A^+ = \frac{1}{2}(|A| + A)$, $A^- = \frac{1}{2}(|A| - A)$, so that $A = A^+ - A^-$, $|A| = A^+ + A^-$.

In particular, this notation will be used for $n \times n$ matrices M and $n \times 1$ matrices, i.e., vectors $u \in \mathbb{R}^n$, except that the null vector will usually be denoted by o. In addition, $I = I_n$ denotes the $n \times n$ unit matrix and $e = (e_i) \in \mathbb{R}^n$ is the vector with $e_i = 1$ $(i = 1, 2, \ldots, n)$. (The dimension n will be defined in the context.) Besides the relations $u \geqslant o$ and $u \succcurlyeq o$ we shall use *strong order relations* $u \gneqq o$ having the property (I.1.6).

For $z \succ o$, $u \in \mathbb{R}^n$, and $M \in \mathbb{R}^{n,n}$:

$$\|u\|_z = \inf\{\alpha \geqslant 0 : |u| \leqslant \alpha z\} = \max\{|u_i|/z_i : i = 1, 2, \ldots, n\},$$

$$\|M\|_z = \inf\{q \geqslant 0 : \|Mu\|_z \leqslant q\|u\|_z \text{ for } u \in \mathbb{R}^n\}$$

$$= \inf\{q \geqslant 0 : |M|z \leqslant qz\} = \max\left\{\sum_k |m_{ik}|z_k/z_i : i = 1, 2, \ldots, n\right\}.$$

Moreover, $\|M\|_\infty = \|M\|_e$ and $\|M\|_1 = \|M^{\mathrm{T}}\|_\infty$.

A matrix $M \in \mathbb{R}^{n,n}$ is *inverse-positive* or *strictly inverse-positive* if M^{-1} exists and $M^{-1} \geqslant O$ or $M^{-1} \succ O$, respectively (cf. Proposition 1.1). Obviously, these conditions hold, if and only if the transposed matrix M^{T} satisfies $(M^{\mathrm{T}})^{-1} \geqslant O$ or $(M^{\mathrm{T}})^{-1} \succ O$, respectively. Consequently, all conditions for inverse-positivity and strict inverse-positivity derived below may also be applied to M^{T} instead of M.

A matrix $M \in \mathbb{R}^{n,n}$ is said to be *off-diagonally negative* provided $m_{ik} \leqslant 0$ for $i \neq k$, and it is *diagonally positive* (*diagonally strictly positive*) if all diagonal elements satisfy $m_{ii} \geqslant 0$ ($m_{ii} > 0$). We denote the set of all off-diagonally negative square matrices by $\mathbf{Z}$ and the set of all inverse-positive matrices $M \in \mathbf{Z}$ by $\mathbf{M}$. A Z-*matrix* is a matrix $M \in \mathbf{Z}$, and an M-*matrix* is a matrix $M \in \mathbf{M}$. (Concerning this notation, see the notes at the end of this section.) A Z-matrix is said to be *diagonally dominant* (*weakly diagonally dominant*) if $Me \succ o$ ($Me > o$).

A matrix $M \in \mathbb{R}^{n,n}$ is called *reducible* if there exist nonempty disjoint index sets J_0, J_1 with $\{1, 2, \ldots, n\} = J_0 \cup J_1$ such that $m_{ik} = 0$ for all (i, k) with $i \in J_0$, $k \in J_1$. In other words, *M is reducible if and only if there exists a permutation matrix P such that $A = PMP^{\mathrm{T}}$ has the form*

$$A = \begin{pmatrix} A_{11} & 0 \\ A_{21} & A_{22} \end{pmatrix} \quad \text{with} \quad A_{11} \in \mathbb{R}^{r,r}, \quad A_{22} \in \mathbb{R}^{n-r,n-r}, \quad \text{and} \quad 0 < r < n.$$

Otherwise, M is called *irreducible*. It can be shown that *M is irreducible if and only if, for each ordered pair of indices i, k with $i \neq k$, $1 \leqslant i, k \leqslant n$, there exist pairwise different indices l_j ($j = 0, 1, 2, \ldots, p$) such that*

$$l_0 = i, \qquad l_p = k, \qquad 1 \leqslant l_j \leqslant n, \qquad \prod_{j=1}^{p} m_{l_{j-1}, l_j} \neq 0. \tag{2.1}$$

Problems

2.1 Prove the equivalence of the two characterizations of an irreducible matrix.

2.2 Prove that a positive matrix $T \in \mathbb{R}^{n,n}$ is irreducible if and only if $I + T + \cdots + T^{n-1} \succ O$.

2.3 Prove that $M \in \mathbb{R}^{n,n}$ is inverse-positive if and only if the equation $Mu = r$ has a solution $u \geqslant o$ for each $r \geqslant o$.

2.2 General results for square matrices

For matrices $M \in \mathbb{R}^{n,n}$ as considered here, the results of Section 1 can be reformulated so that they assume a simpler form.

Theorem 2.1. *A matrix $M \in \mathbb{R}^{n,n}$ is inverse-positive if and only if the following two conditions are satisfied:*

I. *For all* $u \in \mathbb{R}^n$, *the inequalities* $u \geqslant o$, $Mu \succ o$ *imply* $u \succ o$.
II. *There exists a vector* $z \in \mathbb{R}^n$ *with* $z \geqslant o$, $Mz \succ o$.

This statement follows immediately from the monotonicity theorem and Theorem 1.6(i), when the strict order $\preccurlyeq$ is used in place of the strong order $\overset{\prime}{\leqslant}$ which occurs in those theorems.

More detailed results are obtained by considering an arbitrary strong order in $(\mathbb{R}^n, \leqslant)$. Let $\overset{\prime}{\leqslant}$ be a given order relation in $\mathbb{R}^n$ which is stronger than the natural order.

Theorem 2.2. *Suppose that* M *is pre-inverse-positive with respect to the strong order* $\overset{\prime}{\leqslant}$, *that means, for all* $u \in \mathbb{R}^n$,

$$\left.\begin{array}{r} u \geqslant o \\ Mu \overset{\prime}{>} o \end{array}\right\} \quad \Rightarrow \quad u \succ o. \tag{2.2}$$

Then the following four properties are equivalent:

(i) M *is inverse-positive;*
(ii) M *is inverse-positive and for all* $u \in \mathbb{R}^n$

$$Mu \overset{\prime}{\geqslant} o \quad \Rightarrow \quad u \succcurlyeq o; \tag{2.3}$$

(iii) *there exists a vector* $z \in \mathbb{R}^n$ *with*

$$z \geqslant o, \qquad Mz \overset{\prime}{>} o; \tag{2.4}$$

(iv) *there exists a vector* $z \in \mathbb{R}^n$ *with* $z \succ o$, $Mz \succ o$.

Proof. (ii) $\Rightarrow$ (i) and (iv) $\Rightarrow$ (iii) are trivial. (i) $\Rightarrow$ (ii) follows from the implications $Mu = o \Rightarrow u = o$ and $Mu \overset{\prime}{>} o \Rightarrow Mu \geqslant o \Rightarrow u \geqslant o$, and from the assumption of the theorem. (iii) $\Rightarrow$ (i), as a consequence of the monotonicity theorem. To prove (i) $\Rightarrow$ (iv), let z be the solution of $Mz = e \succ o$ with inverse-positive M. Then $z \geqslant o$, so that $z \succ o$ by the above assumption. □

If the matrix M satisfies (2.2), then the inverse-positivity can be established by constructing a vector z with property (2.4). The actual construction of z may be avoided by using the following result. Here a set $\mathscr{K} \subset \mathbb{R}^{n,n}$ is called *arcwise connected* if to each pair $M_0, M_1 \subset \mathscr{K}$ there exists a family of matrices $M(t) \subset \mathscr{K}$ $(0 \leqslant t \leqslant 1)$ such that $M(0) = M_0$, $M(1) = M_1$, and all elements $m_{ik}(t)$ of $M(t)$ depend continuously on t. If all elements $m_{ik}(t)$ are piecewise linear functions (polygons), then $\mathscr{K}$ is called *piecewise linearly connected.*

Theorem 2.3. *Let* $\mathscr{K}$ *be an arcwise connected set of weakly pre-inverse-positive invertible matrices* $M \in \mathbb{R}^{n,n}$.

Then, if at least one matrix $M \in \mathscr{K}$ *is inverse-positive, all matrices* $M \in \mathscr{K}$ *are inverse-positive.*

Proof. It suffices to verify assumptions (iii) and (iv) of the continuation theorem 1.8. Since $e \in M\mathbb{R}^n$ for invertible M, (iii) is obvious. By standard theorems of matrix theory, the solution $z(t)$ of $M(t)z(t) = e$ depends continuously on t if the elements of $M(t) \in \mathscr{K}$ depend continuously on t. Thus, (iv) holds also. □

Notice, that this theorem can also be used to find sets $\mathscr{K}$ of matrices M with property (2.3), due to the equivalence stated in Theorem 2.2.

While in the above theorem all occurring matrices are supposed to be invertible, the following theorem is concerned with singular matrices which can be approximated by inverse-positive matrices.

Let $\Lambda \subset \mathbb{R}$ denote a set which has a point of accumulation $\lambda_0 \in \Lambda$, and let $\{M(\lambda) : \lambda \in \Lambda\}$ be a family of matrices in $\mathbb{R}^{n,n}$ which is continuous at $\lambda = \lambda_0$.

Theorem 2.4. *Suppose that the following three conditions are satisfied.*

(i) *$M(\lambda)$ is inverse-positive for all $\lambda \in \Lambda$ with $\lambda \neq \lambda_0$.*
(ii) *$M(\lambda_0)$ is singular.*
(iii) *For all $u \in \mathbb{R}^n$, the inequalities $u > o$, $M(\lambda_0)u \geqslant o$ imply $u \succ o$.*

Then there exists a vector $\varphi \succ o$ in $\mathbb{R}^n$ such that $M(\lambda_0)\varphi = o$ and that each vector u with $M(\lambda_0)u = o$ is a multiple of φ.

Proof. For each $\lambda \in \Lambda$ with $\lambda \neq \lambda_0$ there exists a vector $z(\lambda) \succ o$ with $M(\lambda)z(\lambda) \succ o$, because of assumption (i) and Theorems 1.6 and 2.2. Without loss of generality, we may assume that $\|z(\lambda)\|_2 = 1$. Then there exists a sequence $\{\lambda_m\}$ with limit λ_0 such that the sequence $\{z(\lambda_m)\}$ converges toward a vector φ. Obviously, $M(\lambda_0)\varphi \geqslant o$, $\varphi \geqslant o$, and $\|\varphi\|_2 = 1$, so that $\varphi > o$.

Thus, because of (iii), all assumptions of Theorem 1.9 are satisfied for $M = M(\lambda_0)$, with $z = \varphi$ in (1.8). Therefore, properties (i′) and (ii′) in Theorem 1.9 hold, since $M(\lambda_0)$ is singular. □

Problems

The following statements are to be proved.

2.4 If $A, B \in \mathbb{R}^{n,n}$ are inverse-positive and $B \leqslant A$, then $O \leqslant A^{-1} \leqslant B^{-1}$.
2.5 Suppose that for $M \in \mathbb{R}^{n,n}$ there exists a vector $z \succ o$ with $Mz \succ o$. Then M is inverse-positive if and only if $\|u\|_z \leqslant \|Mu\|_{Mz}$ for all $u \in \mathbb{R}^n$.
2.6 If $A, B, C \in \mathbb{R}^{n,n}$ such that A and C are inverse-positive and $A \leqslant B \leqslant C$, then B is inverse-positive.
2.7 If $A, B \in \mathbb{R}^{n,n}$ are inverse-positive and $A \leqslant B$, then the matrices $A + B$, $B^{-1}A$, AB^{-1} are inverse-positive, and $B^{-1}A \leqslant I$.
2.8 Suppose that for a given $M \in \mathbb{R}^{n,n}$ with $n > 1$ there exists a vector $z > o$ with $Mz \geqslant o$, but that for any matrix $\tilde{M}$ which is obtained from M by omitting at least one column no vector $\zeta > o$ with $\tilde{M}\zeta \geqslant o$ exists. Prove that under this

assumption for all $u \in \mathbb{R}^n$, the inequalities $u > o$, $Mu \geqslant o$ imply $u \succ o$. Then apply Theorem 1.9.

2.3 Off-diagonally negative matrices

When we want to apply the results of the preceding sections we need sufficient conditions for the property (2.2) of pre-inverse-positivity. If $M \in \mathbf{Z}$, such conditions are easily obtained for a series of strong order relations $\stackrel{\cdot}{\leqslant}$. This can be considered as the main reason why an extended theory of inverse-positivity has been developed for Z-matrices, as presented in this section. Here and in the following sections, the dimension n shall be fixed. That means all occurring matrices are $n \times n$ matrices and all occurring vectors are n-vectors, unless otherwise stated.

Proposition 2.5. (i) *Each $M \in \mathbf{Z}$ is weakly pre-inverse-positive, i.e., the inequalities $u \geqslant o$, $Mu \succ o$ imply $u \succ o$.*

(ii) *If M is irreducible and $M \in \mathbf{Z}$, then for all $u \in \mathbb{R}^n$*

$$\left.\begin{array}{r} u > o \\ Mu \geqslant o \end{array}\right\} \quad \Rightarrow \quad u \succ o.$$

In particular, each irreducible $M \in \mathbf{Z}$ is strictly pre-inverse-positive, i.e., the inequalities $u \geqslant o$, $Mu > o$ imply $u \succ o$.

Proof. (i) Let $u \in \mathbb{R}^n$ satisfy $u \geqslant o$ and $Mu \succ o$, but $u \not\succ o$. Then $u_j = 0$ for some component u_j and, consequently, $(Mu)_j = \sum_{k \neq j} m_{jk} u_k \leqslant 0$, which contradicts $Mu \succ o$.

(ii) Here, it suffices to show that $(Mu)_j < 0$ for some index j, if $u > o$ and $u \not\succ o$. Let J_0 and J_1 denote the sets of indices such that $u_j = 0$ for $j \in J_0$ and $u_k > 0$ for $k \in J_1$. None of these sets is empty, and since M is irreducible, there exist indices $j \in J_0$ and $k \in J_1$ with $m_{jk} < 0$. Thus, $(Mu)_j \leqslant m_{jk} u_k < 0$. □

Proposition 2.5 and Theorem 2.2 yield

Theorem 2.6. (i) *A matrix $M \in \mathbf{Z}$ is inverse-positive if and only if $z \geqslant o$, $Mz \succ o$ for some vector z.*

(ii) *An irreducible matrix $M \in \mathbf{Z}$ is inverse-positive if and only if $z \geqslant o$, $Mz > o$ for some vector z. Then M is also strictly inverse-positive.*

Example 2.1. *Each (lower or upper) triangular Z-matrix $M \in \mathbb{R}^{n,n}$ with $m_{ii} > 0$ $(i = 1, 2, \ldots, n)$ is inverse-positive.* To prove this for a lower triangular matrix, choose $z = (z_i)$ with $z_i = \varepsilon^{n-i}$ and small enough $\varepsilon > 0$. □

The following corollaries are immediate consequences of the above theorem.

Corollary 2.6a. (i) *Each $M \in \mathbf{M}$ is strictly diagonally positive.*

(ii) *If $M \in \mathbf{M}$ and $D = (d_i \, \delta_{ij})$ is positive, then $M + D \in \mathbf{M}$.*

Corollary 2.6b. *A matrix $M \in \mathbf{Z}$ (an irreducible matrix $M \in \mathbf{Z}$) belongs to* $\mathbf{M}$ *if and only if there exists a diagonal matrix $D \in \mathbf{M}$ such that $D^{-1}MD$ is diagonally dominant (weakly diagonally dominant).*

In particular, $M \in \mathbf{Z}$ is inverse-positive, if M is diagonally dominant. Also, an irreducible $M \in \mathbf{Z}$ is strictly inverse-positive if M is weakly diagonally dominant.

Example 2.2. The tridiagonal $n \times n$ matrix

$$M = \begin{bmatrix} 2 & -1 & & & & & \\ -1 & 2 & -1 & & & 0 & \\ & -1 & 2 & -1 & & & \\ & & & \ddots & & & \\ & 0 & & & -1 & 2 & -1 \\ & & & & & -1 & 2 \end{bmatrix}$$

is irreducible and weakly diagonally dominant. Thus, M has an inverse $M^{-1} = (\mu_{ik})$ such that $\mu_{ik} > 0$ for all its elements.

According to Theorem 2.2 there exists also a vector $z \succ o$ with $Mz \succ o$. For example, $z = (z_i)$ with $z_i = \sin i\pi h$ satisfies these inequalities, since $Mz = 2(1 - \cos \pi h)z$ with $h = (n+1)^{-1}$. □

There are interesting Z-matrices which are neither irreducible nor diagonally dominant. A typical example is the tridiagonal $n \times n$ matrix

$$M = \begin{bmatrix} 1 & 0 & & & & \\ -1 & 2 & -1 & & 0 & \\ & -1 & 2 & -1 & & \\ & & \ddots & \ddots & \ddots & \\ & 0 & & -1 & 2 & -1 \\ & & & & 0 & 1 \end{bmatrix} \tag{2.5}$$

The question arises whether a strong order $\leqslant$ can be found such that $Me \succ o$ implies inverse-positivity for such matrices. According to Theorem 2.2, it suffices to verify (2.2).

Proposition 2.7a. *Suppose that $M \in \mathbb{R}^{n,n}$ belongs to* $\mathbf{Z}$, *and that there exist index sets J_1, J_2 with the following properties:*

(i) $J_1 \cup J_2 = \{1, 2, \ldots, n\}$, $J_1 \cap J_2 = \varnothing$, $J_1 \neq \varnothing$.

(ii) *For each $i \in J_2$ there exist a $k \in J_1$ and pairwise different indices l_j $(j = 0, 1, \ldots, p)$ with property* (2.1).

Then the matrix M is pre-inverse-positive with respect to the strong order $\overset{\scriptscriptstyle d}{\leqslant}$ defined by

$$u \,\overset{\scriptscriptstyle d}{>}\, o \quad\Leftrightarrow\quad u \geqslant o \quad \text{and} \quad u_i > 0 \quad \text{for all} \quad i \in J_1. \tag{2.6}$$

The proof is left to the reader.

Example 2.3a. The matrix in (2.5) satisfies the above condition for $J_1 = \{1, n\}$. Moreover, we have $Me \,\overset{\scriptscriptstyle d}{>}\, o$ for the corresponding strong order relation. Thus, M is inverse-positive, according to Theorem 2.2. □

Proposition 2.7b. *A Z-matrix $M \in \mathbb{R}^{n,n}$ is inverse-positive if and only if there exists a vector $z \in \mathbb{R}^n$ such that $z \geqslant o$, $Mz \geqslant o$, and $\sum_{k=1}^{i} m_{ik} z_k > 0$ $(i = 1, 2, \ldots, n)$.*

Proof. The necessity of the condition is clear (cf. Theorem 2.6(i)). The sufficiency will be derived from Proposition 2.7a and Theorem 2.2.

Let J_1 denote the set of all indices i such that $(Mz)_i > 0$. To verify condition (ii) of Proposition 2.7a choose an index $i = l_0 \in J_2$. Since $\sum_{k=i+1}^{n} m_{ik} z_k = -\sum_{k=1}^{i} m_{ik} z_k < 0$, there exists at least one index $l_1 > l_0$ with $m_{l_0 l_1} \neq 0$. If $l_1 \in J_1$, then (ii) holds with $l_1 = k$. If $l_1 \in J_2$, the process is continued with l_1 in place of l_0. In this way one obtains indices l_j $(j = 0, 1, \ldots, p)$ such that (2.1) holds and $k = l_p \in J_1$. The process ends with an index in J_1, since $l_{j-1} < l_j$ and $n \in J_1$.

Thus, the matrix M is pre-inverse-positive with respect to the strong order defined by (2.6). Since also (2.4) holds, M is inverse-positive, according to Theorem 2.2. □

Example 2.3b. For the matrix in (2.5), the vector $z = e$ satisfies the inequalities required in Proposition 2.7b. □

There are a series of possibilities to characterize (or define) the matrix class **M**. For example, one may use the equivalence statement of Theorem 2.6(i). Another interesting description of the class **M** is given in the following theorem. Further possibilities will be discussed in the subsequent section.

Theorem 2.8. $\mathbf{M} \cap \mathbb{R}^{n,n}$ *is the largest piecewise linearly connected set of invertible $M \in \mathbf{Z} \cap \mathbb{R}^{n,n}$ which contains a diagonal matrix $D \in \mathbf{M} \cap \mathbb{R}^{n,n}$.*

Proof. In this proof, we write **M** and **Z** instead of $\mathbf{M} \cap \mathbb{R}^{n,n}$ and $\mathbf{Z} \cap \mathbb{R}^{n,n}$, respectively. Let $\mathscr{K}$ denote the largest set with the properties described in the theorem. It is clear that this set exists, since each pair D_1, D_2 of diagonal matrices in **M** is linearly connected within **M**. The relation $\mathscr{K} \subset \mathbf{M}$ follows from Theorem 2.3. On the other hand, if $M \in \mathbf{M}$, then each matrix $M_t = (1-t)M + tD$ with $t \in [0, 1]$ and given diagonal $D \in \mathbf{M}$ belongs to **M**, since $M_t \in \mathbf{Z}$ and $M_t z \succ o$ for any element $z \geqslant o$ with $Mz \succ o$ (see Theorem 2.6(i)). Thus, M is linearly connected with $D \in \mathbf{M}$, so that $M \in \mathscr{K}$, by definition of $\mathscr{K}$. □

While the preceding theorem characterizes the whole set **M**, the following theorem describes an important subset.

Theorem 2.9. *A symmetric $M \in \mathbf{Z}$ is inverse-positive if and only if M is positive definite.*

Proof. Let $\mathscr{K}$ denote the convex set of all positive definite symmetric $M \in \mathbf{Z}$. This set contains a diagonal matrix $D \in \mathbf{M}$. Therefore Theorem 2.3 implies $\mathscr{K} \subset \mathbf{M}$. On the other hand, if M is symmetric and belongs to **M**, then all matrices $(1 - t)M + tD\ (0 \leqslant t \leqslant 1)$ have these properties also. Moreover, all of these matrices are positive definite, since otherwise one of them would be positive semidefinite but singular. Thus, $M \in \mathscr{K}$. □

Notice that this theorem involves two different kinds of order relations. This is even more clearly seen if we use the following phrasing. *A symmetric Z-matrix has a positive inverse if and only if it has a positive definite inverse.*

The above theorem and Theorem 2.6(ii) immediately yield

Corollary 2.9a. *An irreducible symmetric $M \in \mathbf{Z}$ is strictly inverse-positive if and only if M is positive definite.*

Besides (regular) M-matrices also certain singular limits of M-matrices are of interest. An $n \times n$ matrix M will be called an M_0-*matrix* if M is singular and if $M + \lambda I$ is an M-matrix for each $\lambda > 0$. We derive from Theorem 2.4:

Theorem 2.10. *If M is an irreducible M_0-matrix, then the equation $Mu = o$ has a solution $u^* \succ o$ and all solutions of this equation are multiples of u^*.*

We conclude this section by applying Theorems 2.2 and 2.10 and the result of Example 2.1 to some simple economic models.

Example 2.4 (A *production model*). Z-matrices and M-matrices occur, for example, in certain models in economy. Suppose that n manufacturers $\mu_1, \mu_2, \ldots, \mu_n$ produce n goods $\gamma_1, \gamma_2, \ldots, \gamma_n$, where μ_i produces only γ_i. Assume, moreover, that the production of one unit of γ_i requires $c_{ij} \geqslant 0$ units of γ_j and that u_i units of γ_i are produced within one year. $C = (c_{ij}) \in \mathbb{R}^{n,n}$ is called the *consumption matrix* and $u = (u_i) \in \mathbb{R}^n$ the *production vector*.

The total amount of the n goods produced in one year is described by the row vector $u^{\mathrm{T}} - u^{\mathrm{T}}C = u^{\mathrm{T}}(I - C)$. Now, if there exists an outside demand for these n goods which is described by the demand vector $d \geqslant o$, the question arises whether there exists a production vector $u \geqslant o$ which meets the demand without any surplus, that means $u^{\mathrm{T}}(I - C) = d^{\mathrm{T}}$, or $Mu = d$ with $M = I - C^{\mathrm{T}}$.

Obviously M is a Z-matrix. An application of Theorem 2.6(i) shows that *the following two properties are equivalent.* (i) *To each demand $d \geqslant o$ there exists a production $u \geqslant o$ which meets the demand without surplus.* (ii) *There exists a production $u \geqslant o$ such that an amount >0 is produced of each good.*

The second part of Theorem 2.6 can also be applied. Observe that M is reducible if and only if there exists a "subgroup" of p of the manufacturers with $1 \leqslant p < n$ such that these manufacturers for their production need only goods produced by this subgroup. □

Example 2.5 (A special *production model*). We treat a model similar to that in Example 2.4. Now we consider a single plant which produces parts $\gamma_1, \gamma_2, \ldots, \gamma_n$. It is assumed that these parts are listed in such an order that only parts γ_k with $k \leqslant i$ are needed in the production of γ_i. In this case we arrive again at an equation $Mu = d$ with $M = I - C^{\mathrm{T}}$, as in Example 2.4, where now, however, the production matrix C is lower triangular and hence M is upper triangular. From the statement in Example 2.1 we thus obtain the following result. *If the (trivial) assumption is satisfied that for the production of one unit of γ_i less than one unit of γ_i is used ($i = 1, 2, \ldots, n$), then to each demand $d \geqslant o$ there exists a production $u \geqslant o$ which meets the demand without any surplus.* □

Example 2.6 (A *model of exchange*). Suppose that an economy of n manufacturers $\mu_1, \mu_2, \ldots, \mu_n$ produces n goods $\gamma_1, \gamma_2, \ldots, \gamma_n$, where only μ_i produces γ_i. Assume that in one year exactly one unit of γ_i is produced, that μ_i consumes the amount a_{ij} of γ_j, where $0 \leqslant a_{ij} \leqslant 1$, and that all goods produced are also consumed by the economy. Then all row sums $\sum_{j=1}^{n} a_{ij}$ (and also all column sums) equal 1.

If u_i is the price for one unit of γ_i, then μ_i pays $\sum_{j=1}^{n} a_{ij}u_j$ for the goods he consumes and receives u_i for the unit of γ_i he produces. In a stable economy these two amounts must be equal, i.e., $Mu = o$ for $u = (u_i)$ and $M = I - A$. We are asking whether a price vector $u > o$ exists such that this equation holds.

M is a Z-matrix with $Me = o$. Consequently, M is singular; but $\lambda I + M$ is an M-matrix for each $\lambda > 0$, since $(\lambda I + M)e = \lambda e \succ o$. Finally observe that the matrix M is irreducible if and only if there exists no "subgroup" of p manufacturers ($1 \leqslant p < n$) which consume only goods produced by this subgroup. *If this situation holds, then there exists a price vector $u^* \succ o$ such that the economy is stable,* by Theorem 2.10. Moreover, each other positive price vector for which the economy is stable is a multiple of u^*. □

Problems

2.9 Prove Proposition 2.7a.

2.10 Let J_1, J_2, J_3 denote disjoint subsets of $J = \{1, 2, \ldots, n\}$ such that $J_1 \cup J_2 \cup J_3 = J$. Define $U \mathrel{\dot{>}} o$ by: $U > o$, $U_j > 0$ for some $j \in J_2$, $U_i > 0$ for all $i \in J_3$. Find conditions such that M is pre-inverse-positive with respect to $\mathrel{\dot{\leqslant}}$.

2.11 For a Z-matrix M let there exist a permutation matrix P such that $A = PMP^{\mathrm{T}}$ has the following properties:

$$A = \begin{bmatrix} A_{11} & A_{12} \\ A_{21} & A_{22} \end{bmatrix} \quad \text{with} \quad \begin{cases} \text{diagonally strictly positive } A_{11} \in \mathbb{R}^{p,p}, \\ \text{irreducible } A_{22} \in \mathbb{R}^{n-p,n-p},\ A_{21} \neq O. \end{cases}$$

Find a strong order $\preceq$ as in (2.6) such that M is pre-inverse-positive with respect to that order. Apply the result to the matrix in (2.5).

2.12 Generalize the result at the end of Example 2.6 to the case in which M is not irreducible. Use the fact that by suitable permutations M can be transformed into a block-lower triangular matrix with irreducible diagonal blocks. Formulate conditions such that all solutions $u > o$ also satisfy $u \succ o$.

2.13 In this problem, M-matrices are characterized by properties of submatrices and subdeterminants. A *principal submatrix* of M is a matrix $A = (a_{ik}) \in \mathbb{R}^{p,p}$ with $1 \leqslant p \leqslant n$; $a_{ik} = m_{q_i q_k}$, $1 \leqslant q_1 < q_2 < \cdots < q_p \leqslant n$; a *principal minor* is the determinant of a principal submatrix. It shall be proved that *a Z-matrix is inverse-positive if and only if one of the following three conditions hold*:

(i) *Each principal minor of M is >0.*

(ii) *All eigenvalues of each principal submatrix of M have a real part >0.*

(iii) *There exist a lower triangular matrix $L \in \mathbf{M}$ and an upper triangular matrix $U \in \mathbf{M}$ such that $M = LU$.*

The proof may be done in the following way: (a) (i) $\Rightarrow$ (iii) by the theory of the Gauß elimination procedure; (b) (iii) $\Rightarrow$ $M \in \mathbf{M}$; (c) estimating appropriately one derives from $(m_{ii} - \lambda)\varphi_i = \sum_{k \neq i} m_{ik}\varphi_k$ for $|\varphi_i| = \|\varphi\|_\infty$ that each eigenvalue of M has a real part >0, if $Me \succ o$ (here, the theorem of Gerschgorin [1931] may also be used); (d) $M \in \mathbf{M}$, $Me \succ o \Rightarrow$ (ii); (e) $M \in \mathbf{M} \Rightarrow$ (ii); (f) (ii) $\Rightarrow$ (i).

2.4 Eigenvalue problems

The theory of the matrix classes $\mathbf{M}$ and $\mathbf{Z}$ is closely related to the theory of eigenvalue problems. Some of the relations will be discussed in this section. First we consider an eigenvalue problem

$$(\lambda D + M)\varphi = o \quad \text{with} \quad M \in \mathbf{Z} \quad \text{and diagonal} \quad D \in \mathbf{M}. \tag{2.7}$$

Theorem 2.11. *The eigenvalue problem* (2.7) *has a real eigenvalue λ_0 such that* (*for $\lambda \in \mathbb{R}$*) *the matrix $\lambda D + M$ is inverse-positive if and only if $\lambda > \lambda_0$.*

Proof. To prove the above statement, Theorem 2.3 could be used. Here, however, we prefer the following direct proof.

Since $(\lambda D + M)e \succ o$ for large enough λ, we have $\lambda D + M \in \mathbf{M}$ for these λ. Also, if $\lambda D + M \in \mathbf{M}$, then $\lambda' D + M \in \mathbf{M}$ for all $\lambda' \geqslant \lambda$ (cf. Corollary 2.6a). On the other hand, the element $-e < o$ satisfies $(\mu D + M)(-e) \geqslant o$ for some $\mu < 0$, so that $\mu D + M \notin \mathbf{M}$. Consequently, $\lambda_0 = \inf\{\lambda \in \mathbb{R} : \lambda D + M \in \mathbf{M}\}$ is finite, and it remains only to be shown that λ_0 is an eigenvalue.

Suppose $\lambda_0 D + M$ were invertible, so that $(\lambda_0 D + M)^{-1} \geqslant O$ by reasons of continuity, i.e., $\lambda_0 D + M \in \mathbf{M}$. Let then $z \geqslant o$ be the solution of $(\lambda_0 D + M)z = e$. Because $e \succ \varepsilon D z$ for some $\varepsilon > 0$, we would have $(M + (\lambda_0 - \varepsilon)D)z \succ o$, which implies that $M + (\lambda_0 - \varepsilon)D \in \mathbf{M}$, contrary to the definition of λ_0. □

In Theorem 2.11 it was *assumed* that $M \in \mathbf{Z}$. However, properties of the type described in that theorem can also be used to *characterize* the matrix classes $\mathbf{Z}$ and $\mathbf{M}$, as we shall see in what follows. This characterization will lead to the definition of an operator class $\mathbf{Z}$ in Section 4.2. As before, let $D \in \mathbf{M}$ be a given diagonal matrix.

Theorem 2.12. *For $M \in \mathbb{R}^{n,n}$, the following two properties are equivalent.*

(i) $M \in \mathbf{Z}$.
(ii) *$M + \lambda D$ is weakly pre-inverse-positive for all sufficiently large λ.*

Also, the following two properties are equivalent.

(i′) *M is irreducible and $M \in \mathbf{Z}$.*
(ii′) *$M + \lambda D$ is strictly pre-inverse-positive for all sufficiently large λ.*

Proof. Suppose that M has property (ii). Then, since $(\lambda D + M)e \succ o$ for large enough λ, the matrix $M + \lambda D$ is inverse-positive for those λ. That means

$$(M + \lambda D)^{-1} = \mu D^{-1}(I - \mu A + \mu^2 A^2 - + \cdots) \geqslant O$$

with $\mu = \lambda^{-1}$ and $A = MD^{-1}$, for all sufficiently large $\lambda > 0$. This inequality can only be true for all small enough $\mu > 0$, if the factor $-A$ of μ has all its off-diagonal elements $\geqslant 0$. That implies $M \in \mathbf{Z}$. Thus, (ii) ⇒ (i), while (i) ⇒ (ii) is obvious.

(i′) ⇒ (ii′) follows from Proposition 2.5(ii), since $\lambda D + M$ is irreducible if M is irreducible.

Now let (ii′) hold, so that $M \in \mathbf{Z}$, because (ii′) ⇒ (ii) ⇒ (i). Suppose that M is reducible and denote by J_0, J_1 the corresponding sets in the definition of a reducible matrix at the end of Section 2.1. Choose $u \in \mathbb{R}^n$ with $u_i = 0$ for $i \in J_0$, $u_i = 1$ for $i \in J_1$. Then $u \geqslant o$, $u \not\succ o$, and $(M + \lambda D)u > o$ for large enough λ. This contradicts (ii′), so that M is irreducible. □

Corollary 2.12a. *A matrix $M \in \mathbb{R}^{n,n}$ belongs to $\mathbf{M}$ if and only if M is inverse-positive and $M + \lambda D$ is inverse-positive for all sufficiently large λ.*

Proof. Let $M + \lambda D$ be inverse-positive for all $\lambda \geqslant \lambda_1$. Then, by Theorem 2.1, $M + \lambda D$ is weakly pre-inverse-positive for all such λ, so that $M \in \mathbf{Z}$, by Theorem 2.12. Thus, $M \in \mathbf{M}$, if M is also inverse-positive.

On the other hand, if $M \in \mathbf{M}$, then M has all the properties mentioned in the corollary (cf. Corollary 2.6a). □

Example 2.7. The matrix

$$\hat{M} + \lambda I \qquad \text{with} \quad \hat{M} = \begin{bmatrix} 5 & -4 & 1 & & & & \\ -4 & 6 & -4 & 1 & & 0 & \\ 1 & -4 & 6 & -4 & 1 & & \\ & \ddots & \ddots & \ddots & \ddots & \ddots & \\ & & 1 & -4 & 6 & -4 & 1 \\ & 0 & & 1 & -4 & 6 & -4 \\ & & & & 1 & -4 & 5 \end{bmatrix}$$

is inverse-positive for $\lambda = 0$, since $\hat{M} = M^2$ with M in Example 2.2 and, therefore, $\hat{M}^{-1} = (M^{-1})^2 \geqslant O$. However, since $\hat{M} \notin \mathbf{Z}$, there exist arbitrarily large λ such that $\hat{M} + \lambda I$ is not inverse-positive. □

Since a matrix $T \in \mathbb{R}^{n,n}$ is irreducible if and only if $M := -|T|$ is irreducible, Theorem 2.12 can be used to characterize irreducible matrices in the following way.

Corollary 2.12b. *A matrix $T \in \mathbb{R}^{n,n}$ is irreducible if and only if $\lambda I - |T|$ is strictly pre-inverse-positive for all sufficiently large λ.*

Now we consider a more general eigenvalue problem

$$(\lambda A - B)\varphi = o \tag{2.8}$$

where the pair (A, B) of $n \times n$ matrices A, B satisfies the following

Condition (R): A *is inverse-positive and* $-A^{-1}B \in \mathbf{Z}$.

For example, Condition (R) holds if A is inverse-positive and $A^{-1}B \geqslant O$. In particular, this is true if A is inverse-positive and $B \geqslant O$.

By $\sigma = \sigma(A, B)$ we denote the largest real eigenvalue of problem (2.8) (provided such eigenvalue exists).

Proposition 2.13. *If A and B satisfy Condition* (R), *the following statements hold for each $\lambda \in \mathbb{R}$.*

(i) *The matrix $\lambda A - B$ is weakly pre-inverse-positive.*
(ii) *The matrix $\lambda A - B$ is inverse-positive if and only if $\lambda I - A^{-1}B$ is inverse-positive.*

Proof. (i) The inequalities $u \geqslant o$, $(\lambda A - B)u \succ o$ imply $(\lambda I - A^{-1}B)u \succ o$, so that $u \succ o$, since $\lambda I - A^{-1}B \in \mathbf{Z}$.

(ii) If $\lambda I - A^{-1}B$ is inverse-positive, then

$$(\lambda A - B)u \geqslant o \Rightarrow (\lambda I - A^{-1}B)u \geqslant o \Rightarrow u \geqslant o,$$

so that $\lambda A - B$ is also inverse-positive.

On the other hand, if $\lambda A - B$ is inverse-positive, then $(\lambda A - B)z = e$ for some $z \geqslant o$ and, consequently, $(\lambda I - A^{-1}B)z = A^{-1}e \succ o$, so that $\lambda I - A^{-1}B$ is also inverse-positive, by Theorem 2.6(i). □

Theorem 2.14. *Let A and B satisfy Condition* (R). *Then the eigenvalue problem* (2.8) *has a largest real eigenvalue σ, which has the following properties*:

(i) *$\lambda A - B$ is inverse-positive if and only if $\lambda > \sigma$,*

(ii) *$\sigma = \inf\{\lambda \in \mathbb{R} : (\lambda A - B)z \succ o$ for some $z \geqslant o\}$.*

Proof. Since problem (2.8) is equivalent to problem (2.7) with $D = I$ and $M = -A^{-1}B$, statement (i) is derived from Theorem 2.11 and Proposition 2.13(ii). Moreover, statement (ii) follows from the equivalence of the properties in Theorem 2.2. □

Corollary 2.14a. *If A is inverse-positive and $A^{-1}B \geqslant O$, then $\sigma(A, B)$ equals the spectral radius $\rho(A^{-1}B)$ of $A^{-1}B$.*

Proof. For each $\lambda > \sigma = \sigma(A, B)$, the matrix $\lambda I - A^{-1}B$ is inverse-positive, so that $\lambda z \succ A^{-1}Bz$ for some $z \succ o$. Since $A^{-1}B \geqslant O$, these inequalities yield $\rho(A^{-1}B) \leqslant \|A^{-1}B\|_z < \lambda$. Consequently, $\rho(A^{-1}B) \leqslant \sigma(A, B)$, while $\sigma(A, B) \leqslant \rho(A^{-1}B)$ is obvious. □

The above results yield the following corollary which generalizes Theorem 2.9.

Corollary 2.14b. *For $M \in Z$ the following three properties are equivalent.*

(i) *M is inverse-positive.*

(ii) *$\lambda > 0$ for each real eigenvalue λ of M.*

(iii) *$\operatorname{Re} \lambda > 0$ for each complex eigenvalue λ of M.*

Proof. (i) $\Leftrightarrow$ (ii) according to Theorem 2.11. (iii) $\Rightarrow$ (ii) is obvious. To prove that (ii) $\Rightarrow$ (iii) we write $M = pI - B$ with $p > 0$ and $B \geqslant O$. Clearly, (ii) $\Leftrightarrow \sigma(I, B) < p \Leftrightarrow \rho(B) < p$ (cf. Corollary 2.14a). The last inequality, however, implies (iii). □

Example 2.8. Let $A = \hat{M}$, where $\hat{M}$ denotes the inverse-positive matrix in Example 2.7. This matrix has a smallest eigenvalue $\mu_0 = 4(1 - \cos \pi h)^2$ with $h = (n + 1)^{-1}$. Thus, $\sigma = \mu_0^{-1}$ is the largest eigenvalue of $(\lambda \hat{M} - I)\varphi = o$ and hence $\lambda \hat{M} - I$ is inverse-positive if and only if $\lambda > \mu_0^{-1}$. Consequently, for $\mu > 0$, $\hat{M} - \mu I$ is inverse-positive if and only if $\mu < \mu_0$. □

Example 2.9. Let again $A = \hat{M} = M^2$ with $\hat{M}$ in Example 2.7 and M in Example 2.2. Moreover, let $B = -\alpha\hat{M}C_1 + \beta MC_2 + \gamma C_3$ with $C_1 \in \mathbf{Z}$, $C_2 \geqslant O$, $C_3 \geqslant O$, and constants $\alpha, \beta, \gamma \geqslant 0$. Since $M \in \mathbf{M}$, we have $A^{-1} \geqslant O$ and $-A^{-1}B \in \mathbf{Z}$, so that Condition (R) holds. Thus, $\lambda A - B$ is inverse-positive if and only if $\lambda > \sigma(A, B)$. If $\alpha = 0$, for instance, we have $\sigma(A, B) = \rho(A^{-1}B)$. □

Example 2.10. The statements of Theorem 2.14 become false when $A \in \mathbf{M}$ and $-B \in \mathbf{Z}$ are assumed instead of Condition (R). One easily calculates that the inverse of

$$\lambda\begin{bmatrix} 2 & -1 \\ -1 & 2 \end{bmatrix} + \begin{bmatrix} 1 & 0 \\ 0 & 1 \end{bmatrix}$$

is positive if and only if either $\lambda \geqslant 0$ or $-1 < \lambda \leqslant -\frac{1}{2}$, while the above matrix is singular for $\lambda = -1$ and $\lambda = -\frac{1}{3}$. □

The above results can be used to estimate the largest eigenvalue σ.

Theorem 2.15. *For a pair of matrices* (A, B) *satisfying Condition* (R), *the following statements hold* (*with* $\sigma = \sigma(A, B)$):

(i) *If* $(pA - B)z \geqslant o$ *for some* $z \succ o$, *then* $p \geqslant \sigma$.
(ii) *If* $(pA - B)z \leqslant o$ *for some* $z \succ 0$, *then* $p \leqslant \sigma$.
(iii) *If* $(\hat{A}, \hat{B})$ *is a further pair of matrices which also satisfy Condition* (R) *and if*

$$(\sigma + \varepsilon)(A - \hat{A}) - (B - \hat{B}) \leqslant O \tag{2.9}$$

for small enough $\varepsilon > 0$, *then* $\sigma(\hat{A}, \hat{B}) \leqslant \sigma(A, B)$.

Proof. (i) For each $\varepsilon > 0$, we have $((p + \varepsilon)I - A^{-1}B)z \succ o$, so that $(p + \varepsilon)I - A^{-1}B \in \mathbf{M}$. Therefore, $p + \varepsilon > \sigma$ and hence $p \geqslant \sigma$.
(ii) Obviously, $pA - B$ is not inverse-positive, so that $p \leqslant \sigma$, by Theorem 2.14(i).
(iii) For each $\varepsilon > 0$, there exists a vector $z \succ o$ with $((\sigma + \varepsilon)A - B)z \succ o$. Because of (2.9) this vector satisfies also $((\sigma + \varepsilon)\hat{A} - \hat{B})z \succ o$, if ε is small enough. Consequently, $\sigma + \varepsilon \geqslant \sigma(\hat{A}, \hat{B})$ and hence $\sigma \geqslant \sigma(\hat{A}, \hat{B})$. □

Properties (i) and (ii) are often applied in the following way:

Corollary 2.15a. *Suppose that* A *and* B *satisfy Condition* (R), *and that* $Az \succ o$ *for some* $z \succ o$. *Then*

$$q \leqslant \sigma(A, B) \leqslant p \tag{2.10}$$

with $q = \min_i \mu_i$, $p = \max_i \mu_i$, $\mu_i = (Bz)_i/(Az)_i$ $(i = 1, 2, \ldots, n)$.

Example 2.11. We consider the eigenvalue problem $(\lambda A - B)\varphi = o$, where A equals M in Example 2.2, and $B = CD$ with tridiagonal C and diagonal D given by $c_{i-1,i} = 1, c_{ii} = 10, c_{i,i+1} = 1, d_i = 1 + ih, h = (n+1)^{-1}$.
For $n = 7$, the iterative procedure

$$\varphi^{(0)} = e, \qquad A\psi^{(\nu+1)} = B\varphi^{(\nu)}, \qquad \varphi^{(\nu+1)} = \psi^{(\nu+1)}/\psi_4^{(\nu+1)}$$

yields a vector $z = \psi^{(4)}$ such that $(Bz)_i/(Az)_i = 117 + \delta_i$, where the first two decimals of the numbers δ_i $(i = 1, 2, \ldots, 7)$ are given by 0.30, 0.33, 0.35, 0.34, 0.28, 0.18, 0.08. Since $A^{-1}B \geqslant O$, we obtain from (2.10) that $117.08 \leqslant \rho(A^{-1}B) \leqslant 117.36$. □

While so far we have been concerned with the largest eigenvalue of problem (2.8), we turn now to the investigation of the corresponding eigenspace. Instead of Condition (R), we now need the following stronger

Condition (R′): *A is inverse-positive,* $-A^{-1}B \in \mathbf{Z}$, *and* (*with* $\sigma = \sigma(A, B)$) *for all* $u \in \mathbb{R}^n$

$$u > o, \quad (\sigma A - B)u \geqslant o \quad \Rightarrow \quad u \succ o. \tag{2.11}$$

Obviously, if this implication holds, then $\sigma A - B$ is strictly pre-inverse-positive. For example, Condition (R′) is satisfied in the following two special cases (see Proposition 2.5(ii)):

(a) Condition (R) holds and $A^{-1}B$ is irreducible,
(b) $A \in \mathbf{M}$, $B \geqslant O$, and $\sigma A - B$ is irreducible.

Theorem 2.16. *If A and B satisfy Condition* (R′), *then*

(i) $\sigma(A, B)$ *is a simple eigenvalue of problem* (2.8),
(ii) *there exists an eigenvector* $\varphi \succ o$ *corresponding to the eigenvalue* $\sigma(A, B)$,

$$\text{(iii)} \quad \sigma(A, B) = \begin{cases} \min\{\lambda \in \mathbb{R} : (\lambda A - B)z \geqslant o \text{ for some } z \succ o\}, \\ \max\{\lambda \in \mathbb{R} : (\lambda A - B)z \leqslant o \text{ for some } z \succ o\}. \end{cases}$$

Proof. These results follow from Theorem 2.4 applied to $M(\lambda) = \lambda A - B$, $\lambda_0 = \sigma(A, B)$, and $\Lambda = [\lambda_0, \infty)$, by observing Theorems 2.14 and 2.15. □

Theorem 2.16 serves as a convenient tool for deriving further comparison statements.

Theorem 2.17. *Suppose that* (A, B) *and* $(\hat{A}, \hat{B})$ *are pairs of matrices in* $\mathbb{R}^{n,n}$ *which both satisfy Condition* (R′). *Let*

$$\sigma = \sigma(A, B), \qquad \hat{\sigma} = \sigma(\hat{A}, \hat{B}), \qquad C = \sigma(A - \hat{A}) - (B - \hat{B}).$$

Then $C \geqslant O \Rightarrow \sigma \leqslant \hat{\sigma}$; $C \leqslant O \Rightarrow \sigma \geqslant \hat{\sigma}$; $C < O \Rightarrow \sigma > \hat{\sigma}$.

Proof. According to Theorem 2.16 there exists a vector $\varphi \succ o$ such that $o = (\sigma A - B)\varphi = (\sigma \hat{A} - \hat{B})\varphi + C\varphi$. For $C \geqslant O$ $(\leqslant O)$, the inequality $(\sigma \hat{A} - \hat{B})\varphi \leqslant o$ $(\geqslant o)$ yields $\sigma \leqslant \hat{\sigma}$ $(\sigma \geqslant \hat{\sigma})$, according to Theorem 2.15.

Now suppose that $C < O$ and $\sigma = \hat{\sigma}$. Then $\hat{\sigma}\hat{A} - \hat{B}$ is inverse-positive, since this matrix is strictly pre-inverse-positive and $(\hat{\sigma}\hat{A} - \hat{B})\varphi > o$ (cf. Theorem 2.2). This contradicts the definition of $\hat{\sigma}$, so that $\sigma > \hat{\sigma}$. □

Corollary 2.17a. *If the assumptions of Theorem* 2.17 *are satisfied and if* $A = \hat{A}$, *then* $\hat{B} < B \Rightarrow \hat{\sigma} < \sigma$.

Corollary 2.17b. *If the assumptions of Theorem* 2.17 *are satisfied and if* $A - \hat{A} = B - \hat{B} > O$, *then either* $\hat{\sigma} < \sigma < 1$, *or* $\hat{\sigma} = \sigma = 1$, *or* $1 < \sigma < \hat{\sigma}$.

Proof. Theorem 2.17 immediately yields, $\sigma - 1 \geqslant 0 \Rightarrow \sigma \leqslant \hat{\sigma}$, and $\sigma - 1 \leqslant 0 \Rightarrow \sigma \geqslant \hat{\sigma}$, and $\sigma - 1 < 0 \Rightarrow \sigma > \hat{\sigma}$. By exchanging the roles of (A, B) and $(\hat{A}, \hat{B})$ in Theorem 2.17, we also obtain $\hat{\sigma} - 1 > 0 \Rightarrow \hat{\sigma} > \sigma$. These implications together yield the statements of the corollary. □

The main part of the Perron–Frobenius theory on positive matrices is contained in the preceding results, or can be derived from them by continuity arguments.

Theorem 2.18 (*Theorem of Perron and Frobenius*). *The spectral radius of an (irreducible) matrix* $B \geqslant O$ *in* $\mathbb{R}^{n,n}$ *is a (simple nonzero) eigenvalue of that matrix, and there exists a corresponding (strictly) positive eigenvector. Moreover, if both* B *and* $\hat{B}$ *are irreducible, then* $O \leqslant \hat{B} < B \Rightarrow \rho(\hat{B}) < \rho(B)$.

Problems

In the next three problems, let $\stackrel{\prime}{\leqslant}$ be a given strong order and let (A, B) denote a pair of matrices which satisfy Condition (R). The term pre-inverse-positive is understood with respect to $\stackrel{\prime}{\leqslant}$. Prove the following statements.

2.14 For each $\lambda \in \mathbb{R}$, $\lambda A - B$ is pre-inverse-positive if A is pre-inverse-positive.

2.15 If $\lambda A - B$ is pre-inverse-positive for each $\lambda \in \mathbb{R}$, then statement (ii) in Theorem 2.14 can be replaced by $\sigma = \inf\{\lambda \in \mathbb{R} : (\lambda A - B)z \stackrel{\prime}{>} o$ for some $z \geqslant o\}$.

2.16 If $pA - B$ is pre-inverse-positive and $(pA - B)z \stackrel{\prime}{>} o$ for some $z \succ o$, then $p > \sigma(A, B)$.

2.17 Let $B \in \mathbb{R}^{n,n}$ be a tridiagonal matrix with $b_{ii} = 0$, $0 < a \leqslant b_{i,i+1} \leqslant b$, $a \leqslant b_{i+1,i} \leqslant b$. To obtain upper and lower bounds for $\rho(B)$, apply Corollary 2.15a with $z = (z_i)$, $z_i = x_i(1 - x_i)$, $x_i = ih$, $h = (n + 1)^{-1}$.

2.18 Prove Theorem 2.18.

2.5 Iterative procedures

2.5.1 *General remarks*

The theory of M-matrices can be used to investigate certain iterative procedures of n linear algebraic equations for n unknowns. Let $Mu = s$ be such a system. An ordered pair (A, B) of matrices is called a *splitting of* M if A is nonsingular and $M = A - B$. We shall often simply say that $M = A - B$ is a splitting. *The iterative procedure corresponding to the splitting* $M = A - B$ is given by

$$Au_{\nu+1} = Bu_\nu + s \qquad (\nu = 0, 1, 2, \ldots), \qquad u_0 \in \mathbb{R}^n. \tag{2.12}$$

If the sequence $\{u_\nu\}$ converges for given u_0, then its limit is a solution of the given equation. We shall say that *the procedure* (*corresponding to the splitting* $M = A - B$) *converges* if and only if the sequence $\{u_\nu\}$ converges for arbitrary u_0. This property holds if and only if the spectral radius $\rho(A^{-1}B)$ of the *iteration matrix* $A^{-1}B$ satisfies $\rho(A^{-1}B) < 1$. Obviously, this condition does not depend on the vector s. (In this section lower indices 0 and ν for vectors are iteration indices; they do not indicate components of the vectors.)

For given $M \in \mathbb{R}^{n,n}$, the matrices $D(M)$, $L(M)$, and $U(M)$ are defined as follows. Let

$$M = D(M) - L(M) - U(M),$$

where $D(M) = (m_{ii}\,\delta_{ik})$ is diagonal, $L(M)$ is strictly lower triangular, and $U(M)$ is strictly upper triangular (all diagonal elements of $L(M)$ and $U(M)$ equal 0). When no misunderstanding is likely, we shall simply write D, L, U.

As typical examples, we consider the *Jacobi procedure* (= *total step procedure* = *Gesamtschrittverfahren*) described by $A = D$, $B = L + U$, and the *Gauß–Seidel procedure* (= *single step procedure* = *point successive method* = *Einzelschrittverfahren*) described by $A = D - L$, $B = U$. For both these methods the matrix D is supposed to be nonsingular. Most of the results for these methods also hold for corresponding relaxation methods, such as the SOR-*method* and the SSOR-*method*, provided the relaxation parameter ω belongs to a certain interval (see the Notes). However, this interval, in general, does not contain the optimal parameter which yields the fastest method. For this reason, and since the basic ideas are more easily explained for the simpler methods of Jacobi and Gauß–Seidel, we shall not consider relaxation methods in greater detail.

The following criteria play an important role in the convergence of the above procedures and similar procedures. We note the

row-sum criterion: $(|D| - |L| - |U|)e \succ o$,
weak row-sum criterion: $(|D| - |L| - |U|)e > o$,
column-sum criterion: $e^{\mathrm{T}}(|D| - |L| - |U|) \succ o$,
weak column-sum criterion: $e^{\mathrm{T}}(|D| - |L| - |U|) > o$.

Obviously, for a diagonally positive Z-matrix, these criteria are equivalent to, respectively, $Me \succ o$, $Me > o$, $M^{\mathrm{T}}e \succ o$, and $M^{\mathrm{T}}e > o$. Corresponding generalized criteria are obtained when e is replaced by an arbitrary vector $z \succ o$.

2.5.2 *Regular splittings*

A splitting $M = A - B$ will be called *regular* if A is inverse-positive and $A^{-1}B \geqslant O$. Obviously, then A and B satisfy Condition (R) of Section 2.4, so that the results of that section apply. In particular, $M = A - B$ is a regular splitting if A is inverse-positive and $B \geqslant O$. The more general condition $A^{-1}B \geqslant O$ holds if and only if for all $u, v \in \mathbb{R}^n$

$$Au = Bv, \quad v \geqslant o \quad \Rightarrow \quad u \geqslant o. \tag{2.13}$$

Theorem 2.19. *For a regular splitting $M = A - B$ of a given matrix M, the following three properties are equivalent:*

(i) *The iterative procedure corresponding to the splitting converges.*
(ii) *M is inverse-positive.*
(iii) *$z \geqslant o$, $Mz \succ o$ for some $z \in \mathbb{R}^n$.*

Proof. According to Theorem 2.14 and Corollary 2.14a the inequality $\rho(A^{-1}B) < 1$ is equivalent to each of the properties (ii) and (iii). □

Corollary 2.19a. *For strictly diagonally positive $M \in \mathbf{Z}$, the Jacobi procedure converges if and only if M is inverse-positive. The same statement holds for the Gauß–Seidel procedure.*

Proof. Both procedures correspond to regular splittings. □

The above theorem shows that *each criterion for the inverse-positivity of M is also a criterion for the convergence of each iterative procedure corresponding to a regular splitting $M = A - B$.* A series of criteria for inverse-positivity has been given in previous sections.

In particular, for diagonally positive $M \in \mathbf{Z}$, the row–sum criterion and the column–sum criterion are sufficient for the convergence of each procedure corresponding to a regular splitting. The same statement holds for the generalized row–sum criterion and column–sum criterion. If, in addition, M is irreducible, the corresponding weak criteria are also sufficient. Furthermore, we recall that each positive definite symmetric $M \in \mathbf{Z}$ is inverse-positive.

While the convergence of an iterative procedure corresponding to a regular splitting is independent of the special choice of the splitting, the *ratio of convergence* $-\log \rho(A^{-1}B)$, in general, *does* depend on the splitting $M = A - B$. The same is true for norms of the iteration matrix. Such norms are used for sufficient convergence criteria and error estimations. We collect some related results.

Theorem 2.20. *Suppose that both $M = A_1 - B_1$ and $M = A_2 - B_2$ are regular splittings, and let $T_i = A_i^{-1}B_i$, $\rho_i = \rho(T_i)$ $(i = 1, 2)$.*

(i) *If $(pA_1 - B_1)z \geqslant o$ for some $z \succ o$, then $\|T_1\|_z \leqslant p$ and, moreover,*

$$(p - 1)(B_1 - B_2) \leqslant O \quad \Rightarrow \quad \|T_2\|_z \leqslant p.$$

(ii) *If $B_1 - B_2 > O$ and if both matrices $\rho_i A_i - B_i$ $(i = 1, 2)$ have the property* (2.11) *(in place of $\sigma A - B$), then*

$$\text{either} \quad \rho_2 < \rho_1 < 1, \quad \text{or} \quad \rho_1 = \rho_2 = 1, \quad \text{or} \quad 1 < \rho_1 < \rho_2.$$

Proof. (i) The inequality $pA_1 z - B_1 z \geqslant o$ implies $T_1 z \leqslant pz$, so that $\|T_1\|_z \leqslant p$, since T_1 is positive. For $(p - 1)(B_1 - B_2) \leqslant O$ we have also $pA_2 z - B_2 z \geqslant o$ and hence $\|T_2\|_z \leqslant p$. (ii) follows from Corollary 2.17b. □

Example 2.12. Let $M = D - L - U \in \mathbf{Z}$ with $D \in \mathbf{M}$ and consider the Gauß–Seidel procedure, that means, choose $A = D - L$ and $B = U$. In order to obtain the norm of the corresponding iteration matrix $T = A^{-1}B$, one may proceed as follows. Calculate Ue and solve the triangular system $(D - L)v = Ue$ for v. Then $p = \max_i v_i$ is the smallest number with $(pI - T)e \geqslant o$. Therefore, $\|T\|_e = p$.

For example, if M is the matrix in Example 2.2, then $v_i = 1 - 2^{-i}$ $(i = 1, 2, \ldots, n - 1)$ and $v_n = \frac{1}{2}v_{n-1}$, so that $p = 1 - 2^{-n+1} < 1$. □

Example 2.13. Let M be as in Example 2.12 and consider again the Gauß–Seidel procedure. Now we want to calculate the norm $\|T\|_1$. Solve the triangular system $(D - L)^{\mathrm{T}}z = e$ for z and compute $w = U^{\mathrm{T}}z$. Then, $p = \max_i w_i$ is the smallest number with

$$(pI - U^{\mathrm{T}}(D - L^{\mathrm{T}})^{-1})e = (p(D - L^{\mathrm{T}}) - U^{\mathrm{T}})z \geqslant o.$$

Therefore, $\|T\|_1 = \|(D - L)^{-1}U\|_1 = \|U^{\mathrm{T}}(D - L^{\mathrm{T}})^{-1}\|_\infty = p$.

Observe that the inequality $(p(D - L^{\mathrm{T}}) - U^{\mathrm{T}})z \geqslant o$ also yields

$$\|(D - L^{\mathrm{T}})^{-1}U^{\mathrm{T}}\|_z \leqslant p. \quad \square$$

Corollary 2.20a. *Suppose that M is a diagonally strictly positive Z-matrix. Let T_{J} and T_{G} denote the iteration matrices of the Jacobi procedure and the Gauß–Seidel procedure, respectively, and let $\rho_{\mathrm{J}} = \rho(T_{\mathrm{J}})$, $\rho_{\mathrm{G}} = \rho(T_{\mathrm{G}})$.*

(i) *If $\|T_{\mathrm{J}}\|_z \leqslant 1$ for some $z \succ o$, then $\|T_{\mathrm{G}}\|_z \leqslant \|T_{\mathrm{J}}\|_z$.*

(ii) *If M is irreducible, then either $\rho_{\mathrm{G}} < \rho_{\mathrm{J}} < 1$, or $\rho_{\mathrm{G}} = \rho_{\mathrm{J}} = 1$, or $1 < \rho_{\mathrm{J}} < \rho_{\mathrm{G}}$.*

Proof. The statements follow from Theorem 2.20 applied to $A_1 = D$, $B_1 = L + U$, $A_2 = D - L$, $B_2 = U$, where $M = D - L - U$. □

2.5.3 *Majorization*

The theory of the preceding section can also be used to obtain results on iterative procedures corresponding to nonregular splittings. One derives sufficient convergence criteria which are useful in some cases.

Theorem 2.21. *For a given splitting $M = A - B$, let A_0, B_0 denote matrices such that*

(i) $|A^{-1}B| \leqslant A_0^{-1}B_0$, (2.14)
(ii) $M_0 = A_0 - B_0$ *is inverse-positive,*
(iii) $M_0 = A_0 - B_0$ *is a regular splitting of* M_0.

Then the iterative procedure corresponding to the splitting $M = A - B$ converges. Moreover, $\rho(A^{-1}B) \leqslant \rho(A_0^{-1}B_0)$ and

$$\|A^{-1}B\|_z \leqslant \|A_0^{-1}B_0\|_z \qquad \textit{for each} \quad z \succ o. \tag{2.15}$$

The proof is left to the reader. (See Problems 2.22–2.24.)

In order to obtain a bound for the norm of $A^{-1}B$, one may apply the results of Section 2.5.2 to $M_0 = A_0 - B_0$ and then use the estimation (2.15). Obviously, the relation (2.14) holds if $|A^{-1}| \leqslant A_0^{-1}$ and $|B| \leqslant B_0$.

For example, *the inequality $|A^{-1}| \leqslant A_0^{-1}$ is satisfied for*

$$A_0 = D_0 - C_0 \qquad \textit{with} \quad D_0 = |D(A)|, \qquad C_0 = |L(A) + U(A)|,$$

if this matrix A_0 is inverse-positive. This follows from

$$A_0^{-1} = \sum_{k=0}^{\infty} (D_0^{-1}C_0)^k D_0^{-1} \geqslant \left| \sum_{k=0}^{\infty} [D^{-1}(A)(L(A) + U(A))]^k D^{-1}(A) \right| = |A^{-1}|.$$

Thus, when we use $A_0 = |D|$, $B_0 = |L + U|$ for the Jacobi procedure, and $A_0 = |D| - |L|$, $B_0 = |U|$ for the Gauß–Seidel procedure (where $M = D - L - U$), we obtain

Corollary 2.21a. *The Jacobi procedure and the Gauß–Seidel procedure for the equation $Mu = s$ converge if $M_0 = |D| - |L| - |U|$ is inverse-positive. In particular, this is true if the (generalized) row–sum criterion or the (generalized) column–sum criterion is satisfied. If, in addition, M is irreducible, the corresponding weak critera are also sufficient for convergence.*

2.5.4 *Monotonic approximations*

We turn now to iterative procedures for two sequences of vectors which not only approximate a solution of the given equation $Mu = s$, but yield also bounds for this solution. The starting vectors of the iterative procedure have to satisfy conditions which are closely related to the inverse-positivity of a certain matrix $\hat{M}$.

Theorem 2.22. *Suppose that $M = A - B$ is a regular splitting and that there exist vectors $x_0, y_0 \in \mathbb{R}^n$ with*

$$Mx_0 \leqslant s \leqslant My_0, \qquad x_0 \leqslant y_0.$$

Then the sequences $\{x_\nu\}$ and $\{y_\nu\}$ defined by

$$Ax_{\nu+1} = Bx_\nu + s, \qquad Ay_{\nu+1} = By_\nu + s \qquad (\nu = 0, 1, 2 \ldots)$$

converge toward limits x^* and y^*, respectively, and

$$x_0 \leqslant x_1 \leqslant \cdots \leqslant x_\nu \leqslant x^* \leqslant y^* \leqslant y_\nu \leqslant \cdots \leqslant y_1 \leqslant y_0. \tag{2.16}$$

Proof. The iterative procedures can be written as $x_{\nu+1} = Tx_\nu$, $y_{\nu+1} = Ty_\nu$ with $Tu = A^{-1}Bu + A^{-1}s$. Since $A^{-1}B \geqslant O$, the results stated follow from Theorem I.3.4 (see also Remark 3). □

The above theorem can be generalized to certain nonregular splittings.

Theorem 2.23. *Let $A, B, A_1, A_2, B_1, B_2 \in \mathbb{R}^{n,n}$ such that*

$$M = A - B \quad \text{with} \quad A = A_1 - A_2, \qquad B = B_1 - B_2, \tag{2.17}$$

and define $\tilde{M}, \tilde{A}, \tilde{B} \in \mathbb{R}^{2n,2n}$ by

$$\tilde{M} = \tilde{A} - \tilde{B} \quad \text{with} \quad \tilde{A} = \begin{pmatrix} A_1 & A_2 \\ A_2 & A_1 \end{pmatrix}, \qquad \tilde{B} = \begin{pmatrix} B_1 & B_2 \\ B_2 & B_1 \end{pmatrix}. \tag{2.18}$$

Assume that the following two conditions are satisfied.

(i) *$\tilde{M} = \tilde{A} - \tilde{B}$ is a regular splitting.*
(ii) *There exist vectors $x_0, y_0 \in \mathbb{R}^n$ with*

$$\begin{gathered} A_1x_0 - A_2y_0 - B_1x_0 + B_2y_0 \leqslant s \leqslant A_1y_0 - A_2x_0 - B_1y_0 + B_2x_0, \\ x_0 \leqslant y_0. \end{gathered} \tag{2.19}$$

Then the sequences $\{x_\nu\}$, $\{y_\nu\}$ defined by

$$\begin{aligned} A_1x_{\nu+1} - A_2y_{\nu+1} &= B_1x_\nu - B_2y_\nu + s, \\ -A_2x_{\nu+1} + A_1y_{\nu+1} &= -B_2x_\nu + B_1y_\nu + s \end{aligned}$$

$(\nu = 0, 1, 2, \ldots)$ *converge toward limits x^*, y^* and the inclusion statement* (2.16) *holds.*

Proof. This result follows when Theorem 2.22 is applied to the procedures

$$\tilde{A}\tilde{x}_{\nu+1} = \tilde{B}\tilde{x}_\nu + \tilde{s}, \qquad \tilde{A}\tilde{y}_{\nu+1} = \tilde{B}\tilde{y}_\nu + \tilde{s},$$

with

$$\tilde{s} = \begin{pmatrix} -s \\ s \end{pmatrix}, \qquad \tilde{x}_\nu = \begin{pmatrix} -y_\nu \\ x_\nu \end{pmatrix}, \qquad \tilde{y}_\nu = \begin{pmatrix} -x_\nu \\ y_\nu \end{pmatrix}. \quad □$$

Introducing

$$u_\nu = \tfrac{1}{2}(x_\nu + y_\nu), \qquad z_\nu = \tfrac{1}{2}(y_\nu - x_\nu), \qquad u^* = \tfrac{1}{2}(x^* + y^*),$$

we can reformulate Theorem 2.23 in the following way.

Corollary 2.23a. *Suppose that assumption (i) of Theorem 2.23 is satisfied, and that there exist vectors $u_0, z_0 \in \mathbb{R}^n$ with*

$$|Mu_0 - s| \leqslant \hat{M}z_0, \qquad z_0 \geqslant o, \tag{2.20}$$

where

$$\hat{M} = \hat{A} - \hat{B}, \qquad \hat{A} = A_1 + A_2, \qquad \hat{B} = B_1 + B_2. \tag{2.21}$$

Then, the sequences $\{u_\nu\}$, $\{z_\nu\}$ defined by

$$Au_{\nu+1} = Bu_\nu + s, \qquad \hat{A}z_{\nu+1} = \hat{B}z_\nu \qquad (\nu = 0, 1, \ldots)$$

converge. Moreover, the sequence $\{z_\nu\}$ is antitonic and

$$|u_\nu - u_{\nu+1}| \leqslant z_\nu - z_{\nu+1}, \qquad |u_\nu - u^*| \leqslant z_\nu \qquad (\nu = 0, 1, 2, \ldots)$$

with $u^* = \lim u_\nu$.

Obviously, Theorem 2.22 is contained in Theorem 2.23 as a special case where $A = A_1$ and $B = B_1$. Consequently, if $M = A - B$ is a regular splitting, then Corollary 2.23a holds with $\hat{A} = A$, $\hat{B} = B$, and $\hat{M} = M$.

Now we shall describe assumption (i) of Theorem 2.23 in a different way, and also formulate some special cases. In the following propositions, the occurring matrices shall be related as above (see (2.17), (2.18), and (2.21)).

Proposition 2.24. (i) *$\tilde{M} = \tilde{A} - \tilde{B}$ is a regular splitting if and only if the following two properties hold:*

(a) $u, v \in \mathbb{R}^n$, $|Au| \leqslant \hat{A}v \Rightarrow |u| \leqslant v$.
(b) $x, y, u, v \in \mathbb{R}^n$, $Au = Bx$, $\hat{A}v = \hat{B}y$, $|x| \leqslant y \Rightarrow |u| \leqslant v$.

(ii) *If $\tilde{M} = \tilde{A} - \tilde{B}$ is a regular splitting, then $\hat{M} = \hat{A} - \hat{B}$ is a regular splitting.*

Proof. (i) By applying $\tilde{A}$ to the vector $\binom{-u+v}{u+v}$, it is easily seen that property (a) is equivalent to the inverse-positivity of $\tilde{A}$. In a similar way, one can show that property (b) is equivalent to (2.13) applied to $\tilde{A}$, $\tilde{B}$ instead of A, B. The proof of (ii) is left to the reader. □

Proposition 2.25. *Let $A, B \in \mathbb{R}^{n,n}$ be given matrices with $M = A - B$.*

(i) *If A is inverse-positive, then the choice $A_1 = A$, $A_2 = O$, $B_1 = B^+$, $B_2 = B^-$, yields a regular splitting $\tilde{M} = \tilde{A} - \tilde{B}$.*

(ii) *The choice*

$$A_1 = D(A) - (L(A) + U(A))^+, \qquad B_1 = B^+,$$
$$A_2 = -(L(A) + U(A))^-, \qquad B_2 = B^-,$$

yields a regular splitting $\tilde{M} = \tilde{A} - \tilde{B}$ *if and only if* $\hat{A} = D(A) - |L(A) + U(A)|$ *is inverse-positive.*

The proof is left to the reader.

In applying the preceding results, the problem arises of how to obtain suitable vectors x_0, y_0 or u_0, z_0. Such vectors shall be called *admissible starting vectors* for the corresponding iterative procedures provided the inequalities (2.19) or (2.20) hold. In this section, we shall discuss only the theoretical conditions under which admissible starting vectors exist. Since all conditions on $x_0 = u_0 - z_0$, $y_0 = u_0 + z_0$ can easily be obtained from corresponding (equivalent) conditions on u_0, z_0, the following proposition is formulated only for the latter vectors.

Proposition 2.26. *If assumption* (i) *of Theorem* 2.23 *is satisfied, the following statements hold.*

(i) *There exist admissible starting vectors* u_0, z_0 *with* $\hat{M}z_0 \succ o$ *if and only if* $\hat{M}$ *is inverse-positive.*

(ii) *Let* $\hat{M}$ *be strictly pre-inverse-positive. Then, there exist admissable starting vectors* u_0, z_0 *with* $\hat{M}z_0 > o$ *if and only if* $\hat{M}$ *is inverse-positive.*

Proof. (i) According to Proposition 2.24(ii), $\hat{M} = \hat{A} - \hat{B}$ is a regular splitting, so that the matrix $\hat{M}$ is weakly pre-inverse-positive. Thus, if a vector $z_0 \geqslant o$ with $\hat{M}z_0 \succ o$ exists, the matrix $\hat{M}$ is inverse-positive. On the other hand, if $\hat{M}$ is inverse-positive, there exists a vector $z \geqslant o$ with $Mz \succ o$, so that $z_0 = mz$ satisfies (2.20) for given u_0 and large enough m. (ii) is proved in a similar way. □

Unless the vector u_0 satisfies at least one of the given n algebraic equations (described by $Mu = s$) exactly, each vector z_0 with (2.20) has the property $\hat{M}z_0 \succ o$. Thus, *exceptional cases excluded, there exist admissible starting vectors if and only if* $\hat{M}$ *is inverse-positive.* According to Theorem 2.14, this property is equivalent to the condition $\rho(\hat{A}^{-1}\hat{B}) < 1$, which implies the convergence $z_\nu \to o$.

Problems

2.19 The *point successive relaxation method* is defined by (2.12) with $A = \omega^{-1}D - L$, $B = (\omega^{-1} - 1)D + U$, $\omega \neq 0$. Prove that the statement in Corollary 2.19a holds also for this procedure, if $0 < \omega \leqslant 1$. What is the situation for $\omega > 1$?

2.20 Use Theorem 2.20(ii) for comparing point successive relaxation methods with different parameters $\omega \in (0, 1]$.
2.21 (2.13) means that $Au \in C \Rightarrow u \geqslant o$, where $C = \{w \in \mathbb{R}^n : w = Bv, v \geqslant o\}$. Try to apply Theorem 1.2 to this type of inverse-positivity.
2.22 Prove $\rho(T) = \min\{\| T \|_z : z \succ o\}$ for positive $T \in \mathbb{R}^{n,n}$ by first verifying this relation for positive irreducible T.
2.23 For $T_1, T_2 \in \mathbb{R}^{n,n}$ derive $\rho(T_1) \leqslant \rho(T_2)$ from $|T_1| \leqslant T_2$.
2.24 Prove Theorem 2.21, observing the preceding two problems.
2.25 Prove Proposition 2.25.

3 DIFFERENTIAL OPERATORS OF THE SECOND ORDER, POINTWISE INEQUALITIES

The abstract theory of Section 1 will here be applied to second order linear differential operators related to boundary value problems. In particular, inverse-positivity statements of the following type will be derived:

$$\left.\begin{aligned} L[u](x) &\geqslant 0 \quad (0 < x < 1) \\ B_0[u] &\geqslant 0 \\ B_1[u] &\geqslant 0 \end{aligned}\right\} \Rightarrow \quad u(x) \geqslant 0 \quad (0 \leqslant x \leqslant 1). \tag{3.1}$$

Here L denotes a (formal) differential operator in the usual sense and B_0, B_1 are boundary operators. These three terms will be combined in one operator $M = (L, B_0, B_1)$ such that (3.1) is equivalent to the inverse-positivity of M, i.e., $Mu \geqslant o \Rightarrow u \geqslant o$. Note that then $Mu = r \in \mathbb{R}[0, 1]$ constitutes a boundary value problem.

In many of the results below no regularity conditions on the coefficients of L are required. In cases, however, where due to "smooth" coefficients a suitable existence theory for boundary value problems $Mu = r$ is available, a richer theory on inverse-positivity can be developed. We shall, in particular, often consider *regular operators* M, as defined in Section 3.1.

Section 3.2 mainly provides sufficient conditions and necessary conditions for (3.1) in terms of differential inequalities for a *majorizing function* z. In Section 3.3 sufficient conditions on the coefficients of the operators L, B_0, B_1 are formulated such that a majorizing function exists. The actual construction of such a function can often be avoided by considering connected sets of operators, as in Section 3.4. This method leads to several characterizations of the set of all regular inverse-positive operators M. Section 3.5 is concerned with property (3.1) under constraints such as $B_0[u] = B_1[u] = 0$.

Often, the inequality $Mu \geqslant o$ implies stronger positivity properties of u than $u \geqslant o$. Corresponding results are contained in Sections 3.6 and 3.7. For regular operators, Theorem 3.17 is the strongest and most general statement.

An operator M is inverse-positive if and only if M has a positive inverse. Under suitable regularity conditions, the inverse is an integral operator with a Green function $\mathscr{G}$ as its kernel. Section 3.8 is concerned with the relation between inverse-positivity of M and the positivity (and similar properties) of $\mathscr{G}$.

The results on inverse-positivity can be used for obtaining other interesting statements. For example, boundary maximum principles are formulated in Sections 3.2, 3.6, and 3.7. For regular operators, we show in Sections 3.9 and 3.10 in which way statements on inverse-positivity, in particular Theorem 3.17, can serve as a tool for deriving oscillation theorems and results on eigenvalue problems.

It is only for simplicity of notation that we mainly consider the special interval $[0, 1]$ in (3.1). Analogous results for an arbitrary compact interval $[\alpha, \beta]$ are easily derived from the theory below. In many cases one may simply replace $[0, 1]$ by $[\alpha, \beta]$. We shall sometimes make use of this fact. Differential operators M on noncompact intervals, such as $[0, \infty)$ and $(-\infty, \infty)$, are studied in Section 3.11. In many cases, one can either develop a theory similar to that for operators on compact intervals, or obtain results by using the theorems for compact intervals.

Throughout this Section 3 we consider only functions u in (3.1) such that all derivatives $u'(x)$, $u''(x)$ which occur are defined in the usual sense. Functions with weaker smoothness properties will be considered in the theory for nonlinear operators in Chapter III.

3.1 Assumptions and notation

Our main interest is in operators $M: R \to S$ of the form

$$Mu(x) = \begin{cases} L[u](x) & \text{for} \quad 0 < x < 1 \\ B_0[u] & \text{for} \quad x = 0 \\ B_1[u] & \text{for} \quad x = 1, \end{cases} \tag{3.2}$$

where

$$\begin{aligned} L[u](x) &= -a(x)u'' + b(x)u' + c(x)u, \\ B_0[u] &= -\alpha_0 u'(0) + \gamma_0 u(0), \qquad B_1[u] = \alpha_1 u'(1) + \gamma_1 u(1); \\ &\quad a, b, c \in \mathbb{R}(0, 1); \qquad \alpha_0, \gamma_0, \alpha_1, \gamma_1 \in \mathbb{R}; \end{aligned} \tag{3.3}$$

with

$$\begin{aligned} &a(x) \geqslant 0 \quad (0 < x < 1), \qquad \alpha_0 \geqslant 0, \qquad \alpha_1 \geqslant 0, \\ &\gamma_0 > 0 \quad \text{if} \quad \alpha_0 = 0, \qquad \gamma_1 > 0 \quad \text{if} \quad \alpha_1 = 0. \end{aligned} \tag{3.4}$$

In the following, unless stated differently, M is an operator of this type, $S = \mathbb{R}[0, 1]$, and R denotes an arbitrary linear space with $C_2[0, 1] \subset R \subset R_M$.

Here R_M is the set of all $u \in C_0[0, 1] \cap C_1(0, 1)$ such that all derivatives exist which occur in $Mu(x)$. That means, $u''(x)$ exists for each $x \in (0, 1)$ with $a(x) \neq 0$, $u'(0)$ exists (as one-sided derivative) if $\alpha_0 \neq 0$, and $u'(1)$ exists if $\alpha_1 \neq 0$. (For $u \in R_M$ the preceding notation is to be interpreted in an appropriate way. For example, when $\alpha_0 = 0$, the term $\alpha_0 u'(0)$ stands for the value 0.)

For convenience, the notation $M = (L, B_0, B_1)$ and $Mu = (L[u](x), B_0[u], B_1[u])$ will be used, in particular, when special operators are described. For example, the meaning of $Mu = (-u'' + u, u(0), u(1))$ is obvious. Furthermore, we shall use the notation $M = (a(x), b(x), c(x), \alpha_0, \gamma_0, \alpha_1, \gamma_1)$ when we want to indicate the coefficients which determine the operator.

Both spaces R and S are endowed with the natural order relation. Thus, throughout Section 3 the following notation is used:

$$\begin{aligned} u \geqslant o \quad &\Leftrightarrow \quad u(x) \geqslant 0 \quad (0 \leqslant x \leqslant 1); \\ u = o \quad &\Leftrightarrow \quad u(x) = 0 \quad (0 \leqslant x \leqslant 1); \\ u > o \quad &\Leftrightarrow \quad u \geqslant o, \quad u(\xi) > 0 \quad \text{for some} \quad \xi \in [0, 1]. \end{aligned}$$

The corresponding strict order on R is given by

$$u \succ o \quad \Leftrightarrow \quad u(x) > 0 \quad (0 \leqslant x \leqslant 1).$$

Whenever other order relations or order relations on intervals different from $[0, 1]$ occur, we shall point this out explicitly.

According to our notation an operator $M = (L, B_0, B_1)$ is *inverse-positive* if for all $u \in R$, $Mu \geqslant o \Rightarrow u \geqslant o$, which means, explicitly, that (3.1) holds for all $u \in R$. Sometimes, when the choice of R is essential, we shall use the term *inverse-positive on R*.

An operator $M = (L, B_0, B_1)$ is called *regular*, if

$$a, b, c \in C_0[0, 1] \qquad \text{and} \qquad a(x) > 0 \qquad (0 \leqslant x \leqslant 1). \tag{3.5}$$

A regular operator is said to be *formally symmetric* or *formally self-adjoint if* $L[u] = -(au')' + cu$ with $a \in C_1[0, 1]$. To each regular $M = (L, B_0, B_1)$ we adjoin a formally symmetric regular operator $M_s = (L_s, B_0, B_1)$, where

$$L_s[u] = -(pu')' + qu = dL[u] \tag{3.6}$$

with

$$p = \exp\left[-\int_{x_0}^{x} b(t)a^{-1}(t)\, dt\right], \qquad q = cd, \qquad d = a^{-1}p \tag{3.7}$$

and some $x_0 \in [0, 1]$. Since $d(x) > 0$ $(0 \leqslant x \leqslant 1)$, it would theoretically suffice to consider only regular operators which are formally self-adjoint. For example, a regular operator M *is inverse-positive if and only if* M_s *is inverse-positive.*

Our concept includes also differential operators of the first order

$$Mu(x) = \begin{cases} u' + c(x)u & \text{for} \quad 0 < x \leqslant 1 \\ u(0) & \text{for} \quad x = 0 \end{cases} \tag{3.8}$$

defined on $R = C_0[0, 1] \cap C_1(0, 1]$, where $c \in \mathbb{R}(0, 1]$. This operator has the form (3.2) with $\alpha_1 = 1$, $\gamma_1 = c(1)$.

3.2 Basic results on inverse-positivity

For a fixed operator $M = (L, B_0, B_1): R \to S$, as described above, we formulate conditions such that M is inverse-positive or has similar properties. These conditions are formulated in terms of differential inequalities for a function $z \in R$.

Proposition 3.1. *The given operator M is pre-inverse-positive with respect to the order relation $\lessdot$ defined in S by*

$$U \gtrdot o \quad \Leftrightarrow \quad U(x) > 0 \quad (0 \leqslant x \leqslant 1),$$

that means, $u \geqslant o$ and $Mu \gtrdot o$ imply $u \succ o$.

Proof. It suffices to show that for each $\xi \in [0, 1]$

$$u \geqslant o, \quad u(\xi) = 0 \quad \Rightarrow \quad Mu(\xi) \leqslant 0, \tag{3.9}$$

since then the assumption of Proposition 1.3 is satisfied for $\ell U = U(\xi)$.

First, let $\xi \in (0, 1)$. Since u assumes its minimum at ξ, the derivatives satisfy the relations $u(\xi) = u'(\xi) = 0$ and $u''(\xi) \geqslant 0$ (provided $u''(\xi)$ exists). Therefore, $Mu(\xi) = -a(\xi)u''(\xi) \leqslant 0$. For $\xi = 0$ we have $u(0) = 0$ and $u'(0) \geqslant 0$ (provided $u'(0)$ exists), so that $Mu(0) = -\alpha_0 u'(0) \leqslant 0$. The case $\xi = 1$ is treated analogously. □

From this proposition and the monotonicity theorem the following result is immediately derived.

Theorem 3.2. *If there exists a function $z \in R$ such that*

$$z \geqslant o, \qquad Mz(x) > 0 \qquad (0 \leqslant x \leqslant 1), \tag{3.10}$$

then M is inverse-positive.

A function $z \in R$ with this property (3.10) will be called a *majorizing function for M*. (This definition is consistent with that of a majorizing element in Section 1.2, since (3.10) implies $z \succ o$.)

In general, the existence of a majorizing function is also necessary. If for any $r(x) > 0$ $(0 \leqslant x \leqslant 1)$ the problem $Mu = r$ has a solution $z \in R$, then this solution satisfies (3.10), provided M is inverse-positive. For a regular inverse-positive M, the equation $Mu = r$ with $r(x) \equiv 1$ has a solution $z \in C_2[0, 1]$. This yields

Theorem 3.3. *A regular operator $M = (L, B_0, B_1)$ is inverse-positive if and only if there exists a majorizing function for M.*

Moreover, a regular operator M is inverse-positive on R_M if and only if M is inverse-positive on $C_2[0, 1]$.

Before we discuss these results, we shall consider some modifications, which will be used later. Theorem 3.2 can also be derived from Theorem 1.7, as the proof of Proposition 3.1 shows. One chooses $\mathscr{L}$ to be the set $\{\ell_x : 0 \leqslant x \leqslant 1\}$ of all point functionals $\ell_x u = u(x)$ on R, $\mathscr{L}_1 = \mathscr{L}$, and $\mathscr{L}_2 = \varnothing$. Then the implication (3.9) is equivalent to (1.3) with $\ell = \ell_\xi$ and $\ell U = U(\xi)$. Moreover, with these notations, Corollary 1.7a yields the following result.

Corollary 3.2a. *Let the assumption of Theorem 3.2 be satisfied. If $Mu \geqslant o$ and $Mu(\xi) > 0$ for a given $u \in R$ and some $\xi \in [0, 1]$, then $u \geqslant o$ and $u(\xi) > 0$.*

This statement remains true if for all ξ with $(Mu)(\xi) > 0$ only the inequality $Mz(\xi) \geqslant 0$ is required instead of $Mz(\xi) > 0$, and if $z \succ o$.

In particular, we obtain

Theorem 3.4. *If there exists a function $z \in R$ such that*

$$z \succ o, \qquad L[z](x) \geqslant 0 \quad (0 < x < 1), \qquad B_0[z] > 0, \qquad B_1[z] > 0, \tag{3.11}$$

then for all $u \in R$

$$\left.\begin{aligned} L[u](x) > 0 \quad (0 < x < 1) \\ B_0[u] \geqslant 0 \\ B_1[u] \geqslant 0 \end{aligned}\right\} \Rightarrow u(x) > 0 \quad (0 < x < 1). \tag{3.12}$$

According to these results, the inverse-positivity of M and similar properties can be established by solving certain (differential) inequalities such as (3.10) and (3.11).

Example 3.1. (a) If $c(x) > 0\ (0 < x < 1)$, $\gamma_0 > 0$, $\gamma_1 > 0$, then $z(x) \equiv 1$ satisfies the inequalities (3.10) and, therefore, M is inverse-positive.

(b) If

$$c(x) \geqslant 0 \qquad (0 < x < 1), \qquad \gamma_0 > 0, \qquad \gamma_1 > 0, \tag{3.13}$$

then $z(x) \equiv 1$ satisfies the inequalities (3.11) and, therefore, statement (3.12) is true. □

The condition $c(x) > 0$ in part (a) of Example 3.1 is rather restrictive. In particular, this inequality is not satisfied for the simple operator $Mu = (-u'', u(0), u(1))$. The conditions in part (b) hold for many more operators which are of practical interest. However, property (3.12) does not imply inverse-positivity. Neither do the inequalities (3.13) alone imply inverse-positivity.

Example 3.2. The inequalities (3.13) hold for $M = (L, B_0, B_1)$ with

$$L[u](x) = -(1 - 2x)^2 u'' - 4(1 - 2x)u', \quad B_0[u] = u(0), \quad B_1[u] = u(1).$$

However, this operator is not inverse-positive, since $u = 1 - 8x(1 - x)$ satisfies $Mu \geqslant o$, but $u(\frac{1}{2}) < 0$. □

In Section 3.3 we shall derive results on inverse-positivity by actually constructing suitable functions z. In particular, we shall show that M is inverse-positive if (3.13) holds and the coefficients $a(x)$ and $b(x)$ satisfy certain boundedness conditions or sign conditions. These conditions on $a(x)$ and $b(x)$ may be relaxed when the following result is used, where

$$B_0[u](x) = -\alpha_0 u'(x) + \gamma_0 u(x), \qquad B_1[u](x) = \alpha_1 u'(x) + \gamma_1 u(x). \quad (3.14)$$

Theorem 3.5. *Suppose that the following assumptions* (i) *and* (ii) *are satisfied.*

(i) *There exists a function* $z_0 \in C_1[0, 1] \cap D_2(0, 1)$ *such that*

$$z_0(x) > 0 \quad (0 \leqslant x \leqslant 1), \qquad L[z_0](x) \geqslant 0 \quad (0 < x < 1),$$
$$B_0[z_0] > 0, \qquad B_1[z_0] > 0.$$

(ii) *For each subinterval* $[\alpha, \beta] \subset (0, 1)$ *with* $\alpha < \beta$, *there exists a function* $z \in C_1[\alpha, \beta] \cap D_2(\alpha, \beta)$ *such that*

$$z(x) \geqslant 0 \quad (\alpha \leqslant x \leqslant \beta), \qquad L[z](x) > 0 \quad (\alpha < x < \beta),$$
$$B_0[z](\alpha) > 0, \qquad B_1[z](\beta) > 0.$$

Then $M = (L, B_0, B_1)$ *is inverse-positive.*

Proof. Let $u \in R$ satisfy $Mu(x) \geqslant 0$ $(0 \leqslant x \leqslant 1)$. For reasons of continuity, there exist numbers α_m, β_m $(m = 1, 2, \ldots)$ such that $\alpha_{m+1} < \alpha_m < \beta_m < \beta_{m+1}$; $\alpha_m \to 0$, $\beta_m \to 1$, for $m \to \infty$; and $B_0[u + m^{-1}z_0](\alpha_m) \geqslant 0$, $B_1[u + m^{-1}z_0](\beta_m) \geqslant 0$. Moreover, we have $L[u + m^{-1}z_0](x) \geqslant 0$ $(\alpha_m < x < \beta_m)$.

If we now apply Theorem 3.2 to $u + m^{-1}z_0$ instead of u and to the interval $[\alpha_m, \beta_m]$ instead of $[0, 1]$, we conclude that $u(x) + m^{-1}z_0(x) \geqslant 0$ $(\alpha_m \leqslant x \leqslant \beta_m)$. Since this is true for all m, the desired inequality $u(x) \geqslant 0$ $(0 \leqslant x \leqslant 1)$ follows. □

Example 3.2 has shown that conditions (3.13) are not sufficient for M to be inverse-positive. They are not necessary either, as seen in the following example.

Example 3.3. Let $Mu = (-u'' + c(x)u, u(0), u(1))$. *If* $c(x) > -\pi^2 + \delta$ $(0 < x < 1)$ *for some* $\delta > 0$, *then* M *is inverse-positive*, because for $\varepsilon > 0$ small enough $z = \cos(\pi - \varepsilon)(x - \frac{1}{2})$ is a majorizing function. For $c(x) \equiv -\pi^2$, however, the operator M is not inverse-positive, since $M\varphi = o$ for $\varphi = \alpha \sin \pi x$ with arbitrary $\alpha \in \mathbb{R}$. □

Example 3.4. The operator (3.8) is inverse-positive if $c(x) > -\rho(x)$ $(0 < x \leqslant 1)$ for some $\rho \in C_0(0, 1]$ such that $\int_0^x \rho(t)\,dt$ $(0 < x \leqslant 1)$ exists. Under this condition, the function $z(x) = \exp\{\int_0^x \rho(t)\,dt\}$ satisfies (3.10). The condition on c cannot be dropped completely. For if $c(x) = -2x^{-1}$ $(0 < x \leqslant 1)$, then $u = -x^2$ satisfies $Mu \geqslant o$, but $u < o$. □

Example 3.5. The operator $Mu = (-u'' + u, -u'(0) + \gamma u(0), u'(1) + \gamma u(1))$ is inverse-positive if $\gamma > -\frac{4}{9}$. This is shown by using $z = \alpha - x(1 - x)$ with $1 + \alpha\gamma > 0$, $\alpha > \frac{9}{4}$. □

In several of the preceding examples we proved inverse-positivity by constructing a suitable function z, and we shall continue to do this in a more systematical way in the subsequent section. The question arises, exactly under what conditions on $c(x)$, γ_0, γ_1 (and the other coefficients) do there exist suitable functions z. This question will be answered in Section 3.4.

The property of inverse-positivity is sometimes referred to as a boundary maximum principle, the reason being that, in a special case, inverse-positivity indeed is equivalent to a boundary maximum principle in the usual sense. This special case is described by

$$Mu(x) = L[u](x) \quad \text{for} \quad 0 < x < 1, \qquad Mu(0) = u(0), \qquad Mu(1) = u(1). \tag{3.15}$$

For this operator, we may use the corresponding space $R = R_M$, or simply $R = C_0[0, 1] \cap C_2(0, 1)$.

Theorem 3.6. *Let there exist a function $z_0 \in R$ such that*

$$z_0(x) > 0 \qquad (0 \leqslant x \leqslant 1), \qquad L[z_0](x) \geqslant 0 \qquad (0 < x < 1). \tag{3.16}$$

Then the following two properties are equivalent:

(i) *The operator M in* (3.15) *is inverse-positive.*

(ii) *For each $u \in R$, the inequality $L[u](x) \leqslant 0$ $(0 < x < 1)$ implies that $u(x) \leqslant \mu z_0(x)$ $(0 \leqslant x \leqslant 1)$ with*

$$\mu = \max\{0, \mu_0\}, \qquad \mu_0 = \max\{u(0)/z_0(0), u(1)/z_0(1)\}.$$

When $L[z_0](x) = 0$ $(0 < x < 1)$, the number μ in (ii) *can be replaced by $\mu = \mu_0$.*

Proof. (i) ⇒ (ii) follows when we apply (3.1) to $v = -u + \mu z_0$ instead of u. To derive (3.1) from (ii), apply (ii) to $-u$. □

For the case $z_0(x) \equiv 1$, property (ii) is called the *weak boundary maximum principle*. For the general case, (ii) is the *generalized weak boundary maximum principle*. For example, $z_0(x) \equiv 1$ can be used if (3.13) holds.

Notice that, for the operator (3.15), condition (3.16) is equivalent to assumption (i) in Theorem 3.5. Thus the corresponding boundary maximum principle holds if assumption (ii) of Theorem 3.5 is satisfied.

Example 3.6. We have $L[u] := -(1 - \sin \pi x)u'' + \pi^2 u \leqslant 0\,(0 < x < 1)$ for $u = \sin \pi x - 1$. Since $Mu = (L[u], u(0), u(1))$ is inverse-positive and $z_0(x) \equiv 1$ satisfies (3.16), Theorem 3.6 asserts that $u(x) \leqslant \max\{0, u(0), u(1)\} = 0\,(0 \leqslant x \leqslant 1)$. The considered function u attains this value 0 only at $x = \frac{1}{2}$, while $u(0) < 0$ and $u(1) < 0$. This shows that, in general, the constant μ cannot be replaced by μ_0. □

Problems

3.1 Suppose that $M = (L, B_0, B_1)$ is regular and inverse-positive. Prove that then $\hat{M} = (\hat{L}, \hat{B}_0, \hat{B}_1)$ with $\hat{L}[u](x) = L[u](x) + d(x)u(x)$, $\hat{B}_0[u] = B_0[u] + \beta_0 u(0)$, $\hat{B}_1[u] = B_1[u] + \beta_1 u(1)$, $d(x) \geqslant 0\,(0 < x < 1)$, $\beta_0 \geqslant 0$, $\beta_1 \geqslant 0$ is also inverse-positive.

3.2 Suppose that $M = (L, B_0, B_1)$ is regular and d, β_0, β_1 are chosen as in Problem 3.1. Show that M is not inverse-positive, if the eigenvalue problem $L[\varphi] = \lambda\, d\varphi\,(0 < x < 1)$, $B_0[\varphi] = \lambda\beta_0\varphi(0)$, $B_1[\varphi] = \lambda\beta_1\varphi(1)$ has an eigenvalue $\lambda \leqslant 0$.

3.3 Prove that for $c \in C_0[0, 1]$ with $c(x) < -\pi^2$ $(0 \leqslant x \leqslant 1)$ no function $z \in C_2[0, 1]$ exists such that $z \geqslant o$, $z(0) > 0$, $z(1) > 0$, $-z''(x) + c(x)z(x) > 0$ $(0 < x < 1)$.

3.3 Construction of majorizing functions

In order to apply the results of Section 3.2 to some concrete operator $M = (L, B_0, B_1)$, one may either actually construct a suitable function z or verify the existence of such z by other means. The second possibility will be pursued in Section 3.4. In what follows we shall list some simple monotonicity statements which are gained by using specific functions z that satisfy the required inequalities.

Theorem 3.7. *Let*

$$c(x) \geqslant 0 \qquad (0 < x < 1), \qquad \gamma_0 \geqslant 0, \qquad \gamma_1 \geqslant 0, \tag{3.17}$$

and assume that one of the following five conditions is satisfied, where the occurring inequalities for a, b, c are supposed to hold for all $x \in (0, 1)$.

(i) $c(x) > 0$, $\gamma_0 > 0$, $\gamma_1 > 0$;
(ii) $a(x) + c(x) > 0$, $|b(x)| \leqslant \hat{b}a(x)$ *with some* $\hat{b} \geqslant 0$, $\gamma_0 + \gamma_1 > 0$;
(iii) $a(x) + c(x) > 0$, $b(x)\operatorname{sgn}(x - x_0) \leqslant \hat{b}a(x)$ *with some* $x_0, \hat{b} \in \mathbb{R}$,

$$\gamma_0 > 0 \quad \textit{for} \quad x_0 > 0, \qquad \gamma_1 > 0 \quad \textit{for} \quad x_0 < 1;$$

(iv) $a(x) + b(x) + c(x) > 0$, $b(x) \geqslant 0$, $\gamma_0 > 0$;
(v) $a(x) - b(x) + c(x) > 0$, $b(x) \leqslant 0$, $\gamma_1 > 0$.

Then M is inverse-positive.

Proof. For each of the five cases, we show that a function z exists which satisfies the inequalities (3.10). For (i) we choose $z(x) \equiv 1$. Case (ii) is included in (iii). In cases (iii)–(v), we find constants ξ, γ, and $\rho \geqslant 0$ such that the required inequalities hold for

$$z(x) = \gamma - \int_\xi^x E(s) \int_\xi^s E^{-1}(t)\, dt\, ds \qquad \text{with} \quad E(x) = \exp \rho|x - \xi|. \tag{3.18}$$

A formal calculation yields

$$z'(x) = -E(x) \int_\xi^x E^{-1}(t)\, dt, \qquad L[z] = a + \{a\rho - b \operatorname{sgn}(x - \xi)\}|z'| + cz.$$

Choose

$$\rho = \hat{b} \quad \text{and} \quad \begin{cases} \xi = x_0 & \text{for} \quad 0 < x_0 < 1 \\ \xi < 0 & \text{for} \quad x_0 \leqslant 0 \\ \xi > 1 & \text{for} \quad x_0 \geqslant 1 \end{cases} \qquad \text{for case (iii),}$$

$$\rho = 0 \quad \text{and} \quad \xi > 1 \qquad \text{for case (iv);}$$

$$\rho = 0 \quad \text{and} \quad \xi < -1 \qquad \text{for case (v).}$$

Then, for large enough $\gamma > 0$, $z(x)$ has all required properties. □

The following example will explain condition (iii), and it will show again that *the inequalities* (3.17) *alone are not sufficient for inverse-positivity.*

Example 3.7. We consider two operators $M = (L, B_0, B_1)$ and $\hat{M} = (\hat{L}, B_0, B_1)$ given by

$$L[u] = -u'' + b(x)u', \qquad \hat{L}[u] = -u'' + \hat{b}(x)u',$$
$$B_0[u] = u(0), \qquad B_1[u] = u(1),$$

with $b(x) = -\hat{b}(x) = b_0(1 - 2x)^{-1}$ for $x \neq \frac{1}{2}$, $b(\frac{1}{2}) = \hat{b}(\frac{1}{2}) = 0$, and $b_0 \geqslant 6$.

The first operator M is inverse-positive, since

$$b(x) \operatorname{sgn}(x - \tfrac{1}{2}) = -b_0|1 - 2x|^{-1} \leqslant 0$$

for $x \neq \frac{1}{2}$, and hence condition (iii) is satisfied for $x_0 = \frac{1}{2}$, $\hat{b} = 0$. The second operator $\hat{M}$, however, is not inverse-positive, since $u = 2(2x - 1)^4 - 1$ satisfies $Mu \geqslant o$, but $u(\frac{1}{2}) < 0$. Consequently, condition (iii) does not hold. Here, the quotient $\hat{b}(x) : a(x) = b_0(1 - 2x)^{-1}$ has the "wrong sign." For

$b_0 = 6$ the operator $\hat{M}$ is not even one-to-one, since $Mv = o$ for $v = u - 1$ with u mentioned above. □

The assumptions on the coefficients a, b, c in conditions (ii) and (iii) of Theorem 3.7 can be relaxed further.

Corollary 3.7a. *Theorem* 3.7 *remains true, if* (iii) *is replaced by*

(iii′) $a(x) + c(x) > 0\ (0 < x < 1),\ \gamma_0 > 0,\ \gamma_1 > 0,$

and if for each subinterval $[\alpha, \beta] \subset (0, 1)$ *with* $\alpha < \beta$ *there exist a real* x_0 *and a real* $\hat{b}$ *such that* $b(x)\operatorname{sgn}(x - x_0) \leqslant \hat{b}a(x)\ (\alpha < x < \beta)$.

Proof. Now the inequalities in condition (i) of Theorem 3.5 hold for $z_0(x) \equiv 1$. Moreover, for each subinterval $[\alpha, \beta] \subset (0, 1)$, there exists a function z which satisfies the inequalities in condition (ii) of that theorem. Choose z in (3.18) with $\rho = \hat{b}$ and $\xi = x_0$. □

As we have already seen in Example 3.3, the condition $c(x) \geqslant 0\ (0 < x < 1)$ is not necessary for inverse-positivity. A more careful examination of $Mz(x)$ with z as in (3.18) shows that this function z, in general, is a majorizing function also if $c(x)$ assumes certain values <0. This fact is used in the following example.

Example 3.8. Now let $a, b, c \in C_2[0, \infty)$ and $a(x) > 0\ (0 \leqslant x < \infty)$, so that $L[u](x)$ is defined for $u \in C_2(0, \infty)$ and all $x > 0$. Moreover, suppose that $\gamma_0 > 0, \gamma_1 > 0$ and define $B_1[u](x)$ as in (3.14). Then *there exists an* $l > 0$ *such that for all* $u \in C_1[0, l] \cap C_2(0, l)$

$$\left.\begin{array}{r} L[u](x) \geqslant 0 \quad (0 < x < l) \\ B_0[u] \geqslant 0 \\ B_1[u](l) \geqslant 0 \end{array}\right\} \Rightarrow u(x) \geqslant 0 \quad (0 \leqslant x \leqslant l). \tag{3.19}$$

To prove this statement we replace the variable x by xl. Then it turns out that the above property holds if and only if the following operator M_l is inverse-positive:

$$M_l u(x) = \begin{cases} -a(xl)u''(x) + lb(xl)u'(x) + l^2c(xl)u(x) & \text{for } 0 < x < 1 \\ -\alpha_0 u'(0) + l\gamma_0 u(0) & \text{for } x = 0 \\ \alpha_1 u'(1) + l\gamma_1 u(1) & \text{for } x = 1. \end{cases} \tag{3.20}$$

For small enough $l > 0$ the inverse-positivity of M_l follows from Theorem 3.2, when we use a function z as in (3.18). □

A similar result on *local inverse-positivity* is provided by the following theorem.

Theorem 3.8. *Suppose that*

$$a(x) > 0, \qquad b(x)\,\mathrm{sgn}(x - x_0) \leqslant \hat{b}a(x), \qquad c(x) \geqslant \hat{c}a(x) \qquad (0 < x < 1),$$

for some $x_0, \hat{b}, \hat{c} \in \mathbb{R}$.

Then there exists a constant $\delta > 0$ *such that for any interval* $[\alpha, \beta] \subset [0, 1]$ *of length* $0 < \beta - \alpha \leqslant \delta$ *and all* $u \in C_0[\alpha, \beta] \cap C_2(\alpha, \beta)$

$$\left.\begin{array}{ll} L[u](x) \geqslant 0 & (\alpha < x < \beta) \\ u(\alpha) \geqslant 0, & u(\beta) \geqslant 0 \end{array}\right\} \Rightarrow \quad u(x) \geqslant 0 \quad (\alpha \leqslant x \leqslant \beta).$$

Proof. Given a subinterval $[\alpha, \beta] \subset [0, 1]$, introduce a new independent variable $t = (\beta - \alpha)^{-1}(x - \alpha)$, so that the above inverse-positivity statement for $[\alpha, \beta]$ is transformed into a statement for an operator $\tilde{M}u = (\tilde{L}[u], u(0), u(1))$ on $[0, 1]$. Then show that the function z which has been used in the proof of Theorem 3.7 is a majorizing function for $\tilde{M}$, provided $\delta = \beta - \alpha$ is small enough. □

Problems

3.4 Prove that condition (ii) in Theorem 3.7 can be replaced by (ii′) $a(x) + c(x) > 0, |b(x)| \leqslant \beta(x)a(x)$ for some $\beta \in C_0(0, 1)$ such that $\int_0^1 \beta(t)\,dt$ converges, $\gamma_0 + \gamma_1 > 0$. Hint: use z in (3.18) with $E(x) = \exp|\int_\xi^x \rho(t)\,dt|$ and suitable ξ and $\rho(x)$.

3.5 May condition (iii) in Theorem 3.7 be relaxed in a corresponding way?

3.6 The boundary value problem

$$-u'' + b(x)u' - 630u = 0, \qquad u(0) = 0, \qquad u(1) = 5;$$

$$b(x) = -\left[90 \,\mathrm{ctg}\,\frac{\pi}{6}(1 + x) + 60 \,\mathrm{tg}\,\frac{\pi}{6}(1 + x)\right],$$

with $-190.53 \leqslant -110\sqrt{3} = \min b(x) \leqslant \max b(x) = -60\sqrt{6} \leqslant -147$ arises when considering stress distributions in a spherical membrane (Russell and Shampine [1972]). This problem can be written as $Mu = r$ with an operator as in (3.2). Show that M is inverse-positive, by constructing a majorizing function of the form $z(x) = \exp(-\kappa x)$ with a constant κ.

3.4 Sets of inverse-positive regular differential operators

In this section the application of the abstract continuation theorem shall be explained. We describe sets $\mathcal{M}$ of regular differential operators $M = (L, B_0, B_1)$ such that all $M \in \mathcal{M}$ are inverse-positive. In each of the following results, the main assumption to be verified is that all $M \in \mathcal{M}$ are one-to-one. Functions z as in Theorem 3.2 need not be constructed.

Here we suppose that $R = C_2[0, 1]$. This means no loss of generality, according to Theorem 3.3. In Sections 3.6 and 3.7 it will be shown that each inverse-positive regular operator has even stronger monotonicity properties. Thus, the results of this section are also useful for proving such stronger properties.

A set $\mathscr{M}$ of (regular) operators is called *arcwise connected* if for each pair of operators M_0, M_1 in $\mathscr{M}$ there exists a *connecting family* $\{M(t): 0 \leqslant t \leqslant 1\}$ of operators $M(t) = (a(x, t), b(x, t), c(x, t), \alpha_0(t), \gamma_0(t), \alpha_1(t), \gamma_1(t)) \in \mathscr{M}$ such that $M(0) = M_0$, $M(1) = M_1$ holds and the coefficients of $M(t)$ are continuous on $[0, 1] \times [0, 1]$ and $[0, 1]$, respectively. $\mathscr{M}$ is called *piecewise linearly connected* if $M(t)$ can be choosen such that the coefficients are piecewise linear (polygons) in t. For example, any convex set is piecewise linearly connected.

Theorem 3.9. *Let $\mathscr{M}$ be an arcwise connected set of regular differential operators M such that the following conditions are satisfied.*

(α) *For each $M \in \mathscr{M}$ the homogeneous boundary value problem $Mu = o$ has only the trivial solution $u = o$.*

(β) *There exists at least one inverse-positive $M \in \mathscr{M}$.*

Then all operators $M \in \mathscr{M}$ are inverse-positive.

Remark. According to Theorem 3.7(i), assumption (β) is satisfied if there exists an $M \in \mathscr{M}$ with

$$c(x) > 0 \qquad (0 < x < 1), \qquad \gamma_0 > 0, \qquad \gamma_1 > 0. \tag{3.21}$$

Proof of the theorem. Let the order relation $\leqslant$ in S be defined as in Proposition 3.1. The above statement follows from Theorem 1.8, if we show that assumptions (i)–(iv) of that theorem are satisfied.

Property (i) follows from Proposition 3.1. If (α) holds, then each $M \in \mathscr{M}$ is one-to-one and its range contains the function $r(x) = 1$ $(0 \leqslant x \leqslant 1)$, so that (ii) and (iii) are satisfied. For given $M_0, M_1 \in \mathscr{M}$, let $\{M(t): 0 \leqslant t \leqslant 1\} \subset \mathscr{M}$ be a connecting family as described above. Because of the continuity of the coefficients of these operators $M(t)$, the solutions $z(t)$ of $M(t)z(t) = r$ are o-continuous on $[0, 1]$. Thus, (iv) is also true. □

Example 3.9 (Continuation of Example 3.8). Let the assumptions of Example 3.8 be satisfied. For small enough $l > 0$ the operator M_l in (3.20) is inverse-positive and hence invertible. Consequently, there exists a greatest l_0 with $0 < l_0 \leqslant \infty$ such that M_l is invertible for all $l < l_0$. The set $\mathscr{M} = \{M_l: 0 < l < l_0\}$ satisfies all assumptions of Theorem 3.9. Therefore, all $M_l \in \mathscr{M}$ are inverse-positive. We reformulate this result:

The statement (3.19) *holds for all* $l \in (0, l_0)$, *where* $l_0 > 0$ *is the smallest of all values* $l > 0$ *such that the problem*

$$L[u](x) = 0 \qquad (0 < x < l), \qquad B_0[u] = 0, \qquad B_1[u](l) = 0$$

has a nontrivial solution, or $l_0 = \infty$, *if no such* l *exists.* □

An immediate consequence of Theorem 3.9 is

Corollary 3.9a. *Let* $M = (a, b, c, \alpha_0, \gamma_0, \alpha_1, \gamma_1)$ *be regular and let* $\hat{c} \in C_0[0, 1]$, $\hat{\gamma}_0 \in \mathbb{R}$, $\hat{\gamma}_1 \in \mathbb{R}$ *satisfy*

$$c(x) \leqslant \hat{c}(x) > 0 \qquad (0 < x < 1), \qquad \gamma_0 \leqslant \hat{\gamma}_0 > 0, \qquad \gamma_1 \leqslant \hat{\gamma}_1 > 0. \tag{3.22}$$

Suppose that for $0 \leqslant t \leqslant 1$ *each boundary value problem*

$$\begin{aligned} -au'' + bu' + [tc + (1 - t)\hat{c}]u &= 0 \qquad (0 < x < 1), \\ -\alpha_0 u'(0) + [t\gamma_0 + (1 - t)\hat{\gamma}_0]u(0) &= 0, \\ \alpha_1 u'(1) + [t\gamma_1 + (1 - t)\hat{\gamma}_1]u(1) &= 0 \end{aligned}$$

has only the trivial solution $u = o$.

Then the operator M *is inverse-positive.*

While the above theorem yields sufficient conditions for operators M to be inverse-positive, the following theorem characterizes the set of *all* (regular) inverse-positive differential operators $M = (L, B_0, B_1)$.

Theorem 3.10. *Let* $\mathscr{K}$ *denote the largest piecewise linearly connected set of invertible regular operators* M *which contains all regular operators* M *with* $c(x) > 0$ $(0 < x < 1)$, $\gamma_0 > 0$, $\gamma_1 > 0$.

Then a regular operator M *is inverse-positive if and only if* $M \in \mathscr{K}$.

Proof. To justify the above definition of $\mathscr{K}$, we observe that the set of all operators M with property (3.21) is convex.

According to Theorem 3.9, each $M \in \mathscr{K}$ is inverse-positive. Thus, we have only to show that each regular inverse-positive operator $M = (a, b, c, \alpha_0, \gamma_0, \alpha_1, \gamma_1)$ belongs to $\mathscr{K}$. For given M consider a regular operator $\hat{M} = (a, b, \hat{c}, \alpha_0, \hat{\gamma}_0, \alpha_1, \hat{\gamma}_1)$ such that (3.22) is satisfied and define

$$\mathscr{M} = \{M(t) = (1 - t)M + t\hat{M} : 0 \leqslant t \leqslant 1\}. \tag{3.23}$$

For each $M(t)$ the solution $z \in R$ of $Mz(x) = 1$ $(0 \leqslant x \leqslant 1)$ satisfies $z \geqslant o$ and $M(t)z = Mz + t(\hat{M} - M)z \geqslant Mz \succ o$, so that all these operators $M(t)$ are inverse-positive.

Since $\hat{M}$ belongs to $\mathscr{K}$ and M is linearly connected with $\hat{M}$ by the family $\mathscr{M}$, the operator M belongs also to $\mathscr{K}$. Otherwise, $\mathscr{K}$ would not be the largest set as described in the theorem. □

Often one may not want to consider all regular operators $M = (L, B_0, B_1)$, but only a certain subset. The following corollary is proved in the same way as the above theorem.

Corollary 3.10a. *Suppose that $\mathscr{M}_0$ is a set of regular operators $M = (L, B_0, B_1)$ with the following two properties.*

The set of all $M \in \mathscr{M}_0$ with (3.21) is piecewise linearly connected.

For any $M = (a, b, c, \alpha_0, \gamma_0, \alpha_1, \gamma_1) \in \mathscr{M}_0$, there exists an $\hat{M} = (a, b, \hat{c}, \alpha_0, \hat{\gamma}_0, \alpha_1, \hat{\gamma}_1) \in \mathscr{M}_0$ with (3.22) such that the set (3.23) belongs to $\mathscr{M}_0$.

Under these conditions Theorem 3.10 *remains true if the term "regular operator M" is replaced anywhere by "operator $M \in \mathscr{M}_0$."*

Example 3.10. Consider the set $\mathscr{M}_0$ of all operators $Mu = (-u'' + cu, u(0), u'(1) + \gamma u(1))$ with $c \in \mathbb{R}$ and $\gamma \in \mathbb{R}$. Each $M \in \mathscr{M}_0$ is described by a point in the (c, γ)-plane, $M = M_{(c,\gamma)}$.

The operator $M_{(c,\gamma)}$ is one-to-one if and only if (c, γ) does not belong to one of the infinitely many eigencurves $\Gamma_1, \Gamma_2, \ldots$ of the two-parameter eigenvalue problem $M_{(c,\gamma)}u = o$. The eigencurve Γ_1, which passes through $(c, \gamma) = (0, -1)$ and $(c, \gamma) = (-\pi^2/4, 0)$ and approaches $c = -\pi^2$ for $\gamma \to +\infty$, is the boundary of an unbounded domain Ω in the (c, γ)-plane such that $M_{(c,\gamma)}$ is invertible for all $(c, \gamma) \in \Omega$. In particular $(1, 1) \in \Omega$. Thus, according to Corollary 3.10a, the set of all inverse-positive $M \in \mathscr{M}_0$ is equivalent to the set of all $M_{(c,\gamma)}$ with $(c, \gamma) \in \Omega$. □

The next theorem characterizes inverse-positive regular operators $M = (L, B_0, B_1)$ by positive definiteness. This characterization is closely related to the theory of eigenvalue problems. (See Problem 3.11 for a different proof of the main part of the theorem below.) Similar results will be derived by a different approach in Section III.3.

To each regular $M = (L, B_0, B_1)$ and each $g \in C_0[0, 1]$ with $g(x) > 0$ $(0 \leqslant x \leqslant 1)$, we adjoin an operator $A: D \to C_0[0, 1]$ which is defined by

$$Au(x) = g^{-1}(x)L[u](x), \qquad D = \{u \in C_2[0, 1] : B_0[u] = B_1[u] = 0\}. \tag{3.24}$$

Obviously, $Au(x) = k^{-1}(x)L_s[u](x)$ with $k = gpa^{-1}$ and L_s and p defined in (3.6), (3.7). This operator A is symmetric with respect to the inner product

$$\langle u, v \rangle = \int_0^1 k(x)u(x)v(x)\, dx, \tag{3.25}$$

that is, $\langle Au, v \rangle = \langle u, Av \rangle$ for all $u, v \in D$.

The given operator M will be called *positive definite* if and only if A is positive definite with respect to the above inner product, i.e., $\int_0^1 L_s[u]u\, dx > 0$

for all $u \neq o$ in D. (Observe that the latter property does not depend on the choice of g.) If M is positive definite, then $\langle Au, u\rangle \geqslant \kappa\langle u, u\rangle$ for some $\kappa > 0$. Moreover, the largest such constant κ is the smallest eigenvalue of A, according to the eigenvalue theory of differential operators.

For $u \in D$ we obtain by partial integration that $\langle Au, u\rangle = J[u]$, where

$$J[u] = \int_0^1 (pu'^2 + qu^2)\,dx + \beta_0[u] + \beta_1[u], \tag{3.26}$$

$$\beta_0[u] = \begin{cases} 0 & \text{for} \quad \alpha_0 = 0 \\ \alpha_0^{-1}\gamma_0 u^2(0)p(0) & \text{for} \quad \alpha_0 > 0, \end{cases}$$

$$\beta_1[u] = \begin{cases} 0 & \text{for} \quad \alpha_1 = 0 \\ \alpha_1^{-1}\gamma_1 u^2(1)p(1) & \text{for} \quad \alpha_1 > 0. \end{cases}$$

Theorem 3.11. *For a regular operator M the following three properties are equivalent:*

(i) *M is inverse-positive,*
(ii) *M is positive definite,*
(iii) *$J[u] > 0$ for all $u \in C_2[0, 1]$ which do not vanish identically and satisfy $u(0) = 0$ if $\alpha_0 = 0$, $u(1) = 0$ if $\alpha_1 = 0$.*

Proof. It is sufficient to consider only formally symmetric operators $M = (L, B_0, B_1)$, where $a = p$, $b = -a'$. We shall only carry out the proof for the case $\alpha_0 > 0$, $\alpha_1 = 0$. All other cases are treated similarly. For the considered case we shall even assume $\alpha_0 = 1$, $\alpha_1 = 0$, $\gamma_1 = 1$, without loss of generality.

Obviously, (iii) $\Rightarrow$ (ii). Now suppose that (ii) holds, so that $J[u] \geqslant \kappa\langle u, u\rangle$ for some $\kappa > 0$ and all $u \in D$. If (iii) is not true, then $J[v] < \kappa\langle v, v\rangle$ for some $v \in C_2[0, 1]$ with $v(1) = 0$. One can modify this function v in a sufficiently small neighborhood of $x = 0$ such that the modified function u belongs to D and still $J[u] < \kappa\langle u, u\rangle$. This contradicts (ii).

For proving (iii) $\Rightarrow$ (i), we apply Theorem 3.9. Let $\mathscr{M}_0$ denote the set of all positive definite regular $M = (L, B_0, B_1)$ with $\alpha_0 = 1, \alpha_1 = 0, \gamma_1 = 1$. This set is convex, since $J[u]$ depends linearly on $p = a$, $q = c$, and γ_0. Each positive definite operator is invertible. Moreover, $M \in \mathscr{M}_0$ is inverse-positive for $\gamma_0 > 0$, $c(x) > 0$ $(0 \leqslant x \leqslant 1)$. Thus, all $M \in \mathscr{M}_0$ are inverse-positive, by Theorem 3.9.

To prove (i) $\Rightarrow$ (ii), let $M = (a, b, c, 1, \gamma_0, 0, 1)$ be inverse-positive. Moreover, consider an operator $\hat{M} = (a, b, \hat{c}, 1, \hat{\gamma}_0, 0, 1)$ with $c(x) \leqslant \hat{c}(x) > 0$ $(0 \leqslant x \leqslant 1)$ and $\gamma_0 \leqslant \hat{\gamma}_0 > 0$, so that $\hat{M} \in \mathscr{M}_0$. Then all operators $M(t)$ which belong to $\mathscr{M}$ in (3.23) are inverse-positive, as shown in the proof of Theorem 3.10.

Let $J_t[u]$ be the functional (3.26) which corresponds to $M(t)$. Then $\mu(t) := \min\{J_t[u]/\langle u, u\rangle : u \in D_t,\ u \neq o\}$ depends continuously on t. Since

$\mu(1) > 0$, we have either $\mu(t) > 0$ $(0 \leqslant x \leqslant 1)$, or $\mu(t) = 0$ for some $t \in [0, 1)$. The second case cannot occur, because $\mu(t) = 0$ implies that $M(t)$ is not invertible. Thus, $\mu(0) > 0$, so that M is positive definite. □

3.5 Inverse-positivity under constraints

We take up the question under what conditions the inverse-positivity statement (3.1) holds if the functions u are subjected to certain constraints. Here we shall only consider the case that all functions u satisfy the boundary conditions $B_0[u] = B_1[u] = 0$.

Let M be a given operator (3.2), but assume now that

$$a, b, c \in \mathbb{R}[0, 1] \qquad \text{and} \qquad a(x) \geqslant 0 \qquad (0 \leqslant x \leqslant 1).$$

(We still require $\alpha_0 \geqslant 0$, $\alpha_1 \geqslant 0$; now these conditions do not mean any loss of generality.)

Define an operator $\overline{M}: \overline{R} \to S = \mathbb{R}[0, 1]$ by

$$\overline{M}u(x) = L[u](x) \qquad (0 \leqslant x \leqslant 1),$$

where $\overline{R}$ denotes the set of all functions $u \in D_2[0, 1]$ which satisfy the boundary conditions $B_0[u] = B_1[u] = 0$. This operator $\overline{M}$ is inverse-positive if and only if for all $u \in D_2[0, 1]$

$$L[u] \geqslant o, \quad B_0[u] = B_1[u] = 0 \quad \Rightarrow \quad u \geqslant o. \tag{3.27}$$

We shall see in Section 3.8 that a regular operator M is inverse-positive if and only if the corresponding operator $\overline{M}$ is inverse-positive. To prove this fact, the following results will be needed.

Lemma 3.12. *The operator $\overline{M}: \overline{R} \to S$ is pre-inverse-positive with respect to the strong order in S defined by $U \succ o \Leftrightarrow U(x) > 0$ $(0 \leqslant x \leqslant 1)$.*

Proof. In $\overline{R}$, the relation $u \succ o$ is equivalent to the following inequalities:

$$\begin{array}{llll} u(x) > 0 & (0 < x < 1), & & \\ u'(0) > 0 & \text{if } \alpha_0 = 0, & u(0) > 0 & \text{if } \alpha_0 > 0, \\ u'(1) < 0 & \text{if } \alpha_1 = 0, & u(1) > 0 & \text{if } \alpha_1 > 0. \end{array} \tag{3.28}$$

Thus, if $u \geqslant o$ and $u \not\succ o$ for some $u \in \overline{R}$, one of these inequalities is wrong. In each case one obtains $u(\xi) = u'(\xi) = 0$ for some $\xi \in [0, 1]$. For example, if $\alpha_0 > 0$ and $u(0) = 0$, we have $u'(0) = -\alpha_0^{-1} B_0[u] = 0$. If $u(\xi) = u'(\xi) = 0$ and $u \geqslant o$, then also $u''(\xi) \geqslant 0$. These relations together imply $L[u](\xi) \leqslant 0$, so that $\overline{M}u \not\succ o$. □

This lemma and the monotonicity theorem yield the following result.

Theorem 3.13. *If there exists a function $z \geqslant o$ in $D_2[0, 1]$ such that*

$$L[z](x) > 0 \qquad (0 \leqslant x \leqslant 1), \qquad B_0[z] = B_1[z] = 0, \tag{3.29}$$

then $\overline{M}$ is inverse-positive, i.e., (3.27) is true.

In general, it will be more difficult to construct a function z as in (3.29), than to construct one which satisfies $Mz(x) > 0$ $(0 \leqslant x \leqslant 1)$. However, the conditions (3.29) yield some new possibilities.

Example 3.11. Suppose that the eigenvalue problem

$$\tilde{L}[\varphi](x) = \lambda\varphi(x) \qquad (0 \leqslant x \leqslant 1), \qquad B_0[\varphi] = B_1[\varphi] = 0,$$

with

$$\tilde{L}[\varphi] = -a\varphi'' + b\varphi'$$

has an eigenvalue λ_1 which corresponds to an eigenfunction $\varphi_1 \succ o$. When $\alpha_0 > 0$ and $\alpha_1 > 0$, the function $z = \varphi_1$ satisfies (3.29) if $c(x) > -\lambda_1$ $(0 \leqslant x \leqslant 1)$. If $\alpha_0\alpha_1 = 0$, so that $\tilde{L}[\varphi_1]$ vanishes at a boundary point, one may modify φ_1 slightly for obtaining a suitable z.

For example, suppose that $\alpha_0 = \alpha_1 = 0$, $a(x) \geqslant \kappa > 0$ $(0 \leqslant x \leqslant 1)$, and the coefficients a, b, c are bounded. Then, if $c(x) > -\lambda_1 + \delta$ $(0 \leqslant x \leqslant 1)$ for some $\delta > 0$, the inequalities (3.29) are satisfied with $z = \varphi_1 - \varepsilon x^2(1-x)^2$ and sufficiently small $\varepsilon > 0$. □

Example 3.12. *The operator $\overline{M}$ is inverse-positive if the coefficients a, b are bounded and if $c(x) \geqslant \kappa$ $(0 \leqslant x \leqslant 1)$ for some sufficiently large constant κ.*

For proving this, find a polygon $g(x)$ such that $B_0[g] = B_1[g] = 0$ and $g(x) > 0$ $(0 \leqslant x \leqslant 1)$. Then construct $z \in C_2[0, 1]$ such that $z(x) > 0$ $(0 \leqslant x \leqslant 1)$ and $z(x) = g(x)$ for all $x \in [0, 1]$, except in sufficiently small neighborhoods of the "jump points" of $g'(x)$. This function z satisfies (3.29) if κ is large enough. □

Problems

3.7 Let $\overline{R}$ denote the set of all $u \in C_0[0, 1] \cap D_2(0, 1]$ such that $B_1[u] = 0$ and $u'(0)$ exists when $\alpha_0 > 0$. Define an operator $\overline{M}: \overline{R} \to \mathbb{R}[0, 1]$ by $\overline{M}u(x) = L[u](x)$ for $0 < x \leqslant 1$, $\overline{M}u(0) = B_0[u]$; where now $a, b, c \in \mathbb{R}[0, 1]$ and $a(x) \geqslant 0$ $(0 < x \leqslant 1)$. Prove the following statement. *The operator $\overline{M}$ is inverse-positive if there exists a function $z \in \overline{R}$ such that $z \geqslant o$ and $\overline{M}z(x) > 0$ $(0 \leqslant x \leqslant 1)$.*

3.8 The monotonicity theorem can also be applied to treat periodic boundary conditions. Let $\overline{R}$ be the set of all $u \in D_2[0, 1]$ such that $u(0) = u(1)$. Define an operator $\overline{M}: \overline{R} \to S = \mathbb{R}[0, 1]$ by $\overline{M}u(x) = L[u](x)$ for $0 \leqslant x < 1$, $Mu(1) = -u'(0) + u'(1)$, where now $a, b, c \in \mathbb{R}[0, 1]$ and $a(x) \geqslant 0$ $(0 \leqslant x \leqslant 1)$. Prove the following statement. *The operator* $\overline{M}$ *is inverse-positive if there exists a function* $z \in \bar{R}$ *such that* $z \geqslant o$, $L[z](x) > 0$ $(0 \leqslant x < 1)$ *and* $-z'(0) + z'(1) \geqslant 0$.

3.6 Stronger results on positivity in the interior

Most operators $M = (L, B_0, B_1)$ which are inverse-positive have even stronger monotonicity properties. Such properties will be formulated in the following theorem, from which, for example, strong boundary maximum principles can be derived. These investigations are continued in the following section.

Theorem 3.14. (A) *The operator* $M = (L, B_0, B_1)$ *is inverse-positive if the following two conditions are satisfied*:

I. *For each* $\xi \in (0, 1)$ *and all* $u \in R$

$$\left.\begin{array}{ll} u(x) \geqslant 0 \quad (0 < x < 1), & u(\xi) = 0 \\ L[u](x) \geqslant 0 \quad (0 < x < 1) & \end{array}\right\} \Rightarrow u(x) = 0 \quad (0 < x < 1). \tag{3.30}$$

II. *There exists a function* $z \in R$ *such that*

$$z \geqslant o, \qquad L[z](x) \geqslant 0 \qquad (0 < x < 1), \qquad B_0[z] > 0, \qquad B_1[z] > 0.$$

(B) *If* M *is inverse-positive and condition* I *holds, then for all* $u \in R$ *the inequality* $Mu \geqslant o$ *implies that* $u = o$ *or* $u(x) > 0$ $(0 < x < 1)$.

Proof. This result may either be derived immediately from the monotonicity theorem, or from Theorem 1.7, which itself follows from the monotonicity theorem. When the latter theorem is used for the proof, one defines the strong order in S by $U \succ o \Leftrightarrow U(x) \geqslant 0\,(0 < x < 1), U(0) > 0, U(1) > 0$.

To derive the result from Theorem 1.7, let

$$\mathscr{L} = \{\ell_x : 0 \leqslant x \leqslant 1\} \qquad \text{with} \quad \ell_x u = u(x),$$

$$\mathscr{L}_1 = \{\ell_0, \ell_1\}, \qquad \mathscr{L}_2 = \{\ell_x : 0 < x < 1\}.$$

Then condition Ia of Theorem 1.7 is satisfied, since (3.9) holds for $\xi = 0$ and $\xi = 1$. Moreover, it is immediate that the assumptions of Theorem 3.14 are sufficient for assumptions Ib and II of Theorem 1.7 and that the statement of the above theorem follows. $\square$

Corollary 3.14a. *Condition* I *of Theorem* 3.14 *is satisfied if the following conditions hold*:

(i) $a(x) > 0\,(0 < x < 1)$,

(ii) *for each interval* $[\alpha, \beta] \subset (0, 1)$, *there exist constants* $\hat{b}, \hat{c}$ *such that* $|b(x)| \leqslant \hat{b}a(x), |c(x)| \leqslant \hat{c}a(x) \; (\alpha \leqslant x \leqslant \beta)$.

Proof. Suppose that condition I is not satisfied, so that

$$u \geqslant o, \qquad u(\xi) = 0, \qquad L[u](x) \geqslant 0 \qquad (0 < x < 1), \qquad u(\eta) > 0 \quad (3.31)$$

for some $u \in R$ and some $\xi, \eta \in (0, 1)$. First, let $\xi < \eta$. Without loss of generality, we may also assume that $\eta = \xi + \varepsilon$ with some arbitrarily small $\varepsilon > 0$. We choose ε so small that a function $z \in C_2[\xi, \eta]$ with $z(x) > 0$, $L[z](x) \geqslant 0$ $(\xi \leqslant x \leqslant \eta)$ exists. Such a function can be constructed as in the proof of Theorem 3.8.

After having fixed ε, we find a function $\varphi \in C_2[\xi, \eta]$ such that $\varphi(\xi) = 0$, $\varphi'(\xi) > 0$, $\varphi(x) > 0$ $(\xi < x \leqslant \eta)$, $L[\varphi](x) < 0$ $(\xi < x < \eta)$. Because of the preceding assumptions, $\varphi(x) = \exp \kappa(x - \xi) - 1$ satisfies these relations for large enough κ. Finally, for given φ a sufficiently small $\mu > 0$ is chosen such that $v = u - \mu\varphi$ satisfies $v(\xi) = 0$, $v(\eta) > 0$, $L[v](x) > 0$ $(\xi < x < \eta)$. These relations yield $v(x) > 0$ $(\xi < x < \eta)$, according to Theorem 3.4, applied to the interval $[\xi, \eta]$ in place of $[0, 1]$. The properties of v together yield $v'(\xi) \geqslant 0$, so that

$$u'(\xi) = v'(\xi) + \mu\varphi'(\xi) \geqslant \mu\varphi'(\xi) > 0. \tag{3.32}$$

However, the inequality $u'(\xi) > 0$ cannot hold, since u attains its minimum at ξ. Therefore, (3.31) cannot be true for $\eta > \xi$. For $\eta < \xi$, one proceeds similarly, using $\psi(x) = \exp \kappa(\xi - x) - 1$ instead of φ. □

Remark. Obviously, the assumptions of this corollary can be generalized. Instead of requiring the preceding conditions on the coefficients a, b, c, one may assume that functions z, φ, ψ as used in the proof exist.

Under the stronger assumptions that

$$\begin{gathered} R = C_1[0, 1] \cap C_2(0, 1); \qquad a, b, c \in C_0[0, 1]; \\ a(x) > 0 \qquad (0 < x < 1) \end{gathered} \tag{3.33}$$

we shall provide a second proof. The ideas involved will later be used in other connections.

Condition I of Theorem 3.14 holds if and only if for given u, ξ, and η there exists a functional $\ell U = U(\zeta)$ with $\zeta \in (0, 1)$ such that the relations $u \geqslant o$, $u(\xi) = 0$, $u(\eta) > 0$ together imply $\ell Mu < 0$. We could not use this fact in the above proof since, in general, ζ is unknown. In the following proof, an integral functional $\ell \geqslant o$ will actually be constructed such that the above implication holds and $\ell Mu < 0$ contradicts $L[u](x) \geqslant 0$ $(0 < x < 1)$.

Second proof (under the assumption (3.33)). Suppose again (3.31) holds. We shall present two possibilities for choosing a suitable functional ℓ. First, let

$$\ell U = \int_{\xi}^{\eta} \psi(x)E(x)U(x)\,dx \qquad \text{when} \quad \xi < \eta,$$

$$\ell U = \int_{\eta}^{\xi} \varphi(x)E(x)U(x)\,dx \qquad \text{when} \quad \xi > \eta,$$

for $U \in MR$, where

$$E = a^{-1}p, \qquad p = \exp\left\{-\int_{\xi}^{x} b(t)a^{-1}(t)\,dt\right\},$$

$$\varphi = \exp \kappa(x-\eta) - 1, \qquad \psi = \exp \kappa(\eta - x) - 1$$

and $\kappa > 0$ so large that $L[\varphi](x) \leqslant 0$, $L[\psi](x) \leqslant 0$ $(\xi \leqslant x \leqslant \eta)$. We obtain $\ell Mu < 0$ by partial integration. For example, when $\xi < \eta$,

$$\ell Mu = p(\eta)\psi'(\eta)u(\eta) + \int_{\xi}^{\eta} E(x)L[\psi](x)u(x)\,dx < 0,$$

since $\psi'(\eta) < 0$. Observe that $EL[u]$ has the self-adjoint form.

In order to describe a second possible choice of ℓ, we assume, without loss of generality, that $u(x) > 0$ for $\xi - \mu \leqslant x < \xi$, or $u(x) > 0$ for $\xi < x \leqslant \xi + \mu$, where $\mu > 0$ is sufficiently small. Then let $(\alpha, \beta) = (\xi - \varepsilon, \xi + \varepsilon)$ with arbitrarily small $\varepsilon \in (0, \mu]$ and define for $U \in MR$

$$\ell U = \int_{\alpha}^{\beta} g(x)E(x)U(x)\,dx \qquad \text{with} \quad g(x) = \begin{cases} \varphi(x)\psi(\xi) & \text{for} \quad \alpha \leqslant x \leqslant \xi \\ \varphi(\xi)\psi(x) & \text{for} \quad \xi \leqslant x \leqslant \beta, \end{cases} \tag{3.34}$$

where now $\varphi(x) = \exp \kappa(x - \alpha) - 1$, $\psi(x) = \exp \kappa(\beta - x) - 1$, and κ so large that $L[\varphi](x) \leqslant 0$ $(\alpha \leqslant x \leqslant \xi)$, $L[\psi](x) \leqslant 0$ $(\xi \leqslant x \leqslant \beta)$. Because of the special properties of φ and ψ, we obtain again $\ell Mu < 0$ by partial integration.

If the operator M is regular and $R = C_2[0, 1]$, we can even choose a functional ℓ as in (3.34) with $\alpha = 0$, $\beta = 1$, and κ so large that $L[\varphi](x) < 0$, $L[\psi](x) < 0$ for $0 \leqslant x \leqslant 1$. This means, instead of a "local" integral functional, we can use a "global" integral functional. This functional does not depend on η, i.e., we can prove that

$$u > o, \quad \ell u = 0 \quad \Rightarrow \quad \ell_{\ell} Mu < 0,$$

where $\ell u = u(\xi)$ and $\ell = \ell_{\ell}$ depends only on ℓ. (Compare this property with (1.3), (1.4), and (3.9).) □

The stronger monotonicity statements of Theorem 3.14 yield also stronger maximum principles, as stated in the following theorem. This theorem is proved in essentially the same way as Theorem 3.6. Property (i) of the following theorem is equivalent to condition I in Theorem 3.14; Property (ii) is called the *(generalized) strong maximum principle.*

Theorem 3.15. *Let* $R = C_0[0, 1] \cap C_2(0, 1)$ *and suppose there exists a function* $z \in R$ *such that* $z(x) > 0$ $(0 \leqslant x \leqslant 1)$, $L[z](x) \geqslant 0$ $(0 < x < 1)$.

Then the following two properties are equivalent:

(i) *For all* $u \in R$, *the inequalities* $u \geqslant o$ *and* $L[u](x) \geqslant 0$ $(0 < x < 1)$ *imply that either* $u = o$ *or* $u(x) > 0$ $(0 < x < 1)$.

(ii) *For all* $u \in R$, *the inequality* $L[u](x) \leqslant 0$ $(0 < x < 1)$ *implies that either* $u = \mu z$ *or* $u(x) < \mu z(x)$ $(0 < x < 1)$, *where* $\mu = \max\{0, \mu_0\}$, $\mu_0 = \max\{u(0)/z(0), u(1)/z(1)\}$.

When $L[z](x) = 0$ $(0 < x < 1)$, *the number* μ *in* (ii) *can be replaced by* $\mu = \mu_0$.

3.7 Strictly inverse-positive operators

As we have seen in the preceding section, the inequality $Mu \geqslant o$ often implies the positivity statement $u(x) > 0$ $(0 < x < 1)$. Now we shall derive even stronger positivity results which also describe the behavior of u at the endpoints 0 and 1. These results include statements on strict inverse-positivity as defined in Section 1.1. We shall have to require a stronger type of pre-inverse-positivity, while at the same time the assumptions on the occurring function z can be weakened. This is important for several applications in the following sections.

Throughout this section, R will denote the set $D_2[0, 1]$ of twice differentiable functions on $[0, 1]$, if not otherwise specified.

Theorem 3.16. (A) *The operator* $M = (L, B_0, B_1)$ *is inverse-positive if the following two conditions are satisfied.*

I. *For each* $\xi \in [0, 1]$ *and all* $u \in R$

$$\left.\begin{array}{ll} u \geqslant o, & u(\xi) = u'(\xi) = 0 \\ L[u](x) \geqslant 0 & (0 < x < 1) \end{array}\right\} \quad \Rightarrow \quad u = o. \tag{3.35}$$

II. *There exists a function* $z \in R$ *such that*

$$z \geqslant o, \qquad Mz \geqslant o, \qquad Mz(\zeta) > 0 \qquad \text{for some} \quad \zeta \in [0, 1];$$
$$z(0) > 0 \quad \text{if} \quad \alpha_0 = 0, \qquad z(1) > 0 \quad \text{if} \quad \alpha_1 = 0. \tag{3.36}$$

(B) *If M is inverse-positive and if M satisfies condition* I, *then the following property* (P) *holds:*

(P) *For all* $u \in R$, $Mu \geqslant o$ *implies that either* $u = o$ *or the following inequalities are satisfied*:

$$\left.\begin{array}{llll} u(x) > 0 & (0 < x < 1), & & \\ u(0) > 0 & \text{if } \alpha_0 > 0, & u'(0) > 0 & \text{if } u(0) = 0, \\ u(1) > 0 & \text{if } \alpha_1 > 0, & u'(1) < 0 & \text{if } u(1) = 0. \end{array}\right\} \quad (3.37)$$

Proof. This result will be derived from Theorem 1.7, where $\mathscr{L} = \{\ell_x : 0 \leqslant x \leqslant 1\}$ with $\ell_x u = u(x)$. We have to distinguish several cases. In each case, we choose appropriate sets $\mathscr{L}_1$, $\mathscr{L}_2$ and verify condition I of Theorem 1.7, while it will be immediately clear that condition II is also satisfied.

(a) Let $\alpha_0 > 0$, $\alpha_1 > 0$. We choose $\mathscr{L}_1 = \{\ell_\zeta\}$ and $\mathscr{L}_2 = \mathscr{L}$. For $\ell = \ell_\zeta$ the implication (1.3) holds with $\not{\ell} U = U(\zeta)$, since (3.9) is true for $\xi = \zeta$.

In order to prove (1.4) for each $\ell_\xi \in \mathscr{L}$, let $u \in R$ satisfy $u > o$, $Mu \geqslant o$, and $u(\xi) = 0$. Then $u'(\xi) = 0$ holds also. For example, we have $0 \leqslant u'(0) = -\alpha_0^{-1} B_0[u] \leqslant 0$ for $\xi = 0$. Thus condition I implies that $u > o$ is false. This verifies condition Ib of Theorem 1.7.

(b) Let $\alpha_0 = \alpha_1 = 0$. Then $\gamma_0^{-1} B_0[z] = z(0) > 0$ and $\gamma_1^{-1} B_1[z] = z(1) > 0$, so that all assumptions of Theorem 3.14 are satisfied. That means, we can proceed as in the proof of that theorem; condition I of Theorem 1.7 holds for $\mathscr{L}_1 = \{\ell_0, \ell_1\}$, $\mathscr{L}_2 = \{\ell_x : 0 < x < 1\}$. Consequently, the inequalities $Mu \geqslant o$ and $u \neq o$ imply $u(x) > 0$ $(0 < x < 1)$. Moreover, if $u \neq o$ and $u(0) = 0$ $(u(1) = 0)$, then $u'(0) > 0$ $(u'(1) < 0)$, by condition I of this theorem.

(c) For the remaining cases we proceed similarly and choose

$$\begin{array}{llll} \mathscr{L}_1 = \{\ell_0\}, & \mathscr{L}_2 = \{\ell_x : 0 < x \leqslant 1\} & \text{if } \alpha_0 = 0, & \alpha_1 > 0, \\ \mathscr{L}_1 = \{\ell_1\}, & \mathscr{L}_2 = \{\ell_x : 0 \leqslant x < 1\} & \text{if } \alpha_0 > 0, & \alpha_1 = 0. \end{array} \quad \square$$

Corollary 3.16a. *Condition* I *of Theorem* 3.16 *is satisfied if*

$$a(x) > 0, \qquad |b(x)| \leqslant \hat{b}a(x), \qquad |c(x)| \leqslant \hat{c}a(x) \qquad (0 < x < 1) \qquad (3.38)$$

for some constants $\hat{b} \geqslant 0$, $\hat{c} \geqslant 0$.

Proof. Let (3.31) hold for some $u \in R$ and some $\xi, \eta \in [0, 1]$. First, it is shown as in the proof of Corollary 3.14a that ξ cannot belong to $(0, 1)$. For $\xi = 0$ it suffices to establish $u'(\xi) > 0$. This inequality is derived in the same way as (3.32) in the proof of Corollary 3.14a. For $\xi = 1$ one proceeds similarly. $\square$

If M is regular and $R = C_2[0, 1]$, the result can also be derived as in the "second proof" of Corollary 3.14a. Now the occurring functionals are also defined for $\xi = 0$ and $\xi = 1$, respectively.

Example 3.13. *Property* (P) *of Theorem* 3.16 *holds for each operator* $M = (L, B_0, B_1)$ *which satisfies* (3.38) *and*

$$c(x) \geqslant 0 \qquad (0 < x < 1), \qquad \gamma_0 \geqslant 0, \qquad \gamma_1 \geqslant 0,$$
$$c(\zeta) + \gamma_0 + \gamma_1 > 0 \qquad \textit{for some} \quad \zeta \in (0, 1).$$

Notice that $z(x) \equiv 1$ satisfies condition II of Theorem 3.16. □

Under suitable assumptions, condition (3.36) in Theorem 3.16 may be dropped. Also, the statement of the theorem can be generalized by including the possibility that $Mz = o$. For simplicity, the following result is only formulated for regular operators. (See, however, the subsequent corollary.)

Theorem 3.17. *Let* $M = (L, B_0, B_1)$ *be a regular operator such that* $z > o$, $Mz \geqslant o$ *for some* $z \in R = D_2[0, 1]$.
Under these assumptions, the following statements hold:

(i) *If* $Mz > o$, *then* M *has property* (P) *in Theorem* 3.16.
(ii) *If* $Mz = o$, *then each solution* $u \in R$ *of* $Mu = o$ *is a multiple of* z, *and* z *satisfies the inequalities* (3.37) *in property* (P).

Proof. We shall derive the above result from Theorem 1.9, where again, we have to distinguish several cases. Observe that condition I of Theorem 3.16 is satisfied, according to Corollary 3.16a.

(a) Let $\alpha_0 > 0, \alpha_1 > 0$. We have shown in the proof of Theorem 3.16 (part a) that (1.4) holds for all $\ell \in \mathscr{L} = \{\ell_x : 0 \leqslant x \leqslant 1\}$. However, this property is equivalent to (1.7) in Theorem 1.9, so that the above statements follow from that theorem, since $u \succ o \Leftrightarrow u(x) > 0 \; (0 \leqslant x \leqslant 1)$.
(b) Let $\alpha_0 = \alpha_1 = 0$ and $z(0) > 0, z(1) > 0$. Then the above statement (i) follows from Theorem 3.16. Statement (ii) is also true, since $Mz \neq o$.
(c) Let $\alpha_0 = \alpha_1 = 0$ and $z(0) = z(1) = 0$. Here we shall apply Theorem 1.9 to the corresponding operator $\overline{M} : \overline{R} \to S$ in Section 3.5, where $\overline{R} = \{u \in D_2[0, 1] : u(0) = u(1) = 0\}$. The positive cone $\overline{K} = \{u \in \overline{R} : u \geqslant o\}$ in $\overline{R}$ is completely described by $\mathscr{L} = \{\ell_x : 0 < x < 1\} \cup \{\ell_{(0)}, \ell_{(1)}\}$ with $\ell_{(0)}u = u'(0)$, $\ell_{(1)}u = -u'(1)$. Now $u \succ o$ is defined by (3.28).

Since condition I of Theorem 3.16 is satisfied, the implication (1.4) holds for each $\ell \in \mathscr{L}$, with $\overline{M}$ in place of M and $\overline{R}$ in place of R. The element Mz vanishes if and only if $\overline{M}z$ vanishes. Thus, statement (ii) follows immediately from Theorem 1.9(ii′). Moreover, if $Mz > o$, then $\overline{M}z > o$, so that $\overline{M}u \geqslant o \Rightarrow u \geqslant o$. In this case, let $\bar{\zeta} \geqslant o$ be the solution in $\overline{R}$ of $L[\bar{\zeta}](x) = 1 \; (0 \leqslant x \leqslant 1)$. Then the function $\zeta = \bar{\zeta} + \varepsilon$ satisfies $M\zeta(x) > 0 \; (0 \leqslant x \leqslant 1)$ if $\varepsilon > 0$ is small enough. Consequently, M is inverse-positive and statement (i) follows from Theorem 3.16B.

(d) All remaining cases are treated by properly combining the methods which have been used above. For example, if $\alpha_0 > 0$, $\alpha_1 = 0$, $z(1) = 0$, one considers an operator $\overline{M}$ as in Problem 3.7. □

The above proof shows that the conditions on M can be relaxed.

Corollary 3.17a. *The statements of Theorem* 3.17 *remain true if the assumption that M be regular is replaced by the following two conditions:*

(i) *There exist constants* $\hat{a}, \hat{b}, \hat{c}$ *with*

$$a(x) \geqslant \hat{a} > 0, \qquad |b(x)| \leqslant \hat{b}, \qquad |c(x)| \leqslant \hat{c} \qquad (0 < x < 1).$$

(ii) *If M is invertible, there exists a function* $\bar{\zeta} \in D_2[0, 1]$ *with* $L[\bar{\zeta}](x) \geqslant 1$ $(0 \leqslant x \leqslant 1)$ *and*

$$\begin{aligned} \bar{\zeta}(0) &= 0 \quad \text{if} \quad \alpha_0 = 0, \qquad B_0[\bar{\zeta}] \geqslant 1 \quad \textit{if} \quad \alpha_0 > 0, \\ \bar{\zeta}(1) &= 0 \quad \text{if} \quad \alpha_1 = 0, \qquad B_1[\bar{\zeta}] \geqslant 1 \quad \textit{if} \quad \alpha_1 > 0. \end{aligned}$$

When $\alpha_0 > 0$ and $\alpha_1 > 0$, property (P) means that the operator $M = (L, B_0, B_1)$ is strictly inverse-positive. If one of the numbers α_0, α_1 equals zero, then a regular operator $M: R \to S$ cannot be strictly inverse-positive. For example, if M is inverse-positive and $\alpha_0 = 0$, then the solution of $L[u](x) = 1$ $(0 < x < 1)$, $u(0) = 0$, $B_1[u] = 1$ is not a strictly positive element of R. However, property (P) includes the statement that the "restricted" operator $\overline{M}: \overline{R} \to S$ defined in Section 3.5 is strictly inverse-positive, whether or not the numbers α_0, α_1 vanish.

Property I in Theorem 3.16 can be considered as a certain type of pre-inverse-positivity. Again, this pre-inverse-positivity is equivalent to a boundary maximum principle as formulated in the following theorem. This theorem is proved similarly as Theorem 3.6.

Theorem 3.18. *If there exists a function* $z \in R$ *with* $z(x) > 0$, $L[z](x) \geqslant 0$ $(0 < x < 1)$, *then the following properties are equivalent:*

(i) *Condition* I of *Theorem* 3.16 *holds.*

(ii) *For all* $u \in R$ *the inequality* $L[u](x) \leqslant 0$ $(0 < x < 1)$ *implies that either* $u = \mu z$ *or* $u(x) < \mu z(x)$ $(0 < x < 1)$ *and*

$$u'(0) < \mu z'(0) \quad \textit{if} \quad u(0) = \mu z(0), \qquad u'(1) > \mu z'(1) \quad \textit{if} \quad u(1) = \mu z(1),$$

where $\mu = \max\{0, \mu_0\}$, $\mu_0 = \max\{u(0)/z(0), u(1)/z(1)\}$.

When $L[z](x) = 0$ $(0 < x < 1)$ *the number* μ *in (ii) can be replaced by* $\mu = \mu_0$.

Problems

3.9 Modify the assumptions of Corollary 3.17a such that the implication in condition I of Theorem 3.16 holds for $\xi \in [0, 1)$.

3.10 What statement can be proved if the implication in condition I of Theorem 3.16 holds for all ξ which belong to certain intervals contained in $[0, 1]$ and if a suitable condition II is satisfied? What is a suitable condition II?

3.8 Equivalence statements

Theorem 3.17 of the preceding section can be used to prove a series of different results for regular operators. Several of these results will be derived in the following four sections.

In Section 3.5 we considered monotonicity statements for functions u which satisfy boundary conditions. The question arises whether such statements hold under less strong conditions than those which are required for inverse-positivity of M. An answer to this question is given in the following theorem.

Theorem 3.19. *For a regular operator* $M = (L, B_0, B_1)$ *the following five statements are equivalent* (*where conditions* (i)–(iv) *are interpreted to hold for all* $u \in R = C_2[0, 1]$).

(i) $Mu \geqslant o \Rightarrow u \geqslant o$,
(ii) $L[u](x) \geqslant 0\ (0 < x < 1),\ B_0[u] = 0,\ B_1[u] = 0 \Rightarrow u \geqslant o$,
(iii) $L[u](x) = 0\ (0 < x < 1),\ B_0[u] \geqslant 0,\ B_1[u] = 0 \Rightarrow u \geqslant o$,
(iv) $L[u](x) = 0\ (0 < x < 1),\ B_0[u] = 0,\ B_1[u] \geqslant 0 \Rightarrow u \geqslant o$.
(v) *M has Property* (P) *in Theorem* 3.16.

Proof. Obviously, it suffices to show that (v) follows from each of the other four properties.

Suppose that one of the implications (i)–(iv) holds (for all $u \in R$) and, for the moment, call this implication (π). Then define $z \in R$ by $Mz = r$ where $r(x) = 1\ (0 < x < 1)$ if $L[u](x) \geqslant 0\ (0 < x < 1)$ is assumed in the premise of (π), $r(x) = 0\ (0 < x < 1)$ otherwise; $r(0) = 1$ if $B_0[u] \geqslant 0$ is assumed in the premise of (π), $r(0) = 0$ otherwise; and $r(1)$ is defined analogously. This function satisfies $z \geqslant o$, $Mz > o$. Thus, according to Theorem 3.17, the operator M has property (P). □

Now suppose that the regular operator $M = (L, B_0, B_1)$ is invertible. Then since $R = C_2[0, 1]$, the range of M consists of all $U \in \mathbb{R}[0, 1]$ such that $U(x) = \bar{U}(x)\ (0 < x < 1)$ with some $\bar{U} \in C_0[0, 1]$. Moreover, for such $U \in S$, $M^{-1}U$ assumes the form $M^{-1}U(x) = U(0)\psi(x) + U(1)\varphi(x) + \int_0^1 \mathcal{G}(x, \xi)\bar{U}(\xi)\, d\xi$, where $\varphi, \psi \in R$ satisfy

$$L[\varphi](x) = 0 \qquad (0 < x < 1), \qquad B_0[\varphi] = 0, \qquad B_1[\varphi] = 1;$$

$$L[\psi](x) = 0 \qquad (0 < x < 1), \qquad B_0[\psi] = 1, \qquad B_1[\psi] = 0$$

and $\mathscr{G}$ denotes the Green function

$$\mathscr{G}(x, \xi) = -\frac{1}{a(\xi)W(\xi)}\begin{cases}\varphi(x)\psi(\xi) & \text{for } 0 \leqslant x \leqslant \xi \leqslant 1\\ \varphi(\xi)\psi(x) & \text{for } 0 \leqslant \xi \leqslant x \leqslant 1\end{cases}$$

$$\text{with} \quad W = \varphi\psi' - \varphi'\psi.$$

Because of this representation of $M^{-1}U$, Theorem 3.19 immediately yields the following result, where $\mathscr{G} \geqslant o$ means that $\mathscr{G}(x, \xi) \geqslant 0$ for all $x, \xi \in [0, 1]$.

Corollary 3.19a. *An invertible regular operator $M = (L, B_0, B_1)$ is inverse-positive if and only if one of the following four statements holds:*

(i) $\varphi \geqslant o, \psi \geqslant o, \mathscr{G} \geqslant o$;
(ii) $\mathscr{G} \geqslant o$;
(iii) $\varphi \geqslant o$;
(iv) $\psi \geqslant o$.

Moreover, from the equivalence of (i) and (v) in Theorem 3.19 one easily derives stronger statements on φ, ψ, and $\mathscr{G}$, such as $\varphi(x) > 0$ $(0 < x \leqslant 1)$, $\mathscr{G}_\xi(x, 0) > 0$ $(+0 \leqslant x < 1)$ if $\alpha_0 = 0$, etc.

3.9 Oscillation and separation of zeros

Now we shall study the oscillatory behavior of solutions of the homogeneous equation

$$L[u](x) := -a(x)u'' + b(x)u' + c(x)u = 0.$$

As before, we shall mostly consider the interval $[0, 1]$. However, since other intervals occur also, let us assume, for simplicity, that

$$a, b, c \in C_0(-\infty, \infty), \qquad a(x) > 0 \qquad (-\infty < x < \infty). \tag{3.39}$$

The results are derived from Theorem 3.17 applied to the operator $Mu = (L[u](x), u(0), u(1)) : C_2[0, 1] \to \mathbb{R}[0, 1]$.

Proposition 3.20. *Let $v, w \in C_2[0, 1]$ satisfy*

$$\begin{aligned} v > o; \quad & L[v](x) \leqslant 0 \quad (0 < x < 1), \quad v(0) = 0, \quad v(1) = 0;\\ & L[w](x) \geqslant 0 \quad (0 < x < 1), \quad w(0) > 0. \end{aligned}$$

Then $w(\xi) < 0$ for some $\xi \in (0, 1)$ and hence $w(\eta) = 0$ for some $\eta \in (0, 1)$.

Proof. Define M as above and suppose that $w \geqslant o$. Then also $Mw > o$, so that M is inverse-positive, according to Theorem 3.17. Consequently, the inequality $Mv \leqslant o$ implies $v \leqslant o$, which contradicts the assumption $v > o$. □

Example 3.14. Let $L[u] = -u'' - 10u$. Then, $v = \sin \pi x$ and $w = 1 - 8x^2$ satisfy the required inequalities. Obviously, the above statements hold with $\eta = 1 : 2\sqrt{2}$. □

Theorem 3.21 (*Sturmian oscillation theorem*). *Suppose that $v, w \in C_2(I)$ are linearly independent solutions of the differential equation $L[u](x) = 0$ $(x \in I)$, where I denotes any given interval and* (3.39) *is satisfied.*

Then the zeros of v and w separate each other, that is, between two consecutive zeros of v in I there exists exactly one zero of w, and vice versa.

Proof. Notice first that the zeros of v or w do not have a finite point of accumulation.

Let $\alpha, \beta \in I$ denote two subsequent zeros of v and suppose that $v(x) > 0$ $(\alpha < x < \beta)$. (Otherwise, consider $-v$.) If $w(\alpha) = 0$, there would exist constants c_1, c_2 with $|c_1| + |c_2| > 0$ such that $y := c_1 v + c_2 w$ satisfies $y'(\alpha) = 0$, $y(\alpha) = 0$. This would imply $y(x) \equiv 0$ on I, which is not possible, since v and w are linearly independent. Therefore, we may suppose $w(\alpha) > 0$. (Otherwise, consider $-w$.) Then $w(\xi) = 0$ for a $\xi \in (\alpha, \beta)$, by Proposition 3.20 applied to $[\alpha, \beta]$ in place of $[0, 1]$. □

Example 3.15. (a) The functions $v = \sin x$ and $w = \cos x$ are solutions of $-u'' - u = 0$ on $(-\infty, \infty)$. These functions show the "typical" oscillatory behavior explained in the above theorem.

(b) The functions $v = \sinh x$ and $w = \cosh x$ are solutions of $-u'' + u = 0$ on $(-\infty, \infty)$. None of these functions has two zeros. Here, the statement of Theorem 3.21 is empty, but nevertheless true.

(c) For $L[u] = -xu'' + u'$ the assumptions of Theorem 3.21 are not satisfied on $[-1, 1]$. Moreover, the statement of this theorem is false, since $v = 1 - x^2$ vanishes at $x = -1$ and $x = 1$, but $w(x) \equiv 1$ has no zero. □

The above theorem describes the relation between the oscillatory behavior of two solutions on any interval I. The question whether such solutions "oscillate" at all on a fixed interval is important for inverse-positivity.

An arbitrary interval I is called an *interval of nonoscillation* for L, if each solution $u \in C_2(I)$ of $L[u](x) = 0$ either vanishes identically or has at most one zero in I. It is easily seen that $Mu = (L[u](x), u(0), u(1))$ is invertible if $[0, 1]$ is an interval of nonoscillation for L.

Theorem 3.22. *The regular operator $Mu = (L[u](x), u(0), u(1))$ is inverse-positive if and only if $[0, 1]$ is an interval of nonoscillation for L.*

Proof. If $[0, 1]$ is an interval of nonoscillation, then the solution $z \in C_2[0, 1]$ of $L[z](x) = 0$ $(0 < x < 1)$, $z(0) = 0$, $z(1) = 1$ satisfies $z(x) > 0$ $(0 < x \leqslant 1)$. Thus, M is inverse-positive, according to Theorem 3.17.

Suppose now that M is inverse-positive and that there exists a nontrivial solution $v \in C_2[0, 1]$ of $L[v](x) = 0$ $(0 < x < 1)$ which has two zeros in $[0, 1]$.

If $v(0) = 0$, then $v(1) \neq 0$, because M is invertible. Without loss of generality, we may assume $v(1) > 0$, so that $Mv > 0$ and $v(x) > 0$ $(0 < x \leqslant 1)$, by Theorem 3.17.

If $v(0) \neq 0$, we choose $w \in C_2[0, 1]$ such that $L[w](x) = 0$ $(0 < x < 1)$, $w(0) = -v(0)$, $w'(0) \neq -v'(0)$. The functions v and w are linearly independent and $y = v + w$ satisfies $L[y](x) = 0$ $(0 < x < 1)$, $y(0) = 0$. Thus, we have $y(x) \neq 0$ $(0 < x \leqslant 1)$, by the arguments applied to v in the last paragraph. On the other hand, the zeros of v and w separate each other (cf. Theorem 3.21), so that y changes its sign on $(0, 1)$. Consequently, v cannot have two zeros. □

Example 3.16. Each nontrivial solution of $L[u](x) = -u'' - u = 0$ can be written as $u = \alpha \cos(x - \gamma)$ with $\alpha \neq 0$. Any two consecutive zeros of such a function have the distance π. Thus, each interval $[\alpha, \beta]$ of length $\beta - \alpha < \pi$ is an interval of nonoscillation for L. Consequently, $Mu = (-u'' - u, u(\alpha), u(\beta))$ is inverse-positive on each such interval. (Apply Theorem 3.22 to $[\alpha, \beta]$ instead of $[0, 1]$.) □

3.10 Eigenvalue problems

We consider the eigenvalue problem

$$L[\varphi](x) = \lambda g(x)\varphi(x) \qquad (0 < x < 1), \qquad B_0[\varphi] = 0, \qquad B_1[\varphi] = 0 \tag{3.40}$$

for $\varphi \in C_2[0, 1]$, where $M = (L, B_0, B_1)$ is a regular operator and g denotes a continuous function on $[0, 1]$ with $g(x) > 0$ $(0 \leqslant x \leqslant 1)$. We shall prove the existence of a lowest eigenvalue with a positive eigenfunction and derive bounds for this eigenvalue. Moreover, the set of all inverse-positive regular operators will be characterized in a way different from that in Section 3.4. Let $R = C_2[0, 1]$ throughout this section.

For each real λ, we define an operator $M_\lambda: R \to S$ by

$$M_\lambda = M - \lambda D \qquad \text{with} \quad Du(x) = \begin{cases} g(x)u(x) & \text{for} \quad 0 < x < 1, \\ 0 & \text{for} \quad x = 0, \\ 0 & \text{for} \quad x = 1. \end{cases} \tag{3.41}$$

Obviously, the above eigenvalue problem can then be written as $M_\lambda \varphi = o$.

Proposition 3.23. *The eigenvalue problem* (3.40) *has a smallest eigenvalue* $\lambda_0 > -\infty$. *The operator* M_λ *is inverse-positive if and only of* $\lambda < \lambda_0$.

Proof. Consider the set Λ of all $\lambda \in \mathbb{R}$ such that M_λ is inverse-positive. We shall show that (a) $\Lambda \neq \varnothing$, (b) Λ is bounded above, (c) $\Lambda = (-\infty, \lambda_0)$ for some eigenvalue λ_0. Then λ_0 is the smallest eigenvalue, since M_λ is invertible for $\lambda < \lambda_0$.

(a) There exists a function $z \in C_2[0, 1]$ such that $z(x) > 0$ $(0 \leqslant x \leqslant 1)$, $B_0[z] > 0$, $B_1[z] > 0$ (for example, such a function z can be constructed similarly as in Example 3.12). Consequently, $M_\lambda z(x) > 0$ $(0 \leqslant x \leqslant 1)$ for $\lambda < 0$ with sufficiently large $|\lambda|$. These λ belong to Λ (cf. Theorem 3.17).

(b) Choose any $\mu \in \Lambda$ and any $r \in C_2[0, 1]$ such that $r(x) > 0$ $(0 < x < 1)$, $r(0) = 0$ if $\alpha_0 = 0$, $r(1) = 0$ if $\alpha_1 = 0$. Then define $v \in C_2[0, 1]$ to be the solution of $L[v] - gv = r(x)$ $(0 < x < 1)$, $B_0[v] = B_1[v] = 0$. This function satisfies the inequalities (3.37), since M_μ has property (P). Therefore, we have $L[-v] - \lambda g(-v) = -r(x) + (\lambda - \mu)g(x)v(x) > 0$ $(0 < x < 1)$ for large enough λ, so that $M_\lambda(-v) \geqslant o$. For these λ, M_λ cannot be inverse-positive, since $-v \leqslant o$. Consequently, Λ is bounded above.

(c) Let $\lambda \in \Lambda$ and suppose that $z \in C_2[0, 1]$ satisfies $z \geqslant o$, $M_\lambda z > o$. Then for each $\mu < \lambda$ we have also $M_\mu z > o$, so that $\mu \in \Lambda$. Consequently, all $\lambda < \lambda_0 := \sup \Lambda$ belong to Λ.

Suppose now that M_{λ_0} is invertible and let $\mathcal{M} = \{M_\lambda : \lambda_0 - \varepsilon \leqslant \lambda \leqslant \lambda_0\}$ with some $\varepsilon > 0$. This set satisfies the assumptions of Theorem 3.9, so that M_{λ_0} is inverse-positive. Therefore, the solution $\zeta \in C_2[0, 1]$ of $M_{\lambda_0}\zeta(x) = 1$ $(0 \leqslant x \leqslant 1)$ satisfies $\zeta \geqslant o$. Moreover, we have $M_\lambda \zeta(x) > 0$ $(0 \leqslant x \leqslant 1)$ for all λ in a sufficiently small neighborhood of λ_0. Consequently, this neighborhood belongs to Λ, which contradicts the definition of λ_0. □

The given eigenvalue problem (3.40) can be written as $A\varphi = \lambda\varphi$ with A defined in (3.24). Since A is symmetric with respect to the inner product (3.25), it means no loss of generality considering only real-valued eigenfunctions. Even if we considered complex-valued eigenfunctions φ in (3.40), we would obtain the same (real) eigenvalues.

The following theorem extends Proposition 3.23.

Theorem 3.24. *Suppose that* $M = (L, B_0, B_1)$ *is regular and* $g \in C_0[0, 1]$ *satisfies* $g(x) > 0$ $(0 \leqslant x \leqslant 1)$.

Then problem (3.40) *has a smallest eigenvalue* λ_0, *which has the following properties*:

(i) λ_0 *is a simple eigenvalue and there exists a corresponding eigenfunction* φ *which has the positivity properties* (3.37) (*with* φ *in place of* u);

(ii) *the operator* M_λ *in* (3.41) *is inverse-positive if and only if* $\lambda < \lambda_0$.

Proof. According to Proposition 3.23 it suffices to prove statement (i). Consider any $\lambda \in R$ and define $\varphi = \varphi_\lambda \in C_2[0, 1]$ to be the solution of the initial value problem

$$L[\varphi](x) - \lambda g(x)\varphi = 0 \qquad (0 < x < 1), \qquad B_0[\varphi] = 0,$$
$$\varphi(0) = 1 \quad \text{if} \quad \alpha_0 > 0, \qquad \varphi'(0) = 1 \quad \text{if} \quad \alpha_0 = 0.$$

For $\lambda < \lambda_0$ the operator M is inverse-positive, so that $M_\lambda \varphi_\lambda > o$ implies $\varphi_\lambda > o$. By taking the limit $\lambda \to \lambda_0 - 0$, we obtain $\varphi_{\lambda_0} > o$ and $B_1[\varphi_{\lambda_0}] \geqslant 0$, since φ_λ and φ'_λ depend continuously on λ. The inequality $B_1[\varphi_{\lambda_0}] > 0$ cannot hold, since $M_{\lambda_0}\varphi_{\lambda_0} > o$ implies that M_{λ_0} is inverse-positive (cf. Theorem 3.17). Thus, $z = \varphi_{\lambda_0}$ satisfies $z > o$ and $Mz = o$, so that (i) follows from Theorem 3.17. □

Corollary 3.24a. (i) *If* $z \in C_2[0, 1]$ *satisfies*

$$z > o, \qquad L[z](x) \leqslant \mu g(x)z(x) \qquad (0 < x < 1),$$
$$B_0[z] \leqslant 0, \qquad B_1[z] \leqslant 0$$

for some real μ, *then problem* (3.40) *has an eigenvalue* $\lambda \leqslant \mu$.

(ii) *If* $z \in C_2[0, 1]$ *satisfies*

$$z > o, \qquad L[z](x) \geqslant \nu g(x)z(x) \qquad (0 < x < 1),$$
$$B_0[z] \geqslant 0, \qquad B_1[z] \geqslant 0$$

for some real ν, *then each eigenvalue* λ *of* (3.40) *satisfies* $\lambda \geqslant \nu$.

Proof. (i) The inequalities $M_\mu(-z) \geqslant o$ and $-z < o$ show that M_μ is not inverse-positive, so that $\mu \geqslant \lambda_0$, by the preceding theorem.

(ii) From $M_\nu z \geqslant o$ and $z > o$ the inequality $M_\mu z > o$ follows for all $\mu < \nu$, so that $\nu \geqslant \lambda_0$. □

This corollary may be applied in the following way. *Choose a function* $z \in C_2[0, 1]$ *such that* $z(x) > 0$ $(0 < x < 1)$, $B_0[z] = B_1[z] = 0$. *Then the smallest eigenvalue* λ_0 *of* (3.40) *satisfies*

$$\inf\left\{\frac{L[z](x)}{g(x)z(x)} : 0 < x < 1\right\} \leqslant \lambda_0 \leqslant \sup\left\{\frac{L[z](x)}{g(x)z(x)} : 0 < x < 1\right\}. \tag{3.42}$$

Problem

3.11 Show with Theorem 3.24 that the properties (i) and (ii) in Theorem 3.11 are equivalent.

3.11 Differential operators on noncompact intervals

Many results of the preceding sections carry over to problems where the considered functions u are only defined on a noncompact interval. Here we shall treat such problems in two ways. In the first method, we introduce generalized boundary operators and define an operator M by a formula such as (3.2). Then the monotonicity theorem yields results which correspond to Theorems 3.2 and 3.14. In the second method—the method of absorption—the given noncompact interval is absorbed by compact intervals to which then the theory of the previous sections may be applied. Here results are obtained which resemble Theorem 3.5. (For further results see Sections III.2.3.5 and IV.2.1.4.)

3.11.1 The method of generalized boundary operators

For explaining the main ideas we take up the following special case. Let $[0, l)$ be the given interval where $0 < l \leqslant \infty$. Suppose that L and B have the same meaning as in Section 3.1, except that now $a, b, c \in \mathbb{R}(0, l)$ and (3.4) is replaced by $a(x) \geqslant 0$ $(0 < x < l)$, $\alpha_0 \geqslant 0$, $\gamma_0 > 0$ if $\alpha_0 = 0$. We are interested in conditions such that for suitable functions u the inequalities $L[u](x) \geqslant 0$ $(0 < x < l)$ and $B_0[u] \geqslant 0$ imply $u(x) \geqslant 0$ $(0 \leqslant x < l)$, or stronger positivity properties of u.

Suppose that $p \in \mathbb{R}[0, l)$ is a preassigned function with $0 < \varepsilon \leqslant p(x)$ $(0 < x < l)$ for some constant ε. Then define R to be the space of all functions $u \in C_0[0, l) \cap C_2(0, l)$ such that $u'(0)$ exists if $\alpha_0 > 0$ and

$$u(x) = p(x)(\delta + o(1)) \qquad \text{for} \quad x \to l, \tag{3.43}$$

where δ denotes some constant, which may depend on u. Define

$$B_1[u] = \lim_{x \to l} p^{-1}(x)u(x) \tag{3.44}$$

and $M = (L, B_0, B_1): R \to S = \mathbb{R}[0, l]$ by

$$\begin{aligned} Mu(x) &= L[u](x) \qquad \text{for} \quad 0 < x < l, \\ Mu(0) &= B_0[u], \qquad Mu(l) = B_1[u]. \end{aligned} \tag{3.45}$$

Theorem 3.25. *If there exists a function $z \in R$ such that*

$$z(x) \geqslant 0 \qquad (0 \leqslant x < l), \qquad Mz(x) > 0 \qquad (0 \leqslant x \leqslant l), \tag{3.46}$$

then the operator M is inverse-positive, that is, for all $u \in R$

$$Mu(x) \geqslant 0 \qquad (0 \leqslant x \leqslant l) \quad \Rightarrow \quad u(x) \geqslant 0 \qquad (0 \leqslant x < l).$$

Proof. We define $U \succ o$ by $U(x) > 0$ $(0 \leqslant x \leqslant l)$ and derive this theorem from Theorem 1.2 by verifying condition I. Let $u \in R$ satisfy $u \geqslant o$ and $Mu \succ o$, that is, $u(x) \geqslant 0$ $(0 \leqslant x < l)$, $Mu(x) > 0$ $(0 \leqslant x < l)$, $B_1[u] > 0$. As in the proof of Proposition 3.1, the first two of these inequalities imply $u(x) > 0$ $(0 \leqslant x < l)$. Thus, it suffices to show that

$$u(x) > 0 \quad (0 \leqslant x < l), \quad B_1[u] > 0 \quad \Rightarrow \quad u \succ o. \tag{3.47}$$

To see this, consider an arbitrary $v \in R$. Since $B_1[n_1 u - v] > 0$ for some $n_1 \in \mathbb{N}$, we have

$$\begin{aligned} (n_1 u - v)(x) &\geqslant \varepsilon p^{-1}(x)(n_1 u - v)(x) \\ &\geqslant (\varepsilon/2)B_1[n_1 u - v] > 0 \qquad \text{for} \quad \alpha < x < l \end{aligned}$$

and some $\alpha \in (0, l)$. On the other hand, $(n_2 u - v)(x) \geqslant 0$ for $0 \leqslant x \leqslant \alpha$ and sufficiently large $n_2 \in \mathbb{N}$. These inequalities together yield $nu - v \geqslant o$ for large enough $n \in \mathbb{N}$. Hence $u \succ o$. □

Example 3.17. Let $c(x) > 0$ $(0 < x < l)$ and $\gamma_0 > 0$ and choose $p(x) \equiv 1$. Then $z(x) \equiv 1$ satisfies the inequalities (3.46). Thus, for all $u \in R$,

$$\left.\begin{aligned} L[u](x) \geqslant 0 \quad (0 < x < l) \\ B_0[u] \geqslant 0 \\ \lim_{x\to l} u(x) \geqslant 0 \end{aligned}\right\} \Rightarrow \quad u(x) \geqslant 0 \quad (0 \leqslant x < l). \quad \square$$

Condition (3.46) often is overly restrictive. For example, there is no $z \in C_2(0, \infty)$ such that $z(x) \geqslant 0$, $-z''(x) > 0$ for $0 < x < \infty$, and $z(x) \to 0$ for $x \to \infty$. Therefore, we are interested in relaxing this condition.

Theorem 3.26. *The statements of Theorem* 3.14 *and Corollary* 3.14a *hold also for the operator M in* (3.45) (*when the endpoint* $x = 1$ *is replaced by* $x = l$).

Proof. This result is proved in very much the same way as the theorem and corollary mentioned, except that (3.47) is needed to verify condition I of Theorem 1.2. □

Example 3.18. *Suppose that* $a, b, c \in C_0(0, l)$, $\gamma_0 > 0$ *and* $a(x) > 0$, $c(x) \geqslant 0$ $(0 < x < l)$. *Then for each* $u \in C_1[0, l) \cap C_2(0, l)$

$$\left.\begin{aligned} L[u](x) \geqslant 0 \quad (0 < x < l) \\ B_0[u] \geqslant 0 \\ \lim_{x\to l} u(x) \geqslant 0 \end{aligned}\right\} \Rightarrow \left\{\begin{aligned} &\textit{either} \quad u(x) = 0 \quad (0 \leqslant x < l) \\ &\textit{or} \quad\quad u(x) > 0 \quad (0 < x < l). \end{aligned}\right.$$

For proving this statement choose $p(x) \equiv 1$ and $z(x) \equiv 1$. □

Example 3.19. *Let* $k > 0$. *Then for each* $u \in C_0[0, \infty) \cap C_2(0, \infty)$,

$$\left.\begin{aligned} &(-u'' + k^2 u)(x) \geqslant 0 \quad (0 < x < \infty) \\ &u(0) \geqslant 0 \\ &\lim_{x\to\infty} u(x) \exp(-kx) \geqslant 0 \end{aligned}\right\} \Rightarrow \left\{\begin{aligned} &\textit{either} \quad u(x) = 0 \\ &\qquad\qquad (0 \leqslant x < \infty) \\ &\textit{or} \qquad u(x) > 0 \\ &\qquad\qquad (0 < x < \infty). \end{aligned}\right.$$

This statement follows from Theorem 3.26, when $z(x) = p(x) = \exp(kx)$ is used. □

The above theory can be generalized in several ways. For example, boundary operators B_1 other than (3.44) can be used. The only essential condition on B_1 is that (3.47) holds. Moreover, with essentially the same ideas one can also treat problems on open intervals with generalized boundary operators at both endpoints. The generalization is straightforward.

3.11.2 Approximation by compact intervals

This method, too, will be explained by treating a special case. Now let $(-\infty, \infty)$ be the given interval. Suppose that $\{\alpha_m\}$, $\{\beta_m\}$ are given sequences with $\alpha_{m+1} < \alpha_m < \beta_m < \beta_{m+1}$ $(m = 1, 2, \ldots)$, and $\alpha_m \to -\infty$, $\beta_m \to \infty$ $(m \to \infty)$.

Theorem 3.27. *Suppose that the following conditions hold:*

(i) *For each interval* $[\alpha_m, \beta_m]$ $(m = 1, 2, \ldots)$ *and all* $u \in C_2[\alpha_m, \beta_m]$

$$\left.\begin{array}{ll} L[u](x) \geqslant 0 & (\alpha_m < x < \beta_m) \\ u(\alpha_m) \geqslant 0, & u(\beta_m) \geqslant 0 \end{array}\right\} \Rightarrow u(x) \geqslant 0 \quad (\alpha_m \leqslant x \leqslant \beta_m).$$

(ii) *There exist functions* $z_m \in C_0[\alpha_m, \beta_m] \cap C_2(\alpha_m, \beta_m)$ $(m = 1, 2, \ldots)$ *and a function* $z \in \mathbb{R}(-\infty, \infty)$ *such that*

$$0 < z_m(x) \leqslant z(x) \qquad (\alpha_m \leqslant x \leqslant \beta_m), \qquad L[z_m](x) \geqslant 0 \qquad (\alpha_m < x < \beta_m).$$

Then for each $u \in C_2(-\infty, \infty)$ *the relations*

$$\begin{array}{lll} & L[u](x) \geqslant 0 & \text{for} \quad -\infty < x < \infty, \\ \limsup_{m \to \infty} \gamma_m \geqslant 0 & \text{with} & \gamma_m = \min\{u(\alpha_m) : z_m(\alpha_m),\, u(\beta_m) : z_m(\beta_m)\} \end{array} \tag{3.48}$$

imply that $u(x) \geqslant 0$ $(-\infty < x < \infty)$.

Proof. Let u satisfy the inequalities in the premise of the statement. Without loss of generality, we may assume that the sequence $\{\gamma_m\}$ converges. In this case set $v_m = u + \mu_m z_m$ with $\mu_m = \max\{0, -\gamma_m\}$. Then property (i), applied to v_m instead of u, yields $v_m(x) \geqslant 0$ $(\alpha_m \leqslant x \leqslant \beta_m)$, that is, $u(x) \geqslant -\mu_m z_m(x) \geqslant -\mu_m z(x)$ $(\alpha_m \leqslant x \leqslant \beta_m)$. Since $\mu_m \to 0$, we reach the desired conclusion that $u(x) \geqslant 0$ $(-\infty < x < \infty)$. □

This theorem, too, can be generalized in many ways. Suppose for example, that a stronger condition than (i) is assumed where $u(x) \geqslant 0$ $(\alpha_m \leqslant x \leqslant \beta_m)$ is replaced by the property that either $u(x) \equiv 0$ or $u(x) > 0$ $(\alpha_m < x < \beta_m)$. Then the inequality $u(x) \geqslant 0$ $(-\infty < x < \infty)$ in the statement of the theorem may be replaced by the statement that either $u(x) \equiv 0$ or $u(x) > 0$ $(-\infty < x < \infty)$.

To verify condition (i) or the stronger condition just mentioned, the results of the preceding sections may be used. For example, the following corollary can be derived with Theorem 3.14 and Corollary 3.14a.

Corollary 3.27a. *If* $a, b, c \in C_0(-\infty, \infty)$ *and* $a(x) > 0$ $(-\infty < x < \infty)$, *then condition* (i) *of Theorem* 3.27 *may be dropped. Moreover the inequalities* (3.48) *imply that either* $u(x) = 0$ $(-\infty < x < \infty)$ *or* $u(x) > 0$ $(-\infty < x < \infty)$.

Example 3.20. *Let $k > 0$. Then for each $u \in C_2(-\infty, \infty)$*

$$\left.\begin{array}{l} -u'' + k^2 u \geqslant 0 \quad (-\infty < x < \infty) \\ \\ \limsup\limits_{|x| \to \infty} u(x) \exp(-k|x|) \geqslant 0 \end{array}\right\} \Rightarrow \left\{\begin{array}{ll} \text{either} & u(x) = 0 \\ & (-\infty < x < \infty) \\ \text{or} & u(x) > 0 \\ & (-\infty < x < \infty). \end{array}\right.$$

This statement follows from Theorem 3.27 and Corollary 3.27a, when $z_m(x) = z(x) = \cosh kx$ is used. The sequences $\{\alpha_m\}$ and $\{\beta_m\}$ may be chosen arbitrarily. □

Problems

3.12 Consider the following generalization of B_1 in (3.44). Let α_1 and γ_1 be given functions on $[0, l)$ with $\alpha_1(x) \geqslant 0$ $(0 \leqslant x < l)$ and $B_1[u](x) = \alpha_1(x)u'(x) + \gamma_1(x)u(x)$. Define R as in Section 3.11.1, with (3.43) replaced by the condition that $B_1[u] = \lim_{x \to l} B_1[u](x)$ exists. Show that (3.47) holds under these conditions.

3.13 Prove that Theorem 3.25 holds for M in (3.45) with B_1 defined as in the preceding problem.

4 ABSTRACT THEORY (CONTINUATION)

We continue our theory on inverse-positive linear operators M.

Section 4.1 is mainly concerned with proving the pre-inverse-positivity of M, which is required in the monotonicity theorem. We have seen in Sections 2 and 3, that this property can easily be established for Z-matrices and certain differential operators of the second order (at least for certain strong order relations $U \succ o$). For other operators, such as differential operators of higher order, however, pre-inverse-positivity holds only under rather restrictive conditions and it is much more complicated to verify. In such cases the *method of reduction* explained in Section 4.1 often is a useful tool.

The Z-matrices in Section 2 and the second order differential operators in Section 3, besides both being pre-inverse-positive, have a series of other properties in common which are related to inverse-positivity. We mention, for example, the relation between eigenvalue theory and inverse-positivity. Introducing *Z-operators* (and Z-pairs of operators) in Section 4.2, we formulate properties which constitute the underlying reason for the analogous behavior of Z-matrices and second order differential operators.

More generally, we shall introduce *Z(W)-operators* (and Z(W)-pairs) which depend on the order cone W of a given strong order $\leqslant$. The theory of these operators generalizes large parts of the theory of Z-matrices, irreducible Z-matrices, second order differential operators, etc. The common basic property of all these operators is described by condition (4.22), which is stronger than pre-inverse-positivity. We point out that the terms introduced are also closely related to the term *quasi-monotone*, which is often used for vector-valued differential operators, and to the notion of an *irreducible operator.*

In Section 4.3, we consider operators $\lambda I - T$ with positive T as a special case of Z-operators, also introducing the term of a *W-irreducible operator.*

In the general assumptions of Section 2, it was required that the space $(R, \leqslant)$ contain a strictly positive element. Section 4.4 is concerned with spaces $(R, \leqslant, K)$ such that $K^c = \varnothing$. Here, from conditions similar to those in the monotonicity theorem, a certain *restricted inverse-positivity* is derived.

4.1 The method of reduction

We introduce the *method of reduction* as a tool in applying the monotonicity theorem. The simple idea behind the method is to transform the given linear operator $M: R \to S$ into a linear operator PM for which the assumptions of Theorem 1.2 are more easily verified. We shall here describe in an abstract form several ways to perform such a reduction (Sections 4.1.1, 4.1.2, and 4.1.4). Section 4.1.3 provides an application to a differential operator of the fourth order.

We assume in the following that R and S satisfy the general assumptions (A_1), (A_2) in Section 1.1, so that, in particular, a strong order $\lessdot$ is defined in S.

4.1.1 Global reduction

Let $(Y, \leqslant)$ denote a further ordered linear space and assume that in this space, too, a strong linear order $\lessdot$ is given. Moreover, let $P: S \to Y$ be a given linear operator such that for all $U \in S$

$$U \geqslant o \quad \Rightarrow \quad PU \geqslant o; \qquad U \gtrdot o \quad \Rightarrow \quad PU \gtrdot o. \tag{4.1}$$

For convenience, we use the same symbols $\leqslant$, $\lessdot$ for order relations in different spaces. It will always be clear to which spaces the occurring elements belong and, thus, which order relation is meant. The term pre-inverse-positive shall, in the following, always mean pre-inverse-positive with respect to the strong order $\lessdot$ (defined in the range of the operator considered).

Theorem 4.1. *If*

I. *$PMu \not\gtrdot o$ for each $u \in R$ with $u \geqslant o$ and $u \not\succ o$,*
then M is pre-inverse-positive. If, in addition,

II. *$Mz \gtrdot o$ or $PMz \gtrdot o$ for some $z \geqslant o$ in R, then M is inverse-positive and for $u \in R$, $Mu \gtrdot o \Rightarrow u \succ o$.*

Proof. Since $Mu \geqslant o \Rightarrow PMu \geqslant o$ and $Mu \succ o \Rightarrow PMu \succ o$, the result follows from Theorem 1.2 applied to PM. $\square$

The theorem essentially describes the *reduction method*: Find an operator $P: S \to Y$ *and* a strong order relation in Y such that the assumptions of the theorem hold, i.e., PM is pre-inverse-positive (with respect to the strong order in Y) and a suitable z exists.

In general, the main purpose of the reduction method is to show that M is pre-inverse-positive. Since both strong orders in S and Y are involved here, reduction means not only the choice of an operator P, but also that of suitable strong orders.

Example 4.1. Let the sets R, S, and Y all equal the vector space $\mathbb{R}^4$ with $\leqslant$ denoting the natural order relation. However, we shall distinguish between S and Y by defining

$$U \succ o \quad \Leftrightarrow \quad U > o, \qquad \text{for} \quad U \in S; \qquad y \succ o \quad \Leftrightarrow \quad y \succ o, \qquad \text{for} \quad y \in Y.$$

We have

$$PM = \begin{bmatrix} 5 & 0 & 0 & 0 \\ -5 & 5 & 0 & 0 \\ 0 & -5 & 5 & 0 \\ 0 & 0 & -5 & 5 \end{bmatrix}$$

for

$$M = \begin{bmatrix} 3 & -1 & 0 & 0 \\ -3 & 3 & -1 & 0 \\ 1 & -3 & 3 & -1 \\ 0 & 1 & -3 & 2 \end{bmatrix}, \qquad P = \begin{bmatrix} 4 & 3 & 2 & 1 \\ 3 & 6 & 4 & 2 \\ 2 & 4 & 6 & 3 \\ 1 & 2 & 3 & 4 \end{bmatrix}.$$

The matrix PM satisfies assumption I of Theorem 4.1, since it is a Z-matrix. To verify assumption II, we can, for example, show that $Mz > o$ for $z^{\mathrm{T}} = (1, 3, 6, 10)$ or show that $PMz \succ o$ for $z^{\mathrm{T}} = (1, 2, 3, 4)$. Theorem 4.1 then yields that M is inverse-positive and, moreover, $Mu > o \Rightarrow u \succ o$. Consequently, *$M$ is strictly inverse-positive.*

Observe that the strict pre-inverse-positivity of M here follows from the weak pre-inverse-positivity of PM, since $U > o \Rightarrow PU \succ o$. $\square$

If P satisfies (4.1) and if

$$PM = M_1 - N \tag{4.2}$$

with a linear operator $M_1: R \to Y$ and a positive linear operator $N: R \to Y$ then *M is pre-inverse-positive, if M_1 is pre-inverse-positive,* since then PM has this property, too.

A reduction may also be achieved through a (multiplicative and additive) splitting

$$M = LM_1 - Q. \tag{4.3}$$

Corollary 4.1a. *Let* (4.3) *hold with a positive linear operator* $Q: R \to S$, *a pre-inverse-positive linear operator* $M_1: R \to Y$ *and an inverse-positive linear operator* L *such that* $LY = S$ *and for all* $y \in Y$, *the inequalities* $Ly \gtrdot o$ *and* $y \geqslant o$ *imply* $y \gtrdot 0$.

Then M *is pre-inverse-positive.*

Proof. Applying $P = L^{-1} \geqslant O$ to (4.3) yields (4.2) with $N = PQ \geqslant O$. Moreover, if $U = Ly \gtrdot o$, then $Ly \geqslant o$ and hence $y \geqslant o$, so that $PU = y \gtrdot o$ by the preceding assumptions. Thus, the second implication in (4.1) holds also, so that M is pre-inverse-positive according to Theorem 4.1. □

Example 4.1 (continued). We have $M = LM_1$ with

$$L = \begin{bmatrix} 2 & -1 & 0 & 0 \\ -1 & 2 & -1 & 0 \\ 0 & -1 & 2 & -1 \\ 0 & 0 & -1 & 2 \end{bmatrix} \quad \text{and} \quad M_1 = \begin{bmatrix} 1 & 0 & 0 & 0 \\ -1 & 1 & 0 & 0 \\ 0 & -1 & 1 & 0 \\ 0 & 0 & -1 & 1 \end{bmatrix}.$$

All assumptions of Corollary 4.1a are satisfied. Observe that L is strictly pre-inverse-positive, so that the inequalities $Ly > o$, $y \geqslant o$ imply $y \succ o$ (see Example 2.2). □

A simple but sometimes quite useful application of the reduction method is described in the next two corollaries (see also Problem 4.1).

Corollary 4.1b. *Suppose that* $M = D + A - B - G$ *with positive linear operators* D, A, B, G *such that* D *and* $D - B$ *are inverse-positive,* $DR = S$, $(D - B)R = S$, *and* $A \leqslant BD^{-1}G$. *Then* M *is weakly pre-inverse-positive.*

Proof. Here, (4.3) holds with $L = D - B$, $M_1 = I - D^{-1}G$, and $Q = BD^{-1}G - A \geqslant O$. Observe that M_1 and L are weakly pre-inverse-positive, due to Proposition 1.4 and Theorem 1.6(i). □

In particular, this statement may be applied to $n \times n$ matrices, as in the next result, where $\leqslant$ denotes the natural order relation in $\mathbb{R}^n$.

Corollary 4.1c. *Let* M *denote a diagonally strictly positive* $n \times n$ *matrix and* $D = (m_{ii}\,\delta_{ij})$. *Suppose there exist matrices* B, G *with the following properties:* $(M - D)^- = B + G$, $B \geqslant O$, $G \geqslant O$, $D - B$ *is inverse-positive, and to each* $m_{ij} > 0$ *with* $i \neq j$ *an index* k *exists for which* $m_{ij} m_{kk} < b_{ik} g_{kj}$. *Then* M *is weakly pre-inverse-positive.*

Example 4.2. For

$$M = \begin{bmatrix} 1 & & & & & & \\ -1 & 2 & -1 & & & & \\ 1 & -16 & 30 & -16 & 1 & & \\ & 1 & -16 & 30 & -16 & 1 & \\ & & 1 & -16 & 30 & -16 & 1 \\ & & & & -1 & 2 & -1 \\ & & & & & & 1 \end{bmatrix},$$

$$B = \begin{bmatrix} 0 & & & & & & \\ 0 & 0 & & & & & \\ 0 & 8 & 0 & 16 & & & \\ & & 8 & 0 & 8 & & \\ & & & 16 & 0 & 8 & \\ & & & & & 0 & 0 \\ & & & & & & 0 \end{bmatrix}, \qquad C = \begin{bmatrix} 0 & & & & & & \\ 1 & 0 & 1 & & & & \\ 0 & 8 & 0 & 0 & & & \\ & & 8 & 0 & 8 & & \\ & & & 0 & 0 & 8 & \\ & & & & 1 & 0 & 1 \\ & & & & & & 0 \end{bmatrix},$$

the assumptions of Corollary 4.1c are satisfied. Choose $k = i + 1$ for elements $m_{i,i+2} = 1$, for example. □

Problems

4.1 Generalize Corollary 4.1b in the following way. Suppose that $\preccurlyeq_1$ is a strong order in S, $\preccurlyeq_2$ a strong order in R, and that $U \succ_1 o \Rightarrow (D - B)^{-1}U \succ_2 o; u \succ_2 o \Rightarrow (D^{-1}G + I)u \succ o$. Prove that then M is pre-inverse-positive with respect to $\preccurlyeq_1$.

4.2 If M is inverse-positive and (4.2) holds with positive N or (4.3) holds with positive Q and inverse-positive L, then for all $u \in R$, $Mu \geqslant o$ implies that $u \geqslant o$ and $M_1 u \geqslant o$. This property can be considered as inverse-positivity of M with respect to an order relation in R given by the two inequalities on the right-hand side of the above implication. Apply the monotonicity theorem to this type of inverse-positivity.

4.1.2 Local reduction

To prove that M is pre-inverse-positive, certain conditions have to be verified for each $v \in \partial K$ (i.e., $v \in R$ with $v \geqslant o$, $v \not\succ o$). Sometimes, a reduction for proving the pre-inverse-positivity is carried out more easily by using a *family* of operators $P(v): S \to Y(v)$ with $v \in \partial K$ as parameter. We assume that in each of the linear spaces $Y = Y(v)$ which will occur, a linear order $\leqslant$ and a further stronger linear order $\preccurlyeq$ are given, which, of course, may depend on the parameter v. Again, whenever a symbol $\leqslant$ or $\prec$ occurs in the following, it will be clear from the context to which space this relation belongs.

Theorem 4.2. *Suppose that the operator M has the following two properties:*

I. *To each $v \in R$ with $v \geqslant o$ and $v \not> o$, a linear operator $P = P(v): S \to Y = Y(v)$ exists such that* (4.1) *holds and $P(v)Mv \not\succ o$.*
II. *There exists a $z \geqslant o$ in R such that $Mz \succ o$, or $P(v)Mz \succ o$ for all operators $P(v)$ occurring in* I.

Then M is inverse-positive and $Mu \succ o \Rightarrow u > o$, for all $u \in R$.

Proof. Let $\hat{Y}$ be the product space $\hat{Y} = \prod_{v \in \partial K} Y(v)$ with elements $\hat{y} = \{y(v)\}$ and define $\hat{y} \geqslant o$ and $\hat{y} \succ o$ componentwise, so that, for example, $\hat{y} \succ o \Leftrightarrow y(v) \succ o$ for all $v \in \partial K$. Moreover, define $\hat{P}: S \to \hat{Y}$ by $\hat{P}U = \{P(v)U\}$. Then the preceding statements follow from Theorem 4.1 with $P = \hat{P}$. □

In particular, all spaces $Y(v)$ may equal $\mathbb{R}$, in which case we obtain from Theorem 4.2 the following result, which essentially was already obtained in Section 1 (cf. Proposition 1.3 and Theorem 1.7).

Corollary 4.2a. *Suppose that for each $v \in R$ with $v \geqslant o$ and $v \not> o$ a positive linear functional ℓ on S exists such that $\ell Mv \leqslant 0$ and $U \succ o$ implies $\ell U > 0$, for all $U \in S$. Moreover, let a function $z \in R$ exist which satisfies $z \geqslant o$, $Mz \succ o$.*
Then M is inverse-positive and for $u \in R$, $Mu \succ o \Rightarrow u > o$.

4.1.3 A differential operator of the fourth order

We shall illustrate the reduction method by applying it to a particular differential operator of the fourth order. The results obtained are formulated in Propositions 4.3 and 4.4. We shall first discuss the problem in an elementary way and then describe several possibilities to carry out a reduction. The emphasis is not only on the results themselves but as well on the methods to obtain them. The ideas involved can also be used to derive results on more general fourth order operators (see Section 6.1).
The boundary value problem

$$L[u](x) = r(x) \qquad (0 \leqslant x \leqslant 1), \qquad u(0) = u''(0) = u(1) = u''(1) = 0 \\ \text{with} \quad L[u](x) = u^{\text{iv}}(x) + cu(x) \tag{4.4}$$

describes the bending of a beam which is attached to an elastic support; u denotes the deviation of the beam under the (continuous) load function r, and c is an elasticity constant. (Later we shall also consider functions $c \in C_0[0, 1]$.) Our question is: *For which c does a load $r \geqslant o$ always result in a deviation*

$u \geqslant o$? In other words, for which c is the following operator M inverse-positive:

$$M: R \to S, \qquad R = \{u \in C_4[0, 1] : u(0) = u''(0) = u(1) = u''(1) = 0\},$$
$$S = C_0[0, 1], \qquad Mu(x) = L[u](x) \qquad (0 \leqslant x \leqslant 1). \tag{4.5}$$

Before we apply our theory (to the more general case where $c = c(x)$), we shall first discuss the preceding simple problem in a more direct way. This will help us in understanding the meaning of the assumptions in Propositions 4.3 and 4.4.

If M is invertible, then the inverse is an integral operator with the Green function corresponding to M as its kernel. *M is inverse-positive if and only if the Green function exists and is positive* ($\geqslant 0$). For the above operator M the Green function $\mathscr{G}(x, \xi)$ can be calculated explicitly. In particular, one sees by formal calculations that $\mathscr{G}(x, \xi) = \mathscr{G}(x, \xi, c)$ *(exists and) satisfies*

$$\mathscr{G}(x, \xi, c) \geqslant 0 \qquad (0 \leqslant x, \xi \leqslant 1) \qquad \textit{if and only if} \quad -\pi^4 < c \leqslant c_0, \tag{4.6}$$

where $c_0 = 4\kappa^4 \approx 951$ with κ the smallest solution of $\tan\kappa = \tanh\kappa$ satisfying $\kappa > 0$. For example,

$$\mathscr{G}(x, 1 - x, 4\kappa^4) = \frac{2}{\kappa}\frac{\sin\kappa\cosh\kappa - \cos\kappa\sinh\kappa}{\cosh^2\kappa - \cos^2\kappa}x^2 + O(x^3) \quad \text{for} \quad x \to 0.$$

Since π^4 is the smallest eigenvalue of the problem $u^{\mathrm{iv}} = \lambda u$ associated with the preceding boundary conditions, the operator M is not invertible for $c = -\pi^4$, but it is invertible for all $c > -\pi^4$. This explains the lower bound $-\pi^4$ for c in (4.6); the upper bound c_0, however, has to be explained differently.

We point out that the necessary and sufficient condition in (4.6) for the inverse-positivity of the fourth order operator M differs from corresponding conditions for second order operators in an essential way (cf. Proposition 3.23).

We shall describe the behavior of $\mathscr{G}(x, \xi, c)$ in the neighborhood of $c = c_0$ in more detail. If $-\pi^4 < c < c_0$, one calculates $G(x, \xi, c) > 0$ $(0 < x, \xi < 1)$ and $\varphi(x) > 0$ $(0 < x < 1)$, $\varphi(1) = 0$, $\varphi'(1) < 0$ for $\varphi(x) = \mathscr{G}_\xi(x, 0)$. If c varies continuously from values $< c_0$ to values $> c_0$, the function $\mathscr{G}(x, \xi, c)$ changes from a function $\geqslant 0$ to a function which assumes also negative values. For $c = c_0$ all the preceding inequalities remain true except that now $\varphi'(1) = 0$. For c in a right-hand neighborhood of c_0, one has $\varphi'(1) > 0$, so that $\mathscr{G}(x, \xi, c)$ assumes negative values near $(x, \xi) = (1, 0)$ (and, analogously, near $(x, \xi) = (0, 1)$).

Let us explain these facts somewhat differently without using the Green function. For each $c > c_0$ there exists a load $r \geqslant o$ with deviation $u \ngeqslant o$. It turns out that for c only "slightly larger" than c_0 the load r has to be "concentrated near one end" of the beam in order to achieve values $u(x) < 0$, which then occur "near the other end."

To study the "limit behavior" for $c \to c_0 - 0$, we consider the functions Φ, Ψ which for $c > -\pi^4$ are defined by

$$\begin{aligned} &L[\Phi](x) = 0 \quad (0 \leqslant x \leqslant 1), \\ &\qquad \Phi(0) = 0, \quad \Phi''(0) = -1, \quad \Phi(1) = 0, \quad \Phi''(1) = 0, \\ &L[\Psi](x) = 0 \quad (0 \leqslant x \leqslant 1), \\ &\qquad \Psi(0) = 0, \quad \Psi''(0) = 0, \quad \Psi(1) = 0, \quad \Psi''(1) = -1. \end{aligned} \tag{4.7}$$

For $-\pi^4 < c \leqslant c_0$ both these functions satisfy $\Phi \geqslant o$, $\Psi \geqslant o$ since they are limits of solutions u_n of boundary value problems (4.4) with loads $r_n \geqslant o$ $(n \to \infty)$. For example, consider $r_n(x) = 6n^2(1 - nx)$ for $0 \leqslant nx \leqslant 1$, $r_n(x) = 0$ otherwise. The momentum of this load with respect to $x = 0$ equals 1, i.e., $\int_0^1 x r_n(x)\, dx = 1$. For $n \to \infty$ the loads become concentrated more and more near $x = 0$, and the corresponding solutions u_n converge toward the above function Φ. This function represents the deviation of the beam which results from a momentum $-\Phi''(0) = 1$ at $x = 0$ with no load present.

By elementary calculations one obtains

$$\operatorname{sgn} \Phi'(1) = -\operatorname{sgn}(c_0 - c), \qquad \operatorname{sgn} \Psi'(0) = \operatorname{sgn}(c_0 - c) \tag{4.8}$$

for $-\pi^4 < c \leqslant c_0 + \varepsilon$ with sufficiently small $\varepsilon > 0$. Consequently, Φ and Ψ assume negative values for c in a right-hand neighborhood of c_0.

This shows that, for the particular operator M considered here, the inverse-positivity is closely related to the behavior of the functions Φ, Ψ. We call Φ and Ψ the *limit functions corresponding to M*. Because of (4.8), the upper bound c_0 in (4.6) has the property that $c = c_0$ is an eigenvalue of the problem

$$\begin{aligned} &\Phi^{\text{iv}}(x) + c\Phi(x) = 0 \quad (0 \leqslant x \leqslant 1), \\ &\qquad \Phi(0) = 0, \quad \Phi(1) = \Phi'(1) = \Phi''(1) = 0, \end{aligned} \tag{4.9}$$

with c considered as parameter, as well as of

$$\begin{aligned} &\Psi^{\text{iv}}(x) + c\Psi(x) = 0 \quad (0 \leqslant x \leqslant 1), \\ &\qquad \Psi(0) = \Psi'(0) = \Psi''(0) = 0, \quad \Psi(1) = 0. \end{aligned} \tag{4.10}$$

(Obviously, for constant c, (4.9) can be transformed into (4.10) by defining $\Psi(x) = \Phi(1 - x)$.)

One can show that the functions Φ, Ψ in (4.7) satisfy $\mathscr{G}_\xi(x, 0) = \Phi(x)$, $-\mathscr{G}_\xi(x, 1) = \Psi(x)$.

In the following, we shall apply global reduction as well as local reduction to the operator M in (4.5), where now c denotes a function in $C_0[0, 1]$.

First we shall reduce this operator M to a differential operator of the second order by applying a positive integral operator P. Define P on $S = C_0[0, 1]$ by

$$PU(x) = \int_0^1 \Gamma(x, t)U(t)\,dt, \qquad \Gamma(x, t) = \begin{cases} \psi(x)\varphi(t) & \text{for } 0 \leqslant x \leqslant t \leqslant 1 \\ \psi(t)\varphi(x) & \text{for } 0 \leqslant t \leqslant x \leqslant 1 \end{cases} \tag{4.11}$$

with

$$\varphi, \psi \in C_4[0, 1], \qquad \varphi(x) > 0, \qquad \psi(x) > 0 \qquad (0 < x < 1). \tag{4.12}$$

For $u \in R$ one obtains by partial integration, using the boundary conditions,

$$PMu(x) = M_1u(x) - Nu(x) \tag{4.13}$$

with

$$\begin{aligned} M_1u(x) = &-p(x)u''(x) + p'(x)u'(x) + q(x)u(x) \\ &- \varphi(x)(\psi(0)u'''(0) + \psi''(0)u'(0)) \\ &+ \psi(x)(\varphi(1)u'''(1) + \varphi''(1)u'(1)), \end{aligned}$$

$$p = \varphi\psi' - \varphi'\psi, \qquad q = \varphi'''\psi - \varphi\psi''',$$

$$Nu(x) = -\int_0^x \varphi(x)L[\psi](t)u(t)\,dt - \int_x^1 \psi(x)L[\varphi](t)u(t)\,dt.$$

For each $u \geqslant o$ in R we have $Nu(x) \geqslant 0\,(0 \leqslant x \leqslant 1)$, provided the relations

$$L[\varphi](x) \leqslant 0, \qquad L[\psi](x) \leqslant 0 \qquad (0 \leqslant x \leqslant 1) \tag{4.14}$$

hold, which will be assumed in the following.

In order to apply the results of Section 4.1.1, we have to determine a suitable space Y and strong order relations in S and Y. Let

$$Y = \{y \in C_1[0, 1] : y(0) = y(1) = 0\},$$

$$U \succ o \;\Leftrightarrow\; U > 0, \quad \text{for } U \in S; \qquad y \succ o \;\Leftrightarrow\; y \succ o, \quad \text{for } y \in Y, \tag{4.15}$$

i.e., $y \succ o \Leftrightarrow y(x) > 0$ $(0 < x < 1)$, $y'(0) > 0$, $y'(1) < 0$. Then $M_1R \subset Y$, $PS \subset Y$, and (4.1) hold if

$$\varphi(1) = 0, \qquad \psi(0) = 0 \tag{4.16}$$

and

$$\varphi'(1) < 0, \qquad \psi'(0) > 0, \tag{4.17}$$

besides the conditions already imposed on φ, ψ.

In order to prove that $M: R \to S$ is (strictly) pre-inverse-positive, it suffices to show that $M_1: R \to Y$ is (weakly) pre-inverse-positive. M_1 has this property if

$$p = \varphi\psi' - \varphi'\psi \geqslant o, \qquad \varphi''(1) \geqslant 0, \qquad \psi''(0) \geqslant 0. \tag{4.18}$$

To prove this, let $u \in R$ satisfy $u \geqslant o$, $M_1 u \succ o$ but $u \not\succ o$. Then $u(\xi) = 0$ for some $\xi \in (0, 1)$ or $u'(0) = 0$ or $u'(1) = 0$. The first case can be excluded, since $u(\xi) = u'(\xi) = 0$, $u''(\xi) \geqslant 0$, $u'(0) \geqslant 0$, $u'(1) \leqslant 0$ and the inequalities (4.17) contradict $M_1 u(\xi) > 0$. If $u'(0) = 0$ then $u'''(0) \geqslant 0$, since $u \in R$ satisfies also $u''(0) = 0$. Now $(M_1 u)'(0) = -\varphi(0)\psi'(0)u'''(0) + \psi'(0)\varphi''(1)u'(1) \leqslant 0$ under the assumptions imposed on φ and ψ, while $M_1 u \succ 0 \Rightarrow (M_1 u)'(0) > 0$. The case $\xi = 1$ is treated similarly. □

Summarizing the requirements on φ and ψ we obtain the following statement.

Proposition 4.3. *If there exist functions* $\varphi, \psi \in C_4[0, 1]$ *such that* (4.12), (4.14), (4.16), (4.17), *and* (4.18) *hold, then the operator* M *in* (4.5) (*with* $c \in C_0[0, 1]$) *is strictly pre-inverse-positive. If, in addition, a function* $z \in R$ *satisfying* $z \geqslant o$, $Mz > o$ *exists, then* M *is strictly inverse-positive.*

For $c(x) > -\pi^4$ $(0 \leqslant x \leqslant 1)$, the function $z = \sin \pi x$ satisfies the required assumptions. Thus, it remains to choose suitable φ, ψ. We shall give some examples.

1. For the very simple choice $\varphi(x) = 1 - x$, $\psi(x) = x$, (4.14) holds for $c \leqslant o$, while all other assumptions on φ, ψ are satisfied for arbitrary $c \in C_0[0, 1]$.
2. Next we try with polynomials $\varphi(x) = \varepsilon(1 - x) + (1 - \varepsilon)(1 - x)^2 - (1 - x)^4$, $\psi(x) = \varepsilon x + (1 - \varepsilon)x^2 - x^4$. If $c(x) < 96$ $(0 \leqslant x \leqslant 1)$, these functions satisfy all assumptions for sufficiently small $\varepsilon > 0$.
3. Finally, we shall use the limit functions Φ, Ψ defined by (4.7), where now $L[u](x) = u^{\text{iv}}(x) + c(x)u(x)$ with given $c \in C_0[0, 1]$. For $c(x) > -\pi^4$ $(0 \leqslant x \leqslant 1)$ the operator M is positive definite and hence invertible, so that functions Φ, Ψ satisfying (4.7) exist. Choosing $\varphi = \Phi$, $\psi = \Psi$, we see that M *is inverse-positive if* $c(x) > -\pi^4$ $(0 \leqslant x \leqslant 1)$, $\Phi(x) > 0$ *and* $\Psi(x) > 0$ $(0 < x < 1)$, $p \geqslant o$, $\Phi'(1) < 0$, *and* $\Psi'(0) > 0$.

For constant c these inequalities hold if $-\pi^4 < c < c_0$; they are not true, however, for $c = c_0$ (cf. (4.8)).

With the *global reduction* carried out above, we were not able to show that M is inverse-positive for $c(x) \equiv c_0$. We could treat this case by choosing the space Y and the functions φ, ψ differently. Instead, we shall apply *local reduction*, again assuming $c(x) > -\pi^4$ $(0 \leqslant x \leqslant 1)$.

There are several possible ways to proceed. We shall here apply Corollary 4.2a. Let $v \in R$ satisfy $v \geqslant o$ and $v \not> o$. If $v(\xi) = 0$ for some $\xi \in (0, 1)$, we choose

$$\ell U = \int_0^1 g(t)U(t)\,dt \qquad \text{with} \quad g(t) = \Gamma(\xi, t) \quad \text{and} \quad \Gamma \text{ in (4.11).} \qquad (4.19)$$

(In other words: $\ell U = PU(\xi)$.) Using the arguments employed above, we establish that $U > o$ implies $\ell U > 0$, and that $\ell Mv \leqslant 0$, provided the relations (4.12), (4.14), (4.16), (4.18) hold. Here, (4.17) is not needed.

If $v(x) > 0$ $(0 < x < 1)$ and $v'(0) = 0$, we choose $\ell U = \int_0^1 \varphi(t)U(t)\,dt$ with $\varphi \in C_4[0, 1]$. If φ has the properties formulated for φ in (4.12), (4.14), (4.16), and (4.18), we see again that $U > o$ implies $\ell U > 0$ and obtain $\ell Mv \leqslant 0$ by partial integration. The case $v'(1) = 0$ is treated analogously using a function ψ with the properties formulated for ψ in the same formulas.

Now let $\varphi = \Phi_0$, $\psi = \Psi_0$ with Φ_0, Ψ_0 denoting the limit functions which correspond to $c(x) \equiv c_0$. Then, for $-\pi^4 < c(x) \leqslant c_0$ all assumptions imposed on φ and ψ in the course of the above local reduction are satisfied, so that the following result is proved.

Proposition 4.4. *The operator M in* (4.5) *with $c \in C_0[0, 1]$ is strictly inverse-positive if $-\pi^4 < c(x) \leqslant c_0$ $(0 \leqslant x \leqslant 1)$, where $c_0 = 4\kappa^4$ and κ is the smallest solution of* $\tan \kappa = \tanh \kappa$ *satisfying* $\kappa > 0$.

4.2 Z-operators and M-operators

Z-matrices and M-matrices as considered in Section 2 can be characterized in a coordinate-free way (cf. Theorem 2.12 and Corollary 2.12a). This allows one to generalize these terms by introducing Z-operators and M-operators. For such operators a theory analogous to that for matrices can be developed. More generally, we shall investigate certain classes $\mathbf{Z}(W)$ of pairs of operators (M, N), thus obtaining results, for example, on eigenvalue problems of the form $(M + \lambda N)\varphi = o$.

The abstract theory is developed in Section 4.2.1, while Section 4.2.2 contains an application to vector-valued differential operators. The terms introduced are closely related to the concept of irreducibility which will be discussed in Section 4.3.

4.2.1 *Abstract operators*

Let $(R, \leqslant)$ and $(S, \leqslant)$ be Archimedian ordered linear spaces with order cones K and C, respectively, and suppose that $\lessdot$ denotes a strong order in $(S, \leqslant)$ with order cone W. Assume, moreover, that

$$R \subset S, \qquad K = R \cap C, \qquad R \cap C^c \neq \varnothing.$$

Then also $K^c \neq \varnothing$, since $R \cap C^c \subset K^c$. All topological terms in S (and R) shall be defined with respect to the order topology, which is induced by a norm $\| \ \|_w$ with $w \in C^c = C^i$.

We shall investigate certain pairs (M, N) of operators $M \in L(R, S)$, $N \in L(S, S)$, always assuming in the following that

(i) $N \geqslant O$,
(ii) there exists an element $\zeta \in R$ with

$$\zeta \geqslant o, \qquad (M + \lambda N)\zeta \succ o \qquad \text{for some} \quad \lambda > 0. \tag{4.20}$$

In particular, we shall consider the special case where

$$N = I, \qquad \text{i.e.,} \quad Nu = u \quad \text{for} \quad u \in S. \tag{4.21}$$

Here, both conditions (i) and (ii) hold, with (4.20) being satisfied for each $\zeta \in R \cap C^i$. Due to this fact, the theory below becomes somewhat simpler for this special case. Consideration of more general operator pairs is motivated, for example, by the applications in Section 4.2.2.

The class of operator pairs now introduced depends on the strong order $\overset{\text{d}}{\leqslant}$ in S, i.e., on the cone W. Define $\mathbf{Z}(W)$ to be the class of all pairs (M, N) of linear operators $M: R \to S$, $N: S \to S$ satisfying (i), (ii) such that the following condition holds:

For all sufficiently large $\lambda \in \mathbb{R}$ and all $u \in R$

$$\left.\begin{array}{r} u \geqslant o \\ (M + \lambda N)u \succ o \end{array}\right\} \quad \Rightarrow \quad u \succ o. \tag{4.22}$$

A pair $(M, N) \in \mathbf{Z}(W)$ will also be called a Z(W)*-pair*. Moreover, if $R \subset S$ and $Nu = u$ for $u \in R$, as in the special case (4.21), then M is said to be a *Z(W)-operator* if and only if (M, I) is a Z(W)-pair. An M(W)*-operator* is an inverse-positive Z(W)-operator.

For the important case that W equals the cone $C_s = \{U \in S : U \succcurlyeq o\}$ of the strict order in S, we use a simpler notation leaving out the variable W. That is, we define $\mathbf{Z} = \mathbf{Z}(C_s)$ and speak of Z-pairs, Z-operators, and M-operators if $W = C_s$.

According to Theorem 2.12 and Corollary 2.12a, Z-matrices (M-matrices) are special cases of Z-operators (M-operators), and irreducible Z-matrices (irreducible M-matrices) are special cases of Z(C)-operators (M(C)-operators).

Proposition 4.5. *Let (M, N) be a Z(W)-pair. Then $M + \lambda N$ is W-pre-inverse-positive for all $\lambda \in \mathbb{R}$. Moreover, $M + \lambda N$ is inverse-positive for all sufficiently large $\lambda \in \mathbb{R}$.*

The proof is left to the reader.

In the following we shall generalize mainly those results on M-matrices which are related to eigenvalue theory. We consider the eigenvalue problem

$$(M + \lambda N)\varphi = o \tag{4.23}$$

in the real linear space R, so that an eigenvalue is a real number, by definition.

Of course, under the weak assumptions made so far, there need not exist any eigenvalue. Instead of the "largest eigenvalue" of problem (4.23) we shall, therefore, consider a certain quantity κ defined by "spectral properties" of (M, N). For most results, however, it will be explicitly required that κ be an eigenvalue.

For a Z(W)-pair (M, N) let $\kappa = \kappa(M, N) \in \{\mathbb{R}, -\infty, +\infty\}$ denote the infimum of all $\mu \in \mathbb{R}$ such that for all $\lambda > \mu$,

$$\begin{aligned} &\text{(a)} \quad M + \lambda N \text{ is one-to-one,} \\ &\text{(b)} \quad (M + \lambda N)R = S, \\ &\text{(c)} \quad (M + \lambda N)^{-1}N : S \to S \text{ is bounded.} \end{aligned} \tag{4.24}$$

Theorem 4.6. *If $(M, N) \in \mathbf{Z}(W)$ and $\kappa < \infty$, then the operator $M + \lambda N$ is inverse-positive for each $\lambda > \kappa$.*

Proof. We apply Theorem 1.8 to the set $\mathscr{M}$ of all operators $M + \lambda N$ with $\lambda > \kappa$. Obviously, conditions (i)–(iii) of the theorem are satisfied (observe Proposition 4.5). Because of the boundedness of the operators $(M + \lambda N)^{-1}N$, one shows by standard estimations that for fixed $r \succ o$ in S the solutions $z(\lambda)$ of $(M + \lambda N)z(\lambda) = r$ depend continuously on λ. Consequently, property (iv) of Theorem 1.8 holds also.

Finally, according to Proposition 4.5 there exists an operator in $\mathscr{M}$ which is inverse-positive. Thus, by Theorem 1.8, all operators in $\mathscr{M}$ are inverse-positive. □

Since we shall assume in the theorems below that κ is an eigenvalue of problem (4.23), we state the following sufficient condition.

Proposition 4.7. *Suppose that S is a Banach space, $\kappa < \infty$, and that for some $\mu > \kappa$ the operator $T = (M + \mu N)^{-1}N : S \to S$ is compact and has a spectral radius $\rho(T) > 0$.*

Then κ is finite and there exists a $\varphi > o$ in R such that $(M + \kappa N)\varphi = o$.

Proof. The positive and compact operator T has an eigenelement $\varphi > o$ corresponding to the eigenvalue $\Lambda = \rho(T) > 0$ (cf. Theorem I.3.3). A formal calculation yields $(M + \lambda_0 N)\varphi = o$ with $\lambda_0 = \mu - \Lambda^{-1}$. By definition of κ, $\lambda_0 \leqslant \kappa$. Thus, to prove $\lambda_0 = \kappa$, it suffices to show that properties (a)–(c) in (4.24) hold for $\lambda_0 < \lambda < \mu$. For such a value λ, however, an equation $(M + \lambda N)u = v$ is equivalent to $u - Au = w$ with $A = (\mu - \lambda)T$ and $w = (M + \mu N)^{-1}v$. Since $\rho(A) = (\mu - \lambda)\Lambda < 1$, the latter equation has a unique solution u for each $w \in R$. Consequently, $M + \lambda N$ is one-to-one and maps R onto S. Moreover, the operator $(M + \lambda N)^{-1}N = (\mu - \lambda)^{-1}(I - A)^{-1}A$ is bounded. □

A series of equivalence statements similar to those for matrices can be derived, if κ is an eigenvalue.

Theorem 4.8. *Let $(M, N) \in \mathbf{Z}(W)$ and suppose that $\kappa = \kappa(M, N)$ is an eigenvalue of problem* (4.23).

Then the following five properties are equivalent for each $\lambda \in \mathbb{R}$:

(i) *$M + \lambda N$ is inverse-positive and $(M + \lambda N)v \succ o$ for some $v \in R$.*
(ii) *$u \in R$, $(M + \lambda N)u \geqslant o \Rightarrow u \geqslant o$; and $(M + \lambda N)v \succ o$ for some $v \in R$.*
(iii) *$z \geqslant o$, $(M + \lambda N)z \succ o$ for some $z \in R$.*
(iv) *$z \succ o$, $(M + \lambda N)z \succ 0$ for some $z \in R$.*
(v) *$\lambda > \kappa(M, N)$.*

Proof. (i) ⇒ (ii), as a consequence of Proposition 4.5 (cf. Corollary 1.2a). (ii) ⇒ (iii) is obvious. (iii) ⇒ (i) follows from the monotonicity theorem and Proposition 4.5. (iii) ⇒ (iv) by Proposition 4.5; (iv) ⇒ (iii) is obvious.

(ii) ⇒ (v): If $\lambda \leqslant \kappa$, then $v \succ o$ and $(M + \kappa N)v \succ o$, so that $M + \kappa N$ is inverse-positive by Theorem 1.2. This contradicts κ being an eigenvalue.

(v) ⇒ (iii): Since the range of $M + \lambda N$ equals S and $r \succ o$ for some $r \in S$, there exists a $z \in R$ with $(M + \lambda N)z = r \succ o$. This element satisfies $z \geqslant o$ because $M + \lambda N$ is inverse-positive by Theorem 4.6. □

Corollary 4.8a. *Under the assumptions of the preceding theorem, the following estimates for κ hold.*

(i) *If $(M + \lambda N)z \leqslant o$ for some $z \succ o$, then $\lambda \leqslant \kappa(M, N)$.*
(ii) *If $(M + \lambda N)z \geqslant o$, $Nz \succ o$ for some $z \succ o$, then $\lambda \geqslant \kappa(M, N)$.*

The proof is left to the reader. (Compare the proof of Theorem 2.15.)

We are particularly interested in conditions such that $\lambda > \kappa$ is equivalent to the inverse-positivity of $M + \lambda N$. The following two theorems provide such conditions.

Theorem 4.9. *Suppose that $(M, N) \in \mathbf{Z}(W)$, $\kappa = \kappa(M, N)$ is an eigenvalue of problem* (4.23), *and $(M + \lambda N)R \cap W \neq \{o\}$ for each (real) λ which is not an eigenvalue of problem* (4.23).

Then the operator $M + \lambda N$ is inverse-positive if and only if $\lambda > \kappa(M, N)$. Moreover,

$$\kappa(M, N) = \inf\{\lambda \in \mathbb{R} : (M + \lambda N)z \succ o \text{ for some } z \succ o\}. \tag{4.25}$$

Proof. By Theorem 4.6, $M + \lambda N$ is inverse-positive for $\lambda > \kappa$. On the other hand, if $M + \lambda N$ is inverse-positive, then λ is not an eigenvalue, so that property (i) of Theorem 4.8 holds; $\lambda > \kappa$ follows from the equivalence of (i) and (v). Finally, the equivalence of (iv) and (v) yields (4.25). □

Theorem 4.10. *Suppose that* $(M, N) \in \mathbf{Z}(W)$, κ *is finite, and*

$$(M + \kappa N)\varphi = o \qquad \text{for some} \quad \varphi > o. \tag{4.26}$$

Then $M + \lambda N$ *is inverse-positive if and only if* $\lambda > \kappa$. *Moreover,* (4.25) *holds.*

Proof. For $\lambda > \kappa$, the inverse-positivity of $M + \lambda N$ is derived from Theorem 4.6. For $\lambda \leqslant \kappa$, the operator $M + \lambda N$ cannot be inverse-positive, since $(M + \lambda N)(-\varphi) \geqslant o$, but $-\varphi < o$. Statement (4.25) follows from the equivalence of properties (iv) and (v) in Theorem 4.8. □

In the eigenvalue theory of positive operators one is interested in conditions such that there exists a *strictly* positive eigenelement φ corresponding to the largest (real) eigenvalue. We shall here show how $\varphi > o \Rightarrow \varphi \succ o$ can be derived for φ in (4.26), if (M, N) belongs to the class $\mathbf{Z}(C)$. For such operators M, N the results of Section 1.5 on noninvertible operators can be applied to $M + \lambda N$.

Proposition 4.11. *Suppose that* $(M, N) \in \mathbf{Z}(C)$ *and* $Nu > o$ *for all* $u > o$ *in* R. *Then for each* $\lambda \in \mathbb{R}$ *and all* $u \in R$

$$u > o, \quad (M + \lambda N)u \geqslant o \quad \Rightarrow \quad u \succ o.$$

Proof. The inequalities $u > o$ and $(M + \lambda N)u \geqslant o$ imply $(M + \lambda' N)u > o$, for all $\lambda' > \lambda$. Thus, $u \succ o$ follows, since $(M, N) \in \mathbf{Z}(C)$. □

Theorem 4.12. *Suppose that* $(M, N) \in \mathbf{Z}(C)$, $Nu > o$ *for all* $u > o$ *in* R, κ *is finite, and* $(M + \kappa N)\varphi = o$ *for some* $\varphi > o$.
Then κ *is a simple eigenvalue of problem* (4.23) *and* $\varphi \succ o$. *Moreover,*

$$\kappa = \min\{\lambda \in \mathbb{R} : (M + \lambda N)z \geqslant o \text{ for some } z \succ o\}$$
$$\kappa = \max\{\lambda \in \mathbb{R} : (M + \lambda N)z \leqslant o \text{ for some } z \succ o\}.$$

Proof. We apply Theorem 1.9 to $M + \kappa N$ (in place of M). According to Proposition 4.11, assumption I is satisfied. Moreover, assumption II holds with $z = \varphi$. Thus, κ is simple and $\varphi \succ o$. The preceding formulas for κ then follow from Corollary 4.8a. □

Finally, we shall describe certain sufficient conditions for (M, N) being a Z(W)-pair. These conditions are particularly useful to prove that $(M, N) \in \mathbf{Z}$. We assume that $\mathscr{L}$ is a given set of linear functionals $\ell > o$ on R which describes the cone K completely (see Section I.1.3) and that $\mathscr{F}$ is a nonempty set of linear functionals $\ell \geqslant o$ on S such that

$$U \succ o \quad \Leftrightarrow \quad \begin{cases} U \geqslant o \\ \ell U > 0 \qquad \text{for all} \quad \ell \in \mathscr{F}. \end{cases} \tag{4.27}$$

Proposition 4.13a. *If to each $\ell \in \mathscr{L}$ there exists an $\ell = \ell_\ell \in \mathscr{F}$ such that for all $u \in R$*

$$\left.\begin{aligned} u \geqslant o \\ \ell u = 0 \end{aligned}\right\} \quad \Rightarrow \quad \begin{cases} \ell Mu \leqslant 0 \\ \ell Nu = 0, \end{cases} \tag{4.28}$$

then (4.22) *holds for all $\lambda \in \mathbb{R}$ and $u \in R$.*

The proof is obvious.

Formula (4.27) may be used to *define* the strong order $\prec$ by choosing $\mathscr{F}$ to be the set of all $\ell = \ell_\ell$ ($\ell \in \mathscr{L}$) which occur in (4.28).

Let us consider a special case, where it is convenient to redefine some terms. Suppose that (4.21) holds and denote by $\mathscr{L}$ a set of linear functionals $\ell > o$ on S, but use the letter $\mathscr{L}$, for simplicity, also for the set of the restrictions of these functionals to R. Then, as before, assume that $\mathscr{L}$ describes the cone in R completely. For $\ell = \ell$ property (4.28) is here equivalent to

$$u \geqslant o, \quad \ell u = 0 \quad \Rightarrow \quad \ell Mu \leqslant 0. \tag{4.29}$$

A linear operator M such that this implication holds for all $u \in R$ and all $\ell \in \mathscr{L}$ will be called *$\mathscr{L}$-quasi-antitone*. (Compare the analogous definition for nonlinear operators in Section IV.2.1.3.) For such operators we obtain the following statement under the above assumptions.

Proposition 4.13b. *An $\mathscr{L}$-quasi-antitone operator M is a Z(W)-operator, where the cone W corresponds to the relation $U \succ o$ defined by $U \geqslant o$ and $\ell U > 0$ for all $\ell \in \mathscr{L}$. If, in particular, the set $\mathscr{L}$ also describes the cone in S completely, then M is a Z-operator.*

4.2.2 Quasi-antitone vector-valued differential operators

Many results of Section 3 can be carried over to certain vector-valued differential operators. (See, for example, Section IV.2.3.) Here we shall apply the theory of the preceding section to such operators under the condition that they have sufficiently smooth coefficients.

Let $R = C_2^n[0, 1]$ and S be the space of all $U \in \mathbb{R}^n[0, 1]$ such that the restriction of U to $(0, 1)$ can be extended to a function in $C_0^n[0, 1]$. In both spaces, $\leqslant$ shall denote the natural order relation. Then $U \in S$ is strictly positive if and only if $\inf\{U_i(x) : 0 \leqslant x \leqslant 1; i = 1, \ldots, n\} > 0$.

We consider operators $M \in L[R, S]$, $N \in L[S, S]$ given by

$$Mu(x) = \begin{cases} L[u](x) & \text{for } 0 < x < 1 \\ B_0[u] & \text{for } x = 0 \\ B_1[u] & \text{for } x = 1, \end{cases}$$
$$Nu(x) = \begin{cases} g(x)u(x) & \text{for } 0 < x < 1 \\ g^0 u(0) & \text{for } x = 0 \\ g^1 u(1) & \text{for } x = 1, \end{cases} \tag{4.30}$$

where

$$L[u](x) = -a(x)u'' + b(x)u' + c(x)u,$$
$$B_0[u] = -\alpha^0 u'(0) + \gamma^0 u(0), \qquad B_1[u] = \alpha^1 u'(1) + \gamma^1 u(1),$$

and the occurring quantities have the following meaning.

For each $x \in [0, 1]$, $a(x)$, $b(x)$, and $g(x)$ are diagonal matrices (in $\mathbb{R}^{n,n}$) with diagonal elements $a_i(x)$, $b_i(x)$, and $g_i(x)$, respectively, such that $a_i(x) > 0$, $g_i(x) > 0$ $(i = 1, 2, \ldots, n)$, and $c(x)$ is a Z-matrix (in $\mathbb{R}^{n,n}$). These matrices are supposed to depend continuously on x. Moreover, α^0, α^1, g^0, g^1 are positive diagonal matrices (in $\mathbb{R}^{n,n}$) with diagonal elements α_i^0, g_i^0, α_i^1, g_i^1 $(\geqslant 0)$ and $\gamma^0 = (\gamma_{ik}^0)$, $\gamma^1 = (\gamma_{ik}^1)$ are Z-matrices (in $\mathbb{R}^{n,n}$). In the course of the proof of Proposition 4.14 we shall see that for each $\lambda \in \mathbb{R}$ the operator $M + \lambda N$ is $\mathscr{L}$-quasi-antitone with $\mathscr{L}$ being the set of all point functionals $\ell_{x,i} u = u_i(x)$ $(0 \leqslant x \leqslant 1; i = 1, 2, \ldots, n)$.

We shall furthermore assume that for $i = 1, 2, \ldots, n$

$$g_i^0 > 0 \quad \text{if } \sum_k \gamma_{ik}^0 \leqslant 0, \qquad g_i^1 > 0 \quad \text{if } \sum_k \gamma_{ik}^1 \leqslant 0. \tag{4.31}$$

Then we have $(M + \lambda N)\zeta \succ o$ for $\zeta = (\zeta_i(x))$ with all components $\zeta_i(x) \equiv 1$ and sufficiently large $\lambda > 0$. Consequently, assumption (ii) at the beginning of Section 4.2.1 is satisfied for an arbitrary strong order relation $\leqslant$ in S, while $N \geqslant O$ in (i) is obvious.

Here, we consider more general operators N than $Nu = u$, in order to obtain results on more general classes of eigenvalue problems $(M + \lambda N)\varphi = o$. For example, if $g^0 = g^1 = O$, i.e., $Nu(0) = Nu(1) = o$, the eigenvalue λ does not occur in the boundary conditions.

We are interested in sufficient conditions such that (M, N) is a $Z(W)$-pair corresponding to a strong ordering $\leqslant$ with cone W, and consider first the case that $W = C_s$, i.e., $U \succ o \Leftrightarrow U \succ o$.

Proposition 4.14. *Each pair (M, N) of the above kind is a Z-pair.*

Proof. Using the methods in the proof of Proposition 3.1, one verifies that (4.28) holds for $\ell u = u_j(\xi)$, $\ell U = U_j(\xi)$, $\xi \in [0, 1]$, $j = 1, 2, \ldots, n$. □

If the operator $M + \lambda N$ is one-to-one for some λ, then the inverse $(M + \lambda N)^{-1}$ is given by a vector-valued integral operator with the Green matrix as its kernel. One shows that this inverse is defined on all of S and that $(M + \lambda N)^{-1}N \in L(S, S)$ is a compact operator with spectral radius >0 (the order topology being given by $\|u\| = \sup\{|u_i(x)| : 0 \leqslant x \leqslant 1;\ i = 1, 2, \ldots, n\}$). Hence, by Proposition 4.7, $\kappa = \kappa(M, N)$ is the largest (real) eigenvalue of the problem

$$\begin{aligned} L[\varphi](x) + \lambda g(x)\varphi = o \qquad (0 < x < 1), \\ B_0[\varphi] + \lambda g^0\varphi = o, \qquad B_1[\varphi] + \lambda g^1\varphi = o. \end{aligned} \tag{4.32}$$

Moreover, for each λ which is not an eigenvalue, the range S of $M + \lambda N$ contains a strictly positive element.

Thus, we can apply many results of the preceding section. For example, the five properties in Theorem 4.8 are equivalent for the differential operators M, N and $U \succcurlyeq o \Leftrightarrow U \succ o$. Moreover, the estimates in Corollary 4.8a can be applied, and (4.25) holds. In particular, we obtain

Theorem 4.15. *The differential operator M is inverse-positive if and only if each (real) eigenvalue λ of the eigenvalue problem* (4.32) *satisfies $\lambda < 0$.*

The following simple example will explain the theory.

Example 4.3. Let $n = 2$ and

$$L[u](x) = -u''(x), \qquad B_0[u] = \gamma u(0), \qquad B_1[u] = \gamma u(1),$$

$$\gamma = \begin{pmatrix} 1 & -\delta \\ -\delta & 1 \end{pmatrix} \quad \text{with} \quad \delta > 0, \tag{4.33}$$

$$g(x) = I_2, \qquad g^0 = g^1 = \varepsilon I_2 \qquad \text{with} \quad \varepsilon \geqslant 0.$$

(a) If $\delta < 1$, we may choose $\varepsilon = 0$ (observe (4.31)). For this case, the boundary conditions of the eigenvalue problem (4.32) do not depend on λ, and the eigenvalues are $\lambda_\nu = -\nu^2\pi^2 < 0$ $(\nu = 1, 2, \ldots)$. Thus, the corresponding operator M is inverse-positive.

(b) For arbitrary $\delta > 0$, we may choose $\varepsilon = 1$. Then the numbers λ_ν are again eigenvalues of the corresponding problem (4.32). However, $\lambda = -1 - \delta$ and $\lambda = -1 + \delta$ are also eigenvalues, so that the largest eigenvalue is $\kappa = -1 + \delta$. Thus, the *operator M given by* (4.30), (4.33) *is inverse-positive if and only if $\delta < 1$.* □

Using the results of Sections 3.6 and 3.7, one can also derive sufficient conditions such that $(M, N) \in \mathbf{Z}(W)$ for W different from C_s. For example,

W may be given by

$$U \succ o \quad \Leftrightarrow \quad \begin{cases} U > o \\ U_i(0) > 0 & \text{if } \alpha_i^0 = 0 \quad (i = 1, 2, \ldots, n) \\ U_i(1) > 0 & \text{if } \alpha_i^1 = 0 \quad (i = 1, 2, \ldots, n). \end{cases} \tag{4.34}$$

In the following result we restrict ourselves to the special case in which the cone W so defined equals C.

Proposition 4.16. *Under the assumption*

$$\alpha_i^0 > 0, \qquad \alpha_i^1 > 0 \qquad (i = 1, 2, \ldots, n) \tag{4.35}$$

the operator pair (M, N) *belongs to* $\mathbf{Z}(C)$, *if the matrix* $\gamma^0 + \gamma^1 + \int_0^1 c(x)\, dx$ *is irreducible.*

Proof. We verify (4.22) for the case $U \succ o \Leftrightarrow U > o$. Let $u \in R$ and $\lambda > 0$ satisfy $u \geqslant o$ and $(M + \lambda N)u > o$. Then we have also $(\tilde{M} + \lambda N)u > o$ where $\tilde{M}$ is obtained from M by replacing all off-diagonal elements of $c(x)$, γ^0, and γ^1 by 0. Since the ith component of $(\tilde{M} + \lambda N)u$ depends only on u_i, a linear operator M_i on $C_2[0, 1]$ is defined by $M_i u_i = ((\tilde{M} + \lambda N)u)_i$. Obviously, $M_i u_i(x) \geqslant 0$ $(0 \leqslant x \leqslant 1)$. When we apply Theorem 3.16 and Corollary 3.16a to this operator M_i for sufficiently large λ, we see that either $u_i(x) = 0$ $(0 \leqslant x \leqslant 1)$ or $u_i(x) > 0$ $(0 \leqslant x \leqslant 1)$.

Because of $(\tilde{M} + \lambda N)u > o$, not all components u_i vanish identically. If one of them vanishes identically, then there exist indices j and k such that $u_j(x) = 0$, $u_k(x) > 0$ $(0 \leqslant x \leqslant 1)$ and, moreover, $\gamma_{ik}^0 < 0$, or $\gamma_{ik}^1 < 0$, or $c_{jk}(\xi) < 0$ for some $\xi \in (0, 1)$, since $\gamma^0 + \gamma^1 + \int_0^1 c(x)\, dx$ is irreducible. The above inequalities, however, contradict $((M + \lambda N)u)_j(x) \geqslant 0$ $(0 \leqslant x \leqslant 1)$. Consequently, $u \succ o$. $\square$

If the assumptions of Proposition 4.16 *are satisfied, then the statements of Theorem* 4.12 *are true for the pair of operators* M, N *in* (4.30).

Problems

4.3 Find sufficient conditions such that the operators in (4.30) constitute a pair $(M, N) \in \mathbf{Z}(W)$ with W given by (4.34), without assuming (4.35).

4.4 Let $R = \{u \in C_2^n[0, 1] : u(0) = u(1) = 0\}$, $S = C_0^n[0, 1]$, and $Mu(x) = L[u](x)$, $Nu(x) = g(x)u(x)$ with L and g as in (4.30). Show that $(M, N) \in \mathbf{Z}$. (See Section 3.5.)

4.3 Operators $\lambda I - T$ with positive T

Each Z-matrix M can be written in the form $\lambda I - T$ with a matrix $T \geqslant O$. Thus, operators of this form may be considered as generalizations of Z-matrices. Indeed, under appropriate assumptions, they are Z-operators as defined in Section 4.2. We shall here investigate this special type of Z-operators in somewhat more detail. In particular, conditions on inverse-positivity which involve the spectral radius of T will be discussed. This then leads to results on iterative procedures. Moreover, we shall introduce the concept of W-irreducible positive operators, which generalizes the notion of an irreducible matrix in several ways.

Let $(R, \leqslant, K)$ be an Archimedian ordered linear space which contains an element $\zeta \succ o$. Suppose that R is a Banach space with respect to the norm $\| \ \|_\zeta$. Moreover, let I denote the identity operator in R and $T: R \to R$ a positive linear operator. This operator is bounded, according to Theorem I.1.10(ii). Define $\rho(T)$ to be the spectral radius of T and $\sigma(T)$ to be the supremum of the (real) spectrum of T. If $\sigma(T) > -\infty$, this number belongs to the spectrum. Moreover, $\sigma(T) \leqslant \rho(T)$ by definition of these quantities. (See Section I.3.1.)

4.3.1 Inverse-positive operators $\lambda I - T$

We shall apply the theory of Section 4.2.1 to the operators $M = -T$ and $N = I$. The space $(S, \leqslant)$ occurring in that section will be identified with $(R, \leqslant)$, so that, in particular, $C = K$. Moreover, we choose here $W = K_s$, i.e., $u \gtrdot 0 \Leftrightarrow u \succ o$.

Proposition 4.17. *For each $\lambda \in \mathbb{R}$, $\lambda I - T$ is a Z-operator and $\sigma(T) = \kappa(-T, I)$.*

This statement, which the reader can easily prove, enables one to apply the theory in Section 4.2.1. We shall here formulate explicitly only a few results which involve the spectral radius $\rho(T)$. Further results, such as estimates of the spectral radius, may then immediately be obtained from the theorems in Section 4.2.1.

Theorem 4.18. (i) $\rho(T) = \sigma(T)$.
(ii) *For $\lambda > \rho(T)$ the operator $\lambda I - T$ is inverse-positive.*
(iii) *$\lambda > \rho(T)$ if and only if $z \geqslant o$, $\lambda z \succ Tz$ for some $z \in R$.*

Proof. For $\lambda > \sigma(T) = \kappa(-T, I)$ the operator $\lambda I - T$ is inverse-positive according to Theorem 4.6, and its range equals R, by the definition of $\sigma(T)$. Consequently, for any $\zeta \succ o$ in R the equation $\lambda z - Tz = \zeta$ has a solution $z \geqslant o$. On the other hand, if $z \geqslant o$ and $\lambda z \succ Tz$, then even $z \succ o$ and hence $\sigma(T) \leqslant \rho(T) \leqslant \|T\|_z < \lambda$. Combining these two arguments we see that $\rho(T) = \sigma(T)$, so that all statements are proved. □

As a simple consequence of statement (iii) we obtain

Corollary 4.18a. *$\lambda > \rho(T)$ if and only if $\|T\|_z < \lambda$ for some $z \succ o$ in R.*

The following example will show that, without further assumptions on T, $\lambda > \rho(T)$ is not necessary for the inverse-positivity of $\lambda I - T$.

Example 4.4. Let $R = l_\infty$ and $T = (t_{ik})$ be the infinite tridiagonal matrix with $t_{i,i+1} = t_{i+1,i} = 1$ and $t_{ik} = 0$ otherwise ($i, k = 1, 2, \ldots$). (We identify such matrices with operators from l_∞ into l_∞.)

(a) For $\lambda > 2$, the inequalities $z \geqslant o$ and $\lambda z \succ Tz$ hold with $z = e = (e_i)$, $e_i = 1$ ($i = 1, 2, \ldots$). Thus $\rho(T) \leqslant 2$, and hence $\lambda I - T$ is inverse-positive.

(b) For $\lambda = 2$, one verifies that $(2I - T)^{-1} = (\mu_{ik})$ with $\mu_{ik} = i$ for $i \leqslant k$, and $\mu_{ik} = k$ for $i > k$. Obviously, this inverse is positive, so that $2I - T$ is inverse-positive. However, the inverse operator is not bounded (and not defined on all of R). Consequently, $\lambda = 2$ belongs to the spectrum of T and $\rho(T) = 2$. According to part (iii) of Theorem 4.18 the range of $2I - T$ then does not contain any element $\succ o$. Observe, however, that $2z - Tz > o$ for $z = e \succ o$. □

A behavior as in the example can, for instance, be excluded for compact operators T. We derive from Proposition 4.7 and Theorem 4.10:

Theorem 4.19. *For compact T, the operator $\lambda I - T$ is inverse-positive if and only if $\lambda > \rho(T)$.*

Example 4.5. Define an operator T on $R = C_0^n[0, 1]$ by

$$(Tu)(x) = \int_0^1 \mathscr{G}(x, \xi) u(\xi)\, d\xi \qquad (0 \leqslant x \leqslant 1), \tag{4.36}$$

in vector notation. We assume that each element $\mathscr{G}_{ik}(x, \xi)$ of the $n \times n$ matrix $\mathscr{G}(x, \xi) = (\mathscr{G}_{ik}(x, \xi))$ is positive and integrable on $[0, 1] \times [0, 1]$, and that for each $\varepsilon > 0$ a $\delta(\varepsilon)$ exists such that

$$\int_0^1 |\mathscr{G}_{ik}(x, \xi) - \mathscr{G}_{ik}(y, \xi)|\, d\xi < \varepsilon \qquad \text{whenever} \quad |x - y| < \delta(\varepsilon),$$
$$0 \leqslant x, y \leqslant 1.$$

Then T maps R into R and is compact, so that Theorems 4.18 and 4.19 can be applied.

If, in particular,

$$\int_0^1 \mathscr{G}(x, \xi)\, d\xi \leqslant \Gamma \qquad (0 \leqslant x \leqslant 1) \tag{4.37}$$

for a (constant) $n \times n$ matrix Γ, then $\lambda I - T$ *is inverse-positive if the matrix* $\lambda I_n - \Gamma$ *is inverse-positive*. To prove this, one verifies $\lambda z \succ Tz$ for $z = (z_i)$

with $z_i(x) \equiv \zeta_i$ and a vector $\zeta \in \mathbb{R}^n$ satisfying $\zeta \geqslant o$, $\lambda\zeta_i > (\Gamma\zeta)_i$ $(i = 1, 2, \ldots, n)$. Such a vector ζ exists, according to Theorem 2.6. □

Example 4.6. Consider the operator (4.36) for $n = 1$ and $\mathscr{G}(x, \xi) = \mathscr{G}(\xi, x) = x(1 - \xi)$ $(0 \leqslant x \leqslant \xi \leqslant 1)$. For $\lambda > \pi^{-2}$ the function $z(x) = \cos(\pi - \varepsilon)(x - \frac{1}{2})$ satisfies $z \geqslant o$ and $\lambda z \succ Tz$, if ε is sufficiently small. Consequently, $\lambda I - T$ is inverse-positive for $\lambda > \pi^{-2}$, and $\rho(T) \leqslant \pi^{-2}$. Since, on the other hand, $T\varphi = \lambda\varphi$ for $\lambda = \pi^{-2}$, $\varphi(x) = \sin \pi x$, we have $\rho(T) = \pi^{-2}$.

Using (4.37) with $\Gamma = \frac{1}{8}$, the inverse-positivity of $\lambda I - T$ can be established for $\lambda > \frac{1}{8}$. □

Example 4.7. Consider the operator (4.36) under the additional assumptions that all functions $\mathscr{G}_{ik}(x, \xi)$ are bounded above by a constant c and that $\mathscr{G}(x, \xi) = 0$ for $x \leqslant \xi$, so that

$$Tu(x) = \int_0^x \mathscr{G}(x, \xi)u(\xi)\, d\xi.$$

Now for each $\lambda > 0$ a constant $\gamma > 0$ exists such that $z \geqslant o$, $\lambda z \succ Tz$ for $z = (z_i)$, $z_i(x) = \exp \gamma x$ $(i = 1, 2, \ldots, n)$. Consequently, $\rho(T) = 0$, and $\lambda I - T$ is inverse-positive if and only if $\lambda > 0$. □

4.3.2 Iterative procedures

Most results of Section 2.5 on iterative procedures in finite-dimensional spaces can be proved to hold also in certain Banach spaces. The preceding section provided the main tools for such generalizations. Since most of these generalizations are straightforward, we shall only state a few basic results.

Suppose that $(R, \leqslant)$ and $(S, \leqslant)$ have the properties required in the first paragraph of Section 4.2.1, let $\zeta \in C^c$, and assume that $(S, \leqslant)$ is an ordered Banach space (with respect to the order topology induced by $\| \ \|_\zeta$). The results of the preceding section will be applied to an operator $T: S \to S$ (instead of $T: R \to R$).

We consider an iterative procedure.

$$Au_{v+1} = Bu_v + s \qquad (v = 0, 1, 2, \ldots) \tag{4.38}$$

for solving an equation $Mu = s$ with given $s \in S$. Here $M, A \in L(R, S)$ and $B \in L(S, S)$ are operators such that A is inverse-positive, $AR = S$, $A^{-1}B \in L(S, S)$ is positive and compact, $MR = S$ if M is one-to-one.

The procedure (4.38), which can be written as

$$u_{v+1} = Tu_v + r \qquad \text{with} \quad T = A^{-1}B : S \to S, \qquad r = A^{-1}s,$$

will be called *convergent* if $\rho(T) = \rho(A^{-1}B) < 1$.

Theorem 4.20. *Under the preceding assumptions, the following four properties are equivalent:*

(i) *The procedure* (4.38) *is convergent.*
(ii) $\|T\|_z < 1$ *for some* $z \succ o$ *in* S.
(iii) $M = A - B$ *is inverse-positive.*
(iv) $z \geqslant o$, $Mz \succ o$ *for some* $z \in R$.

Proof. We first recall that $U \succ o \Rightarrow A^{-1}U \succ o$ for $U \in S$ (cf. Corollary 1.2a).

(i) $\Leftrightarrow$ (ii) as a consequence of Corollary 4.18a.

(i) $\Rightarrow$ (iii): If $\rho(T) < 1$, then $I - T$ is inverse-positive (cf. Theorem 4.18(ii)). Consequently, $Mu \geqslant o \Rightarrow (I - T)u \geqslant o \Rightarrow u \geqslant o$.

(iii) $\Rightarrow$ (i): If M is inverse-positive, then $Mz = \zeta \succ o$ for some $z \geqslant o$ and hence $(I - T)z = A^{-1}\zeta \succ o$, so that $\rho(T) < 1$ (cf. Theorem 4.18(iii)).

(iii) $\Rightarrow$ (iv): Choose z satisfying $Mz = \zeta$.

(iv) $\Rightarrow$ (iii): The inequalities $u \geqslant o$ and $Mu \succ o$ imply $u \succ A^{-1}Bu \geqslant o$, so that M is weakly pre-inverse-positive. Then apply Theorem 1.2. □

Example 4.8. Let M and N be the differential operators in Section 4.2.2 and consider the splitting $M = A - B$ with $A = M + pN$, $B = pN$ with $p > 0$ so large that A is inverse-positive. Since A^{-1} is defined on all of S and the integral operator $A^{-1}B$ is compact, the preceding theorem yields that *for this splitting the iterative procedure* (4.38) *converges if and only if M is inverse-positive.* □

4.3.3 Irreducible operators

This section is concerned with operators $T: R \to R$ which have the properties described at the beginning of Section 4.3 and, in addition, are irreducible or W-irreducible in a sense defined below.

The notion of an irreducible matrix has been generalized to operators in Banach spaces by several authors in different ways. We do not intend to give a survey of these results here. However, we shall indicate how some of the most important definitions fit into our theory.

In the space $(R, \leqslant, K)$ let a further linear order $\overset{\text{\tiny W}}{\leqslant}$ be given which is stronger than $\leqslant$ and has an order cone W. A (positive linear) operator $T: R \to R$ is said to be *W-irreducible* if for all $u \in R$ and all sufficiently large $\lambda \in \mathbb{R}$

$$\left.\begin{aligned} u &\geqslant o \\ \lambda u - Tu &\overset{\text{\tiny W}}{>} o \end{aligned}\right\} \quad \Rightarrow \quad u \succ o. \tag{4.39}$$

In particular, a K-irreducible operator is simply called *irreducible*. For such an operator T the preceding implication holds with $>$ in place of $\overset{\text{\tiny W}}{>}$.

(Compare the characterization of an irreducible matrix in Corollary 2.12b.)

It is unfortunate that the negating term "irreducible" is generally used for a property with so many "positive" aspects. Neglecting, however, linguistic problems, we have introduced the term W-irreducible which allows one to describe different "degrees of irreducibility." The extreme cases are obtained for $W = K$ and $W = K_s$. Each (positive linear) operator is K_s-irreducible. (Concerning this terminology, see also the Notes.)

As in Section 4.3.1 we can again apply the results of Section 4.2.1 to the operator pair $(M, N) = (-T, I)$. Obviously, for each $\lambda \in \mathbb{R}$, T is W-irreducible if and only if $\lambda I - T$ is a Z(W)-operator, and T is irreducible if and only if $\lambda I - T$ is a Z(K)-operator. Therefore, we derive from Proposition 4.17, Theorem 4.18(i), and Theorem 4.12:

Theorem 4.21. *If T is irreducible and has an eigenelement $\varphi > o$ corresponding to the eigenvalue $\lambda = \rho(T)$, then this eigenvalue is simple and $\varphi \succ o$.*

An eigenelement $\varphi > o$ corresponding to $\lambda = \rho(T)$ exists, for example, if T is compact and $\rho(T) > 0$ (cf. Theorem I.3.3).

The next two theorems provide a series of necessary and sufficient conditions for W-irreducibility and irreducibility. Obviously, each of the equivalent conditions can also be used as a definition.

Let $\mathscr{L}$ be a set of positive linear functionals on R which describes K completely. Moreover, denote by $R_\lambda = R_\lambda(T)$ for $\lambda > \rho(T)$ the resolvent operator $R_\lambda(T) = (\lambda I - T)^{-1} = \sum_{j=0}^{\infty} \lambda^{-(j+1)} T^j$, the series being convergent with respect to the operator norm.

Theorem 4.22. *The following five properties are equivalent.*

(i) *T is W-irreducible.*
(ii) *There exists a $\lambda > \rho(T)$ such that* (4.39) *holds for all $u \in R$.*
(iii) *$v \gtrdot o \Rightarrow R_\lambda(T)v \succ o$, for each $\lambda > \rho(T)$ and all $v \in R$.*
(iv) *There exists a $\lambda > \rho(T)$ such that $v \gtrdot o \Rightarrow R_\lambda(T)v \succ o$, for all $v \in R$.*
(v) *To each $v \gtrdot o$ in R and each $\ell \in \mathscr{L}$ there exists an integer $m \geqslant 0$ such that $\ell T^m v > 0$.*

Proof. (i) $\Rightarrow$ (ii) and (iii) $\Rightarrow$ (iv) are obvious.

(iv) $\Rightarrow$ (v) $\Rightarrow$ (iii) is derived from $\ell R_\lambda(T)v = \sum_{j=0}^{\infty} \lambda^{-(j+1)} \ell T^j v$.

(ii) $\Rightarrow$ (iii): For each $\lambda > \rho(T)$ and each $v \gtrdot o$, $u = R_\lambda(T)v$ satisfies $u \geqslant o$ and $\lambda u - Tu \gtrdot 0$, so that $u \succ o$.

(iii) $\Rightarrow$ (i): Let $\lambda > \rho(T)$, $u \geqslant o$ and $(\lambda I - T)u = v \gtrdot o$. Then $u = R_\lambda(T)v \succ o$. $\square$

Theorem 4.23. *The following four properties are equivalent.*

(α) *T is irreducible.*
(β) *$v > o \Rightarrow TR_\lambda(T)v \succ o$, for each $\lambda > \rho(T)$ and all $v \in R$.*

(γ) *There exists a* $\lambda > \rho(T)$ *such that* $v > o \Rightarrow TR_\lambda(T)v \succ o$, *for all* $v \in R$.

(δ) *To each* $v > o$ *in* R *and each* $\ell \in \mathscr{L}$, *there exists a number* $m \in \mathbb{N}$ *such that* $\ell T^m v > 0$.

Proof. The equivalence of (α), (γ), and (δ) is proved similarly as the equivalence of (iii), (iv), and (v) in the preceding theorem. To complete the proof, it suffices to show that (β) $\Rightarrow$ (iii) $\Rightarrow$ (δ), where now $v \mathrel{\succ\!\!\!\!\raisebox{0.5ex}{\scriptsize$\smallsetminus$}} o \Leftrightarrow v > o$ is defined in (iii). (β) $\Rightarrow$ (iii) is obvious.

To prove (iii) $\Rightarrow$ (δ), suppose that the five (equivalent) properties of Theorem 4.22 hold for $W = K$, and let $v > o$, $\ell \in \mathscr{L}$, and $\lambda > \rho(T)$ be given. Then (iii) yields $R_\lambda v \succ o$ with $R_\lambda = R_\lambda(T)$. Define $w = TR_\lambda v$.

If $w > o$, then $\ell T^k w > 0$ for some integer $k \geqslant 0$ because of (v). This inequality implies $\sum_{j=1}^{\infty} \lambda^{-j} \ell T^{j+k} v > 0$, so that $\ell T^m v > 0$ for some $m \in \mathbb{N}$, as stated in (δ).

Now suppose that $w = R_\lambda Tv = o$ and, consequently, $Tv = o$. If $v \succ o$, this relation can only hold if $T = O$. For, given any $u \in R$, there exist $\alpha, \beta > 0$ with $-\alpha v \leqslant u \leqslant \beta v$, so that $o = -\alpha Tv \leqslant Tu \leqslant \beta Tv = o$. However, the null operator is not irreducible. If $v \not\succ o$, there exists an $\ell \in \mathscr{L}$ with $\ell v = 0$. From this relation and $TR_\lambda v = o$ we conclude that $\ell R_\lambda v = 0$, which contradicts $R_\lambda v \succ o$. Consequently, $w = o$ cannot hold. □

Example 4.9. *T is irreducible, if for each $v > o$ in R a number $m \in \mathbb{N}$ exists such that $T^m v \succ o$.*

Obviously, this property of T implies (δ). □

We shall explain in the notes following this chapter how several of the above properties are related to known definitions of irreducibility and similar properties.

4.4 Inverse-positivity for the case $K^c = \emptyset$

The monotonicity theorem for linear operators $M : R \to S$ will here be generalized to the case in which R need not contain a strictly positive element. Let the spaces $(R, \leqslant)$ and $(S, \leqslant)$ again satisfy the requirements (A_1), (A_2) in Section 1.1 except the condition $K^c \neq \emptyset$.

We assume that $p > o$ is a given element in R, and we shall then use the notation $\succ_p$ and B_p introduced in Section I.2.3 (see (I.2.6)).

Theorem 4.24. *Let the following two conditions be satisfied.*

I. *For each $u \in R$, the inequalities $u \geqslant o$, $Mu \mathrel{\succ\!\!\!\!\raisebox{0.5ex}{\scriptsize$\smallsetminus$}} o$ imply $u \succ_p o$.*
II. *There exists an element $z \in R$ such that $z \geqslant o$, $Mz \mathrel{\succ\!\!\!\!\raisebox{0.5ex}{\scriptsize$\smallsetminus$}} o$.*

Then for each $u \in B_p$,

$$Mu \geqslant o \;\Rightarrow\; u \geqslant o; \qquad Mu \succ o \;\Rightarrow\; u \succ_p o. \tag{4.40}$$

Proof. Suppose that $Mu \geqslant o$, $u \geqslant -\gamma p$ with some $\gamma > 0$, but $u \not\geqslant o$. Then there exists a minimal $\lambda > 0$ such that $v = (1 - \lambda)u + \lambda z \geqslant o$. Because of II, we have $Mv \succ o$, and hence I implies that $v \geqslant \alpha p$ for some $\alpha > 0$. Therefore, $(1 + \varepsilon)[(1 - \mu)u + \mu z] = (1 - \lambda + \varepsilon)u + \lambda z = v + \varepsilon u \geqslant (\alpha - \varepsilon\gamma)p \geqslant o$ for $(1 + \varepsilon)\mu = \lambda$ and $\varepsilon\gamma = \alpha$, which contradicts λ being minimal. Consequently, the first statement in (4.40) holds. If $Mu \succ 0$, then $u \geqslant o$ and hence $u \succ_p o$ by assumption I. □

We shall, in particular, consider the special case in which

$$(R, \leqslant) \subset (S, \leqslant), \qquad M = \lambda I - T, \qquad \lambda > 0, \qquad T \in L(R, S), \qquad T \geqslant O \tag{4.41}$$

with I the identity in S. For this case the preceding theorem yields

Corollary 4.24a. (a) *If* $z \geqslant o$, $\lambda z - Tz \succ_p o$ *for some* $z \in R$, *then*

$$u \in B_p, \quad (\lambda I - T)u \geqslant o \;\Rightarrow\; u \geqslant o. \tag{4.42}$$

(b) *If* $z > o$, $\mu z - Tz \geqslant o$ *for some* $z \in R$ *and some* $\mu \in (0, \lambda)$, *then* (4.42) *holds for* $p = z$.

(c) *If* $z \geqslant o$, $\lambda z - Tz > o$ *for some* $z \in R$, *then* (4.42) *holds for* $p = \lambda z - Tz$.

Example 4.10. Let $T = (t_{ik})$ denote the infinite matrix with elements $t_{ik} = k^{-1}$ for $i < k$, $t_{ik} = 0$ for $i \geqslant k$ $(i, k = 1, 2, \ldots)$. Choose the spaces $R = \{u \in l_2 : Tu \in l_2\}$ and $S = l_2$, furnished with the natural order relation, and consider the operator $M = I - T\colon R \to S$. *This operator is not inverse-positive*, since $Mu = o$ for $u = (u_i)$ with $u_i = \alpha(1 + i)^{-1}$, $\alpha \in \mathbb{R}$. *However, statement* (4.42) *holds for* $p = (p_i)$ *with* $p_i = i^{-3}$ *and* $\lambda = 1$.

To prove this, one verifies $z_i - (Tz)_i \geqslant \frac{1}{4} i^{-3}$ for $z_i = i^{-1}$ and applies Corollary 4.24a. □

Theorem 4.24 and its corollary yield sufficient conditions for a *restricted inverse-positivity*, meaning that (4.40) is only established for all u which belong to a subset of the domain of M. Nevertheless, these results can also be used to prove (unrestricted) inverse-positivity. One has to verify that for each $u \in R$ a strong order $\leqslant$ and an element $p > o$ exist such that the assumptions of Theorem 4.24 are satisfied and $u \in B_p$.

As an example we shall formulate a corresponding result for the case $R = S$ using the following notation. A subset $\mathscr{P} \subset R$ will be called *positive*, $\mathscr{P} \geqslant o$, if $p \geqslant o$ for all $p \in \mathscr{P}$. Moreover, $\mathscr{P} \subset R$ is said to be *strictly positive*, $\mathscr{P} \succ o$, *if* $\mathscr{P} \geqslant o$, $o \notin \mathscr{P}$, and $R = \bigcup \{B_p : p \in \mathscr{P}\}$. With this notation, we obtain from Corollary 4.24a

Theorem 4.25. *Suppose that for the operator $M = \lambda I - T : R \to R$ a set $\mathscr{Z} \subset R$ exists such that*

$$\mathscr{Z} \geqslant o \quad \text{and} \quad M\mathscr{Z} \succ o. \tag{4.43}$$

Then M is inverse-positive.

Under certain conditions, the assumption of the theorem can be verified by constructing a single element $\varphi \in R$, as formulated in the following corollary. Here, the given space $(R, \leqslant)$ is called *directed* if to each $u \in R$ an element $v \in R$, $v > o$ exists such that $u \geqslant -v$.

Corollary 4.25a. *Suppose that the ordered linear space $(R, \leqslant)$ is directed, and that the domain TR is a subset of an ordered linear space $(\hat{R}, \leqslant) \subset (R, \leqslant)$ such that $(\hat{R}, \leqslant)$ contains a strictly positive element φ satisfying*

$$\lambda\varphi - T\varphi \geqslant \nu\varphi \qquad \textit{for some} \quad \nu > 0. \tag{4.44}$$

Then $M = \lambda I - T : R \to R$ is inverse-positive.

Proof. For each $u \in R$ choose $v > o$ in R such that $u \geqslant -v$, and a constant $\delta > 0$ such that $v\varphi - \delta Tv \geqslant o$. Then $z := \varphi + \delta v$ satisfies $z \geqslant o$, and $u \geqslant -v = -\gamma\lambda\,\delta v \geqslant -\gamma(\lambda z - Tz)$ for $\gamma = \lambda^{-1}\delta^{-1}$. Consequently, (4.43) holds for the set $\mathscr{Z}$ of all these elements z $(u \in R)$. □

Example 4.11. Let $R = L_2(0, 1)$ and define $u \geqslant o$ to mean that $u(x) \geqslant 0$ for a.a. $x \in [0, 1]$. We shall conclude from the corollary that *the integral operator $\lambda I - T : L_2(0, 1) \to L_2(0, 1)$ with*

$$Tu(x) = \int_0^1 \mathscr{G}(x, \xi)u(\xi)\,d\xi \qquad \text{and} \quad \mathscr{G}(x, \xi) = \mathscr{G}(\xi, x) = x(1 - \xi) \quad \text{for } x \leqslant \xi$$

is inverse-positive if $\lambda > \pi^{-2}$.

The space $(R, \leqslant)$ is directed since $u \geqslant -|u|$, and the range TR is a subset of the space $\hat{R} = C_0[0, 1]$ furnished with the natural order relation. Finally, if $\lambda = \pi^{-2} + 2\mu$, $0 < \varepsilon < \pi$, $(\pi - \varepsilon)^{-2} - \pi^{-2} \leqslant \mu$, then $\varphi(x) = \cos(\pi - \varepsilon)(x - \frac{1}{2})$ satisfies (4.44) for $\nu = \mu$, and this function is strictly positive as an element of $\hat{R}$. □

In Section 1 we were also concerned with strictly inverse-positive operators. Here this term, in general, does not make sense, since K^c may be empty. One can, however, often prove a stronger type of inverse-positivity which coincides with strict inverse-positivity under suitable assumptions. The following result is again concerned with the case $(R, \leqslant) = (S, \leqslant)$, $M = \lambda I - T$, $\lambda > 0$, $T \geqslant O$ (the proof is left to the reader).

Proposition 4.26. *If $\lambda I - T: R \to R$ is inverse-positive and T has property (δ) in Theorem 4.23 for the set $\mathscr{L}$ of all linear functionals $\ell > o$ on R, then for all $u \in R$*

$$(\lambda I - T)u \geqslant o \Rightarrow \textit{either } u = o \textit{ or } u \textit{ is a nonsupport point of } K. \tag{4.45}$$

Example 4.11 (continued). For this case $\mathscr{L}$ is the set of all functionals $\ell u = (w_\ell, u)$ with $w = w_\ell \in L_2(0, 1)$, $w > o$. Since $(w, Tu) > 0$ for arbitrary $u > o$, $w > o$ in $L_2(0, 1)$, the implication (4.45) holds here for $\lambda > \pi^{-2}$. □

Problems

The following problems are concerned with the special case (4.41) under the general assumptions made above.

4.5 Suppose that $P(t) = \sum_{k=0}^{m} \gamma_k t^k$ is a polynomial with $\gamma_k \geqslant 0$ $(k = 0, 1, \ldots, m)$ such that $P(1) = 1$ and the following conditions are satisfied. (i) For all $u \in R$, $u > o \Rightarrow P(T)u \succ_p o$. (ii) There exists a $z \geqslant o$ in R such that $z \gtrdot Tz$ or $z \gtrdot P(T)z$. Prove that under these assumptions for all $u \in B_p$: $u - Tu \geqslant o \Rightarrow u \geqslant o$.

4.6 Reformulate (i) in the preceding problem for the case $u \gtrdot o \Leftrightarrow u \succ_p o$ (with some $p > o$ in R).

4.7 Generalize Problems 4.5 and 4.6 using a power series $P(t)$.

4.8 Show that a set $\mathscr{P} = \{p\} \subset R$ consisting of a single element p is strictly positive if and only if $p \succ o$.

4.9 Show that Corollary 4.25a remains true, if the assumption $TR \subset \hat{R}$ is replaced by $T^m R \subset \hat{R}$ and, in (4.44), $\lambda I - T$ is replaced by $\lambda^m I - T^m$. (Here $m \geqslant 1$ denotes a fixed integer.)

5 NUMERICAL APPLICATIONS

For certain problems, for instance, boundary value problems of the second order, the theory of inverse-positivity is a convenient tool to estimate the errors of approximate solutions.

In Section 5.1 the convergence of discretization methods is proved by estimating the error by functions of h, where h denotes the discretization parameter. In particular, we treat several difference methods for two-point boundary value problems. The ideas of proof can also be used for partial differential equations.

Section 5.2 is devoted to the problem of a posteriori numerical error estimates. It is a trivial statement that approximate numerical solutions are of no value, if one does not have, in addition, some indication of how accurate the approximations may be. Certainly, a rigorous estimate of the error would be the most satisfying answer. Such an error estimate, however, in general is much more difficult to obtain than the approximate solution itself. We shall not try to discuss here for which problems such error estimates could possibly be carried out. We simply report on some work which has been done for initial value problems of the first order and boundary value problems of the second order. Sections 5.2.2 and 5.2.3 contain corresponding results for linear problems. The methods can be carried over to nonlinear differential equations and systems of such equations. See Section III.6 and, furthermore, Examples V.2.4 and V.4.12 for the most difficult problems treated so far by our methods. In Example V.2.4 we shall make some remarks concerning the computing time needed for the error estimates.

One may consider all work on numerical error estimates which one can find in the literature as a beginning, a first step. Certainly much more effort has been put into the development of approximation methods, than into the development of estimation methods. How far one can go in constructing methods for numerical error estimation cannot be answered a priori but only by further investigations. It should also be mentioned that such estimation methods together with comparison techniques as described in Section III.4 may be used to prove the existence of solutions with a certain qualitative behavior in cases where comparison functions cannot be constructed by analytic techniques.

5.1 A convergence theory for difference methods

5.1.1 The difference method

Suppose that an equation

$$Mu = r \tag{5.1}$$

with a linear operator $M : R \to S$ is given, where R and S are linear subspaces of the space $\mathbb{R}_b[\xi, \eta]$ of all bounded real-valued functions on a compact interval $I = [\xi, \eta]$. We assume that for the given $r \in S$ a unique solution $u^* \in R$ exists. Moreover, let $e \in R$ and $e \in S$ (where $e(x) \equiv 1$), so that $u \succ o \Leftrightarrow \inf\{u(x): \xi \leqslant x \leqslant \eta\} > 0$ for $u \in R$ and $u \in S$.

We consider a *difference method* $\mathscr{D}$ for obtaining approximate solutions for the given problem. This method is described by a family of *difference equations*

$$M_h\varphi = N_h r \qquad (h \in H) \tag{5.2}$$

for functions $\varphi = \varphi_h \in R_h := \mathbb{R}(I_h)$ defined on a discrete set $I_h = \{x_0, x_1, \ldots, x_n\} \subset I$. (Of course, the x_i and n may depend on h.) Here $M_h : R_h \to R_h$ and $N_h : S \to R_h$ are linear operators, and H is a sequence of numbers

$h_i > 0$ $(i = 1, 2, \ldots)$ converging to 0. We assume that a constant $\kappa > 0$ exists such that for all $U \in S$ and all $h \in H$

$$U \geqslant e \quad \Rightarrow \quad N_h U \geqslant \kappa^{-1} e_h. \tag{5.3}$$

Here and in the following, u_h denotes the restriction of $u \in \mathbb{R}(I)$ to I_h. Thus, $u_h \in \mathbb{R}(I_h)$ and hence $u_h \geqslant o \Leftrightarrow u(x) \geqslant 0$ for all $x \in I_h$. (By our convention, $\leqslant$ denotes the natural order relation in the function space under consideration.) For convenience, we shall often write $M_h u$ instead of $M_h u_h$, if $u \in \mathbb{R}(I)$.

We shall use the notation

$$\|u\|_J = \inf\{\alpha \geqslant 0 : -\alpha \leqslant u(x) \leqslant \alpha \quad \text{for all} \quad x \in J\}$$

for any set $J \subset I$ and any suitable u, so that, in particular,

$$\|u\|_I = \inf\{\alpha \geqslant 0 : -\alpha e \leqslant u \leqslant \alpha e\} \qquad \text{for} \quad u \in R \quad \text{and} \quad u \in S,$$
$$\|\varphi\|_{I_h} = \inf\{\alpha \geqslant 0 : -\alpha e_h \leqslant \varphi \leqslant \alpha e_h\} \qquad \text{for} \quad \varphi \in R_h.$$

Each operator M_h can be described, in a natural way, by an $(n + 1) \times (n + 1)$ matrix. This matrix is inverse-positive (pre-inverse-positive) if and only if the operator M_h is inverse-positive (pre-inverse-positive). Moreover, the matrix is a Z-matrix if and only if M_h is a Z-operator as defined in Section 4.2.

5.1.2 *Sufficient convergence conditions*

The difference method $\mathscr{D}$ is called *convergent* (more precisely: uniformly convergent on I) if solutions $\varphi = \varphi_h$ of the difference equations (5.2) exist for all sufficiently small $h \in H$ and

$$\|u_h^* - \varphi_h\|_{I_h} \to 0 \qquad \text{for} \quad h \to 0 \quad (h \in H). \tag{5.4}$$

The method has the *order of convergence p (is convergent of order p)* if

$$\|u_h^* - \varphi_h\|_{I_h} \leqslant \gamma h^p \qquad \text{for} \quad h \in H,$$

where $p > 0$, and γ denotes a constant which may depend on u^* and p. The *convergence on a subset* $J \subset I$ and the *order of convergence on* J are defined analogously with I_h replaced by $I_h \cap J$.

As a reasonable assumption on the difference method, we require that equations (5.2) approximate the given problem (5.1) "well enough." This requirement will be formulated as a consistency condition. For fixed $u \in R$, the difference method $\mathscr{D}$ is called *consistent at u* if

$$\|\sigma_h(u)\|_{I_h} \to 0 \qquad \text{for} \quad h \to 0 \quad (h \in H), \qquad \text{where} \quad \sigma_h(u) = M_h u_h - N_h M u.$$

The order of consistency at u is *p*, if

$$\rho_h(u) := \|\sigma_h(u)\|_{I_h} \leqslant \gamma h^p \qquad \text{for} \quad h \in H,$$

where $p > 0$, and $\gamma > 0$ denotes a constant which may depend on u and p.

Of course, the numbers p which occur above are, in general, not uniquely determined by the properties mentioned. This could be taken into account by using terms like "the order of convergence is *at least p*". However, we do not expect any difficulties in using the preceding simpler terminology.

In the convergence proofs the inverse-positivity of the difference operators M_h will be employed. This property can very often be derived using properties of the given operator M.

Proposition 5.1. *Let the following two conditions be satisfied.*

(i) *M_h is weakly pre-inverse-positive for each sufficiently small $h \in H$.*

(ii) *There exists a $z \in R$ such that $z \geqslant o$, $Mz \geqslant e$, and the difference method $\mathscr{D}$ is consistent at z.*

Then M_h is inverse-positive for each sufficiently small $h \in H$.

Proof. For sufficiently small $h \in H$

$$M_h z_h = N_h Mz + \sigma_h(z) \geqslant \kappa^{-1} e_h - \tfrac{1}{2}\kappa^{-1} e_h = \tfrac{1}{2}\kappa^{-1} e_h \succ o, \tag{5.5}$$

so that M_h is inverse-positive by Theorem 2.1. □

Theorem 5.2 (*Convergence theorem*). *Suppose that assumptions* (i), (ii) *of Proposition* 5.1 *are satisfied and the difference method $\mathscr{D}$ is consistent at the solution u^*.*

Then Method $\mathscr{D}$ is convergent, and for sufficiently small $h \in H$

$$\|u_h^* - \varphi_h\|_{I_h} \leqslant 2\kappa\rho_h(u^*)\,\|z\|_I. \tag{5.6}$$

Proof. Let $h \in H$ be so small that the arguments in the proof of Proposition 5.1 are valid. Then equation (5.2) has a unique solution φ_h, since M_h is inverse-positive and hence invertible. Moreover, we estimate

$$M_h(u_h^* - \varphi_h) = \sigma_h(u^*) \leqslant \rho_h e_h \leqslant \rho_h 2\kappa M_h z_h \qquad \text{with} \quad \rho_h = \rho_h(u^*),$$

using (5.5), and obtain from this relation $u_h^* - \varphi_h \leqslant \rho_h 2\kappa z_h$. Analogously, the inequality $-\rho_h 2\kappa z_h \leqslant u_h^* - \varphi_h$ is derived, so that (5.6) follows when $\|z\|_{I_h} \leqslant \|z\|_I$ is observed. □

Obviously, (5.6) implies that *Method $\mathscr{D}$ is convergent of order p if it is consistent of order p.*

Quite often difference methods are used such that not all mesh points x_i lie in the interval on which the boundary value problem is given (see the cases (5) and (6) below, for example). To treat such cases, one could generalize the abstract theory above by not requiring $I_h \subset I$. Instead, we shall extend the given problem to a problem on a larger interval. Then the following result will be used.

Corollary 5.2a. *Under the assumptions of Theorem* 5.2,

$$\|u_h^* - \varphi_h\|_{I_h \cap J} \leqslant 2\kappa \rho_h(u^*)\|z\|_J \qquad \text{for any} \quad J \subset I$$
$$\text{and sufficiently small} \quad h \in H.$$

The above convergence theory can also be formulated somewhat differently, so that the notion of stability is involved. (The proofs of the results below are left to the reader.) The difference method $\mathscr{D}$ is called *stable*, if a constant γ exists such that for all sufficiently small h the operator M_h is invertible and $M_h^{-1} : R_h \to R_h$ has an operator norm $\|M_h^{-1}\| \leqslant \gamma$ (with respect to the norm $\|\varphi_h\|_{I_h}$).

Theorem 5.3. *If the difference method* $\mathscr{D}$ *is stable and consistent at* u^*, *then the method is convergent.*

Of course, in applying this result one has first to prove the stability, i.e., one has to estimate the norms $\|M_h^{-1}\|$. Our aim here is to replace the assumption on these operator norms by simple sufficient conditions, using inverse-positivity.

Proposition 5.4. *The difference method* $\mathscr{D}$ *is stable, if the assumptions of Proposition* 5.1 *are satisfied.*

5.1.3 Boundary value problems

The convergence theory of the preceding section will here be applied to a two-point boundary value problem

$$Mu(x) = r(x) \qquad (0 \leqslant x \leqslant 1) \tag{5.7}$$

with a differential operator M of the form (3.2), (3.4) satisfying the regularity condition (3.5) and a function

$$r \in \mathbb{R}[0, 1] \qquad \text{such that} \quad r(x) = s(x) \quad (0 < x < 1) \\ \text{for some} \quad s \in C_0[0, 1]. \tag{5.8}$$

Observe that $r_0 = r(0)$, $r_1 = r(1)$ are the given inhomogeneous terms in the boundary conditions. Let S denote the space of all functions r of the form (5.8); then $M : R = C_2[0, 1] \to S$. (We shall use the letter r not only for the given function in (5.7), but also for an arbitrary element in S.)

To explain various ideas in applying the abstract theory, we shall consider several special cases (1)–(6). More general problems can be treated in an analogous way. For all considered cases, *the difference method converges, if the differential operator* M *is inverse-positive.*

In the following, the notation of the previous sections will be used. In particular, $I_h = \{x_0, x_1, \ldots, x_n\}$, and $u^* \in C_2[0,1]$ denotes the solution of the given problem (5.7). (If M is inverse-positive, a (unique) solution exists for each $r \in S$.) Moreover, define for $h > 0$

$$L_h[\varphi](x) = a(x)\frac{1}{h^2}[-1 \quad 2 \quad -1]_h \varphi(x)$$

$$+ b(x)\frac{1}{2h}[-1 \quad 0 \quad 1]_h \varphi(x) + c(x)\varphi(x),$$

where $[\alpha\ \beta\ \gamma]_h \varphi(x) = \alpha\varphi(x-h) + \beta\varphi(x) + \gamma\varphi(x+h)$.

(1) *(Dirichlet problem, ordinary difference method.)* Suppose that $B_0[u] = u(0)$, $B_1[u] = u(1)$, and choose for each $h \in H = \{n^{-1} : n = 2, 3, \ldots\}$:

$$x_i = ih = in^{-1} \qquad (i = 0, 1, 2, \ldots, n), \qquad N_h r = r_h$$

$$M_h\varphi(x_i) = L_h[\varphi](x_i) \qquad \text{for} \quad 0 < i < n,$$

$$M_h\varphi(x_0) = \varphi(x_0), \qquad M_h\varphi(x_n) = \varphi(x_n).$$

For the difference method so described, we obtain the following result.

Proposition 5.5. (a) *If the differential operator $M : C_2[0,1] \to \mathbb{R}_b[0,1]$ is inverse-positive, then the difference method is convergent on* $[0,1]$.

(b) *The order of convergence is* 2 *or* 1, *if $u^* \in C_4[0,1]$ or $u^* \in C_3[0,1]$, respectively.*

Proof. For sufficiently small h, the operator $M_h : R_h \to R_h$ is described by a Z-matrix, and hence M_h is weakly pre-inverse-positive, so that condition (i) in Proposition 5.1 holds. Moreover, since M is inverse-positive, there exists a $z \in R$ as required in (ii) (choose z satisfying $Mz = e$). Finally, a Taylor expansion of $\sigma_h(u)(x) = M_h u(x) - (Mu)_h(x)$ at $x_i \in [h, 1-h]$ shows that the difference method is consistent at each $u \in C_2[0,1]$. Moreover, the order of consistency is 2 or 1 if $u \in C_4[0,1]$ or $u \in C_3[0,1]$, respectively. Thus, the statements of this proposition follow from Theorem 5.2. □

(2) *(Dirichlet problem, nonconstant mesh width.)* Let again $B_0[u] = u(0)$, $B_1[u] = u(1)$, but choose now for each $h \in H = \{\frac{1}{2}m^{-1} : m = 2, 3, 4, \ldots\}$

$$x_i = \tfrac{1}{2}ih \qquad \text{for} \quad i = 0, 1, 2, \ldots, 2m;$$

$$x_i = \tfrac{1}{2} + (i - 2m)h \qquad \text{for} \quad i = 2m+1, \ldots, n$$

with $h = \frac{1}{2}m^{-1}$, $n = 3m$, and define

$$M_h\varphi(x_i) = L_{h/2}[\varphi](x_i) \qquad \text{for} \quad 0 < i < 2m,$$

$$M_h\varphi(x_i) = L_h[\varphi](x_i) \qquad \text{for} \quad 2m \leqslant i < n,$$

$$M_h\varphi(x_0) = \varphi(x_0), \qquad M_h\varphi(x_n) = \varphi(x_n), \qquad N_h r = r_h.$$

By a proof similar to that in (1), one shows that *for this difference method Proposition* 5.5 *holds also.*

(3) (*Dirichlet problem, Hermite formula.*) To a boundary value problem (5.7) with

$$L[u](x) = -u''(x) + c(x)u(x), \qquad B_0[u] = u(0), \qquad B_1[u] = u(1),$$

we apply a Hermite formula (Mehrstellenverfahren). We choose H, the points x_i, and the values $M_h\varphi(x_0)$, $M_h\varphi(x_n)$ as in (1), but define now

$$M_h\varphi(x_i) = \frac{1}{h^2}[-1 \quad 2 \quad -1]_h\varphi(x_i) + \frac{1}{12}[1 \quad 10 \quad 1]_h(c\varphi)(x_i)$$

$$\text{for} \quad 0 < i < n,$$

$$N_h r(x_i) = \frac{1}{12}[1 \quad 10 \quad 1]_h s(x_i) \qquad \text{for} \quad 0 < i < n,$$

$$N_h r(x_0) = r_0, \qquad N_h r(x_n) = r_1,$$

for r having the form (5.8).

For this difference method Proposition 5.5 *holds also. Moreover, the order of convergence is* 4 *or* 3, *if* $u^* \in C_6[0, 1]$ *or* $u^* \in C_5[0, 1]$, *respectively.*

Again, this statement is proved by similar means as the result for (1). Now $\sigma_h(u)(x) = h^{-2}[-1 \quad 2 \quad -1]_h u(x) + \frac{1}{12}[1 \quad 10 \quad 1]_h u''(x)$.

(4) (*A method of order* 1 *for Sturmian boundary conditions.*) To describe how more general boundary conditions can be treated, we consider the case where

$$B_0[u] = -\alpha_0 u'(0) + \gamma_0 u(0) \qquad \text{with} \quad \alpha_0 > 0, \qquad B_1[u] = u(1). \quad (5.9)$$

All quantities are chosen as in (1), except that now

$$M_h\varphi(x_0) = -\alpha_0 \frac{1}{h}(\varphi(x_1) - \varphi(x_0)) + \gamma_0 \varphi(x_0).$$

For this difference method, part (a) *of Proposition* 5.5 *holds also. Moreover, the order of convergence is* 1 *if* $u^* \in C_3[0, 1]$.

The proof is the same as in (1), except that for $u^* \in C_4[0, 1]$ we can, in general, establish only the order of consistency 1, because $M_h u_h(0) - Mu(0) = \alpha_0(-\frac{1}{2}hu''(0) + O(h^2))$.

(5) (*A method of order* 2 *for Sturmian boundary conditions.*) Consider again the case (5.9). The difference method in (4) was not quite satisfying,

since for $u^* \in C_4[0, 1]$ only convergence of the order 1 could be established. Therefore, we shall now describe another difference method. Choose

$$x_i = -\frac{h}{2} + ih \qquad (i = 0, 1, 2, \ldots, n) \qquad \text{with} \quad h = (n - \tfrac{1}{2})^{-1},$$

$$M_h\varphi(x_0) = -\alpha_0 \frac{1}{h}[-1 \quad 0 \quad 1]_{h/2}\,\varphi(0) + \frac{1}{2}\gamma_0[1 \quad 0 \quad 1]_{h/2}\,\varphi(0),$$

$$M_h\varphi(x_i) = L_h[\varphi](x_i) \qquad (0 < i < n), \qquad M_h\varphi(x_n) = \varphi(x_n),$$

$$N_h r(x_i) = s(x_i) \qquad \text{for} \quad 0 < i < n, \qquad N_h r(x_0) = r_0, \qquad N_h r(x_n) = r_1.$$

For this difference method Proposition 5.5 *holds also.*

Now, I_h is not a subset of the interval $[0, 1]$ on which the boundary value problem (5.7) is given. Nevertheless, the meaning of the above statement is clear, the convergence on $[0, 1]$ being defined by (5.4) with I_h replaced by $I_h \cap [0, 1]$.

However, for being able to apply Theorem 5.2, we have to incorporate the above difference method into our general scheme. This will be achieved by extending the given problem (5.7) to a problem $\hat{M}\hat{u} = \hat{r}$ on an interval $I = [-\varepsilon, 1]$, with an operator $\hat{M} : C_2[-\varepsilon, 1] \to \mathbb{R}_b[-\varepsilon, 1]$ and some $\varepsilon > 0$. Obviously, $I_h \subset I = [-\varepsilon, 1]$ for all sufficiently small h of the above kind. This extension of the given equation is carried out for formal reasons only.

For inverse-positive M there exists a $z \in C_2[0, 1]$ satisfying $Mz = e$, $z(x) > 0$ $(0 \leqslant x \leqslant 1)$. This function can be extended to a function $\hat{z} \in C_2[-\varepsilon, 1]$ with $\hat{z}(x) > 0$ $(-\varepsilon \leqslant x \leqslant 1)$, where $\varepsilon > 0$ denotes a suitably chosen constant. Obviously, $\alpha\hat{z}(x) \geqslant 1$ $(-\varepsilon \leqslant x \leqslant 1)$ for some $\alpha > 0$.

We define $\hat{M}u(x) = \alpha u(x)$ $(-\varepsilon \leqslant x < 0)$, $\hat{M}u(x) = Mu(x)$ $(0 \leqslant x \leqslant 1)$. This operator $\hat{M}$ maps $\hat{R} = C_2[-\varepsilon, 1]$ into the set $\hat{S}$ of all $\hat{r} \in \mathbb{R}_b[-\varepsilon, 1]$ such that $\hat{r}(x) = r(x)$ $(0 \leqslant x \leqslant 1)$ for some $r \in S$. Obviously, $\hat{M}\hat{z}(x) \geqslant 1$ for $-\varepsilon \leqslant x \leqslant 1$. We also extend N_h to an operator $\hat{N}_h$ on $\hat{S}$ by defining $\hat{N}_h\hat{r} = N_h r$ for $\hat{r}$ as above.

For inverse-positive M the given problem (5.7) has a unique solution $u^* \in C_2[0, 1]$, which can be extended to a function $\hat{u}^* \in C_2[-\varepsilon, 1]$ such that, for $k = 0, 1, 2$, $\hat{u}^* \in C_{2+k}[-\varepsilon, 1]$ whenever $u^* \in C_{2+k}[0, 1]$. Clearly, $\hat{u}^*$ is a solution of the extended equation $Mu = r$, where $r(x) = \alpha u^*(x)$ for $-\varepsilon \leqslant x \leqslant 0$ and $\hat{r}(x)$ coincides on $[0, 1]$ with the given $r(x)$ in (5.7).

Now one derives the preceding convergence statement from Theorem 5.2 applied to $\hat{M}\hat{u} = \hat{r}$ and the difference method $M_h\varphi = \hat{N}_h\hat{r}$, in essentially the same way as in the other cases. □

(6) (*A modification of* (5) *where* M_h *is not a Z-operator.*) Once more, we consider the case (5.9) choosing now

$$x_i = -h + ih \qquad (i = 0, 1, \ldots, n) \qquad \text{with} \quad h = (n-1)^{-1},$$

$$M_h\varphi(x_0) = -\alpha_0 \frac{1}{2h}[-1 \quad 0 \quad 1]_h \varphi(0) + \gamma_0 \varphi(0)$$

and defining all $M_h\varphi(x_i)$ for $i > 0$ and N_h by the formulas used in (5).

For this difference method Proposition 5.5 *holds also.*

The proof is essentially the same as for the case (5), with the following exception. If $\gamma_0 > 0$, then M_h is not a Z-operator for any $h > 0$, so that the (weak) pre-inverse-positivity of M_h is not immediately clear. However, this property of M_h can here be established using the method of reduction explained in Section 4.1. For $P_h : R_h \to R_h$ defined by

$$P_h\varphi(x_0) = 4\varphi(x_0) + \alpha_0 h\varphi(x_2), \qquad P_h\varphi(x_i) = \varphi(x_i) \qquad (i = 1, 2, \ldots, n)$$

and sufficiently small h, P_hM_h is a Z-operator. Consequently, P_hM_h is weakly pre-inverse-positive and hence M_h has this property also (cf. Theorem 4.1).

5.1.4 A generalization and its application

We continue now the abstract convergence theory. To treat more complicated difference methods, Theorem 5.2 will be generalized, replacing condition (ii) (in Proposition 5.1) by a weaker assumption. (Here condition (5.3) will not be needed.)

Theorem 5.6. *Let* M_h *be weakly pre-inverse-positive for sufficiently small* $h \in H$. *Moreover, suppose that an element* $z \in R$ *exists and for all sufficiently small* $h \in H$ *elements* $\psi_h \in R_h$ *exist such that*

$$o \leqslant \psi_h \leqslant z_h, \qquad M_h\psi_h \succ o, \tag{5.10}$$

and

$$|\sigma_h(u^*)| \leqslant \tau(h) M_h \psi_h \qquad \text{with some} \quad \tau \in C_0[0, \infty). \tag{5.11}$$

Then for $h \in H$ *sufficiently small,*

$$\|u_h^* - \varphi_h\|_{I_h} \leqslant \tau(h)\|z\|_I. \tag{5.12}$$

In particular, Method $\mathscr{D}$ *is convergent (convergent of order* p) *if* $\tau(0) = 0$ ($\tau(h) = Ch^p$).

This result is proved similarly as Theorem 5.2. While in the proof of that theorem "bounds" ψ_h of the form $\psi_h = \text{const}\, z_h$ with z in (ii) were used, the functions ψ_h here may be chosen differently. We shall give an example where this greater flexibility is exploited.

(7) (*A method convergent of order* 4 *and consistent of order* 2.) We consider again the problem in (1), but choose now

$$M_h\varphi(x_i) = a(x_i)\frac{1}{12h^2}[1 \quad -16 \quad 30 \quad -16 \quad 1]\varphi(x_i)$$

$$+ b(x_i)\frac{1}{12h}[1 \quad -8 \quad 0 \quad 8 \quad -1]\varphi(x_i) + c(x_i)\varphi(x_i) \qquad \text{for} \quad 2 \leqslant i \leqslant n-2,$$

$$M_h\varphi(x_i) = L_h[\varphi](x_i) \quad \text{for} \quad i = 1, n-1; \qquad M_h\varphi(x_i) = \varphi(x_i) \quad \text{for} \quad i = 0, n.$$

Proposition 5.7. *Suppose the differential operator M is inverse-positive. Then the difference method* (7) *is convergent. Moreover, for $k = 2, 3, 4$, the method has the order of convergence k, if $u^* \in C_{2+k}[0, 1]$.*

This statement will be derived from Theorem 5.6 in two different ways.

First proof (*Method of z-splitting*). In the following arguments we shall always assume that *h is sufficiently small*, without mentioning it each time.

Obviously, M_h is not a Z-operator; but we can prove the weak pre-inverse-positivity of M_h by the method of reduction. By formal calculation, one verifies that $M_h \leqslant \mathscr{L}_h A_h$ for $\mathscr{L}_h : R_h \to R_h$ and $A_h : R_h \to R_h$ defined by

$$\mathscr{L}_h\varphi(x_i) = \begin{cases} \varphi(x_i) & \text{for} \quad i = 0 \quad \text{and} \quad i = n \\ \frac{1}{2}h^{-2}a(x_i)\varphi(x_i) & \text{for} \quad i = 1 \quad \text{and} \quad i = n-1 \\ (\frac{1}{12}ah^{-2}[-1 \quad 8 \quad -1]\varphi + \frac{1}{12}bh^{-1}[-1 \quad 0 \quad 1]\varphi)(x_i) & \\ & \text{for} \quad 2 \leqslant i \leqslant n-2 \end{cases}$$

$$A_h\varphi(x_i) = \begin{cases} \varphi(x_i) & \text{for} \quad i = 0 \quad \text{and} \quad i = n \\ [-1 \quad 8 \quad -1]\varphi(x_i) & \text{for} \quad 1 \leqslant i \leqslant n-1. \end{cases}$$

The operator $\mathscr{L}_h$ is inverse-positive, because it is a Z-operator and $\mathscr{L}_h e_h \succ o$. Since also A_h is a Z-operator, M_h is weakly pre-inverse-positive, according to Corollary 4.1a applied to M_h, $\mathscr{L}_h$, A_h in place of M, L, M_1.

Using Taylor expansions one shows that the difference method has the order of consistency 2 at $u \in C_2[0, 1]$. Moreover, since M is inverse-positive, there exists a $\hat{z} \succ o$ in R such that $M\hat{z} \geqslant e$. Thus the convergence of the difference method already follows from Theorem 5.2. However, to prove also the stated order of convergence, we shall use the more general Theorem 5.6, which allows us to exploit "local" orders of consistency. For simplicity, we shall consider only the case $u^* \in C_6[0, 1]$.

First observe that $M_h \hat{z}_h \geqslant \frac{1}{2} e_h$ for h sufficiently small, due to the consistency of the method. Again using Taylor expansion, we estimate for $u^* \in C_6[0, 1]$

$$|\sigma_h(u^*)| \leqslant C[h^2(e^{(1)} + e^{(n-1)}) + h^4 \hat{e}] \tag{5.13}$$

with $\hat{e} = e^{(2)} + e^{(3)} + \cdots + e^{(n-2)}$ and $e^{(i)} = e_h^{(i)} \in R_h$ defined by $e^{(i)}(x_j) = \delta_{ij}$ $(j = 0, 1, \ldots, n)$. Exploiting this estimate we shall construct suitable $\psi_h \in R_h$ as a sum

$$\psi_h = \psi_h^{(1)} + \psi_h^{(2)} + \psi_h^{(3)} \qquad \text{with} \quad M_h \psi_h^{(1)} \geqslant h^{-2} e^{(1)},$$
$$M_h \psi_h^{(2)} \geqslant h^{-2} e^{(n-1)}, \qquad M_h \psi_h^{(3)} \geqslant e_h.$$

These relations together with (5.13) imply (5.11) for $\tau(h) = Ch^4$. Moreover, if the $\psi_h^{(i)}$ are positive and uniformly bounded (for h sufficiently small), then (5.10) holds for $z = \alpha\hat{z}$ with a large enough α and hence (5.12) yields the order of convergence 4.

Obviously, we may choose $\psi_h^{(3)} = 2\hat{z}_h$. Concerning the construction of $\psi_h^{(1)}$ and $\psi_h^{(2)}$, we consider first the special case $a(x) \equiv 1$, $b(x) \equiv 1$. Here, a formal calculation shows that $M_h \omega(x_i) \geqslant (h^{-2} e^{(1)} + c_h \omega)(x_i)$ $(1 \leqslant i \leqslant n-1)$ for $\omega \in R_h$, $\omega(x_0) = 0$, $\omega(x_i) = 3 - hx_i^{-1}$ otherwise. Consequently, $\psi_h^{(1)} = \omega + \alpha\hat{z}$ has all desired properties, if α is a sufficiently large constant. The function $\psi_h^{(2)}$ can be defined in an analogous way.

In the general case, one can construct $\psi_h^{(1)}$ in essentially the same way using now $\omega(x_i) = \exp(\mu x_i)(3 - hx_i^{-1})$ for $i = 1, 2, \ldots, n$, where μ denotes a suitable constant. □

Second proof (*Perturbation method*). The difference procedure (7) can also be written in the form (5.2) with M_h, N_h replaced by operators $\tilde{M}_h$, $\tilde{N}_h$ which are defined as follows:

$$\tilde{M}_h \varphi(x_i) = M_h \varphi(x_i), \qquad \tilde{N}_h r(x_i) = N_h r(x_i) \qquad \text{for} \quad i \neq 1, n-1;$$
$$\tilde{M}_h \varphi(x_1) = h^2 M_h \varphi(x_1) + \varepsilon a(x_1)\varphi(x_0),$$
$$\tilde{M}_h \varphi(x_{n-1}) = h^2 M_h \varphi(x_{n-1}) + \varepsilon a(x_{n-1})\varphi(x_n),$$
$$\tilde{N}_h r(x_1) = h^2 r(x_1) + \varepsilon a(x_1) r(x_0),$$
$$\tilde{N}_h r(x_{n-1}) = h^2 r(x_{n-1}) + \varepsilon a(x_{n-1}) r(x_n).$$

In the first proof we used a particular reduction method to show that M_h is weakly pre-inverse-positive (for all sufficiently small h). The same arguments apply to an operator which is obtained from M_h when

$$[-1 \quad 2 \quad -1]\varphi(x_1)$$

is replaced by

$$[-(1+\varepsilon) \quad 2 \quad -1]\varphi(x_1)$$

and $[-1 \quad 2 \quad -1]\varphi(x_{n-1})$ is replaced by

$$[-1 \quad 2 \quad -(1+\varepsilon)]\varphi(x_{n-1}),$$

where $\varepsilon > 0$ is a sufficiently small number *independent* of h. Consequently, for this $\varepsilon > 0$ and all sufficiently small h the operator $\tilde{M}_h$ is weakly pre-inverse-positive.

The perturbation by the number ε makes an important difference, when (5.11) is verified. For $u^* \in C_6[0, 1]$ we now obtain $|\tilde{\sigma}_h(u^*)| \leqslant Ch^4(e^{(1)} + e^{(n-1)} + \hat{e})$ instead of (5.13). Therefore, we need only construct a $z \geqslant o$ in R such that $\psi_h = z_h$ satisfies $\tilde{M}_h\psi_h \geqslant e_h$. This is done in the following way.

Let $\hat{z} \succ o$ be the solution of $M\hat{z} = e$. Then $M_h\hat{z}_h \geqslant \frac{1}{2}e_h$ for all sufficiently small h, because of the consistency. From the latter inequality one derives $\tilde{M}_h\hat{z}_h \geqslant \mu e_h$ for a $\mu > 0$ satisfying $\frac{1}{2} \geqslant \mu \leqslant \varepsilon\alpha\beta$, where $0 < \alpha \leqslant a(x)$, $0 < \beta \leqslant \hat{z}(x)$ $(0 \leqslant x \leqslant 1)$. Observe, for instance, that $\tilde{M}_h\hat{z}_h(x_1) \geqslant \frac{1}{2}h^2 + \varepsilon a(x_1)\hat{z}(x_0) \geqslant \mu$. Thus, $z = \mu^{-1}\hat{z}$ has the desired properties. □

5.2 A posteriori error estimates

We shall here describe a method to obtain (two-sided) error bounds for solutions of linear equations $Mu = r$ where M is an inverse-positive operator. After explaining the general approach in an abstract setting (Section 5.2.1) we shall illustrate the method by applying it to linear differential equations (Sections 5.2.2 and 5.2.3). The general ideas involved can also be used for more complicated problems, as indicated at the beginning of Section 5.

5.2.1 *General description of the method*

Suppose that an equation

$$Mu = r \tag{5.14}$$

is given where M denotes a linear operator which maps an ordered linear space $(R, \leqslant)$ into an ordered linear space $(S, \leqslant)$ and $r \in S$. We assume that a solution $u = u^* \in R$ exists and that M is inverse-positive, so that the solution is unique. We are here concerned with methods for providing upper and lower bounds for u^* by using the implications

$$M\varphi \leqslant r \leqslant M\psi \quad \Rightarrow \quad \varphi \leqslant u^* \leqslant \psi. \tag{5.15}$$

If an approximate solution $u_0 \in R$ with *defect* $d[u_0] = r - Mu_0$ has been calculated, one will choose bounds of the form $\varphi = u_0 + v$, $\psi = u_0 + w$ and then use the *principle of error estimation*:

$$Mv \leqslant d[u_0] \leqslant Mw \quad \Rightarrow \quad v \leqslant u^* - u_0 \leqslant w, \tag{5.16}$$

which is equivalent to (5.15).

In order to obtain bounds v, w which are close to the actual error $u^* - u_0$, one could try to choose u_0, v, and w from appropriate finite dimensional spaces and determine the parameters involved such that the inequalities in the premise of (5.16) are satisfied and $w - v$ becomes "small" in some prescribed sense. This method, however, leads to a problem of (linear) optimization which, in general, will be rather complicated to solve. We are interested in much simpler methods. Our goal is not to obtain bounds which are as close as possible to the actual error, but to obtain sufficiently close bounds with as little effort as possible.

We prefer to construct first u_0 such that $d[u_0]$ becomes "minimal" in a sense which has to be described and then to construct the bounds v, w. Of course, the details of the procedure, especially the procedure to minimize the defect, will depend on the type of the problem considered. However, the following general remarks concerning the calculation of the bounds will apply to most cases.

After having computed u_0, one could set up a sophisticated numerical procedure to find elements v, w which yield very accurate estimates of the error $u^* - u_0$. Another possibility would be to choose a very "simple" element z with $z \geqslant o$, $Mz \geqslant o$ and calculate a constant α such that $v = -\alpha z$, $w = \alpha z$ satisfy the inequalities required, where little or no effort is put into the construction of the element z. Of course, for a given u_0, the latter method, in general, will lead to less accurate bounds. Nevertheless, in all our examples this method (or a slight modification of it) is proved to be the most effective. This may be explained by the following arguments.

If bounds of the special form $-v = w = \alpha z$ with $\alpha > 0$ are used, the element z has to satisfy the inequalities $z > o$, $Mz > o$ (except in the trivial case $z = o$). Often it is mainly this requirement, which causes the difference $w - v = 2\alpha z$ to be much larger than the error $u^* - u_0$. In particular, this is true for boundary value problems. Then trying to find an optimal z, in general, is not worthwhile.

As an example, let

$$Mu(x) = -u''(x) \qquad (0 < x < 1), \qquad Mu(0) = u(0), \qquad Mu(1) = u(1),$$

and $(u^* - u_0)(x) = \varepsilon \sin m\pi x$ for some $m \in \mathbb{N}$ and $\varepsilon > 0$. Then the choice $z(x) = x(1 - x)$ results in an estimate $|u^* - u_0|(x) \leqslant \varepsilon\frac{1}{2}m^2\pi^2 x(1 - x)$. Here, it would not help very much to find better functions z with $z > o$, $Mz > o$. For example, one may want to exploit the estimate $4|\sin m\pi x| \leqslant 3 - \cos 2m\pi x$ and choose $z = 6x(1 - x) + m^{-2}\pi^{-2}(1 - \cos 2m\pi x)$. For this function, one calculates $|u^* - u_0|(x) \leqslant \varepsilon\{\frac{3}{8}m^2\pi^2 x(1 - x) + \frac{1}{16}(1 - \cos 2m\pi x)\}$, which presents no significant improvement.

For the boundary value problems which will be considered in Section 5.2.3 the error bounds show a similar behavior; they are larger than the error roughly by a factor αm^2, where α is a constant and m the number of free parameters in determining u_0. This factor can be large, in particular, if the accuracy desired can be reached only for a large m. Nevertheless, using such simple bounds v, w can often be recommended, as the numerical results and the following discussion may show.

Suppose now that for a given approximation u_0 elements v, w satisfying $Mv \leqslant d[u_0] \leqslant Mw$ have been calculated—with considerable numerical effort—such that $v \leqslant u^* - u_0 \leqslant w$ is a very accurate estimate. Then $u_1 = u_0 + \frac{1}{2}(v + w)$ in general, is a much better approximation. For this approximation and $z = \frac{1}{2}(w - v)$ we have $-Mz \leqslant d[u_1] \leqslant Mz$ and $-z \leqslant u^* - u_1 \leqslant z$ and hence also $z \geqslant o$, $Mz \geqslant o$. This estimate for $u^* - u_1$ may be very poor. However, the poor estimate $-z \leqslant u^* - u_1 \leqslant z$ for the error of the better approximation u_1 and the accurate estimate $v \leqslant u^* - u_0 \leqslant w$ for the error of the original approximation yield exactly the same bounds for u^*, namely, $v + u_0 = -z + u_1 \leqslant u^* \leqslant z + u_1 = w + u_0$. Because of these relations we suggest that one does not use much additional effort in constructing bounds, but rather calculate a better approximation in the first place.

As mentioned before, the way to proceed will depend on the type of the problem considered. Sometimes it is advantageous to use bounds of the form $v = -\alpha z$, $w = \beta\zeta$ with two prescribed "simple" functions z, ζ. Then one obtains the following method, which we used for all our special algorithms of error estimation for problems of type (5.14).

Method of Approximation and Estimation. *Choose elements* $\varphi_i \in R$ $(i = 0, 1, 2, \ldots)$ *for constructing an approximate solution* $u_0 = \varphi_0 + \sum_{i=1}^{m} \gamma_i \varphi_i$. *Calculate the constants* γ_i $(i = 1, 2, \ldots, m)$ *such that the defect* $d[u_0] = r - M\varphi_0 - \sum_{i=1}^{m} \gamma_i M\varphi_i$ *becomes "minimal" in a sense which has to be prescribed. Choose appropriate elements* z, ζ *in* R *and calculate constants* α, β *with* $-\alpha Mz \leqslant d[u_0] \leqslant \beta M\zeta$.

Then the error estimate $-\alpha z \leqslant u^* - u_0 \leqslant \beta\zeta$ *holds for the solution* u^* *of the given equation* (5.14).

5.2.2 *An initial value problem*

To obtain *exact* error bounds for solutions of differential equations by the method described in the previous section, one has to estimate the defect $d[u_0](x)$ (very accurately) on the *whole interval* considered. This constitutes a difficult task. Considering a very simple initial value problem, we shall demonstrate how one can proceed. The following method can be generalized to nonlinear equations and systems of nonlinear equations (see Section III.6.2 and Example V.2.4). Although for those problems additional difficulties will have to be overcome, the choice of u_0 and the method to estimate the defect will be essentially the same as explained in this section.

Suppose that a given problem

$$u'(x) + f(x, u(x)) = 0 \qquad (0 < x \leqslant 1), \qquad u(0) = r_0 \tag{5.17}$$

with $f(x, u(x)) = c(x)u(x) - s(x)$, $c \in C_0[0, 1]$, $s \in C_0[0, 1]$ has been "solved numerically," so that approximations $\tilde{u}_i$ for the values $u^*(x_i)$ of the solution at certain points x_i are known, where $0 = x_0 < x_1 < \cdots < x_N = 1$. We want to estimate the errors $u^*(x_i) - \tilde{u}_i$.

The method of error estimation explained in the following can be used for arbitrary approximations $\tilde{u}_i$. However, the estimation method is so constructed that it becomes most efficient for values $\tilde{u}_i$ obtained by a fourth order approximation method, such as the standard Runge–Kutta method.

To apply the estimation principle of the preceding section, we have to find an approximate solution u_0 defined for all $x \in [0, 1]$. We now assume, additionally, that $N = 2\mu$ is even, $x_{2k-1} = \frac{1}{2}(x_{2k-2} + x_{2k})$ for $k = 1, 2, \ldots, \mu$, and that on each interval (x_{2k-2}, x_{2k}) the functions c and s have uniformly continuous derivatives up to the order 5. Then we choose u_0 to be a piecewise polynomial function with the following properties: *on the interval* $I_k = [x_{2k-2}, x_{2k}]$, *the approximate solution* u_0 *coincides with a polynomial of the fifth degree such that*

$$u_0(x_i) = \tilde{u}_i, \qquad u_0'(x_i) = f_i := f(x_i, \tilde{u}_i) \qquad \text{for} \quad i = 2k-2, 2k-1, 2k.$$

Obviously, the given problem can be written as $Mu = r$ with the operator $M: C_1[0, 1] \to \mathbb{R}[0, 1]$ defined by $Mu(x) = u'(x) + c(x)u(x)$ for $0 < x \leqslant 1$, $Mu(0) = u(0)$, and $r(x) = s(x)$ for $0 < x \leqslant 1$, $r(0) = r_0$. This operator M is inverse-positive (cf. Example 3.4), so that for $z \in C_1[0, 1]$

$$\left.\begin{aligned} &|\delta[u_0](x)| \leqslant z'(x) + c(x)z(x) \quad (0 < x \leqslant 1) \\ &|r_0 - u_0(0)| \leqslant z(0) \end{aligned}\right\} \Rightarrow \quad |(u^* - u_0)(x)| \leqslant z(x) \quad (0 \leqslant x \leqslant 1), \tag{5.18}$$

where $\delta[u_0] = -u_0' - f(x, u_0)$ denotes the defect of u_0 with respect to the differential equation.

We shall, however, choose here less smooth functions z such that for $k = 1, 2, \ldots, \mu$:

$$z(x) = z_k(x) \qquad (x_{2(k-1)} < x \leqslant x_{2k}) \qquad \text{with} \quad z_k \in C_1[I_k],$$
$$z_k(x_{2(k-1)}) = \varepsilon_{k-1} \qquad \text{with} \quad z_{k-1}(x_{2(k-1)}) \leqslant \varepsilon_{k-1}$$

(where $z_0(0) := z(0)$). The reader may verify that (5.18) holds also for such a function. Of course now $z'(x)$ denotes a left-hand derivative at the points $x_2, x_4, \ldots$.

The main problem is now to estimate $|\delta[u_0](x)|$ on the entire interval $[0, 1]$. This function oscillates, vanishing at all points x_i. It usually is of small size, but written as a difference of much larger quantities. Consequently, careful estimates will be necessary.

We shall estimate the defect $\delta[u_0]$ on each interval I_k separately. Thus let k be fixed. We then introduce $\delta(t) := \delta[u_0](x)$ with $t = h^{-1}(x - x_{2k-1})$, $2h = x_{2k} - x_{2k-2}$ and write this function as $\delta(t) = P(t) + R(t)$, where

$$P(t) = \tfrac{8}{3}t^2(1 - t^2)\alpha + \tfrac{4}{3}t(1 - t^2)\beta$$

with

$$\alpha = \delta(-\tfrac{1}{2}) + \delta(\tfrac{1}{2}), \qquad \beta = -\delta(-\tfrac{1}{2}) + \delta(\tfrac{1}{2})$$

is an interpolating polynomial for $\delta(t)$ with interpolation points $t = -1$, $-\frac{1}{2}, 0, \frac{1}{2}, 1$. Observe that $\delta(-1) = \delta(0) = \delta(1) = 0$, due to the choice of u_0. Assuming now that c and s have bounded derivatives up to the fifth order on $(x_{2(k-1)}, x_{2k})$, we can write the remainder $R(t)$ as

$$R(t) = -h^5 \frac{1}{5!} t(t^2 - \tfrac{1}{4})(t^2 - 1) f^{[5]}(\xi),$$

where

$$f^{[5]}(x) = d^5/dx^5 f(x, u_0(x)) \qquad \text{and} \qquad \xi \in I_k.$$

Simple estimates yield

$$\begin{aligned} |P(t)| &\leqslant \tfrac{8}{3}t^2(1 - t^2)|\alpha| + \tfrac{4}{3}|t|(1 - t^2)|\beta| \leqslant \tfrac{2}{3}|\alpha| + \tfrac{8}{27}\sqrt{3}|\beta|, \\ |R(t)| &\leqslant h^5 \frac{1}{5!} |t(t^2 - \tfrac{1}{4})(t^2 - 1)|\,|f^{[5]}(\xi)| \\ &\leqslant 0.000946h^5 |f^{[5]}(\xi)|. \end{aligned} \tag{5.19}$$

If the approximations $\tilde{u}_i$ are computed with a fourth order method, we have $P(t) = O(h^4)$ and $R(t) = O(h^5)$ for $h \to 0$ (rounding errors not considered). Consequently, in this case, *a good estimate of* $\delta[u_0](x)$ *on* I_k *can essentially be achieved by calculating the two numbers*

$$\alpha = \alpha_k = \delta[u_0](x_{2k-1} - \tfrac{1}{2}h) + \delta[u_0](x_{2k-1} + \tfrac{1}{2}h)$$

and

$$\beta = \beta_k = -\delta[u_0](x_{2k-1} - \tfrac{1}{2}h) + \delta[u_0](x_{2k-1} + \tfrac{1}{2}h)$$

very accurately and estimating the function $f^{[5]}$ *on* I_k *roughly.*

In order to obtain reliable error bounds, the calculation of $|\alpha| = |\alpha_k|$ and $|\beta| = |\beta_k|$ in (5.19) has to be carried out in such a way that the resulting numbers actually are upper bounds for $|\alpha_k|$ and $|\beta_k|$. This may be considered as applying a type of interval arithmetic.

For being able to estimate $f^{[5]}$ on I_k, this function first has to be expressed in terms of derivatives of "simple" functions, i.e., functions which can easily be estimated from above and below. For example, one can estimate u_0 and its derivatives by linear functions of the six known values x_i, f_i $(i = 2k - 2, 2k - 1, 2k)$, without difficulty. To make the algorithm most effective, one would apply a computer program for calculating the derivative $f^{[5]}$ in terms of functions available on the computer and then apply an interval arithmetic program for estimating the expression so obtained.

We point out that some kind of interval arithmetic has to be applied here at some stage. However, interval calculations are used here only in a very limited way, at the end of the procedure to estimate $\delta[u_0]$. Estimating $\delta[u_0]$ directly by interval arithmetic would not be a feasible approach.

Proceeding in this way one finally obtains a rigorous estimate

$$|\delta[u_0](x)| \leqslant \Gamma_k(x) \qquad (x_{2k-2} \leqslant x \leqslant x_{2k}).$$

Here, $\Gamma_k(x)$ may be a constant or a polynomial, depending on which estimates in (5.19) were used.

Finding an error bound z of the described form poses no major problem. One calculates the functions $z_1, z_2, \ldots, z_\mu$ successively. For example, one may choose

$$z_k(x) = \varepsilon_{k-1} + \int_{x_{2k-2}}^{x} (-\varepsilon_{k-1}c_k + \Gamma_k(\xi)) \exp(c_k(\xi - x))\, d\xi$$

with

$$c_k \leqslant c(x) \qquad \text{for} \quad x_{2k-2} \leqslant x \leqslant x_{2k}.$$

To obtain an upper bound ε_k for $z_k(x_{2k})$ one may replace the exponential function by an approximation such as $1 + c_k(\xi - x_{2k})$, then calculate the integral explicitly obtaining a number γ_k, and finally estimate the "small" difference $|z_k(x_{2k}) - \gamma_k|$, using a formula for the remainder of the Taylor expansion, or the like. In this way, error bounds can be obtained which are globally of the magnitude $O(h^4)$, if the approximations are calculated by a fourth order method. We shall not explain the details. Numerical results which were obtained in this way are provided in the following very simple example. Further examples will be given in Section III.6.2 and Example V.2.4.

Example 5.1. The problem

$$u' - u = -x + (1 + x)^{-1} + (1 + x)^{-2}, \qquad u(0) = 0$$

has the solution $u^*(x) = 1 + x - (1 + x)^{-1}$. Approximate values $\tilde{u}_i \approx u^*(x_i)$ were computed using a fourth order Runge–Kutta method with constant stepsize $h = 0.1$. The errors $\Delta_k = |u^*(x_{2k}) - \tilde{u}_{2k}|$ and the error bounds β_k are shown in Table 1. Here the bounds β_k have the same magnitude as the errors. This is true for small x_{2k}, where the approximations are fairly accurate, as well as for larger x_{2k}, where the approximation methods are unsuitable. The bounds $\tilde{\beta}_k$ for Δ_k were obtained by an improved algorithm which yields estimates $-z \leqslant u^* - u_0 \leqslant \zeta$ such that ζ is not necessarily equal to z. □

TABLE 1

Errors and error bounds for Example 5.1

k	x_{2k}	Error Δ_k	Bound β_k	Bound $\tilde{\beta}_k$
10	2	$1.88E-05$	$2.76E-05$	$1.96E-05$
20	4	$1.39E-04$	$2.05E-04$	$1.46E-04$
30	6	$1.03E-03$	$1.51E-03$	$1.08E-03$
40	8	$7.61E-03$	$1.12E-02$	$8.03E-03$
50	10	$5.62E-02$	$8.26E-02$	$5.97E-02$
60	12	$4.15E-01$	$6.10E-01$	$4.43E-01$
70	14	$3.07E\ \ 00$	$4.51E\ \ 00$	$3.30E\ \ 00$
80	16	$2.27E+01$	$3.33E+01$	$2.45E+01$
90	18	$1.68E+02$	$2.46E+02$	$1.82E+02$
100	20	$1.24E+03$	$1.82E+03$	$1.35E+03$

5.2.3 *Boundary value problems*

The general method of approximation and estimation in Section 5.2.1 will be applied here to boundary value problems

$$L[u](x) := (-au'' + bu' + cu)(x) = s(x) \qquad (0 < x < 1),$$
$$u(0) = r_0, \qquad u(1) = r_1 \tag{5.20}$$

with $a, b, c, s \in C_0[0, 1]$ and $a(x) > 0$ $(0 \leqslant x \leqslant 1)$. (Other linear boundary conditions can be treated analogously.) For a given approximate solution $u_0 \in C_2[0, 1]$ denote by $\delta[u_0] = s - L[u_0]$ the defect with respect to the differential equation. We shall use the following principle of error estimation, which can be derived from Theorem 3.17.

Find $u_0, z \in C_2[0, 1]$ *and* $\alpha \in \mathbb{R}$ *such that* $z \geqslant o$, $\alpha \geqslant 0$, $|u_0(0) - r_0| \leqslant \alpha z(0)$, $|u_0(1) - r_1| \leqslant \alpha z(1)$ *and for* $0 \leqslant x \leqslant 1$

$$|\delta[u_0](x)| \leqslant \alpha L[z](x). \tag{5.21}$$

Then the given problem (5.20) *has a solution* $u^* \in C_2[0, 1]$ *such that* $|u^* - u_0| \leqslant \alpha z$. *(The solution is unique if* $L[z](x) \not\equiv 0$.)

Here we consider only functions u_0 which satisfy the given boundary conditions and apply a procedure consisting of the following three steps:

(1) calculate an approximate solution u_0 with "small" defect $\delta[u_0]$;
(2) choose a suitable *bound function* z satisfying $z \geqslant o$, $Mz \geqslant o$;
(3) verify (5.21) for some $\alpha \geqslant 0$.

In all examples which we treated, finding a bound function z as needed in part (2) of the procedure was no problem. Part (3) of the procedure requires that one find a close upper bound α of the continuous function $|\delta[u_0]|(L[z])^{-1}$. Often $L[z]$ can rather easily be estimated from below, so that then the problem remains to calculate an upper bound of $|\delta[u_0]|$ *on the entire interval.* In the preceding section we solved the analogous problem for a differential equation of the first order, combining analytical tools with a type of interval arithmetic. For boundary value problems one can proceed similarly, dividing the given interval [0,1] into subintervals and estimating the defect $\delta[u_0]$ on each subinterval. The details of the method depend on the nature of the approximation u_0. So far, corresponding estimates were carried out for polynomials u_0. However, we shall not describe this work here, but instead report on estimates which were obtained by verifying the inequality (5.21) for all x in a very fine grid G. More precisely, a constant $\bar{\alpha}$ was calculated such that (5.21) holds for all $x \in G$ and $\bar{\alpha}$ (in place of α); then α was obtained by slightly enlarging $\bar{\alpha}$, say $\alpha = 1.1\bar{\alpha}$. The exact mathematical statement obtained in this way is as follows: *If with the number* α *calculated inequality* (5.21) *holds for all* $x \in [0, 1]$, *then* $|u^* - u_0|(x) \leqslant \alpha z(x)$ $(0 \leqslant x \leqslant 1)$. Thus the problem of estimating an approximate solution of a boundary value problem is reduced to the problem of verifying an inequality for continuous functions.

Concerning step (1) of the preceding procedure, we tested a series of approximation methods which are suitable for the error estimation. For example, we used collocation methods with various sets of collocation points and, moreover, the method of Ritz–Galerkin and the method of least squares with the occurring integrals calculated by various integration formulas. Moreover, several function spaces were used, for example, polynomials and spline functions of various forms. Naturally, one cannot find a simple general procedure for solving "all" problems of type (5.20). Here, we shall only report on a simple Method P described below. This method is limited to problems whose solution do not behave "too badly." On the other hand, in all examples which we treated it turned out that, when the method works, it is by far preferable to all other methods which we tested.

Method P is formulated for problems on the interval $[-1, 1]$ instead of $[0, 1]$; problems on $[0, 1]$ or any other compact interval are first transformed to problems on $[-1, 1]$ in the algorithm. Thus, the boundary conditions are here $u(-1) = r_0$, $u(1) = r_1$; moreover, now $a, b, c, s \in C_0[-1, 1]$.

Method P (Collocation using integrated Legendre polynomials and Chebychev collocation points). Choose

$$\varphi_0 = \tfrac{1}{2}(r_0(1-x) + r_1(1+x)), \qquad \varphi_i(x) = \int_{-1}^{x} L_i(t)\,dt \qquad (i = 1, 2, \ldots, m)$$

where L_i denotes the ith Legendre polynomial (of degree i) and determine the constants γ_i in $u_0 = \varphi_0 + \sum_{i=1}^{m} \gamma_i \varphi_i$ such that $\delta[u_0](x_k) = 0$ $(k = 1, 2, \ldots, m)$ with $x_k = -\cos((2k-1)\pi/2m)$ denoting the zeros of the mth Chebyshev polynomials.

In Examples 5.2–5.6 and the corresponding tables we use the notation

$$\Delta = \|u^* - u_0\|, \qquad \beta = \alpha\|z\|, \qquad \text{where} \quad \alpha = 1.1\bar{\alpha}$$
$$\text{with} \quad \bar{\alpha} = \|\delta[u_0] : L[z]\|$$

and $\|u\| = \{\max |u(x)| : x \in G\}$. Here G is a grid consisting of 501 equidistant points of the interval $[0, 1]$ or $[-1, 1]$ considered, including the endpoints. All approximations u_0 in these examples are obtained by an algorithm based on Method P. Some of the tables also contain the condition number $\kappa = \|A\|_\infty \|A^{-1}\|_\infty$ of the matrix A of the linear equations which determine the parameters γ_i.

Example 5.2 (see Table 2). The problem

$$-u'' + 2xu' + (1 - x^2)u = 1 - x^2, \qquad u(-1) = u(1) = 0$$

TABLE 2

Errors Δ and error bounds β for Example 5.2

m	Δ	β	$\Delta\beta^{-1}$	$\Delta\beta^{-1}m^{-2}$	κ
5	$3.41E-04$	$2.50E-03$	7.3	0.293	23
6	$1.66E-04$	$2.00E-03$	12.0	0.334	38
7	$3.07E-06$	$9.71E-05$	31.6	0.644	59
8	$3.06E-06$	$7.63E-05$	24.9	0.389	87
9	$5.12E-08$	$2.96E-06$	57.8	0.714	122
10	$5.24E-08$	$2.27E-06$	43.3	0.433	165
11	$8.03E-10$	$7.36E-08$	91.6	0.757	217
12	$8.73E-10$	$5.47E-08$	62.6	0.435	279
13	$1.18E-11$	$1.51E-09$	127.2	0.753	352
14	$1.26E-11$	$1.12E-09$	88.5	0.452	437
15	$1.47E-13$	$2.67E-11$	180.8	0.803	534
16	$1.72E-13$	$1.96E-11$	113.5	0.443	644

has the exact solution $u^* = 1 - \exp \frac{1}{2}(x^2 - 1)$. We choose a function z such that $-z'' + 2xz'$ becomes positive: $z = \frac{1}{2}(\exp 1 - \exp x^2)$ with $L[z] = \exp x^2 + (1 - x^2)z$. The error functions $u^* - u_0$ and the corresponding defects $\delta[u_0]$ oscillate, with the number of oscillations increasing with m. Observe that u^* and z are even functions, which explains why Δ and β behave differently for n even or odd. □

Example 5.3 (see Table 3). For problems of the very simple form

$$-u'' + cu = x \qquad \text{with} \quad c = -d^2 < 0 \qquad u(0) = u(1) = 0$$

we explain the influence of the size of the constant c. For $c > -\pi^2$ the exact solution is $u^* = -(c \sin d)^{-1}(\sin dx - x \sin d)$; for $c = -\pi^2$ no solution exists. Table 3 provides results for fixed $m = 12$. □

TABLE 3

Errors Δ and error bounds β for Example 5.3 with $m = 12$

c	$\|u^*\|_\infty$	Δ	β	$\kappa \cdot 100^{-2}$
-9	0.73	$2.09E-12$	$5.89E-08$	33
-9.5	1.72	$9.17E-12$	$1.08E-06$	78
-9.8	9.14	$1.62E-10$	$1.95E-04$	409
-9.86	66.28	$7.06E-09$	$7.71E-02$	2965

Example 5.4 (see Table 4). The boundary value problem

$$-u'' + 400u = -400(\cos \pi x)^2 - 2\pi^2 \cos 2\pi x, \qquad u(0) = u(1) = 0$$

was used by Stoer and Bulirsch [1973] in testing several approximation methods. The exact solution is $u^* = (1 + \exp(-20))^{-1}[\exp(-20(1 - x)) + \exp(-20x)] - \cos^2 \pi x$ with $\|u^*\|_\infty = 0.77$. The error estimates in Table 4 were obtained with $z(x) \equiv 1$. □

TABLE 4

Errors Δ and error bounds β for Example 5.4

m	Δ	β	$\beta : \Delta^{-1}$	$100 \cdot \beta : \Delta^{-1} m^{-2}$	κ
10	$2.82E-03$	$5.63E-03$	2.00	2.00	30
16	$1.81E-06$	$6.44E-06$	3.55	1.39	119
18	$1.05E-07$	$4.43E-07$	4.22	1.30	168
20	$5.10E-09$	$2.54E-08$	4.98	1.25	230
22	$2.11E-10$	$1.23E-09$	5.82	1.20	300
24	$7.50E-12$	$5.05E-11$	6.73	1.17	390
26	$2.38E-13$	$1.81E-12$	7.58	1.12	496

Example 5.5 (see Table 5). The boundary value problem

$$-u'' - [90 \cot \tfrac{1}{6}\pi(1 + x) + 60 \tan \tfrac{1}{6}\pi(1 + x)]u' - 630u = 0,$$
$$u(0) = 0, \qquad u(1) = 5$$

was already considered in Problem 3.6. The solution u^*, which is not known explicitly, increases very fast from $u^*(0) = 0$ to its maximal value ≈ 284, which is attained at approximately $x = 0.022$, and then decreases to the boundary value $u^*(1) = 5$. We calculated error bounds using $z(x) = \exp(-\kappa x)$ and several values of κ. The results in Table 5 correspond to $\kappa = 5.5$. (Slightly better error bounds could be obtained by trying to find an "optimal" κ.) □

TABLE 5

Error bounds β for Example 5.5

m	β	m	β
60	$1.82E - 03$	68	$1.38E - 05$
62	$5.59E - 04$	70	$3.81E - 06$
64	$1.68E - 04$	72	$1.03E - 06$
66	$4.87E - 05$	74	$2.71E - 07$

Example 5.6 (see Table 6). When in the preceding example the factor of u' is replaced by its lower bound -190.53, the following boundary value problem is obtained:

$$-u'' - 190.53u' - 630u = 0, \qquad u(0) = 0, \qquad u(1) = 5.$$

Here the exact solution u^* can be calculated. It behaves similarly to the solution of the preceding problem; its maximal value ≈ 132 is attained at approximately $x = 0.022$. For the error estimation the solution z of $-z'' - 190.53z' - 630z = 1$, $z(0) = z(1) = 0$ was used. □

TABLE 6

Errors Δ and error bounds β for Example 5.6

m	Δ	β	$\kappa \cdot 10^{-3}$
40	$1.47E - 02$	$9.69E - 00$	13.0
50	$8.66E - 05$	$7.54E - 02$	22.9
60	$2.18E - 07$	$2.40E - 04$	37.2
70	$2.44E - 10$	$3.22E - 07$	56.7
74	$3.12E - 11$	$3.54E - 08$	66.2

In calculating error bounds for the examples above rounding errors were not estimated. Let us, therefore, briefly discuss the possible influence of such errors. First, rounding errors occur in calculating the coefficients γ_i. These errors may influence the accuracy of the approximation u_0, but they do not affect the validity of the error estimate. Observe that we estimate $u^* - u_0$ where u_0 is constructed with the numbers γ_i actually calculated. (In our examples, the rounding errors in calculating γ_i also had hardly any effect on the accuracy of u_0.)

Now, if we calculate $u_0(x)$ numerically at some point x, we obtain, in general, a number $\hat{u}_0(x)$ different from $u_0(x)$. Our estimate of $|u^*(x) - u_0(x)|$ makes sense only if $|u_0(x) - \hat{u}_0(x)|$ is reasonably small. The computation of $\hat{u}_0(x)$ essentially requires two steps. First the values of the functions $\varphi_i(x)$ are calculated by suitable recursion formulas. These formulas are numerically very stable. The more "dangerous" step is the second where the sum $\sum \gamma_i \varphi_i(x)$ is calculated. In our examples, we had no real problem with this calculation. For example, a double accuracy computation of the whole procedure essentially gave the same results. Here it is also worth remarking that the numbers $\|\varphi_i\|_\infty$ decrease with increasing i and that, in addition, in our examples the constants γ_i became smaller from a certain index on.

Rounding errors, however, may have influence in computing values of the defect $\delta[u_0](x)$. This term depends on $u_0(x)$, $u_0'(x)$ and $u_0''(x)$, where the latter two numbers are calculated similarly as $u_0(x)$. The rounding errors in these calculations have more influence here, since $\delta[u_0](x)$ is a very small number calculated as a linear combination of much larger numbers. If the rounding errors in calculating $\delta(x) = |\delta[u_0](x)| : L[z](x)$ are larger than this value itself, we obtain a very bad approximation $\hat{\delta}(x)$; but any upper bound for the latter value is also an upper bound for $\delta(x)$, so that the validity of the error estimate is not affected. Rounding errors which are considerably smaller than $\delta(x)$ can be taken into consideration by slightly enlarging the bounds for $\hat{\delta}(x)$. Only if the rounding errors are of about the same size as $\delta(x)$ itself, may $\hat{\delta}(x)$ be considerably smaller than $\delta(x)$, so that we cannot estimate $\delta(x)$. This can only happen, however, if the rounding errors are negative. Now, in our examples, for many values of x the values $\delta(x)$ were very close to $\max\{\delta(x) : x \in G\}$, the number used in the error estimate. Thus, we cannot estimate the maximum of $\delta(x)$ on G only when the rounding errors for those many values of x are about the same size as $\delta(x)$ and, in addition, all negative. Of course, there is no doubt, that an exact estimate of the rounding errors and—even more important—an estimate of $\delta(x)$ on the entire interval $[0, 1]$ is desirable.

Finally, let us discuss the range of the applicability of the simple Method P. According to our experience, Examples 5.5 and 5.6 represent about the limit of what can be treated by this method. We point out, however, that for

Examples 5.1–5.6 Method P was definitely preferable to several approximation procedures using Spline functions which we tested, among them the Ritz–Galerkin method for the usual third order C_2-splines, and the collocation method for piecewise third order Hermite polynomials with two Gaussian collocation points per subinterval.

In Example 5.7 the success of Method P was limited, one reason being that the storage capacity was not large enough. In such a singular perturbation problem with a boundary layer at $x = 1$ one may use piecewise polynomial functions with variable stepsize. The experience with Method P even suggests that one try to approximate the solutions by polynomials as in Method P, except in a small interval containing the boundary layer.

Example 5.7 (see Table 7). The problem

$$-\varepsilon u'' + (2 - x^2)u = 1, \qquad u(-1) = u(1) = 0 \qquad \text{with} \quad \varepsilon = 10^{-8}$$

was treated by Russell and Shampine [1972]. It has an "asymptotic solution" $(2 - x^2)^{-1} - \exp\{-(1 + x)\varepsilon^{-1/2}\} - \exp(x - 1)\varepsilon^{-1/2}$ with maximal values ≈ 1 very close to the boundary points. Method P gave the results in Table 7, where the error bounds are calculated with $z(x) \equiv 1$. (Here, for $m = 150$ or 200 a grid G of 2001 points was used, for $m = 250$ or 300 a grid of 3001 points.) Clearly, this problem is better solved with other methods. Still, we obtained some results even for this singular perturbation problem with Method P, and it is interesting to see in which way the method fails. □

TABLE 7
Error bounds β for Example 5.7

m	β
100	$8.20E-01$
150	$3.74E-01$
200	$1.20E-01$
250	$3.10E-02$
300	$6.50E-03$

6 FURTHER RESULTS

6.1 Differential operators of the fourth order

The theory of inverse-positive differential operators of order higher than 2 is much more complicated than that for second order operators. We already treated a simple example in Section 4.1.3. The reduction method explained there can be carried over to more general operators of the fourth order. We shall briefly describe some results obtained in this way (Theorems 6.1 and 6.2). The construction of the functions which occur in these theorems can be avoided by considering suitable *sets* of operators, similarly as in Theorem 1.8. A corresponding result will be stated without proof (Theorem 6.3). It is related to oscillation theory (Theorem 6.4). (For a different result on fourth order operators see Problem V.4.5.)

For simplicity, we consider only self-adjoint operators M. Let

$$L[u] = (au'')'' - (bu')' + cu \tag{6.1}$$

with $a \in C_2[0, 1]$, $b \in C_1[0, 1]$, $c \in C_0[0, 1]$, $a(x) > 0$ $(0 \leqslant x \leqslant 1)$,

$$U_i[u] = \sum_{k=0}^{\mu_i} \alpha_{ik}(-1)^k u^{[k]}(0), \qquad V_i[u] = \sum_{k=0}^{\nu_i} \beta_{ik}(-1)^k u^{[k]}(1) \qquad (i = 1, 2)$$

with $u^{[0]} = u$, $u^{[1]} = u'$, $u^{[2]} = au''$, $u^{[3]} = (au'')' - bu'$, and

$$0 \leqslant \mu_1 < \mu_2 \leqslant 3, \qquad \alpha_{1,\mu_1} = \alpha_{2,\mu_2} = 1, \qquad \alpha_{2,\mu_1} = 0,$$

$$0 \leqslant \nu_1 < \nu_2 \leqslant 3, \qquad \beta_{1,\nu_1} = \beta_{2,\nu_2} = 1, \qquad \beta_{2,\nu_1} = 0.$$

We define R to be the set of all $u \in C_4[0, 1]$ which satisfy the boundary conditions $U_i[u] = V_i[u] = 0$ $(i = 1, 2)$ and $M: R \to S = C_0[0, 1]$ to be the operator given by $Mu(x) = L[u](x)$ $(0 \leqslant x \leqslant 1)$. We assume that $\int_0^1 Mu(x)v(x)\,dx = \int_0^1 u(x)Mv(x)\,dx$ for all $u, v \in R$.

To describe the strictly positive elements in R (with respect to the natural order relation), we consider integers $\lambda_1, \lambda_2, \omega_1, \omega_2$ such that

$$0 \leqslant \lambda_1 < \lambda_2 \leqslant 3, \qquad \lambda_j \neq \mu_i \qquad (i, j = 1, 2);$$

$$0 \leqslant \omega_1 < \omega_2 \leqslant 3, \qquad \omega_j \neq \nu_i \qquad (i, j = 1, 2).$$

For example, λ_1 is the smallest integer $\geqslant 0$ such that $u^{[\lambda_1]}(0) \neq 0$ for some $u \in R$. A function $u \geqslant o$ in R is strictly positive if and only if $\ell_\xi u > 0$ for all $\xi \in [0, 1]$, where

$$\ell_0 u = u^{[\lambda_1]}(0), \qquad \ell_1 u = u^{[\omega_1]}(1), \qquad \ell_\xi u = u(\xi) \qquad \text{for} \quad 0 < \xi < 1.$$

Theorem 6.1. *The operator $M: R \to S$ is inverse-positive if and only if the following conditions are satisfied.*

I. *For each $u \in R$ which satisfies $u \geqslant o$ and $\ell_\xi u = 0$ for some $\xi \in [0, 1]$ there exists a function $g > o$ in $C_4'[0, 1] \cap C_2[0, 1]$ such that $\int_0^1 g(x)Mu(x)\,dx \leqslant 0$.*

II. *There exists a function $z \in R$ with $z \geqslant o$, $Mz(x) > 0$ $(0 \leqslant x \leqslant 1)$.*

Proof. The sufficiency of the conditions follows from the monotonicity theorem and Proposition 1.3. If M is inverse-positive, then I holds for

$g(x) = G(x, \xi)$ with G the corresponding Green function, and II is satisfied for $z \in R$ with $Mz(x) \equiv 1$. □

To verify assumption I one may construct a family of functions g_ξ $(0 \leqslant \xi \leqslant 1)$ such that $\ell_\xi Mu \leqslant 0$ holds for each $u \in R$ with $u \geqslant o$ and $\ell_\xi u = 0$, where $\ell_\xi U = \int_0^1 g_\xi(t)U(t)\,dt$ defines the functional ℓ_ξ on S. In particular, one may use functions g_ξ of the form

$$g_0 = \varphi; \qquad g_1 = \psi; \qquad \begin{aligned} g_\xi(x) &= \psi(\xi)\varphi(x) \quad &&\text{for} \quad 0 < \xi \leqslant x < 1, \\ g_\xi(x) &= \psi(x)\varphi(\xi) \quad &&\text{for} \quad 0 < x \leqslant \xi < 1 \end{aligned}$$

with positive $\varphi, \psi \in C_4[0, 1]$ (cf. (4.19)). To obtain sufficient conditions, one transforms $\ell_\xi Mu$ into the sum of a second order differential operator and an integral operator evaluated at ξ, by partially integrating four times (cf. (4.13)). Using this form one can formulate sufficient conditions on φ and ψ such that $\ell_\xi Mu \leqslant 0$. In this way the following result is obtained from Theorem 6.1.

Theorem 6.2. *The operator $M : R \to S$ is inverse-positive if assumption* II *of Theorem* 6.1 *is satisfied and if there exist functions $\varphi > o, \psi > o$ in $C_4[0, 1]$ such that*

$$L[\varphi] \leqslant o, \quad U_1[\varphi] \geqslant 0, \quad V_1[\varphi] = 0, \quad V_2[\varphi] \geqslant 0;$$

$$L[\psi] \leqslant o, \quad U_1[\psi] = 0, \quad U_2[\psi] \geqslant 0, \quad V_1[\psi] \geqslant 0;$$

$$(\varphi\psi' - \varphi'\psi) \geqslant o, \qquad (\varphi^2 + \psi^2)(x) > 0 \quad (0 < x < 1).$$

In particular, one may choose $\varphi = \Phi, \psi = \Psi$ with Φ and Ψ defined by

$$L[\Phi] = 0, \quad U_1[\Phi] = 0, \quad U_2[\Phi] = -1, \quad V_1[\Phi] = 0, \quad V_2[\Phi] = 0;$$

$$L[\Psi] = 0, \quad U_1[\Psi] = 0, \quad U_2[\Psi] = 0, \quad V_1[\Psi] = 0, \quad V_2[\Psi] = -1.$$

Such functions exist if M is invertible. We call Φ the *right limit function* and Ψ the *left limit function corresponding to M*. The functions considered in (4.7) are the limit functions for the special operator considered there. We had seen in Section 4.1.3 that some properties of these functions are closely related to the inverse-positivity of the operator. In particular, the upper bound c_0 of the interval $(-\pi^4, c_0]$ of values c for which M in (4.5) is inverse-positive was shown to be an eigenvalue of the homogeneous boundary value problem (4.9) (or (4.10)). We shall now present a more general result, where a set $\mathscr{M}$ of inverse-positive operators M is considered and the "boundary" of this set is determined by the property that certain homogeneous boundary value problems have nontrivial solutions.

Let $\mathscr{M}$ denote a set of operators $M : R_M \to S$ of the form described above. These operators need not have the same domain R, that is, the boundary conditions may also differ. For brevity, we shall write $M = (L, U_1, U_2, V_1, V_2) = (L, U_i, V_i)$. Each $M \in \mathscr{M}$ can also be characterized by the coefficient vector $\not{k} = (a, b, c, \alpha_{11}, \ldots, \beta_{23})$. We assume that $\mathscr{M}$ is piecewise analytically connected, i.e., to each pair M_0, M_1 of operators in $\mathscr{M}$ there exists a connecting family of operators $M_t \in \mathscr{M}$ $(0 \leqslant t \leqslant 1)$ such that the corresponding coefficient vectors $\not{k}_t$ depend piecewise analytically on t. The following theorem is stated without proof.

Theorem 6.3. *Suppose that for each operator* $M = (L, U_i, V_i)$ in $\mathscr{M}$ *none of the following three homogeneous boundary value problems has a nontrivial solution in* $C_4[0, 1]$:

(1) $L[u] = o,\ U_1[u] = U_2[u] = 0,\ V_1[u] = V_2[u] = 0;$
(2) $L[u] = o,\ U_1[u] = U_2[u] = u^{(\lambda_1)}(0) = 0,\ V_1[u] = 0;$
(3) $L[u] = o,\ U_1[u] = 0,\ V_1[u] = V_2[u] = u^{(\omega_1)}(1) = 0.$

Assume, moreover, that at least one operator $M_0 \in \mathscr{M}$ *is inverse-positive and that the limit functions of this operator satisfy* $(\Phi\Psi' - \Phi'\Psi)(x) > 0$ $(0 < x < 1)$.

Then all operators $M \in \mathscr{M}$ *are inverse-positive.*

From this theorem one can derive relations between inverse-positivity and oscillation theory. Consider a fixed operator $M : R \to S$ and define $V_i[u](x)$ by the same formula as $V_i[u]$, except that $u^{[k]}(1)$ is replaced by $u^{[k]}(x)$.

Theorem 6.4. *Suppose that* M *is invertible and for each* $\eta \in (0, 1)$ *none of the following three boundary value problems has a nontrivial solution*:

(1) $L[u](x) = 0\ (0 \leqslant x \leqslant \eta),$
$\quad U_1[u] = U_2[u] = 0, \quad V_1[u](\eta) = V_2[u](\eta) = 0;$
(2) $L[u](x) = 0\ (0 \leqslant x \leqslant \eta),$
$\quad U_1[u] = U_2[u] = u^{(\lambda_1)}(0) = 0, \quad V_1[u](\eta) = 0;$
(3) $L[u](x) = 0\ (0 \leqslant x \leqslant \eta),$
$\quad U_1[u] = 0, \quad V_1[u](\eta) = V_2[u](\eta) = u^{(\omega_1)}(\eta) = 0.$

Moreover, assume that also the problem

$$u^{\text{iv}}(x) = 0 \quad (0 \leqslant x \leqslant 1), \qquad (-1)^{\mu_i} u^{(\mu_i)}(0) = 0, \quad u^{(\nu_i)}(1) = 0 \quad (i = 1, 2)$$

has only the trivial solution.

Then the operator $M = (L, U_i, V_i)$ *is inverse-positive.*

Proof. For fixed $\eta \in (0, 1]$ we introduce a new independent variable $\xi = \eta^{-1}x$ and write $L[u](x)$, $U_i[u]$, $V_i[u](\eta)$ as linear combinations of derivatives of ξ, where $\xi \in [0, 1]$. In this way, we obtain an operator $M_\eta = (L_\eta, U_{i,\eta}, V_{i,\eta})$ where $\eta^4 L[u](x) = L_\eta[u](\xi)$, $\eta^{\mu_i} U_i[u] = U_{i,\eta}[u]$, $\eta^{\nu_i} V_i[u](\eta) = V_{i,\eta}[u]$. If in the formulas for the coefficients of M_η the value η is replaced by 0, we obtain an operator M_0 such that $L_0[u](x)$, $U_{i,0}[u]$, and $V_{i,0}[u]$ equal $u^{\text{iv}}(x)$, $(-1)^{\mu_i} u^{(\mu_i)}(0)$, and $u^{(\nu_i)}(1)$, respectively, except possibly constant factors > 0. For the set $\mathscr{M}$ of all M_η $(0 \leqslant \eta < 1)$ the assumptions of Theorem 6.3 are satisfied. In particular, one shows by elementary calculations that the above operator M_0 has all properties required of M_0 in that theorem. Consequently, all M_η $(0 \leqslant \eta < 1)$ are inverse-positive. The inverse-positivity of $M = M_1$ then follows from the fact that the Green function of M_η depends continuously on η. $\square$

Problem

6.1 Find a set $G \in \mathbb{R}^2$ such that for each $(b, c) \in G$ the operator M determined by $L[u] = u^{\text{iv}} - bu'' + cu$ and the boundary conditions $u(0) = u''(0) = 0$, $u(1) = u'(1) = 0$ is inverse-positive.

6.2 Duality

For many properties which occur in the theory of inverse-positive linear operators one can formulate equivalent dual properties. As examples, we consider here duals for the inverse-positivity itself and for the existence of a majorizing element. In the following, we assume that assumption (A_1) in Section 1.1 is satisfied, that $(S, \leqslant)$ contains a strictly positive element and that $M : R \to S$ denotes a linear operator. The elements of R and S will be denoted by u and U, respectively.

Theorem 6.5. (i) *If the range MR contains a strictly positive element in S, then the following properties P and D are equivalent*:

P: $u \in R$, $Mu \geqslant o \Rightarrow u \geqslant o$.
D: *For each* $\ell > o$ *in* R^* *there exists an* $\mathscr{f} = \mathscr{f}_\ell > o$ *in* S^* *with* $\mathscr{f}M = \ell$.

(ii) *The following properties* II_p *and* II_d *are equivalent*:

II_p: *There exists a* $z \in R$ *with* $z \geqslant o$, $Mz \succ o$.
II_d: $\mathscr{f} \in S^*$, $\mathscr{f}M \leqslant o \Rightarrow \mathscr{f} \not> o$.

Proof. (i) If P holds, then for each $\ell > o$ in R^* $A \cap B_\ell = \varnothing$ where $A = \{(U, t) \in S \times \mathbb{R} : U \geqslant o, t < 0\}$, $B_\ell = \{(U, t) \in S \times \mathbb{R} : U = Mu, t = \ell u, u \in R\}$. Since $A^c = \{(U, t) \in A : U \succ o\} \neq \varnothing$, there exists (to each such ℓ) a functional $(\mathscr{g}, -\mu) \neq o$ on $S \times \mathbb{R}$ with $\mathscr{g} \in S^*$, $\mu \in \mathbb{R}$ such that $(\mathscr{g}M - \mu\ell)u = 0$ for $u \in R$; $\mathscr{g}U - \mu t \geqslant 0$ for $U \geqslant o$, $t < 0$; $\mathscr{g}U - \mu t > 0$ for $U \succ o$, $t < 0$ (cf. Theorem I.2.6). The first of these three relations yields $\mathscr{g}M = \mu\ell$, the second $\mu \geqslant 0$ and $\mathscr{g} \geqslant o$. Here $\mu = 0$ cannot hold, since then $\mathscr{g}M = o$, while $\mathcal{g}Mu_0 > 0$ for strictly positive $Mu_0 \in S$. Thus $\mu > 0$, and $\mathcal{f}M = \ell$, $\mathcal{f} > o$ for $\mathcal{f} = \mu^{-1}\mathcal{g}$.

If D holds, then $Mu \geqslant o$ implies $\mathcal{f}_\ell Mu = \ell u \geqslant 0$ for each $\ell > o$ in R^* and hence $u \geqslant o$ (cf. Proposition I.1.5).

(ii) If II_p holds, but $\mathcal{f}M \leqslant o$, $\mathcal{f} > o$ for some $\mathcal{f} \in S^*$, then $(\mathcal{f}M)z \leqslant 0$ due to $\mathcal{f}M \leqslant o$ and $z \geqslant o$, and $\mathcal{f}(Mz) > 0$ due to $\mathcal{f} > o$ and $Mz \succ o$; contradiction.

If II_d holds, but $MK \cap C^c = \emptyset$, then there exists an $\mathcal{f} \in S^*$ such that $\mathcal{f}U \geqslant 0$ for all $U \geqslant o$, $\mathcal{f}Mu \leqslant o < \mathcal{f}U$ for all $u \geqslant o$ and $U \succ o$, and hence $\mathcal{f}M \leqslant o$, $\mathcal{f} > o$; contradiction. □

Corollary 6.5a. *Suppose that the range MR contains a strictly positive element in S and that there exists an $\mathcal{f} > o$ in S^* such that $\mathcal{f}M \leqslant o$. Then M is not inverse-positive.*

Proof. This follows from the equivalence of II_p and II_d and the fact that here II_p is necessary for inverse-positivity (cf. Theorem 1.6(i)). □

Example 6.1. *Suppose that $R = S = \mathbb{R}^n$ and the $n \times n$ matrix M has an eigenvalue $\lambda < 0$ corresponding to an eigenvector $\varphi > o$. Then M is not inverse-positive.* For proving this for invertible M, apply the corollary to M^{T} instead of M and define $\mathcal{f}M^{\mathrm{T}} = \varphi^{\mathrm{T}}M^{\mathrm{T}}$. □

Problem

6.2 Show that MR contains no strictly positive element in S if and only if $\mathcal{f}M = o$ for some $\mathcal{f} > o$ in S^*.

6.3 A matrix theorem of Lyapunov

As an interesting and "unusual" application of the theory of inverse-positive linear operators, we shall give here a proof of a theorem of Lyapunov, which is important in the stability theory of differential equations.

Theorem 6.6. *For each complex-valued $n \times n$ matrix C, the following two properties are equivalent:*

(i) *Each eigenvalue of C has a real part greater than 0.*

(ii) *There exists a complex-valued positive definite hermitian $n \times n$ matrix Z such that $CZ + ZC^*$ is positive definite.*

Proof. Since the theorem is true for the matrix C if and only if it is true for a matrix λC with $\lambda > 0$, we may assume that $I + C$ is nonsingular, without loss of generality.

We define R to be the set of all complex-valued hermitian $n \times n$ matrices $U, V, \ldots$. This set is an n^2-dimensional real linear space when the sum $U + V$

and the product λU with $\lambda \in \mathbb{R}$ are defined in the usual way. We introduce a linear order in R by writing $U \geqslant O$ for positive semidefinite U. Then the strictly positive elements $U \succ O$ in this space $(R, \leqslant)$ are the positive definite matrices. Moreover, we use $(S, \leqslant) = (R, \leqslant)$ and define $U > O$ by $U \succ O$. Finally, we define a linear operator $M: R \to R = S$ by $MU = CU + UC^*$. Then property (ii) is equivalent to:

(iii) *there exists a $Z \in R$ such that $Z \succ O$ and $MZ \succ O$.*

Since here the general assumptions (A_1) and (A_2) in Section 1.1 are satisfied, we can apply the theory in Sections 1 and 4. The operator M can be written as $M = A - B$ with $AU = (I + C)U(I + C^*)$, $BU = U + CUC^*$. It is easily seen that A is inverse-positive and B is positive. Thus, if follows from the results of Section 1.2 that M is inverse-positive if and only if (iii) holds. (Apply, in particular, Theorems 1.2, 1.6(i), and Proposition 1.4, and observe that $MR = R$ for inverse-positive M.)

Moreover, (iii) holds if and only if

(iv) *there exists a $Z \in R$ with $Z \succ O$, $(I - T)Z \succ O$ where $T = A^{-1}B$.*

For proving this equivalence the inverse-positivity of A is employed. On one hand, $MZ \succ O$ implies $(I - T)Z \succ O$. On the other hand, if (iv) holds, then $I - T$ is inverse-positive and hence M is inverse-positive.

Finally, (iv) is equivalent to $\rho(T) < 1$, according to Theorem 4.18(iii). Thus, it remains to be shown that $\rho(T) < 1$ is equivalent to (i).

Now, let γ_i $(i = 1, 2, \ldots, n)$ denote the eigenvalues of C, i.e., the zeros of the characteristic polynomial of C. Then the eigenvalues of $T = A^{-1}B$ are

$$\lambda_{ij} = (1 + \gamma_i)^{-1}(1 + \bar{\gamma}_j)^{-1}(1 + \gamma_i\bar{\gamma}_j) \qquad (i, j = 1, 2, \ldots, n) \tag{6.2}$$

(see the Notes). Applying Schwartz' inequality one sees that $|\lambda_{ij}|^2 \leqslant \lambda_{ii}\lambda_{jj}$. Consequently, $|\lambda_{ij}| < 1$ holds for all indices i, j if and only if $\lambda_{ii} < 1$ holds for all indices i. These two properties, however, are equivalent to $\rho(T) < 1$ and property (i), respectively. □

NOTES

Section 1

A series of results in "classical" mathematics can be interpreted as results on inverse-positivity. For instance, the maximum principle for elliptic differential equations and the theory of M-matrices belong to this field. The first abstract formulation of this property is due to Collatz [1952] who introduced the concept of "problems of monotonic kind" (Aufgaben monotoner Art) and corresponding "operators of monotonic kind." In particular, he made clear the usefulness of this property in numerical mathematics, where it is now a common tool. In our terminology, a linear operator of monotonic kind is an inverse-positive operator, and a nonlinear operator of monotonic kind is an inverse-monotone operator. (This terminology differs slightly from that introduced by Schröder [1962].) By some authors, an inverse-positive operator is simply called positive. See, for example, Beckenbach and Bellman [1961], where a survey is given.

The above theory on inverse-positive linear operators was developed by Schröder [1961a, 1963, 1965b, 1975a]. For example, the monotonicity theorem is contained in the second paper, while the special case of operators $M = A - B$ with inverse-positive A and positive B was treated in the first. The theory on noninvertible operators in Section 1.5 was given by Schröder [1971]. The class of operators studied there generalizes a matrix class investigated by Fiedler and Pták [1966] (see the Notes to Section 2).

Section 2

The concept of M-matrices has been introduced by Ostrowski [1937, 1956] (who used first the term M-determinant). Ostrowski defined M-matrices by properties of their minors (cf. Problem 2.13). The paper mentioned first deals mainly with determinant theory and contains references to further work on that subject (by Minkowski, Hadamard, Markoff, and others). However, the inverse-positivity of M-matrices was also recognized (Satz II). Applications to iterative procedures were given in the second paper. Earlier results related to the inverse-positivity of M-matrices are due to Stieltjes [1887] and Frobenius [1908, 1912]. Also, see Goheen [1949], de Rham [1952], and Egerváry [1954].

As we have seen, M-matrices can be defined by several different, but equivalent properties. Numerous other equivalences have been derived. Systematical treatments and surveys were first undertaken by Fan [1958] and Fiedler and Pták [1962]. While Fan used topological methods of proof, the proofs of Fiedler and Pták are algebraic. More recent surveys are given by Varga [1976], Plemmons [1977], Schröder [1978]. Concerning numerical applications and the relation to inverse-positive matrices (*matrices of positive type*) see also Collatz [1952, 1964]. Ando [1980] considers mainly aspects of the theory of M-matrices which are not treated here, for example, certain relations (inequalities) between M-matrices and functions of M-matrices. A survey on M-matrices and related concepts is also given in the recent book of Berman and Plemmons [1979].

Fiedler and Pták [1962] considered a series of matrix classes **Z**, **K**, **P**, **S**, . . . , each of which contains the class of all M-matrices. We have adopted the notation of the class **Z**. It should be pointed out, however, that Fiedler and Pták denoted the set of all M-matrices by **K**, while we have used the symbol **M**. In the theory of Fiedler and Pták [1966], the letter **M** denotes the matrix class defined by the properties formulated in Problem 2.8. This class consists exactly of those matrices M which satisfy the assumptions of Theorem 1.9 (with respect to the natural order $\leqslant$). We also remark, that matrices $M \in \mathbf{Z}$ with diagonal elements $m_{ii} > 0$ are called L-matrices by Young [1971]. Moreover, Ostrowski [1937] defined a *Minkowski determinant* to be the determinant of an M-matrix M with $Me \geqslant o$.

The equivalence statements in Theorem 2.6 have been proved by Fan [1958] and Fiedler and Pták [1962]. (However, see also Ostrowski [1956], Hilfssatz H, where a related result is formulated.) The properties of M which are assumed in Proposition 2.7a are closely related to M being of *positive type* in the sense of Bramble and Hubbard [1964a]. Essentially, a matrix of positive type can be characterized as an M-matrix which has properties (i), (ii) in Proposition 2.7a and satisfies $Me > o$. *Matrices of generalized positive type* defined by Varga [1976] can be characterized in the same way with $Mz > o$, $z \succ o$ in place of $Me > o$. The result of Proposition 2.7b is due to Beauwens [1976] and Varga [1976]. Theorem 2.8 is derived from the abstract continuation theorem 1.8. Ostrowski [1956] has already considered sets of M-matrices which depend continuously on a parameter (see Hilfssatz D). For M having off-diagonal elements $m_{ik} < 0$, Theorem 2.9 has been proved by Stieltjes [1887]. For that reason, symmetric M-matrices are often called *Stieltjes matrices* (see Varga [1962]). For Theorem 2.10 see Ostrowski [1937] and Fiedler and Pták [1962]. In the latter paper and that of Fan [1958] also Corollary 2.14b can be found, which result contains Theorem 2.9. Problem 2.13 deals with some further equivalences which are typical for the literature on M-matrices. Concerning proofs for the equivalence of various properties of M-matrices see the survey papers mentioned above and also Kotelyanskii [1952], Schneider [1953], Wong [1955], Burger [1957].

Since each M-matrix can be written as $M = pI - B$ with $p > 0$ and $B \geqslant O$, the eigenvalue theory of positive matrices (and positive operators) is very closely related to the eigenvalue theory of M-matrices. In particular, the theory of Perron [1907] and Frobenius [1912] plays an important role. For other simple proofs of the Perron–Frobenius results, see Wielandt [1950], Householder [1958, 1964], Varga [1962], and Rheinboldt and Vandergraft [1973], for example. For $A = I$ and irreducible $B \geqslant O$, Wielandt [1950] also proved the result of Theorem 2.16(iii). For the same case, the inclusion statement of Corollary 2.15a (with $\sigma(A, B) = \rho(B)$) has been derived by Collatz [1942], who also considered diagonal matrices $A \in \mathbf{M}$. A survey on positive matrices is given by Schaefer [1974].

The characterization of M-matrices in Corollary 2.12a is due to Ostrowski [1956] (see Satz X for the necessity statement, which is the essential part of the result). Corollary 2.12a has been generalized by Willson [1971, 1973], who, for example, considered the inverse-positivity of $AD + B$ with $A, B \in \mathbf{Z}$, $D \in \mathbf{M}$.

Since a diagonally positive Z-matrix satisfies the generalized row–sum criterion if and only if it is inverse-positive, the vast literature on iterative procedures also contains many results on inverse-positivity in a more or less different phrasing. That the row–sum criterion and the column–sum criterion are sufficient for the convergence of the Gauß–Seidel method has already been proved by Nekrasov [1892] and Mehmke [1892], respectively (see also Ostrowski [1956]). These criteria have often been rediscovered. Systematical treatments are due to v. Mises and Pollaczek-Geiringer [1929], Geiringer [1948], Collatz [1950, 1964], Bodewig [1959]. A series of recent papers deal with equivalences to the (generalized) row–sum criterion, in a way similar to the treatment of equivalences for M-matrices (see Walter [1967], Schäfke [1968], Müller [1971], for example, and the book of Bohl [1974], in particular, Satz 6.12.) A thorough description and comparison of these and other related results is given by Varga [1976].

In the theory of iterative procedures, Varga [1960, 1962] introduced the concept of a regular splitting as a splitting $M = A - B$ with inverse-positive A and positive B. This author proved, by estimation of the spectral radius, that each iterative procedure corresponding to such a splitting converges if M is inverse-positive, and he gave a series of comparison results. Example 2.12 yields the convergence criterion of Sassenfeld [1951]. Part (ii) of Corollary 2.20a is known as the theorem of Stein–Rosenberg (see Stein and Rosenberg [1948], and also Kahan [1958]). Matrices $|D| - |L| - |U|$ as in Corollary 2.21a have already been used by Ostrowski [1956] to investigate iterative procedures for systems with a matrix $M = D - L - U$. A matrix $M \in \mathbb{R}^{n,n}$ is called an H-*matrix* if $|D| - |L| - |U|$ is an M-matrix (Ostrowski [1937, 1956]).

For a detailed study of the iterative procedures considered in Section 2.5 the reader is referred to Varga [1962], Young [1971], or Bohl [1974]. A recent survey on the role of M-matrices and certain generalizations of M-matrices in the theory of iterative procedures is given by Beauwens [1979]. The paper of Varga [1976] already mentioned also describes the relation between the concept of M-matrices (H-matrices) and the convergence of relaxation methods such as the SOR-method in form of equivalence statements, for suitable relaxation parameters. In a similar way, the SSOR-method is treated by Alefeld and Varga [1976].

Concerning literature related to Section 2.5.4 see the Notes to Section I.3 and the references given by Collatz [1964] and Bohl [1974]. In particular, the latter book cites literature on the construction of admissible starting vectors for iterative procedures (with works of Albrecht, Bohl, Braess, Schmidt, Schröder, and Schwetlick). Concerning monotonic approximations see also Albrecht [1961, 1962].

Recently, the investigation of M_0-matrices and also certain rectangular matrices related to M-matrices has found increasing interest. (Ostrowski [1956] defined an M-matrix in such a way that it is nonsingular, despite the fact that earlier Ostrowski [1937] also considered "uneigentliche M-Determinanten," which belong to singular matrices. Some authors now use also the term "singular M-matrix.") For M_0-matrices and certain rectangular matrices one is mainly interested in the nature of the solutions of the homogeneous equation $Mu = o$, in the positivity of certain generalized inverses, and in applications to iterative procedures. We mention here the papers of Schneider [1954], Mangasarian [1968], Ben-Israel [1970], Plemmons [1972], Berman and Plemmons [1972]. A series of further references can be found in the survey of Poole and Boullion [1974], and also in the more recent papers of Plemmons [1976], Meyer and Stadelmaier [1978], Neumann and Plemmons [1978], and in the book of Berman and Plemmons [1979].

Z-matrices and M-matrices also occur in economic applications. In particular, Leontief [1951] developed a model of interacting industries—frequently called *input–output analysis*—which involves Z-matrices. For that reason, M-matrices are sometimes called *Leontief matrices.* Examples 2.4, 2.5, and 2.6, which are taken from Johnston *et al.* [1966], describe such applications. The reader may also consult Karlin [1959], Hadley [1962], and Koehler *et al.* [1975], for example. Furthermore, M-matrices occur in the theory of linear complementary problems and generalizations. In this context Cottle and Veinott [1972] provided characterizations of M-matrices which differ from the "usual" ones. See also Cottle [1975] and Kaneko [1978].

Section 3

The main purpose of Section 3 was to derive results on inverse-positive differential operators from the abstract theory in Section 1 and to show how these results can be applied to obtain various other statements. The approach in Section 3.2–3.7 to derive statements on differential inequalities has been used by Schröder [1961a, 1963, 1965b, 1971].

Many of the results were known before, at least in a more or less similar form. Often statements on inverse-positivity are derived from (strong) boundary maximum principles under appropriate conditions on the coefficients. A survey of this approach is given by Protter and Weinberger [1967]. (In particular, see there Theorem 11, which essentially says that $M = (L, B_0, B_1)$ is inverse-positive, if a function z with $z(x) > 0$ $(0 \leqslant x \leqslant 1)$, $Mz > o$ exists.)

There exist various other methods for proving results on inverse-positivity of differential operators. Collatz [1952, 1964] gave a simple proof for the inverse-positivity of regular operators $M = (L, B_0, B_1)$ with $c(x) \geqslant 0$ $(0 \leqslant x \leqslant 1)$. Bellman [1957] proved the nonnegativity of the Green function by variational principles. The results obtained are related to Theorem 3.11. The method used, however, is more closely related to the approach in Section III.3 (see the Notes there). Bellman's result is also described by Beckenbach and Bellman [1961]. These authors give a survey on inverse-positivity (using, however, the term "positivity").

A general theory of positive reproducing kernels has been developed by Aronszajn and Smith [1957]. This theory also yields statements on the positivity of the Green function. In particular, a result such as Theorem 3.11 follows.

An extensive bibliography on the boundary maximum principle for elliptic partial differential operators and related results on differential inequalities is provided by Protter and Weinberger [1967]. Naturally, the results on partial differential operators contain also results on ordinary differential operators. For example, for sufficiently smooth coefficients, Theorem 3.27 is a special case of the Phragmèn–Lindelöff principle in Theorem 19 of Protter and Weinberger [1967]. Moreover, the first proof of Corollary 3.14a uses essential ideas of the proof of the strong boundary maximum principle by Hopf [1927].

The applications of inverse-positivity in Sections 3.9 and 3.10 are to be considered as examples. The assumptions may be relaxed and various other theorems can be obtained in this way.

Results of this type have often been derived by use of the strong boundary maximum principle (see Protter and Weinberger [1967], for example). Inclusion statements for eigenvalues such as (3.42) can, under different assumptions, also be derived from the Courant maximum–minimum principle (see Temple [1929] and Collatz [1963]). Surveys on comparison and oscillation theory are given by Swanson [1968] and Coppel [1971].

Section 4

The method of reduction for abstract operators as described above was introduced by Schröder [1970b]. This paper contained Sections 4.1.1, 4.1.2, and 4.1.4, except Corollaries 4.1b and 4.1c. Corollary 4.1b (together with the result of Problem 4.1) is a generalization of the essential part of a theorem of Lorenz [1977, Satz 3], where sufficient conditions are investigated such that a matrix M is the product of two M-matrices. In particular, Lorenz used the criterion in Corollary 4.1c for proving the inverse-positivity of matrices which describe certain difference methods for second order boundary value problems. Factorizations of matrices for proving inverse-positivity were already employed in a different manner by Bramble and Hubbard [1964b]. These authors used multiplications by positive matrices from the left and right, which is a reduction method, in our terminology.

Concerning the results in Section 4.1.3 see the Notes to Section 6.2. Beyn [1978] investigated the role of weakly majorizing functions for fourth order differential operators, applying a reduction method.

The main part of the theory on Z-operators and M-operators in Section 4.2.1 was published by Schröder [1978] without proofs. Beyn [1978] derived a result which (in our terminology) essentially says that a differential operator of higher order than 2 cannot be a Z-operator.

Section 4.3 provides an application of the theory in Section 4.2.1 to operators $M = \lambda I - T$ with $T \geqslant O$. Naturally, inverse-positive operators of this form can be considered as generalizations of M-matrices, and it has long been recognized that more general positive operators have many properties in common with positive matrices. Thus, the extensive literature on positive operators is closely related to these special generalizations of M-matrices.

As far as our subject is concerned, the main interest in the study of positive operators is in generalizing the Perron–Frobenius theory of positive matrices. In particular, this means the generalization of the existence theory for eigenelements as well as the related comparison theory which deals with estimating and comparing spectral radii, generalizations of the Stein–Rosenberg theorem, etc. For the case $K^c \neq \varnothing$ the tools for extending the corresponding matrix theory are provided above.

Concerning the more theoretical aspects of this theory we mention the works of Krein and Rutman [1948], Krasnosel'skii [1964a], Sawashima [1964], Schaefer [1960, 1974]. The existence of an eigenelement $\varphi > o$ corresponding to $\rho(T) > 0$ can be proved under comparably weak assumptions (see Theorem I.3.3). To make stronger statements on the positivity of φ, additional assumptions have been formulated which can be considered as generalizations of the irreducibility of matrices, or related properties, as primitivity, etc. Krein and Rutman considered *strongly positive operators*, Krasnosel'skii *u_0-positive operators*. As generalizations of irreducible matrices, Schaefer introduced the notion of a *quasi-interior operator*, and Sawashima the notion of a *seminonsupport operator*. Under our assumptions, such operators are defined, respectively, by the properties (γ) and (δ) of Theorem 4.23, the proof of which is taken from Schülgen [1972]. (However, Schaefer and Sawashima were mainly interested in ordered spaces where $K^c = \varnothing$. For such spaces the sign $\succ$ is to be replaced by a weaker relation.) In Section 4 the proof of $\varphi \succ o$ relies on Theorem 1.9.

While the authors mentioned above concentrate on the more theoretical aspects, others also consider applications to estimation theory. The reader may consult Barker [1972], Marek

[1967], Rheinboldt and Vandergraft [1973], Robert [1975], and Vandergraft [1968], to name just some papers, which contain a series of further references. Marek [1967] treats general u_0-positive operators, while the other authors consider the finite-dimensional case with general order cone. For that case, Rheinboldt and Vandergraft [1973] provide a simple approach to the existence and comparison theory. Barker [1972] also investigates the relation between several of the relevant notions for positive operators.

For (finite) matrices and the natural order relation, the generalizations of irreducibility mentioned above reduce to the usual notion. The W-irreducibility in Section 4.3.3 introduced by Schröder [1978] is a more general concept. In some circumstances a still more general property is of interest where (4.39) is replaced by: $u \geqslant o$, $(\lambda I - T)u \succ_1 o \Rightarrow u \succ_2 o$. A matrix (or an operator) $T \geqslant O$ with such a property may be called (W_1, W_2)-*connected*, with W_i denoting the cone corresponding to $\succ_i$.

For the finite-dimensional case with arbitrary (closed) cone, Vandergraft [1976] has considered weaker conditions in place of irreducibility to prove convergence theorems, comparison results, etc. In describing these weaker conditions, the notion of a *face of a cone* K is used, which Vandergraft defines to be a subcone in the boundary of K generated by extremal vectors of K. In particular, Vandergraft defines a relation $u >^T o$, for a given matrix $T \geqslant O$. One can show that this relation induces a strong order as defined above, with order cone W_T. Moreover, each $T \geqslant O$ is W_T-irreducible in the sense of our theory. Notice that Barker and Schneider [1975] define the face of a cone differently, in a way more closely related to the notion of an extremal set. Trottenberg and Winter [1979] give a survey on these concepts and, in particular, carry out a theory on W-irreducible matrices.

More algebraic approaches to the Perron–Frobenius theory are given by Demetrius [1974] and Barker and Schneider [1975]. Schneider and Turner [1972] treat certain nonlinear (homogeneous) operators. Elsner [1970] considers an "inverse problem": Given an operator with certain spectral properties, find an appropriate cone such that the Perron–Frobenius theory applies. Elsner defines irreducibility by property (γ) in Theorem 4.23. A further definition of irreducible operators is given by Stečenko [1966].

In several papers Bohl investigates iterative procedures for abstract linear problems (see the references in Bohl [1974]). In particular, Bohl [1968] proves equivalence statements such as those in Theorem 4.20 for operators $M = I - B$, under less restrictive assumptions on the ordered linear space.

Theorem 4.24 is a special case of Theorem IV.4.3, which was proved by Schröder [1971]. This special case is closely related to an earlier result of Stoss [1970].

Section 5

The majorant method employed in Section 5.1 is the "natural" tool for proving the convergence of difference methods for ordinary and partial differential equations, when the difference equations are described by M-matrices. For this case the method goes back to Gerschgorin [1930]. (For extensions and references see also Collatz [1952. 1960], Babuška *et al.* [1966], Gorenflo [1973].) When difference methods of higher accuracy are used for second order differential equations, the occurring matrices, in general, do not belong to **M**. For such cases Bramble and Hubbard [1964a, b] and Price [1968] gave convergence proofs verifying the inverse-positivity of the matrices and exploiting this property in a different way. Lorentz [1977] developed an interesting method which again uses majorization (in a more sophisticated way) and allows one to prove also the earlier results. In this method essentially two tools are used, the reduction method described by Corollaries 4.1b,c, and a perturbation method explained in the second proof of Proposition 5.7. The method of z-splitting in the first proof of Proposition 5.7 is due to Söllner [1975]. See also Metelmann [1973] for the inverse-positivity of band matrices, and Lorentz [1977] for further references.

That inverse-positivity and inverse-monotonicity (or monotonic type) are appropriate tools for deriving a posteriori error estimates was shown by Collatz [1952]. We cannot possibly attempt to survey the abundance of examples where such methods were applied in the meantime. Our object here is to demonstrate that it may be possible to develop computer algorithms for rigorous error estimation which work for certain classes of problems. We consider the method developed by Schröder [1961b] as a first attempt of this type. Of course, first order (nonlinear) differential equations, as considered in this paper, are not too interesting themselves. The work, however, has been extended in an improved form to systems of first order differential equations by Marcowitz [1973] (the error bounds $\tilde{\beta}$ in Table 1 are taken from this paper). Most of the numerical results in Section 5.2.3 were calculated by G. Hübner on the CDC/CYBER 76 of the Computing Center at the University of Cologne. In these and other calculations the recursion formulas for Legendre polynomials turned out to be numerically stable. Thus, the numerical stability stated in Section 5.2.3 is an experimental result, which confirms experiences of other authors (see, for example, Gautschi [1975]). For algorithms based on iterative procedures see the Notes to Section III.4.5.3. A method for constructing upper and lower bounds for solutions of second order boundary value problems in terms of "finite elements" is described by Nickel [1979].

Section 6

The results of Section 6.1 are contained in a paper of Schröder [1968a] where also nonself-adjoint differential operators of the fourth order, subject to two-point boundary conditions, are treated. The basic results are essentially proved with the method of local reduction explained for an example at the end of Section 4.1.3. In earlier papers Schröder [1965a] obtained results on inverse-positivity by means of a *formal splitting* $L = L_2 L_1$ (or, more generally, $L \leqslant L_2 L_1$) of the given differential operator L into second order operators L_1, L_2. To split L in this way *formally* (that means without considering boundary conditions) is comparably easy; the difficulty is to find splittings which are compatible with the boundary conditions as needed for the theory of inverse-positivity. In many cases "optimal" splittings of the regular operator L involve singular operators L_1 and L_2; for instance, splittings constructed with the limit functions Φ, Ψ defined above. For the *Dirichlet boundary conditions* $u(0) = u'(0) = u(1) = u'(1) = 0$ only such singular splittings exist. As formula (4.3) suggests, this method of splitting is closely related to the reduction method. However, here one cannot simply apply Corollary 4.1a, since L_2 is only a formal operator; one would need adjoin suitable boundary conditions to L_2, dependent on the given boundary conditions for L.

For a differential operator L of the nth order with n given boundary conditions the properties of splittings $L = L_m L_{m-1} \cdots L_1$ into (regular or singular) operators L_k of lower order were thoroughly investigated by Trottenberg [1974, 1975]. He adjoined suitable (generalized) boundary conditions to the formal operators L_k and represented the Green function $\mathscr{G}$ corresponding to L and the given boundary conditions as an iterated kernel constructed with the (generalized formal) Green functions $\mathscr{G}_k$ corresponding to the operators L_k. In this way, Trottenberg not only obtained results on inverse-positivity, but also on the total positivity of $\mathscr{G}$.

For the theory of *total positivity* see, in particular, Krein [1939], Gantmacher and Krein [1960], Karlin [1968]. The total positivity of $\mathscr{G}$ implies $\mathscr{G}(x, \xi) \geqslant 0$ and hence the inverse-positivity of the differential operator. On the other hand, the total positivity of $\mathscr{G}$ implies also that L is *disconjugate on* $[0, 1]$ (or, in other words, $[0, 1]$ is an *interval of nonoscillation for* L, that is, each solution of $L[u] = 0$ has at most $n - 1$ zeros on $[0, 1]$). L is disconjugate on $[0, 1]$ if and only if there exists a formal splitting of L into n (real) differential operators of the first order (Mammana [1931], Pólya [1922], Pólya and Szegö [1925]). The theory of such splittings does not involve boundary conditions and thus is different from the theory mentioned above. (Concerning results on disconjugacy and nonoscillation see the surveys by Coppel [1971], Levin [1969], Swanson [1968], and also Barrett [1962].)

For certain boundary value problems with boundary conditions of *minimal order* total positivity and disconjugacy are very closely related properties. For fourth order operators with two boundary conditions at each endpoint only the Dirichlet conditions mentioned above belong to this class of problems. For these boundary conditions disconjugacy implies inverse-positivity, while the converse is not true (compare the case $c = c_0$ in Section 4.1.3). This result is essentially contained in Theorem 6.4. (For the relations between nonoscillation and sign conditions on the Green function, also for multipoint boundary conditions, see Čičkin [1960, 1962], Levin [1963, 1964, 1969] and the references in the last paper.)

The theory of Aronszajn and Smith [1957] mentioned in the Notes to Section 3 also yields results on inverse-positive differential operators of the fourth order. Zurawski [1975] applied this theory to self-adjoint operators as considered in Section 6.1, using splittings $L = L_2 L_1$, and obtained conditions for inverse-positivity similar to those of Schröder [1968a].

For the results in Section 6.2 and further results of this kind see Schröder [1972b]. The essential tools for deriving such results are separation theorems for convex sets, in particular, conditions such that one set is *openly separated* from the other. Not always is the basic separation theorem (Theorem I.2.8) appropriate. For more general results on separation see Klee [1968, 1969], Rockafellar [1968]. Further duality results for the case of matrices are due to Mangasarian [1968] and Ben-Israel [1970].

In Section 6.3 we have adapted a method of proof which was used by Schneider [1965] to derive a more general theorem. Concerning the derivation of (6.2), see the remarks and references in this paper. Note that for real valued C, the matrix Z in property (ii) of Theorem 6.6 may also be chosen real-valued. A corresponding property (with C replaced by C^T) will be applied in Example V.4.15.

In the following we report briefly on some results on inverse-positive operators in (complex) Hilbert spaces (with inner product $\langle\ ,\ \rangle$).

Aronszajn and Smith [1957] considered a Hilbert space R whose elements are functions u such that the value $u(x)$ at each point x is a continuous linear functional. To such a space there corresponds a *reproducing kernel* $K(x, y)$ such that $u(y) = \langle u, K_y \rangle$ with $K_y(x) = K(x, y)$. The authors proved that $K(x, y) \geqslant 0$ if and only if (i) $\|u\| = \|\bar{u}\|$ for all $u \in R$, (ii) to each $u \in R$ there exists a $\tilde{u} \in R$ such that $\|\tilde{u}\| \leqslant \|u\|$ and $\tilde{u}(x) \geqslant |u(x)|$ for all x. This theory is then extended to a theory on *pseudoreproducing kernels* for μ-measurable functional Hilbert spaces, and the results are applied to prove the positivity of Green functions corresponding to boundary value problems. (See the remarks above.)

A series of authors is concerned with self-adjoint operators M in a Hilbert space $R = L_2(N, d\mu)$ where $\langle N, \mu \rangle$ is a σ-finite measure space. These results are described, and references are given by Reed and Simon [1978]. For example, let M be bounded below and denote by $\tau = -\kappa$ the infimum of its spectrum. Then *$\lambda I + M$ is inverse-positive for all $\lambda > \kappa$ if and only if* $\exp(-tM)$ *is positive for all $t > 0$.* Moreover, in the case that σ is an eigenvalue of M it is shown that: *τ is a simple eigenvalue with an eigenelement $\varphi \succ o$ if and only if for all $u \in R$ and all $\lambda > \kappa$ the inequality $(M + \lambda I)u \succ o$ implies $u \succ o$* (here, $u \succ o \Leftrightarrow u(x) > 0$ for all x). Obviously, such operators M have properties similar to those discussed in Section 4.2.1. For example, (4.22) holds with $u \succ o$ replaced by $u \succ o$; also see Theorem 4.23. Beurling and Deny [1958] and Reed and Simon [1978] prove that *for a positive semidefinite self-adjoint operator M the operator* $\exp(-tM)$ *is positive for all $t > 0$ if and only if $\langle |u|, M|u| \rangle \leqslant \langle u, Mu \rangle$ for all $u \in R$.* The latter condition is closely related to that of Aronszajn and Smith [1957] mentioned above. Moreover, these conditions are also related to properties used by Stieltjes [1887] in the study of positive definite Z-matrices (see the Notes corresponding to Section III.3).

Chapter III

Two-sided bounds for second order differential equations

This chapter is mainly concerned with estimates $\varphi \leqslant u^* \leqslant \psi$ for solutions u^* of (scalar-valued) nonlinear differential equations of the second order. We provide a priori estimates, where the existence of a solution is assumed, and also existence and inclusion statements.

An important concept in the theory below is that of an *inverse-monotone operator* M, for which $Mv \leqslant Mw \Rightarrow v \leqslant w$. This property can immediately be used to obtain a priori estimates, but it is also a useful tool in the existence theory. Theorems on inverse-monotone abstract operators which generalize those for inverse-positive linear operators will be proved in Section IV.2 as special cases of a still more general theory. These theorems could be used to derive some of the results on inverse-monotone differential operators below. We prefer, however, to prove these results here directly. Nevertheless, the concept of an abstract inverse-monotone operator is briefly discussed in Section 1.

Section 2 provides sufficient conditions for second order differential operators M, such that $Mv \leqslant Mw \Rightarrow v \leqslant w$, where all inequalities hold pointwise. In Section 3 such inequalities $v \leqslant w$ are derived from certain weak differential inequalities involving v and w. When the results in these sections are applied to obtain inclusion statements $\varphi \leqslant u^* \leqslant \psi$ for solutions (or weak solutions), the existence of such solutions has to be assumed. The conditions required for such an inclusion to hold then in general imply that there is at most one solution (or at most one solution with certain properties, such as positivity).

In Section 4 we derive sufficient conditions such that the existence of *comparison functions*, that is, functions φ, ψ with $\varphi \leqslant \psi$ and $M\varphi \leqslant o \leqslant M\psi$, implies the existence of a solution u^* of the equation $Mu = o$ with $\varphi \leqslant u^* \leqslant \psi$. Here, the uniqueness of the solution, in general, does not follow. Analogously, Section 5 provides sufficient conditions for the existence of weak solutions u^* and estimates $\varphi \leqslant u^* \leqslant \psi$ by *weak comparison functions.*

Two-sided estimates $\varphi \leqslant u^* \leqslant \psi$ can, for instance, be used to obtain theoretical statements on the qualitative behavior of a solution, such as positivity of the solution, or growth properties for $x \to \infty$, or growth properties in the neighborhood of a singularity, as illustrated in several examples. On the other hand, the inclusion statements can also be used to obtain numerical error estimates, as described in Section 6. In some of the results, not only the solution, but also its derivative is estimated.

Most theorems of this chapter are formulated for problems on the interval [0, 1], or for corresponding operators. Moreover, the main existence statements are proved under certain regularity assumptions on the coefficients. Many of these restrictions are not necessary. First observe that analogous theorems hold for any compact interval; they are easily obtained by a simple transformation. (In many cases one may simply replace [0, 1] by an arbitrary compact interval.) Moreover, many problems on unbounded intervals and problems with singular coefficients can also be treated, either by proceeding analogously, or by using the results for "regular" operators as a tool. We explain several methods in Sections 2.3.5 and 4.4.

The existence proofs in Sections 4 and 5 use the *method of modification*, which essentially consists of the following three steps: (1) formulating a modified problem by applying a truncation operation, (2) proving the existence of a solution u^* of the modified problem, (3) proving inclusion statements for u^* which then imply that u^* is also a solution of the given problem. The third step is here carried out using the theories on inverse-monotonicity in Sections 2 and 3, respectively. For this purpose only the simplest statements on inverse-monotonicity need be applied, because the modified problem is defined accordingly. The existence proof in the second step is separated from the proof of the inclusion, so that any appropriate existence theory may be applied. We use mainly Schauder's fixed-point theorem, but also a theorem on monotone operators (monotone definite operators).

Existence and inclusion statements may also be obtained by applying the theorem of Leray–Schauder or some other theory of topological degree (*degree method*), or by applying the *method of monotone iterations*. These methods are briefly described in Sections 4.5.2 and 4.5.3.

For arbitrary functions $v, w \in C_0[0, 1]$, the inequalities which occur in this chapter are always defined by

$$
\begin{aligned}
v \leqslant w \quad &\Leftrightarrow \quad v(x) \leqslant w(x) \quad \text{for} \quad 0 \leqslant x \leqslant 1; \\
v < w \quad &\Leftrightarrow \quad v \leqslant w, \quad v \neq w; \\
v \prec w \quad &\Leftrightarrow \quad v(x) < w(x) \quad \text{for} \quad 0 \leqslant x \leqslant 1,
\end{aligned}
$$

except when we explicitly give a different definition.

1 THE CONCEPT OF AN INVERSE-MONOTONE OPERATOR

Let M denote a (nonlinear) operator with domain in an ordered linear space $(R, \leqslant)$ and range in an ordered linear space $(S, \leqslant)$, and let D be a subset of its domain. This operator is called *inverse-monotone on D* if for all $v, w \in D$

$$Mv \leqslant Mw \quad \Rightarrow \quad v \leqslant w. \tag{1.1}$$

If the choice of D is clear from the context, we shall simply speak of an *inverse-monotone* operator.

Now it often happens that for a given operator M the implication (1.1) is true for *certain* elements v, w, while it does not hold for *all* v, w in the domain of M. Therefore, we shall assume in our theory that v and w are fixed (though possibly unknown) elements and then derive conditions for (1.1) to hold for those fixed elements. The assumptions required from v and w will also describe *sets* of elements for which (1.1) is true.

As for linear operators, property (1.1) can be used to prove inclusion statements for an element u by

$$M\varphi \leqslant Mu \leqslant M\psi \quad \Rightarrow \quad \varphi \leqslant u \leqslant \psi. \tag{1.2}$$

For example, u may be the unknown solution of an equation $Mu = r$ and φ, ψ may be approximations known by calculation. Then (1.1) has to be applied to the pair $(v, w) = (\varphi, u)$ and to the pair $(v, w) = (u, \psi)$. In the first case, the element v on the left is known, in the second case the element w on the right.

Moreover, (1.1) can be used to prove uniqueness statements. Let $Mu = M\bar{u} = r$ and (1.1) hold for $(v, w) = (\bar{u}, u)$ as well as for $(v, w) = (u, \bar{u})$. Then

(1.2) yields $\bar{u} \leqslant u \leqslant \bar{u}$, consequently $u = \bar{u}$. Here, of course, all elements may be unknown. In particular, we obtain

Proposition 1.1. *If M is inverse-monotone on D, then the restriction of M to D has a monotone inverse.*

This result justifies the term inverse-monotone as defined above. For simplicity, however, results on property (1.1) for fixed v, w will also be referred to as results on inverse-monotone operators.

The results on inverse-monotone differential operators in the succeeding sections will be formulated with the situation in mind that one of the elements v, w may be known, so that then there is no harm when this element occurs in the assumptions. In general, we shall consider w to be this known element. Thus, some of the results should not immediately be applied to the case where v is known in (1.1), but w is unknown. To treat the latter case, one could easily derive analogous results with the same methods. On the other hand, such results can also be obtained by applying the theorems below to an operator $\tilde{M}$ which is obtained from M by a simple transformation.

More generally, one can obtain a series of further results by applying the theory to transformed operators, as described now. For fixed $v, w \in D$, property (1.1) can always be written in the form

$$\tilde{M}\tilde{w} \geqslant o \quad \Rightarrow \quad \tilde{w} \geqslant o \tag{1.3}$$

with $\tilde{w} = w - v$ and a suitable operator $\tilde{M}$ satisfying $\tilde{M}o = o$. Such an operator can be chosen in many different ways. As special examples, we mention the operators

$$\begin{aligned} &\tilde{M}_1 u = M(v + u) - Mv, &&\tilde{M}_2 u = Mw - M(w - u), \\ &\tilde{M}_3 = \tilde{M}_1 + \tilde{M}_2, &&\tilde{M}_4 u = \int_0^1 M'(v + t(w - v))\,dt\,u \end{aligned} \tag{1.4}$$

where, in the last case, the derivative M' is defined in an appropriate way under appropriate conditions which shall not be discussed here.

Applying our theorems to $\tilde{M}_1, o, \tilde{w}$ in place of M, v, w yields exactly the same results. By applying the theorems to $\tilde{M}_2, o, \tilde{w}$ results are obtained where the roles of v and w are exchanged.

Of course, considering the property (1.1) we need not assume that R and S are linear spaces. Also the relations denoted by $\leqslant$ may be preorders or have still weaker properties. These possibilities to weaken the assumptions, however, will not be used in this chapter. (For more general inclusion statements see Chapter IV.)

2 INVERSE-MONOTONE DIFFERENTIAL OPERATORS

In this section we mainly consider differential operators M which correspond to two-point boundary value problems of the second order with separate boundary conditions at $x = 0$ and $x = 1$, such as the operators in (2.1) and (2.15). Operators corresponding to initial value problems of the first order are treated as special cases. We are interested in sufficient conditions such that $Mv \leqslant Mw$ implies $v \leqslant w$ for certain (fixed) functions v, w, where $\leqslant$ denotes the natural (pointwise) order relation.

The very simple Theorem 2.1 in Section 2.1 permits one to prove this property for some interesting cases. However, the more sophisticated Theorem 2.2 in Section 2.2 is in general more appropriate for nonlinear differential operators. On the other hand, Theorem 2.1 is completely adequate to derive the existence and inclusion statements for boundary value problems in Section 4 and also analogous theorems for *systems* of differential equations in Section V.3. The main results in Sections 2.1 and 2.2 can also be obtained from the abstract theorems in Section IV.2. However, by the direct approach used here, various possibilities for relaxing the assumptions are easily recognized.

In Section 2.3 we report on a series of further results, such as inverse-monotonicity for periodic boundary conditions and for differential operators on unbounded domains.

2.1 Simple sufficient conditions

We consider an operator defined by

$$Mu(x) = \begin{cases} -a(x)u''(x) + g(x, u(x), u'(x)) & \text{for} \quad 0 < x < 1 \\ g_0(u(0), u'(0)) & \text{for} \quad x = 0 \\ g_1(u(1), u'(1)) & \text{for} \quad x = 1, \end{cases} \tag{2.1}$$

with given functions $a:(0, 1) \to \mathbb{R}$, $g:(0, 1) \times \mathbb{R} \times \mathbb{R} \to \mathbb{R}$, $g_i:\mathbb{R} \times \mathbb{R} \to \mathbb{R}$ $(i = 0, 1)$, where

$$a(x) \geqslant 0 \qquad (0 < x < 1) \tag{2.2}$$

and

$$\begin{matrix} g_0(y, p_1) \geqslant g_0(y, p_2) \\ g_1(y, p_1) \leqslant g_1(y, p_2) \end{matrix} \quad \text{for} \quad p_1 \leqslant p_2 \qquad (y, p_1, p_2 \in \mathbb{R}). \tag{2.3}$$

Let R be the space of all functions $u \in C_0[0, 1] \cap C_1(0, 1)$ such that all derivatives exist which occur in $Mu(x)$. For example, that means $u''(x)$ must exist for each $x \in (0, 1)$ with $a(x) \neq 0$, and $u'(i)$ must exist if $g_i(y, p)$ depends on p $(i = 0, 1)$. Then the operator M maps R into $S = \mathbb{R}[0, 1]$. (Less smooth functions u will be considered below.)

For example, g_0 and g_1 may have the form of Sturmian boundary operators

$$\begin{aligned} g_0(u(0), u'(0)) &= B_0[u] - r_0 \quad \text{with} \quad B_0[u] = -\alpha_0 u'(0) + \gamma_0 u(0) \\ g_1(u(1), u'(1)) &= B_1[u] - r_1 \quad \text{with} \quad B_1[u] = \alpha_1 u'(1) + \gamma_1 u(1) \end{aligned} \tag{2.4}$$

and real r_i, α_i, γ_i such that $\alpha_0 \geqslant 0$, $\alpha_1 \geqslant 0$, $\alpha_0 + |\gamma_0| > 0$, $\alpha_1 + |\gamma_1| > 0$.

As for linear operators, we shall sometimes use a shorter form to describe a special operator. For example, $Mu = (-u'' + u^3, u(0) - 1, u'(1))$ describes the operator given by (2.1) with $a(x) \equiv 1$, $g(x, y, p) = y^3$, $g_0(y, p) = y - 1$, $g_1(y, p) = p$.

In the following let v and w denote fixed functions in R.

Theorem 2.1. *Let there exist a function $z \succ o$ in R such that*

$$Mw(x) < M(w + \lambda z)(x) \quad \textit{for} \quad 0 \leqslant x \leqslant 1, \quad 0 < \lambda < \infty \tag{2.5}$$

or

$$M(v - \lambda z)(x) < Mv(x) \quad \textit{for} \quad 0 \leqslant x \leqslant 1, \quad 0 < \lambda < \infty. \tag{2.6}$$

Then $Mv \leqslant Mw \Rightarrow v \leqslant w$.

Proof. We only prove the theorem under the condition (2.5). If (2.6) holds, the result is derived similarly. Suppose that $v \nleqslant w$. Then there exists a minimal $\lambda > 0$ such that $v \leqslant w + \lambda z$. Obviously,

$$v(\xi) = (w + \lambda z)(\xi) \quad \text{for some} \quad \xi \in [0, 1]. \tag{2.7}$$

We shall show that

$$Mv(\xi) \geqslant M(w + \lambda z)(\xi), \tag{2.8}$$

which with assumption (2.5) yields $Mv(\xi) > Mw(\xi)$, contradicting the premise $Mv \leqslant Mw$. Hence $v \leqslant w$ must be true if $Mv \leqslant Mw$.

If $\xi \in (0, 1)$, we have

$$\begin{aligned} v'(\xi) &= (w' + \lambda z')(\xi), \quad \text{and} \\ v''(\xi) &\leqslant (w'' + \lambda z'')(\xi) \quad \text{when} \quad a(\xi) \neq 0, \end{aligned} \tag{2.9}$$

so that (2.8) follows with (2.2). If $\xi = 0$ $\{\xi = 1\}$, we have $v'(0) \leqslant w'(0) + \lambda z'(0) \{v'(1) \geqslant w'(1) + \lambda z'(1)\}$, provided these derivatives exist. Thus, (2.8) follows from (2.3). □

Example 2.1. *Let M have the special form*

$$Mu(x) = \begin{cases} -a(x)u''(x) + b(x)u'(x) + f(x, u(x)) & \text{for} \quad 0 < x < 1 \\ B_0[u] & \text{for} \quad x = 0 \\ B_1[u] & \text{for} \quad x = 1, \end{cases} \tag{2.10}$$

with B_0, B_1 *in* (2.4) *and* $f:(0,1)\times\mathbb{R}\to\mathbb{R}$. *Suppose that*

$$f(x, w(x)+y) - f(x, w(x)) \geqslant c(x)y \quad \text{for} \quad 0<x<1, \quad y\geqslant 0$$

or (2.11)

$$f(x, v(x)) - f(x, v(x)-y) \geqslant c(x)y \quad \text{for} \quad 0<x<1, \quad y\geqslant 0$$

with some $c\in\mathbb{R}(0,1)$. *Also suppose that there exists a* $z\in R$ *with*

$$z\geqslant o, \qquad L[z](x)>0 \qquad (0<x<1), \qquad B_0[z]>0, \qquad B_1[z]>0 \tag{2.12}$$

where $L[u] = -au'' + bu' + cu$. *Then* $Mv\leqslant Mw \Rightarrow v\leqslant w$.

This statement follows from Theorem 2.1 by verifying (2.5) or (2.6), respectively. For example, a function z satisfying (2.12) may be constructed with the methods described in Section II.3.3. In particular, the assumptions of Theorem II.3.7 are sufficient for the existence of z. □

Example 2.2. For $Mu = (-u'' + \exp u, u(0) - r_0, u(1) - r_1)$, inequality (2.5) holds with $z(x)\equiv 1$ and arbitrary $w\in R$. Consequently, *the operator* M *is inverse-monotone on* R, *and hence the boundary value problem* $Mu = o$ *has at most one solution.* □

Example 2.3 (see also Example 2.5). Let

$$Mu(x) = \begin{cases} -\varepsilon u'' + u' + f(u) & \text{for} \quad 0<x<1 \\ -u'(0) + \gamma_0 u(0) & \text{for} \quad x=0 \\ u'(1) & \text{for} \quad x=1 \end{cases}$$

$$\text{with} \quad f(y) = -\beta(\eta - y)\exp\frac{-k}{1+y}$$

and $\varepsilon, \beta, \gamma_0, k, \eta$ being constants >0 ($f(-1)=0$). The singular perturbation problem $Mu = o$ arises in chemical reactor theory (Luss [1968], Markus and Amundson [1968]). We shall show that *each solution* $u\in R$ *of* $Mu = o$ *satisfies* $o\leqslant u\leqslant \eta$.

Here, inequality (2.11) holds with $w(x)\equiv\eta$ and $c(x)\equiv 0$. Moreover, in the present case a function z satisfying (2.12) exists, because the assumptions of Theorem II.3.7 are satisfied (for example, (ii) and (iv) hold). Thus, $u\leqslant\eta$ follows from $o\leqslant M\eta$.

For a solution $u\leqslant\eta$, however, we have $f(u(x))\leqslant 0$, so that $M_0 u\geqslant o$, where M_0 is obtained from M by omitting $f(u)$. Using the same function z as above we derive from Theorem II.3.7 that M_0 is inverse-positive on R, so that $u\geqslant o$ follows. □

The smoothness requirements on the occurring functions v, w, and z may be relaxed. We describe here certain weaker assumptions, which are useful, for instance, when the results are applied to treat problems on unbounded intervals (see Problem 2.19 and Example 4.6). Let Γ^+ and Γ^- denote given disjoint sets with $\Gamma^+ \cup \Gamma^- = \Gamma := \{x \in (0, 1) : a(x) \neq 0\}$. The smoothness conditions S_r, S_l, S_m defined below will depend on these sets. Denote by D^+ and D^- the right-hand and left-hand derivative, respectively. For example, $D^+u(x) = \lim_{h \to +0} h^{-1}(u(x + h) - u(x))$. For $x = 0$ $\{x = 1\}$ we shall also write $u'(0)$ $\{u'(1)\}$ instead of $D^+u(0)$ $\{D^-u(1)\}$.

A function $u \in \mathbb{R}[0, 1]$ is said to satisfy *Condition* S_r, if it belongs to $C_0[0, 1]$ and has the following differentiability properties:

(α) $D^+u(x)$ and $D^-u(x)$ exist for $x \in (0, 1)$, and

$$D^+u(x) \leqslant D^-u(x) \qquad \text{for} \quad 0 < x < 1; \tag{2.13}$$

(β) for each $x \in \Gamma^+$ $\{x \in \Gamma^-\}$ such that $u'(x)$ exists, the one-sided derivative $D^+u'(x)$ $\{D^-u'(x)\}$ exists also. (This condition includes the requirement that u' exists in a right $\{$left$\}$ neighborhood of x.)

Condition S_l is the same, with (2.13) replaced by $D^+u(x) \geqslant D^-u(x)$ $(0 < x < 1)$. Also, *Condition* S_m is the same except that $D^+u(x) = D^-u(x)$ $(0 < x < 1)$ is required, which means that u is differentiable on $(0, 1)$.

Where D^+u' or D^-u' exist we shall use the notation

$$u''(x) = D^+u'(x) \quad \text{for} \quad x \in \Gamma^+, \qquad u''(x) = D^-u'(x) \quad \text{for} \quad x \in \Gamma^-. \tag{2.14}$$

Corollary 2.1a. *Suppose that the functions v, w, z do not necessarily belong to R, but that v satisfies Condition S_l and w, z satisfy Condition S_r. Moreover, let these functions have (one-sided) derivatives at $x = 0$ $(x = 1)$, in case $g_0(y, p)$ $(g_1(y, p))$ depends on p.*

Then the statements of Theorem 2.1 remain true provided all occurring inequalities involving M are required to hold only for $x = 0$, $x = 1$, and for those $x \in (0, 1)$ such that all three derivatives $v'(x)$, $w'(x)$ and $z'(x)$ exist.

The corollary is proved in essentially the same way as Theorem 2.1, where now (2.14) is used. Notice that (2.7) cannot hold for any $\xi \in (0, 1)$ where one of the functions v, w, z is not differentiable.

Problems

2.1 Prove that $Mu = (-u'' - \mu u + u^3, u(0), u(1))$ is inverse-monotone for $\mu < \pi^2$. Choose $z(x) = \cos(\pi - \varepsilon)(x - \frac{1}{2})$ with $\varepsilon > 0$ sufficiently small.

2.2 Let $Mu = (-u'' + f(x, u), -u'(0) + \gamma_0 u(0), u'(1) + \gamma_0 u(1))$ with f continuously differentiable and $f_y(x, y) \geqslant c_0 \in \mathbb{R}$. Apply the statement in Example

2.1 to find a domain D in the (c_0, γ_0)-plane such that, for all $(c_0, \gamma_0) \in D$, M is inverse-monotone on $C_2[0, 1]$. (a) Use Theorem II.3.7 for constructing a suitable z. (b) Use also Theorem II.3.24.

2.3 The smoothness requirements on u (and z) in Theorem 2.1 and Corollary 2.1a may be weakened further. As an example, prove that $Mu \geqslant o \Rightarrow u \geqslant o$ for all $u \in R = C_0[0, 1]$ where $Mu = (-u'' + f(x, u), u(0), u(1))$, $0 = f(x, 0) \leqslant f(x, y)$ for $0 < x < 1$, $0 \leqslant y$ and $u''(x)$ is defined by $u''(x) = \liminf_{h \to 0} \delta_h^2(u)(x)$, $\delta_h^2(u)(x) = h^{-2}(u(x - h) - 2u(x) + u(x + h))$.

2.2 More general results

Theorem 2.1 will now be generalized in several ways. We shall consider operators of a more general quasilinear form and require less restrictive assumptions. In particular, the inequalities (2.5), (2.6) will be replaced by weaker requirements which, in general, are more appropriate for nonlinear problems.

Now let

$$Mu(x) = \begin{cases} -a[u](x)u''(x) + g[u](x) & \text{for } 0 < x < 1 \\ g_0(u(0), u'(0)) & \text{for } x = 0 \\ g_1(u(1), u'(1)) & \text{for } x = 1 \end{cases} \tag{2.15}$$

with

$$a[u](x) = a(x, u(x), u'(x)), \qquad g[u](x) = g(x, u(x), u'(x)),$$

where g, g_0, and g_1 satisfy the assumptions in Section 2.1, and $a(x, y, p)$ is a real-valued function on $(0, 1) \times \mathbb{R} \times \mathbb{R}$.

Again, let $v, w \in \mathbb{R}[0, 1]$ denote fixed functions. Instead of a function z and a corresponding linear family λz, we shall now consider a more general family $\mathscr{Z} = \{\mathfrak{z}(\lambda): 0 \leqslant \lambda < \infty\} \subset \mathbb{R}[0, 1]$ with suitable properties. We shall write $\mathfrak{z}(\lambda, x) = \mathfrak{z}(\lambda)(x)$ and always denote by $'$ the derivative with respect to x, so that $\mathfrak{z}'(\lambda, x) = \partial \mathfrak{z}(\lambda, x)/\partial x$, for example.

The family $\mathscr{Z}$ is called *admissible* if $\mathfrak{z}(\lambda, x)$ is continuous for $0 \leqslant x \leqslant 1$, $0 < \lambda$ and $\mathfrak{z}(0, x) = 0$ for $0 \leqslant x \leqslant 1$. Moreover, $\mathscr{Z}$ is said to *majorize a function* $u \in \mathbb{R}[0, 1]$ if $u \leqslant \mathfrak{z}(\lambda)$ for some $\lambda \geqslant 0$. Obviously, $\mathscr{Z}$ majorizes each bounded function u if $\mathfrak{z}(\lambda, x) \to +\infty$ for $\lambda \to +\infty$, uniformly on $[0, 1]$.

The simplest results are obtained, when v, w and all functions $\mathfrak{z}(\lambda)$ belong to $C_1[0, 1] \cap C_2(0, 1)$. However, as before, we shall also allow these functions to have weaker smoothness properties. Define *Conditions* S_r and S_l as in the previous section, where now Γ is the set of all $x \in (0, 1)$ such that $a(x, y, p) \neq 0$ for some $y \in \mathbb{R}$, $p \in \mathbb{R}$.

We assume that v satisfies Condition S_l *and that w satisfies Condition* S_r. Then let Γ_0 denote the set of all $x \in (0, 1)$ such that both $v'(x)$ and $w'(x)$ exist. Moreover, for given $\mathfrak{z}(\lambda)$ with $\lambda > 0$, let Γ_λ be the set of all $x \in (0, 1)$ such that all three derivatives $v'(x)$, $w'(x)$, and $\mathfrak{z}'(\lambda, x)$ exist.

Theorem 2.2. *Suppose that there exists an admissible family* $\{\mathfrak{z}(\lambda): 0 \leqslant \lambda < \infty\}$ *which majorizes* $v - w$, *such that all* $\mathfrak{z}(\lambda)$ $(0 \leqslant \lambda < \infty)$ *satisfy Condition* S_r *and the following conditions hold:*

(i) *The inequalities*

$$a[v](x) \geqslant 0 \tag{2.16}$$

and

$$Mw(x) < M(w + \mathfrak{z}(\lambda))(x) \tag{2.17}$$

are satisfied for each $\lambda > 0$ *and each* $x \in \Gamma_\lambda$ *with*

$$v(x) = w(x) + \mathfrak{z}(\lambda, x), \qquad v'(x) = w'(x) + \mathfrak{z}'(\lambda, x). \tag{2.18}$$

(ii) *The inequality*

$$Mw(x) < M(w + \mathfrak{z}(\lambda))(x) \tag{2.19}$$

is satisfied for each $\lambda > 0$ *and each* $x \in \{0, 1\}$ *with* $v(x) = w(x) + \mathfrak{z}(\lambda, x)$.

If, under these assumptions, $Mv(x) \leqslant Mw(x)$ *for each* $x \in \Gamma_0 \cup \{0, 1\}$, *then* $v \leqslant w$.

This result too is proved in essentially the same way as Theorem 2.1 and Corollary 2.1a. Observe that in the proof of Theorem 2.1 the inequality (2.5) was only used at a point $x = \xi$ where (2.7) holds and in case $0 < \xi < 1$ also the first relation in (2.9) holds. The details of the proof are left to the reader.

Because of the *side conditions* (2.18), the inequality (2.17) may be replaced by $(-a[v]\mathfrak{z}''(\lambda) - (a[v] - a[w])w'' + g[v] - g[w])(x) > 0$. If, in particular, M has the special form (2.1) with $a(x)$ satisfying (2.2), then the above factor of w'' vanishes and inequality (2.17) may also be replaced by $M(v - \mathfrak{z}(\lambda))(x) < Mv(x)$.

Various modifications of the preceding theorem can be obtained by first transforming the given operator, as described at the end of Section 1, and then applying Theorem 2.2 to the transformed operator. For example, if one defines $\tilde{M}u = Mw - M(w - u)$ and then applies Theorem 2.2 to $\tilde{M}$, $w - v, o$ in place of M, w, v, one obtains sufficient conditions where, in place of (2.16), the inequality $a[w](x) \geqslant 0$ occurs.

Example 2.4. *Suppose that M has the form* (2.1), (2.4), *that v and w belong to* $C_2[0, 1]$, *and that* $g(x, y, p)$, $g_y(x, y, p)$, *and* $g_p(x, y, p)$ *are continuous* *on* $[0, 1] \times \mathbb{R} \times \mathbb{R}$. *Define* $L[u] = -au'' + bu' + cu$ *with*

$$b(x) = \int_0^1 g_p(x, (1-t)w + tv, (1-t)w' + tv')\, dt,$$

$$c(x) = \int_0^1 g_y(x, (1-t)w + tv, (1-t)w' + tv')\, dt.$$

If there exists a function $z \in R$ *such that the linear inequalities* (2.12) *are satisfied, then* $Mv \leqslant Mw \Rightarrow v \leqslant w$.

To see this, choose $\mathfrak{z}(\lambda) = \lambda z$ and use

$$\frac{1}{\lambda}\{M(w + \lambda z)(x) - Mw(x)\} = -a(x)z'' + b(x, \lambda)z' + c(x, \lambda)z \quad (0 < x < 1)$$

with

$$b(x, \lambda) = \int_0^1 g_p(x, w + t\lambda z, w' + t\lambda z')\, dt,$$

$$c(x, \lambda) = \int_0^1 g_y(x, w + t\lambda z, w' + t\lambda z')\, dt.$$

Then, exploiting the side condition (2.18), replace λz by $v - w$ and $\lambda z'$ by $v' - w'$ in the formulas for $b(x, \lambda)$ and $c(x, \lambda)$.

In order to verify (2.12), use Theorem II.3.7, for example. □

Example 2.5 (Continuation of Example 2.3). For the problem $Mu = o$ in Example 2.3 we shall now show that *there exists at most one solution, provided*

$$1 + 4\varepsilon c_0 > 0, \qquad \gamma_0 + 2c_0 > 0 \qquad \textit{with} \quad c_0 = \min\{f'(y) : 0 \leqslant y \leqslant \eta\} < 0.$$

It suffices to prove that $Mv \leqslant Mw \Rightarrow v \leqslant w$, for arbitrary $v, w \in R$ with $o \leqslant v \leqslant \eta$, $o \leqslant w \leqslant \eta$. According to Example 2.4, this property holds if there exists a function $z \in R$ which satisfies (2.12) for $a(x) \equiv \varepsilon$, $b(x) \equiv 1$, $c(x) \equiv c_0$, and the given boundary operators. Under the above conditions, one can construct such a function by solving $-\varepsilon z'' + z' + c_0 z = 1$, $-z'(0) + \gamma_0 z(0) = 1$, $z'(1) = 1$.

We prefer, however, to proceed somewhat differently. According to Theorem II.3.3, a function z satisfying (2.12) exists if and only if the linear operator (L, B_0, B_1) is inverse-positive. To verify this property, it suffices, for example, to prove that the solution ψ of the boundary value problem $-\varepsilon\psi'' + \psi' + c_0\psi = o$, $-\psi'(0) + \gamma_0\psi(0) = 1$, $\psi'(1) = 0$ satisfies $\psi \geqslant o$ (cf. Corollary II.3.19a). The latter inequality follows from the required conditions by elementary calculations. □

Example 2.6. *Suppose that M has the form* (2.1), (2.4), *that v and w belong to $C_2[0, 1]$, and that*

$$g(x, w(x) + y, p) - g(x, w(x), p) \geqslant c(x)y \qquad \text{for} \quad 0 < x < 1, \quad 0 \leqslant y, \quad p \in \mathbb{R}$$

with some function $c(x)$. Moreover, for each $N > 0$ let there exist a function $b(x) = b_N(x)$ and a function $z = z_N \in C_2[0, 1]$ such that

$$|g(x, w(x), w'(x) + p) - g(x, w(x), w'(x))| \leqslant b(x)|p| \quad \text{for} \quad 0 < x < 1, \quad |p| < N \tag{2.20}$$

and

$$\begin{aligned} &z(x) > 0 \quad (0 \leqslant x \leqslant 1), \qquad B_0[z] > 0, \qquad B_1[z] > 0, \\ &\hat{L}[z](x) := (-az'' - b|z'| + cz)(x) > 0 \qquad (0 < x < 1). \end{aligned} \tag{2.21}$$

Then $Mv \leqslant Mw \Rightarrow v \leqslant w$.

To prove the statement, choose $N \geqslant |v'(x) - w'(x)|$ $(0 < x < 1)$, define $\mathfrak{z}(\lambda) = \lambda z = \lambda z_N$, and use (2.18) in calculating

$$\begin{aligned} \lambda^{-1}\{M(w + \lambda z)(x) - Mw(x)\} &= -az'' + \lambda^{-1}\{g(x, w, v') - g(x, w, w')\} \\ &\quad + \lambda^{-1}\{g(x, w + \lambda z, w' + \lambda z') \\ &\quad - g(x, w, w' + \lambda z')\} \\ &\geqslant \hat{L}[z](x). \end{aligned}$$

To construct suitable functions z, the methods of Section II.3.3 may be used again. For example, such a $z = z_N$ exists for each $N > 0$ if $a(x) \geqslant a_0 > 0$, $b = b_N$ is bounded, $c(x) \geqslant 0$ $(0 < x < 1)$, $\gamma_0 \geqslant 0$, $\gamma_1 \geqslant 0$, $\gamma_0 + \gamma_1 > 0$, (see Theorem II.3.7, case (ii)). □

Example 2.7. Let $Mu = (-u'' + \exp(-u),\ u(0) - r_0,\ u(1) - r_1)$. We shall show that *$Mv \leqslant Mw$ implies $v \leqslant w$, for all $v, w \in C_2[0, 1]$ with $v \succ -\log \pi^2$, $w \succ -\log \pi^2$.* Thus *the boundary value problem $Mu = o$ has at most one solution $u \in C_2[0, 1]$ satisfying $u \succ -\log \pi^2$.*

We verify (2.17), (2.19) with the method explained in Example 2.4. Using (2.18) one calculates for $0 < x < 1$,

$$\begin{aligned} &M(w + \mathfrak{z}(\lambda))(x) - Mw(x) \\ &\quad = -\mathfrak{z}''(\lambda, x) - \int_0^1 \exp[-(1 - t)w(x) - tv(x)]\, dt\, \mathfrak{z}(\lambda, x). \end{aligned}$$

For the functions v and w considered, the integrand is strictly smaller than π^2. Consequently, $\mathfrak{z}(\lambda, x) = \lambda \cos(\pi - \varepsilon)(x - \frac{1}{2})$ has all required properties if $\varepsilon > 0$ is sufficiently small. □

In the preceding examples, we explained certain general methods for applying Theorem 2.2. In a concrete case one may modify these methods or combine them with other methods as, for instance, in the following example.

Example 2.8. The boundary value problem $Mu = o$ with

$$Mu(x) = \begin{cases} -u''(x) + g(u(x), u'(x)) & \text{for } 0 < x < 1 \\ u(0) & \text{for } x = 0 \\ u'(1) & \text{for } x = 1, \end{cases}$$

$$g(y, p) = -\mu y\sqrt{1 - p^2}$$

describes the bending of a beam under a load $\mu \geqslant 0$ (with x denoting the arclength measured along the beam). A function u shall be called a solution of this problem only if $u \in C_2[0, 1]$ and $|u'| \leqslant 1$.

Since the function g is not determined for all real p, we extend it by defining $g(y; p) = 0$ for $p^2 \geqslant 1$. However, we shall not actually use this extension, but only consider functions $v, w \in C_2[0, 1]$ with $|v'| \leqslant 1, |w'| \leqslant 1$.

We choose $\mathfrak{z}(\lambda) = \lambda z$ with $z(x) = \sin(\pi/2)(1 - \varepsilon)(x + \varepsilon)$ and sufficiently small $\varepsilon > 0$. Then the assumptions of Theorem 2.2 are satisfied if for all $x \in (0, 1)$ and $\lambda > 0$ satisfying (2.18)

$$-z'' - \mu z(1 - w'^2)^{1/2} + \lambda^{-1}\mu v[(1 - w'^2)^{1/2} - (1 - v'^2)^{1/2}] > 0.$$

First suppose that $0 \leqslant \mu < \pi^2/4$. Since the above inequality holds for $v(x) \equiv 0$, we conclude that $o \leqslant Mw \Rightarrow o \leqslant w$. Now let u denote a solution of the given problem and apply the latter implication to $w = u$, as well as to $w = -u$. Then we arrive at the statement that *for* $0 \leqslant \mu < \pi^2/4$ *the given problem has only the trivial solution* $u = o$.

For $\mu \geqslant \pi^2/4$, *each solution* $u \geqslant o$ *of* $Mu = o$ *satisfies* $u \leqslant w$ *with* $w(x) = x$. To prove this, verify the above differential inequality, observing that the term within brackets is positive for $v'(x) = w'(x) + \lambda z'(x)$, since $z'(x) \geqslant 0$. □

Instead of extending the function g in the preceding example, we could have used the following statement which concerns functions a, g with restricted domains. (A corresponding statement holds with respect to the functions g_0, g_1.)

Corollary 2.2a. *Let* $\varphi, \psi, \Phi, \Psi \in \mathbb{R}[0, 1]$ *satisfy* $\varphi \leqslant v \leqslant \psi$, $\varphi \leqslant w \leqslant \psi$, $\Phi \leqslant v' \leqslant \Psi$, *and* $\Phi \leqslant w' \leqslant \Psi$. *Then Theorem* 2.2 *is also true if* $a(x, y, p)$ *and* $g(x, y, p)$ *are only defined for* $0 < x < 1$, $\varphi(x) \leqslant y \leqslant \psi(x)$, $\Phi(x) \leqslant p \leqslant \Psi(x)$.

The following examples will explain the advantage in using a nonlinear family $\mathfrak{z}(\lambda)$. (See also Problem 2.11.)

Example 2.9. *Again let M have the form* (2.1), (2.4) *and* $w \in C_2[0, 1]$. *Suppose that for* $0 < x < 1$

$$g(x, w + y, w' + p) - g(x, w, w' + p) \geqslant c(x)y \quad \text{for} \quad y \geqslant 0, \quad |p| < \varepsilon,$$
$$|g(x, w, w' + p) - g(x, w, w')| \leqslant b(x)|p| \quad \text{for} \quad |p| < \varepsilon, \qquad (2.22)$$

where $b, c \in \mathbb{R}[0, 1]$, $c(x) \geqslant 0\,(0 < x < 1)$, *and* $\varepsilon > 0$ *denotes some (arbitrarily small) constant. Moreover, let there exist a function* $z \in C_2[0, 1]$ *which satisfies* (2.21) *with the above functions* b, c. *Then, for each* $v \in C_2[0, 1]$, $Mv \leqslant Mw \Rightarrow v \leqslant w$.

To prove this statement, define

$$\mathfrak{z}(\lambda, x) = \begin{cases} \lambda z(x) & \text{for} \quad 0 \leqslant \lambda \leqslant \delta \\ \delta z(x) + \lambda - \delta & \text{for} \quad \delta < \lambda \end{cases}$$

with $\delta > 0$ so small that $|\delta z'(x)| < \varepsilon\,(0 \leqslant x \leqslant 1)$. Then use arguments similar to those in Example 2.6. Observe that, by using such a nonlinear family, we are able to replace (2.20) by the *local* condition (2.22). □

Example 2.10. Let $Mu(0) = u(0) - r_0$, $Mu(1) = u(1) - r_1$, and

$$Mu(x) = -d/dx\{h(x)k(u(x))u'(x)\} + g[u](x) \quad \text{for} \quad 0 < x < 1, \qquad (2.23)$$

$$h \in C_1[0, 1], \qquad k \in C_1[0, \infty),$$

$$h(x) \geqslant 0 \quad (0 \leqslant x \leqslant 1), \qquad k(y) > 0 \quad (0 \leqslant y < \infty), \qquad \int_0^\infty k(t)\,dt = \infty.$$

For fixed $w \geqslant o$ *and* $z \succ o$ in $C_2[0, 1]$, we define a family $\{\mathfrak{z}(\lambda): 0 \leqslant \lambda < \infty\}$ by

$$\lambda z(x) = F(w(x) + \mathfrak{z}(\lambda, x)) - F(w(x)) \quad \text{with} \quad F(y) = \int_0^y k(t)\,dt. \qquad (2.24)$$

Then condition (2.19) holds and inequality (2.17) assumes the form $(-\lambda(hz')' + g[w + \mathfrak{z}(\lambda)] - g[w])(x) > 0$, where $w + \mathfrak{z}(\lambda) = F^{-1}(F(w) + \lambda z)$ with F^{-1} the inverse function of F. Now we may try to determine z such that this inequality holds for all $x \in (0, 1)$ and $\lambda > 0$ which satisfy the side conditions (2.18).

For example, if $g[u](x) = g(x, u(x))$ does not depend on $u'(x)$ and if $g(x, w(x) + y) - g(x, w(x)) \geqslant 0$ for $0 < x < 1$, $0 \leqslant y$, and $h(x) > 0$ for $0 \leqslant x \leqslant 1$, then the solution z of the boundary value problem $-(hz')' = 1$, $z(0) = z(1) = 1$ has all desired properties. Thus, $Mv \leqslant Mw \Rightarrow v \leqslant w$ for arbitrary positive $v \in C_2[0, 1]$, by Theorem 2.2. □

As the following example shows, this method of constructing $\mathfrak{z}(\lambda)$ can also be applied when the boundary operators $Mu(0)$, $Mu(1)$ have a different form.

Example 2.11. The boundary value problem

$$-\varepsilon x^{-2}\, d/dx\{x^2 k(u(x))u'(x)\} + \hat{g}(u(x)) = 0 \quad (0 < x < 1),$$
$$u'(0) = 0, \qquad u(1) = 1$$

with $k(y) = 1 + \kappa_0(y + \kappa_1)^{-2}$, $\hat{g}(y) = y(y + \kappa_2)^{-1}$, $\kappa_0 \geqslant 0$, $\kappa_1 > 0$, $\kappa_2 > 0$, $\varepsilon > 0$ describes a diffusion-kinetics enzyme problem (Murray [1968]). One is interested in positive solutions $u \in C_1[0, 1] \cap C_2(0, 1]$.

Define $Mu(x)$ $(0 < x < 1)$ by (2.23) with $h(x) = \varepsilon x^2$ and $g(x, y) = x^2\hat{g}(y)$, but now let $Mu(0) = -k(u(0))u'(0)$, $Mu(1) = u(1) - 1$. Then the given equations are equivalent to $Mu = o$ and, moreover, the method of Example 2.10 can be applied. For fixed $w \geqslant o$ and $z \succ o$ in $C_1[0, 1] \cap C_2(0, 1]$, choose $\mathfrak{z}(\lambda)$ as in (2.24). Then, since $\hat{g}$ is monotone, the inequalities (2.17), (2.19) are satisfied if $-(x^2z')' > 0$ $(0 < x < 1)$, $-z'(0) > 0$, $z(1) > 0$. For example, these conditions hold for $z = 5 - (x + 1)^2$. (Compare Theorem II.3.7(v).) Thus we arrive at the statement that $Mv \leqslant Mw \Rightarrow v \leqslant w$, for arbitrary positive $v, w \in C_1[0, 1] \cap C_2(0, 1]$. Consequently, *the given problem has at most one positive solution* $u \in C_1[0, 1] \cap C_2(0, 1]$. □

Theorems 2.1 and 2.2 may also be applied to first order operators. For example, let

$$Mu(x) = \begin{cases} u'(x) + f(x, u(x)) & \text{for} \quad 0 < x \leqslant 1 \\ u(0) - r_0 & \text{for} \quad x = 0 \end{cases} \tag{2.25}$$

with $f\colon (0, 1] \times \mathbb{R} \to \mathbb{R}$ and $r_0 \in \mathbb{R}$. This operator has the form (2.1) with $a(x) \equiv 0$, $g(x, y, p) = p + f(x, y)$, $g_0(y, p) = y - r_0$, $g_1(y, p) = p + f(1, y)$. Here $R = C_0[0, 1] \cap C_1(0, 1]$.

Example 2.12. *The operator M in (2.25) is inverse-monotone if f satisfies a one-sided local Lipschitz condition of the following form: for each $\eta > 0$ there exists a $\kappa > 0$ such that*

$$f(x, \tilde{y}) - f(x, y) \geqslant -\kappa(\tilde{y} - y) \qquad \textit{for} \quad 0 < x \leqslant 1, \qquad -\eta \leqslant y \leqslant \tilde{y} \leqslant \eta.$$

To prove this apply Theorem 2.2 to an arbitrary fixed pair $v, w \in R$ and verify the assumptions for $\mathfrak{z}(\lambda) = \lambda \exp Nx$, $N > \kappa$, $|v(x)| \leqslant \eta$, $|w(x)| \leqslant \eta$. □

Problems

2.4 Complete the proof of Theorem 2.2.

2.5 Using Theorem 2.2 prove that each positive solution $u \in C_0[0, 1] \cap C_2(0, 1]$ of the boundary value problem $-u'' + 27x^{-1/2}u^{3/2} = 0$ $(0 < x < 1)$,

$u(0) = 1$, $u(1) = 0$ satisfies $u \leqslant w$ with $w(x) = 1$ for $0 \leqslant x \leqslant x_0$, $81x^3w(x) = 16$ for $x_0 \leqslant x \leqslant 1$, where $16 = 81x_0^3$. The above differential equation essentially is the Thomas–Fermi equation.

2.6 Consider the operator (2.15), with g satisfying $0 \leqslant g(x, y_1, 0) < g(x, y_2, 0)$ for $0 < x < 1$, $0 \leqslant y_1 < y_2$, and $Mu(0) = u(0)$, $Mu(1) = u(1)$. Apply Theorem 2.2 with $\mathfrak{z}(\lambda) = \lambda z$ and $z(x) \equiv 1$ to prove the following *boundary maximum principle. If* $u \in C_2[0, 1]$ *satisfies* $a[u](x) \geqslant 0$ *and* $(-a[u]u'' + g[u])(x) \leqslant 0$ $(0 < x < 1)$, *then* $u(x) \leqslant \max(0, u(0), u(1))$ $(0 \leqslant x \leqslant 1)$. Also try to weaken the occurring assumptions by use of the side conditions (2.18).

2.7 The implication $Mv \leqslant Mw \Rightarrow v \leqslant w$ can be written as $\tilde{M}\tilde{w} \geqslant o \Rightarrow \tilde{w} \geqslant o$, with $\tilde{M}u = Mw - M(w - u)$ and $\tilde{w} = w - v$. Apply Theorem 2.2 to $\tilde{M}, o, \tilde{w}$ in place of M, v, w, and formulate the resulting assumptions in terms of the given quantities a, g, v, and w. It will turn out that $a[w](x) \geqslant 0$ has to be required, instead of (2.16), for example.

2.8 Consider M as in (2.1), (2.4), with sufficiently smooth g. The implication $Mv \leqslant Mw \Rightarrow v \leqslant w$ can be written as $\tilde{M}\tilde{w} \geqslant o \Rightarrow \tilde{w} \geqslant o$, with $\tilde{M}u(x) = L[u](x)$ $(0 < x < 1)$, $\tilde{M}u(x) = Mu(x)$ $(x = 0, 1)$, $\tilde{w} = w - v$, *and* L as in Example 2.4. Derive the result of that example by applying Theorem II.3.2 to $\tilde{M}$, assuming $r_0 = r_1 = 0$, without loss of generality.

2.9 Consider the problem in Example 2.11 with the boundary condition $u(1) = 1$ replaced by $\varepsilon u'(1) + \mu(u(1) - 1) = 0$, where $\mu > 0$. (Compare Murray [1968] for this problem.) Try to apply Theorem 2.2 using an approach analogous to that in Example 2.11.

2.10 The problems in Examples 2.10 and 2.11 can also be treated in a formally different way. Introduce a new dependent variable $\hat{u} = \int_0^u k(t)\,dt$ and apply Theorem 2.2 to the transformed operator $\hat{M}$ defined by $\hat{M}\hat{u} = Mu$.

2.11 For the first order operator (2.25) assume that $f(x, w(x) + t) - f(x, w(x)) \geqslant -q(t)$ for $x \in (0, 1]$, $t > 0$ with $v(x) = w(x) + t$, where $q \in C_0[0, \infty)$, $q(0) = 0$, $q(t) > 0$ $(0 < t < \infty)$, $\int_0^1 q^{-1}(t)\,dt = \infty$. Then $Mv \leqslant Mw \Rightarrow v \leqslant w$. Derive this statement from Theorem 2.2 using $\mathfrak{z}(\lambda)$ defined by $\mathfrak{z}(0, x) \equiv 0$, $\int_{\mathfrak{z}(\lambda, x)}^{\lambda} q^{-1}(t)\,dt = (1 + \delta)(1 - x)$ for $\lambda > 0$ with some $\delta > 0$. Observe that $\int_1^\infty q^{-1}(t)\,dt = \infty$ may be assumed without loss of generality.

2.12 For an operator M as in (2.25) the implication $Mv \leqslant Mw \Rightarrow v \leqslant w$ can often be proved for functions v, w which are not continuous. Suppose that $v, w \in C_1[x_0, x_1, x_2, \ldots, x_k, x_{k+1}]$ with $x_0 = 1$, $x_{k+1} = 1$, $u(x_i) = u(x_i - 0) \leqslant u(x_i + 0)$ $(i = 1, 2, \ldots, k)$ for $u = w - v$ and that for each $i \in \{0, 1, \ldots, k\}$ the inequalities $v(x_i) \leqslant w(x_i)$ and $Mv(x) \leqslant Mw(x)$ $(x_i < x \leqslant x_{i+1})$ imply $v(x) \leqslant w(x)$ $(x_i \leqslant x \leqslant x_{i+1})$. Then $Mv \leqslant Mw \Rightarrow v \leqslant w$.

Also show that for M in (2.25) and $v, w \in C_1[0, 1]$ (or less smooth functions) Theorem 2.2 remains true if $\mathfrak{z}(\lambda) = \lambda z$ and $z \in C_1[x_0, x_1, \ldots, x_{k+1}]$ is assumed instead of $z \in C_1[0, 1]$.

2.3 Miscellaneous results

2.3.1 Inverse-monotonicity under constraints

When the functions v, w considered satisfy given linear boundary conditions, the sufficient conditions for inverse-monotonicity may be modified. The theory is analogous to that for linear differential operators in Section II.3.5. Here we consider only the following special case.

Define $\bar{R}$ to be the set of all $u \in C_2[0, 1]$ such that $B_0[u] = B_1[u] = 0$ and let $\bar{M} : \bar{R} \to S$ be given by

$$\bar{M}u(x) = -a(x)u'' + g(x, u(x), u'(x)) \qquad \text{for} \quad 0 \leqslant x \leqslant 1,$$

where $a : [0, 1] \to \mathbb{R}$ with $a(x) \geqslant 0$ $(0 \leqslant x \leqslant 1)$, $g : [0, 1] \times \mathbb{R} \times \mathbb{R} \to \mathbb{R}$ and B_0, B_1 are defined as in (2.4). In this space $\bar{R}$, the strict order relation $u \succ o$ is defined by (II.3.37). The simple proof of the following result is left to the reader.

Theorem 2.3. *The statement of Theorem* 2.1 *holds also for the operator* $\bar{M}$ *(in place of* M*), and* $v, w \in \bar{R}$ *if the strict order relation is defined as above.*

Example 2.13. Let $\bar{M}$ have the special form

$$\bar{M}u(x) = L[u](x) - \mu g(x)u(x) + f(x, u(x)) \qquad (0 \leqslant x \leqslant 1)$$

with L and g as in the eigenvalue problem (II.3.40), $\mu \in \mathbb{R}$, and f satisfying $f(x, y) < f(x, \bar{y})$ for $0 \leqslant x \leqslant 1$, $y < \bar{y}$. *If* $\mu < \lambda_0$ *with* λ_0 *the smallest eigenvalue of problem* (II.3.40), *then* $\bar{M}v \leqslant \bar{M}w \Rightarrow v \leqslant w$, *for all* $v, w \in \bar{R}$. *In particular, if* $f(x, 0) \equiv 0$, *the problem* $\bar{M}u = o$ *has only the trivial solution* $u = o$.

This statement follows from Theorem 2.3 by using as z an eigenfunction φ of the eigenvalue problem (II.3.40) as described in Theorem II.3.24. □

2.3.2 Operators of divergence form

In many concrete problems, the occurring differential operators have the following *divergence form*:

$$Mu(x) = -\frac{d}{dx}\,\mathscr{a}(x, u(x), u'(x)) + \mathscr{g}(x, u(x), u'(x)) \qquad \text{for} \quad 0 < x < 1.$$

Of course, for sufficiently smooth $\mathscr{a}$, such operators can be written as in (2.15), so that the theory of the preceding sections may be applied. In Examples 2.10 and 2.11, for instance, special cases of such operators were treated (see also Problem 2.9). In what follows, however, we shall explain a different method which may yield different results for a concrete operator. Observe, in addition, that the theory on weak differential inequalities in Section 3 is also closely related to our present subject.

Consider the operator

$$Mu(x) = \begin{cases} -\dfrac{d}{dx}\,a(x, u(x), u'(x)) + g(x, u(x)) & \text{for} \quad 0 < x < 1 \\ u(0) - r_0 & \text{for} \quad x = 0 \\ u(1) - r_1 & \text{for} \quad x = 1. \end{cases} \tag{2.26}$$

For simplicity, we require that $a(x, y, p)$ be continuously differentiable on $[0, 1] \times \mathbb{R} \times \mathbb{R}$, $g(x, y)$ be defined on $[0, 1] \times \mathbb{R}$, and the given functions v, w belong to $C_2[0, 1]$.

Theorem 2.4. *If*

$$a_p(x, v(x), p) > 0 \qquad \textit{for} \quad 0 \leqslant x \leqslant 1, \quad p \in \mathbb{R} \tag{2.27}$$

and

$$g(x, v(x)) \geqslant g(x, v(x) - y) \qquad \textit{for} \quad 0 \leqslant x \leqslant 1, \quad 0 \leqslant y, \tag{2.28}$$

then $Mv \leqslant Mw \Rightarrow v \leqslant w$.

Proof. Suppose that $Mv \leqslant Mw$ holds, but $v(\zeta) > w(\zeta)$ for some $\zeta \in (0, 1)$. Then let $[\xi, \eta]$ denote the largest interval such that $\zeta \in (\xi, \eta) \subset (0, 1)$ and $v(x) > w(x)$ for $\xi < x < \eta$. We conclude that $v(\xi) = w(\xi)$, $v(\eta) = w(\eta)$, and consequently $v'(\xi) \geqslant w'(\xi)$, $v'(\eta) \leqslant w'(\eta)$. Partial integration yields

$$\{-a[v](\eta) + a[w](\eta)\} + \{a[v](\xi) - a[w](\xi)\}$$

$$= \int_\xi^\eta \left(-\frac{d}{dx}\,a[v](x) + \frac{d}{dx}\,a[w](x)\right) dx.$$

Because of (2.27), the terms within braces both are positive ($\geqslant 0$). On the other hand, (2.28) implies that the integrand is negative ($\leqslant 0$). Consequently, $a[v](x) = a[w](x)$ ($\xi \leqslant x \leqslant \eta$). Due to (2.27), this inequality can hold at $x = \xi$ only if $v'(\xi) = w'(\xi)$.

In an appropriate neighborhood of $(x, y) = (\xi, v(\xi))$, the equation $a(x, y, p) = a[w](x)$ defines a uniquely determined continuously differentiable function $p = f(x, y)$ with $v'(\xi) = f(\xi, v(\xi))$. Since the initial value problem $u'(x) = f(x, u(x))$, $u(\xi) = v(\xi)$ has only one solution, we conclude that $v(x) = w(x)$ in a right neighborhood of $x = \xi$. This contradicts the definition of ξ. □

The following corollary allows us to modify assumptions (2.27), (2.28). The proof is clear.

Corollary 2.4a. *Theorem 2.4 remains true if the inequality in (2.27) is required only for those $x \in [0, 1]$ such that $v(x) = w(x)$. Moreover, (2.28) need only be required for values $y = v(x) - w(x)$ with $y \geqslant 0$.*

Example 2.14. A lubrication problem (Chandra and Davis [1974], Steinmetz [1974]) is governed by a boundary value problem with Reynold's equation

$$-\varepsilon(h^3(x)uu')' + (h(x)u)' = 0, \qquad u(0) = u(1) = 1.$$

Here, $h \in C_1[0, 1]$, $h(x) > 0$ $(0 \leqslant x \leqslant 1)$, $\varepsilon > 0$. Obviously, this problem can be written in the form $Mu = o$ with M as in (2.26), $a(x, y, p) = \varepsilon h^3(x)yp - h(x)y$, $g(x, y) \equiv 0$. Thus, Theorem 2.4 and Corollary 2.4a yield the following statement.

If $v \succ o$ or $w \succ o$, then $Mv \leqslant Mw \Rightarrow v \leqslant w$. Consequently, if the problem $Mu = o$ has a strictly positive solution $u \in C_2[0, 1]$, then u is the only solution of the problem. In particular, *each solution $u \in C_2[0, 1]$ of $Mu = o$ satisfies $u \leqslant w$ with $w(x) = 1 + \varepsilon^{-1} \int_0^x h^{-2}(s)\, ds$,* since $w \succ o$ and $o \leqslant Mw$. In addition, one shows that $1 \leqslant u$ *in case* $h' \leqslant o$. On the other hand, $u \leqslant 1$ *in case* $h' \geqslant o$. □

2.3.3 Estimating derivatives

Consider a boundary value problem $Mu = o$ with M in (2.15), where $a(x, y, p) \equiv 1$ and $g(x, y, p)$ is defined for all $x \in [0, 1]$ and $y, p \in \mathbb{R}$. Suppose that a solution $u^* \in C_2[0, 1]$ exists and that functions v, w with $v \leqslant u^* \leqslant w$ have been found. Often one would like to have bounds for the derivative $u^{*\prime}$ also. We shall sketch a method for obtaining such bounds. The method may also be adopted to the case of a more general function $a(x, y, p)$. (For estimating derivatives see also Section 4.2.)

We define an operator

$$M_0 u_0(x) = \begin{cases} u_0'(x) - g(x, u^*(x), u_0(x)) & \text{for } \; 0 < x \leqslant 1 \\ u_0(0) - u^{*\prime}(0) & \text{for } \; x = 0, \end{cases}$$

where u^* denotes the unknown but fixed solution which we consider. Obviously, $M_0(u^*)' = o$. The above operator has the form (2.25), so that Theorem 2.2 may be applied to M_0. Thus, if the corresponding assumptions are satisfied, we can conclude that $M_0(u^*)' = o \leqslant M_0 w_0 \Rightarrow u^{*\prime} \leqslant w_0$.

In general, $M_0 w_0$ cannot be calculated, since the function u^* is unknown. However, it often is possible to formulate sufficient conditions for $o \leqslant M_0 w_0$. Since $v \leqslant u^* \leqslant w$, *the inequality* $0 \leqslant M_0 w_0(x)$ $(0 < x \leqslant 1)$ *holds if*

$$0 \leqslant w_0'(x) - g(x, y, w_0(x)) \quad \text{for} \quad 0 < x \leqslant 1, \quad v(x) \leqslant y \leqslant w(x).$$

Moreover, *if a bound* $\eta_0 \geqslant u^{*\prime}(0)$ *is known, then* $0 \leqslant M_0 w_0(0)$ *holds if* $\eta_0 \leqslant w_0(0)$. For example, let $Mu(0) = -\alpha_0 u'(0) + \gamma_0 u(0) - r_0$. *If* $\alpha_0 > 0$ *and* $\gamma_0 \geqslant 0$, *then* $u^{*\prime}(0) \leqslant \alpha_0^{-1}(\gamma_0 w(0) - r_0)$. On the other hand, *if* $\alpha_0 = 0, \gamma_0 = 1$ *and* $w(0) - r_0 = 0$, *then* $u^{*\prime}(0) \leqslant \eta_0 = w'(0)$.

In a similar way, a lower bound v_0 for $u^{*\prime}$ can be obtained. Moreover, one may define M_0 differently, so that the boundary conditions at $x = 1$ are used instead of those at $x = 0$.

2.3.4 *Periodic boundary conditions*

The methods of Sections 2.1 and 2.2 can also be applied to problems with other boundary conditions and corresponding operators. As an example, we shall derive a result which is related to periodic boundary conditions. We shall consider only rather smooth functions v, w. However, the smoothness conditions may be relaxed, in a way similar to that in Section 2.2.

Let R denote the space of all $u \in C_2[0, 1]$ which satisfy the boundary condition $u(0) = u(1)$, and suppose that v, w are fixed functions in R. We consider a differential operator $L[u] = -a[u]u'' + g[u]$ where $a[u](x) = a(x, u(x), u'(x))$, $g[u](x) = g(x, u(x), u'(x))$, and the real-valued functions $a(x, y, p)$, $g(x, y, p)$ are defined on $[0, 1) \times \mathbb{R} \times \mathbb{R}$.

Theorem 2.5. *Suppose that an admissible family* $\{\mathfrak{z}(\lambda): 0 \leqslant \lambda < \infty\} \subset R$ *exists which majorizes* $v - w$, *such that the following conditions hold.*

(i) *The inequalities* $a[v](x) \geqslant 0$ *and* $L[w + \mathfrak{z}(\lambda)](x) > L[w](x)$ *are satisfied for each* $\lambda > 0$ *and each* $x \in [0, 1)$ *with* $v(x) = w(x) + \mathfrak{z}(\lambda, x)$, $v'(x) = w'(x) + \mathfrak{z}'(\lambda, x)$.

(ii) $\mathfrak{z}'(\lambda, 0) \leqslant \mathfrak{z}'(\lambda, 1)$ *for* $\lambda > 0$.

Then

$$\left.\begin{array}{ll} L[v](x) \leqslant L[w](x) & \text{for} \quad 0 \leqslant x < 1 \\ -v'(0) + v'(1) \leqslant -w'(0) + w'(1) & \end{array}\right\} \Rightarrow \quad v \leqslant w. \tag{2.29}$$

Proof. Suppose that the inequalities in the premise of (2.29) hold, but $v \nleqslant w$. Then there exists a smallest number $\lambda_0 > 0$ with $v \leqslant w + \mathfrak{z}(\lambda_0)$, and the function $u = w - v + \mathfrak{z}(\lambda_0)$ satisfies $u(\xi) = 0$ for some $\xi \in [0, 1]$. For the case $\xi \in (0, 1)$ a contradiction is derived in the same way as in the proof of Theorem 2.1.

The equations $u(0) = 0$ and $u(1) = 0$ can only hold simultaneously. If they are satisfied, then $u'(0) \geqslant 0$ and $u'(1) \leqslant 0$. For the case $u'(0) = 0$ we have also $u''(0) \geqslant 0$ and obtain a contradiction in the same way as for $\xi \in (0, 1)$. If $u'(0) > 0$, however, we have $-u'(0) + u'(1) < 0$, while the required inequalities for v and w together with assumption (ii) yield $-u'(0) + u'(1) \geqslant 0$. □

Example 2.15. Let $L[u] = -u'' + g(x, u)$ with continuously differentiable $g(x, y)$ on $[0, 1] \times \mathbb{R}$. *If* $g_y(x, y) > 0$, *then the statement* (2.29) *holds for arbitrary* $v, w \in R$, since $\mathfrak{z}(\lambda, x) \equiv \lambda$ satisfies the required assumptions.

2.3.5 *Noncompact intervals*

For differential operators applied to functions on noncompact intervals, statements like (1.1) can often be derived from corresponding results for compact intervals. To explain the method, we consider an operator M which maps $R = C_2[0, \infty)$ into $S = \mathbb{R}[0, \infty)$. Let

$$Mu(x) = \begin{cases} -a(x)u''(x) + g(x, u(x), u'(x)) & \text{for} \quad 0 < x < \infty \\ -\alpha_0 u'(0) + \gamma_0 u(0) & \text{for} \quad x = 0, \end{cases} \tag{2.30}$$

where $a \in \mathbb{R}(0, \infty)$, $g : (0, \infty) \times \mathbb{R} \times \mathbb{R} \to \mathbb{R}$, $\alpha_0 \in \mathbb{R}$, $\gamma_0 \in \mathbb{R}$, and denote by v, w fixed functions in R. Moreover, let q be a given function satisfying $q(x) > 0$ $(0 \leqslant x < \infty)$; this function will be used to describe the behavior of $w - v$ for $x \to \infty$.

For each $l \in (0, \infty)$ we define an operator M_l mapping $R_l = C_2[0, l]$ into $S_l = \mathbb{R}[0, l]$ by

$$M_l u(x) = Mu(x) \qquad \text{for} \quad 0 \leqslant x < l, \qquad M_l u(l) = u(l).$$

Theorem 2.6. *Suppose that there exist a function* $z \in C_2[0, \infty)$ *and a constant* $\varepsilon_0 > 0$ *such that* $q(x) \leqslant \beta z(x)$ $(0 \leqslant x < \infty)$ *for some* $\beta > 0$ *and the following two conditions are satisfied*:

(i) $Mw(x) \leqslant M(w + \varepsilon z)(x)$ *for* $0 \leqslant x < \infty$ *and* $0 < \varepsilon < \varepsilon_0$,
(ii) *for each* $l \in (0, \infty)$ *and each* $\varepsilon \in (0, \varepsilon_0)$

$$M_l v(x) \leqslant M_l(w + \varepsilon z)(x) \quad (0 \leqslant x \leqslant l) \quad \Rightarrow \quad v(x) \leqslant (w + \varepsilon z)(x) \quad (0 \leqslant x \leqslant l).$$

Then

$$\left.\begin{array}{l} Mv(x) \leqslant Mw(x) \quad (0 \leqslant x < \infty) \\ \limsup\limits_{x \to \infty} [(w - v)/q](x) \geqslant 0 \end{array}\right\} \Rightarrow \quad v(x) \leqslant w(x) \quad (0 \leqslant x < \infty).$$

Proof. Let v and w satisfy the inequalities in the premise of the implication above. Because of the condition for $x \to \infty$, there exist $l_m > 0$ and $\varepsilon_m > 0$ $(m = 1, 2, \ldots)$ such that $l_m \to \infty$, $\varepsilon_m \to 0$, and $v(l_m) \leqslant w(l_m) + \varepsilon_m q(l_m) \leqslant w_m(l_m)$ with $w_m = w + \varepsilon_m \beta z$. Moreover, due to assumption (i), we have $Mv(x) \leqslant Mw(x) \leqslant Mw_m(x)$ for $0 \leqslant x < \infty$ and sufficiently large m, say $m \geqslant N$, and hence (ii) implies $v(x) \leqslant w(x) + \varepsilon_m \beta z(x)$ for $0 \leqslant x \leqslant l_m$ and $m \geqslant N$. Consequently, $v(x) \leqslant w(x)$ $(0 \leqslant x < \infty)$. □

In the above proof we made no use of the special form (2.30) of M. Of course, this form will be used if assumption (ii) above is verified by one of the methods in the preceding sections. For example, the following statement is obtained by applying Theorem 2.2 (to M_l and $[0, l]$ instead of M and $[0, 1]$).

Corollary 2.6a. *Assumption* (ii) *of Theorem* 2.6 *is satisfied if* $a(x) \geqslant 0$ $(0 < x < \infty)$, $\alpha_0 \geqslant 0$, *and the following condition holds*:

For each $l \in (0, \infty)$ *and each* $\varepsilon \in (0, \varepsilon_0)$ *there exists a function* $z = z_{l,\varepsilon} \in C_2[0, l]$ *such that* $z(x) > 0$ $(0 \leqslant x \leqslant l)$, $M_l z(0) > 0$, *and*

$$M_l(w + \varepsilon z)(x) < M_l(w + \lambda z)(x) \tag{2.31}$$

for each $x \in (0, l)$ *and each* $\lambda > \varepsilon$ *which satisfy* $v(x) = (w + \lambda z)(x)$, $v'(x) = (w' + \lambda z')(x)$.

Example 2.16. The boundary value problem

$$-u'' + f(u) = 0, \qquad u(0) = \gamma > 0, \ \lim_{x \to +\infty} u(x) = 0 \qquad \text{with} \quad f(y) = 2 \sinh y$$

arises in the theory of colloids (Alexander and Johnson [1949], Feynman *et al.* [1966], Anderson and Arthurs [1968]). The differential equation and the boundary condition at $x = 0$ can be written as $Mu = r$ with an operator M of the form (2.30). Since $f'(y) > 0$, all assumptions of Theorem 2.6 are satisfied for $q(x) \equiv z(x) \equiv 1$. (Observe that (2.31) holds for $z(x) \equiv 1$ independent of l and ε.) Thus, for functions $v, w \in C_2[0, \infty)$ converging for $x \to \infty$, we obtain

$$\left.\begin{array}{l} Mv(x) \leqslant Mw(x) \quad (0 \leqslant x < \infty) \\ \lim\limits_{x\to\infty} v(x) \leqslant \lim\limits_{x\to\infty} w(x) \end{array}\right\} \quad \Rightarrow \quad v(x) \leqslant w(x) \quad (0 \leqslant x < \infty). \tag{2.32}$$

As a consequence, the given problem has at most one solution $u^* \in C_2[0, \infty)$. Moreover, by applying (2.32) to $v = o$ and $w = u^*$ as well as to $v = u^*$ and $w(x) \equiv \gamma$, we see that $0 \leqslant u^*(x) \leqslant \gamma$ $(0 \leqslant x < \infty)$. Since even $f'(y) \geqslant c^2$ with $c = \sqrt{2}$, the required assumptions are also satisfied for $q(x) = \exp(\sqrt{2}x)$ and $z(x) = q(x) + 2 - (1 + x)^{-1}$. □

Problems

2.13 Prove Theorem 2.3.

2.14 Generalize Theorem 2.2 in an analogous way.

2.15 Apply the statement in Example 2.13 to $\overline{M}u = -u'' - \mu u + u^3$ and the boundary conditions $u(0) = u(1) = 0$.

2.16 Show that Theorem 2.4 remains true if the inequalities in (2.27), (2.28) are replaced by, respectively, $\mathscr{a}_p(x, w(x), w'(x)) > 0$ and $\mathscr{g}(x, w(x) + y) - \mathscr{g}(x, w(x)) \geqslant 0$. These inequalities may again be modified, similarly as in Corollary 2.4a. Describe these possibilities.

2.17 Let $M: C_2(-\infty, \infty) \to \mathbb{R}(-\infty, \infty)$ denote an operator of the form $Mu(x) = -a(x)u''(x) + g(x, u(x), u'(x))$. By applying the methods used in Section 2.3.5 find sufficient conditions on the occurring functions such that $v(x) \leqslant w(x)\,(-\infty < x < \infty)$ follows from $Mv(x) \leqslant Mw(x)\,(-\infty < x < \infty)$, $\limsup_{x\to+\infty}[(w - v)(x)/q_1(x)] \geqslant 0$, $\limsup_{x\to-\infty}[(w - v)(x)/q_2(x)] \geqslant 0$.

2.18 The smoothness requirements on the functions v and w in Theorem 2.6 may be relaxed, similarly as in Theorem 2.2. Carry out the details.

2.19 Apply the result of Problem 2.18 to prove that each positive solution $u^* \in C_0[0, \infty) \cap C_2(0, \infty)$ of the problem $-u'' + x^{-1/2}u^{3/2} = 0$ $(0 < x < \infty)$, $u(0) = 1$, $\lim_{x\to\infty} u(x) = 0$ satisfies $u^*(x) \leqslant w(x)$ $(0 < x < \infty)$, where $w(x) = 1$ for $0 \leqslant x \leqslant x_0$, $w(x) = 144x^{-3}$ for $x_0 < x < \infty$, $x_0^3 = 144$. (Compare Problem 2.5.)

2.20 Generalize Theorem 2.6 by not assuming that a function z as in the theorem exists, but requiring instead that for each $l > 0$ a function z_l on $[0, l]$ with suitable properties exists and all these functions z_l are bounded by some function z. (Compare Theorem II.3.27.)

3 INVERSE-MONOTONICITY WITH WEAK DIFFERENTIAL INEQUALITIES

The theory of this section is applicable to certain problems for which the results on pointwise differential inequalities in Section 2 cannot be applied. For example, using weak differential inequalities one may estimate the solutions of variational problems or solutions of generalized boundary value problems, such as Green functions. On the other hand, the following results also yield a different approach to pointwise inequalities (Section 3.4).

Properties IM* and IP*, in which we are interested, will be formulated in Section 3.1.1. They can be written as an implication of type (1.1) with a suitably defined operator (Section 3.1.2). Applications to generalized boundary value problems and variational problems are briefly discussed in Section 3.1.3. Sections 3.2 and 3.3 yield sufficient conditions for Property IM*. In Section 3.2 close relations to the theory of monotone definite operators become apparent, that is, relations to the theory of monotone operators in the Browder–Minty sense. The conditions of Section 3.3 are more in the spirit of the results of Section 2.2 on pointwise differential inequalities. Section 3.5 is mainly concerned with stronger positivity statements for linear problems.

3.1 Description of the problem

3.1.1 Properties IM* *and* IP*

We shall formulate an inverse-monotonicity property, which involves weak differential inequalities and certain (essential) boundary inequalities. To formulate the weak differential inequalities for a function u, we need only require that u have first derivatives in a generalized sense. More precisely, we shall consider the space H_1 of all functions $u \in C_0[0, 1]$ which have an L_2-derivative $u' = v$, this means $u(x) = u(0) + \int_0^x v(t)\,dt$ for some $v \in L_2(0, 1)$. Notice that, for a.a. $x \in [0, 1]$, $h^{-1}(u(x-h) - u(x)) \to v(x)$ for $h \to 0$. (Similar results are obtained when appropriate subspaces of H_1 are considered, such as the space $C'_{0,1}[0, 1]$ of all functions $u \in C_0[0, 1]$ with piecewise continuous derivatives.)

Concerning the inequalities occurring at the boundary points, we consider four cases which are described by the choice of a *boundary set* β. Let, respectively,

$$\beta = \{0, 1\}, \quad \beta = \{0\}, \quad \beta = \{1\}, \quad \beta = \varnothing \quad \text{for the case} \quad 1, 2, 3, 4. \tag{3.1}$$

In what follows, β shall be fixed. Then define

$$H = H^\beta = \{u \in H_1 : u(x) = 0 \quad \text{for} \quad x \in \beta\}, \qquad K = \{u \in H : u \geqslant o\}.$$

We consider a functional

$$B[u, \eta] = \int_0^1 [a(x, u(x), u'(x))\eta'(x) + g(x, u(x), u'(x))\eta(x)]\,dx$$
$$+ g_0(u(0))\eta(0) + g_1(u(1))\eta(1) \tag{3.2}$$

where g_0, g_1 denote functions in $C_0(-\infty, \infty)$ such that $g_0(y) \equiv 0$ if $0 \in \beta$, $g_1(y) \equiv 0$ if $1 \in \beta$, and the real-valued functions $a(x, y, p)$, $g(x, y, p)$ on $[0, 1] \times \mathbb{R} \times \mathbb{R}$ satisfy the *Caratheodory condition*, which we formulate here for a:

(i) for each fixed $y \in \mathbb{R}$ and $p \in \mathbb{R}$, $a(x, y, p)$ is a measurable function of $x \in [0, 1]$;

(ii) for a.a. $x \in [0, 1]$, $a(x, y, p)$ is a continuous function of $y \in \mathbb{R}$ and $p \in \mathbb{R}$.

Let D be a subset of H_1 such that, for each $u \in D$, the relations $a[u](x) = a(x, u(x), u'(x))$, $g[u](x) = g(x, u(x), u'(x))$ define functions $a[u] \in L_2(0, 1)$, $g[u] \in L_1(0, 1)$. Then $B[u, \eta]$ is defined for all $u \in D$ and $\eta \in H$.

For fixed $v, w \in D$, we are interested in the following property of *generalized inverse-monotonicity*:

Property IM* (*for* v, w):

$$\left.\begin{array}{ll} B[v,\eta] \leqslant B[w,\eta] & \textit{for all} \quad \eta \in K \\ v(x) \leqslant w(x) & \textit{for all} \quad x \in \beta \end{array}\right\} \quad \Rightarrow \quad v \leqslant w.$$

In particular, we shall investigate the linear case where B is a bilinear form

$$B[u,\eta] = \int_0^1 [(au' + b_1 u)\eta' + (b_2 u' + cu)\eta](x)\,dx$$
$$+ \gamma_0 u(0)\eta(0) + \gamma_1 u(1)\eta(1) \tag{3.3}$$

with bounded measurable functions a, b_1, b_2, c and $\gamma_0, \gamma_1 \in \mathbb{R}$. Here, we are interested in Property IP* of *generalized inverse-positivity*.

Property IP*: *For all* $u \in H_1$

$$\left.\begin{array}{ll} B[u,\eta] \geqslant 0 & \textit{for all} \quad \eta \in K \\ u(x) \geqslant 0 & \textit{for all} \quad x \in \beta \end{array}\right\} \quad \Rightarrow \quad u \geqslant o.$$

Obviously, if Property IP* holds, then the corresponding Property IM* holds for arbitrary $v, w \in H_1$.

Observe that for a given functional B as in (3.2) the functions $\mathscr{a}$, $\mathscr{g}$, $\mathscr{g}_0$, and $\mathscr{g}_1$ are not uniquely determined. In particular, the coefficients in (3.3) are not uniquely determined. The integrand may be changed by partial integration. Also, the boundary terms may be modified. For example, $\gamma_1 u(1)\eta(1) = \int_0^1 \gamma_1 (xu\eta)'\,dx$.

3.1.2 A different formulation

Property IM* can also be formulated as inverse-monotonicity of a suitably defined operator $\mathscr{M}$. This different formulation is not needed to derive the sufficient conditions in the subsequent sections. However, it throws some new light on the meaning of those conditions.

We shall use the notation

$$\langle u, v\rangle_0 = \int_0^1 u(x)v(x)\,dx, \qquad \|u\|_0 = [\langle u, u\rangle_0]^{1/2},$$

$$\langle u, v\rangle_1 = \langle u, v\rangle_0 + \langle u', v'\rangle_0, \qquad \|u\|_1 = [\langle u, u\rangle_1]^{1/2}.$$

The set H is a Hilbert space with respect to the inner product $\langle u, v\rangle_1$. For given $u \in D$, a bounded linear functional F on H is defined by $F(\eta) = B[u,\eta]$. According to the theorem of Riesz, there exists a unique $w \in H$ such that $B[u,\eta] = \langle w, \eta\rangle_1$ for all $\eta \in H$. Thus, by $Au = w$ an operator $A: D \to H$ is defined such that

$$B[u,\eta] = \langle Au, \eta\rangle_1 \qquad \text{for} \quad u \in D, \quad \eta \in H. \tag{3.4}$$

In order to define $\mathcal{M}$, we have also to incorporate the boundary terms. For the four cases described in (3.1) let, respectively,

$$\mathcal{S} = H \times \mathbb{R} \times \mathbb{R}, \qquad \mathcal{S} = H \times \mathbb{R}, \qquad \mathcal{S} = H \times \mathbb{R}, \qquad \mathcal{S} = H,$$

and write the elements $\mathcal{U} \in \mathcal{S}$ as, respectively,

$$\mathcal{U} = (U, U_0, U_1), \qquad \mathcal{U} = (U, U_0), \qquad \mathcal{U} = (U, U_1), \qquad \mathcal{U} = U \tag{3.5}$$

with $U \in H$, $U_0 \in \mathbb{R}$, $U_1 \in \mathbb{R}$. Then an operator $\mathcal{M}: D \to \mathcal{S}$ is given by, respectively,

$$\begin{aligned} \mathcal{M}u &= (Au, u(0), u(1)), \qquad & \mathcal{M}u &= (Au, u(0)), \\ \mathcal{M}u &= (Au, u(1)), \qquad & \mathcal{M}u &= Au. \end{aligned} \tag{3.6}$$

Finally, the space $\mathcal{S}$ is furnished with a (linear) order relation $\leqslant$, where $\mathcal{U} \geqslant o$ is defined by

$$\langle U, \eta \rangle_1 \geqslant 0 \qquad \text{for all} \quad \eta \in K,$$
$$U_0 \geqslant 0 \quad \text{if} \quad 0 \in \beta, \qquad U_1 \geqslant 0 \quad \text{if} \quad 1 \in \beta.$$

Then the following statement is obvious.

Proposition 3.1. *Let $\mathcal{M}: (D, \leqslant) \to (\mathcal{S}, \leqslant)$ be defined as above.*
Then, for given $v, w \in D$, Property IM* *holds if and only if $\mathcal{M}v \leqslant \mathcal{M}w \Rightarrow v \leqslant w$. In particular, for a bilinear form B as in* (3.3), *Property* IP* *holds if and only if the* (*linear*) *operator $\mathcal{M}$ is inverse-positive on H_1.*

3.1.3 Generalized boundary value problems

For given B and β as above, consider the following *generalized boundary value problem.* Find $u \in D \subset H_1$ such that

$$\begin{aligned} B[u, \eta] &= F(\eta) \qquad \text{for all} \quad \eta \in H, \\ u(0) = r_0 \quad \text{if} \quad 0 \in \beta, & \qquad u(1) = r_1 \quad \text{if} \quad 1 \in \beta, \end{aligned} \tag{3.7}$$

where $r_0, r_1 \in \mathbb{R}$ and F denotes some linear functional which is bounded in the sense that $|F(\eta)| \leqslant \kappa \|\eta\|_1$ for all $\eta \in H$, with some $\kappa \geqslant 0$.

For later use, we state some facts on the solvability of *linear* generalized boundary value problems. Here, the existence of a solution, in general, follows from the uniqueness.

Theorem 3.2. (i) *Let B denote a bilinear form* (3.3) *with bounded measurable functions a, b_1, b_2, c, such that*

$$a(x) \geqslant a_0 \qquad (0 \leqslant x \leqslant 1) \qquad \textit{for some} \quad a_0 > 0. \tag{3.8}$$

(ii) *Suppose that the homogeneous problem*

$$B[u, \eta] = 0 \qquad \textit{for all} \quad \eta \in H,$$
$$u(0) = 0 \qquad \textit{if} \quad 0 \in \beta, \qquad u(1) = 0 \qquad \textit{if} \quad 1 \in \beta, \tag{3.9}$$

has only the trivial solution $u = o$.

Then, for each bounded linear functional F *on* H *and arbitrary* $r_0, r_1 \in \mathbb{R}$, *the problem* (3.7) *has a unique solution* $u \in H_1$.

This result is proved by using the theorem of Lax–Milgram, Gårding's inequality, the Fredholm alternative, and the fact that the embedding $H \to L_2$ is compact. (Compare, for example, Simader [1972], where a much more general case is treated.) The ideas of the following proof of Gårding's inequality will also be used for further purposes, below.

Lemma 3.3 (Gårding's inequality). *Let* B *satisfy assumption* (i) *of Theorem* 3.2. *Then there exist constants* $c_0 > 0$ *and* $k \geqslant 0$ *such that* $B[\eta, \eta] \geqslant c_0\|\eta'\|_0^2 - k\|\eta\|_0^2$ *for all* $\eta \in H$.

Proof. The preceding inequality is derived by using the following estimates with sufficiently small $\varepsilon > 0$:

$$\int_0^1 a(\eta')^2\,dx \geqslant a_0\|\eta'\|_0^2,$$

$$\int_0^1 |\eta\eta'|\,dx \leqslant \|\eta'\|_0\|\eta\|_0 = 2(\varepsilon\|\eta'\|_0)(\tfrac{1}{2}\varepsilon^{-1}\|\eta\|_0) \leqslant \varepsilon^2\|\eta'\|_0^2 + \tfrac{1}{4}\varepsilon^{-2}\|\eta\|_0^2$$

$$\eta^2(1) = \int_0^1 \frac{d}{dx}(x\eta^2(x))\,dx \leqslant \|\eta\|_0^2 + 2\int_0^1 |\eta\eta'|\,dx,$$

$$\eta^2(0) \leqslant \|\eta\|_0^2 + 2\int_0^1 |\eta\eta'|\,dx. \quad \square$$

In the context of this existence and uniqueness theory, Property IP* may be used to verify assumption (ii) of Theorem 3.2.

Proposition 3.4. *If Property* IP* *holds, then the homogeneous problem* (3.9) *has only the trivial solution.*

For bilinear B, a function $\mathcal{G}(x, \xi)$ on $[0, 1] \times [0, 1]$ is called a *Green function corresponding to* B *and* β, if for all fixed $\xi \in [0, 1]$ $g(x) := \mathcal{G}(x, \xi)$ belongs to H and is a solution of (3.7) with $F(\eta) = \eta(\xi)$, $r_0 = r_1 = 0$. Because this functional F is bounded, Theorem 3.2 can be applied. (Observe that $\xi\eta(\xi) = \int_0^\xi (x\eta)'\,dx$ and $(1 - \xi)\eta(\xi) = \int_\xi^1 ((x - 1)\eta)'\,dx$.) Since F is also positive, we obtain the following result by using Property IP*.

Proposition 3.5. *Under the assumptions of Theorem* 3.2, *there exists a unique Green function* $\mathscr{G}(x, \xi)$, *corresponding to* B *and* β; *this function is positive,* $\mathscr{G}(x, \xi) \geqslant 0$ $(0 \leqslant x, \xi \leqslant 1)$, *if Property* IP* *holds.*

Further applications of generalized inverse-positivity and applications of generalized inverse-monotonicity will be explained in examples, below.

Boundary value problems of the type (3.7) also arise in the theory of variational problems. Consider, for instance, the problem of finding a function $u \in H_1$ which satisfies the essential boundary conditions

$$u(0) = r_0 \quad \text{if} \quad 0 \in \beta, \qquad u(1) = r_1 \quad \text{if} \quad 1 \in \beta \tag{3.10}$$

and minimizes

$$J(u) = \int_0^1 f(x, u(x), u'(x))\, dx + f_0(u(0)) + f_1(u(1)) - F(u), \tag{3.11}$$

that is, $J(u) \leqslant J(u + \eta)$ for all $\eta \in H$.

If f, f_0, and f_1 are sufficiently smooth, necessary conditions are obtained as follows. For a solution u and fixed $\eta \in H$ the function $\mathscr{J}(\varepsilon) = J(u + \varepsilon\eta)$ assumes its minimum at $\varepsilon = 0$, so that $\mathscr{J}'(0) = 0$. The latter relation yields

$$\int_0^1 [f_p(x, u, u')\eta' + f_y(x, u, u')\eta]\, dx + f'_0(u(0))\eta(0) + f'_1(u(1))\eta(1) = F(\eta)$$

and hence u is a solution of a generalized boundary value problem as in (3.7). Thus, *Property* IM* *can be used to estimate solutions of variational problems or to prove uniqueness.*

In particular, let

$$\begin{aligned} f(x, y, p) &= \tfrac{1}{2}a(x)p^2 + b(x)py + \tfrac{1}{2}c(x)y^2, \\ f_0(y) = \tfrac{1}{2}\gamma_0 y^2, &\qquad f_1(y) = \tfrac{1}{2}\gamma_1 y^2 \end{aligned} \tag{3.12}$$

with bounded measurable functions a, b, c. Then the corresponding boundary value problem is linear and we obtain the following statement.

Proposition 3.6. *Suppose that Property* IP* *holds for* B *in* (3.3) *with* $b_1 = b_2 = b$, *that* F *is a positive linear functional on* H *and that* $r_0 \geqslant 0$, $r_1 \geqslant 0$.

Then each solution $u \in H_1$ *of the variational problem corresponding to* (3.10), (3.11), (3.12) *is positive.*

3.2 Monotone definite operators

Throughout this section the general assumptions made in Section 3.1.1 shall be satisfied.

Theorem 3.7. *For given $w \in D$ let*

$$B[w + \eta, \eta] - B[w, \eta] > 0 \tag{3.13}$$

for all $\eta \in K$ such that $\eta > o$, $a[w + \eta]\eta' \in L_1(0, 1)$, $g[w + \eta]\eta \in L_1(0, 1)$.
Then Property IM* *holds for all $v \in D$.*

Proof. For given $u \in H_1$ the functions u^+ and u^- defined by $u^+(x) = \max\{u(x), 0\}$, $u^-(x) = \max\{-u(x), 0\}$ also belong to H_1. One has $u = u^+ - u^-$ and $u' = (u^+)' - (u^-)'$ in the L_2-sense. In particular,

$$(u^+)'(x) = 0 \quad \text{for a.a} \quad x \in [0, 1] \quad \text{with} \quad u(x) \leqslant 0,$$
$$(u^-)'(x) = 0 \quad \text{for a.a} \quad x \in [0, 1] \quad \text{with} \quad u(x) \geqslant 0.$$

Now, suppose that v and w satisfy the inequalities in the premise of Property IM*, but $v \nleqslant w$. Then $\eta := (w - v)^- > o$. Moreover, since $v(x) \leqslant w(x)$ for $x \in \beta$, we have $\eta(x) = 0$ for $x \in \beta$, so that $\eta \in K$. Because of the above relations for L_2-derivatives,

$$\eta'(x) = 0 \quad \text{for a.a.} \quad x \in [0, 1] \quad \text{with} \quad \eta(x) = 0, \tag{3.14}$$

and

$$\begin{aligned} w(x) - v(x) = -\eta(x), \qquad w'(x) - v'(x) = -\eta'(x) \\ \text{for a.a.} \quad x \in [0, 1] \quad \text{with} \quad \eta(x) \neq 0. \end{aligned} \tag{3.15}$$

Thus we obtain $a[v](x)\eta'(x) = a[w - (w - v)](x)\eta'(x) = a[w + \eta](x)\eta'(x)$ for a.a. $x \in [0, 1]$ and, similarly, $g[v](x)\eta(x) = g[w + \eta](x)\eta(x)$, so that $B[v, \eta] = B[w + \eta, \eta]$. On the other hand, the assumption of the theorem yields $B[v, \eta] \leqslant B[w, \eta] < B[w + \eta, \eta]$. Because of this contradiction, $v \leqslant w$ must be true. □

As the proof suggests, the assumptions may be weakened in the following way.

Corollary 3.7a. *For given $v, w \in D$, let $w - v \ngeqslant o$ imply that*

$$B[w + \eta, \eta] - B[w, \eta] > 0 \qquad \textit{for} \quad \eta = (w - v)^-. \tag{3.16}$$

Then Property IM* *holds.*

The *side condition* $\eta = (w - v)^-$ implies (3.15). These relations can be used in a way similar to that in which the conditions (2.18) have been used in Section 2.2. With the aid of (3.15) one may modify the inequality (3.13) and then require the modified condition to hold for all $\eta > o$ in K such that the terms are integrable. For example, (3.13) may be replaced by $B[v, \eta] - B[v - \eta, \eta] > 0$, since the terms are equal for $\eta = (w - v)^-$.

(In the proof of the preceding theorem we could also have used the function

$$\eta(x) = \begin{cases} -(w - v)(x) & \text{for} \quad x \in J \\ 0 & \text{otherwise,} \end{cases}$$

where $J = (\xi, \zeta) \subset (0, 1)$ is some nonempty open interval such that $-(w - v)(x) > 0$ for all $x \in J$, $(w - v)(\xi) = 0$ in case $\xi > 0$, and $(w - v)(\zeta) = 0$ in case $\zeta < 1$. The relations (3.14), (3.15) hold for this function η, too, so that the preceding remarks still apply. In particular, this choice of η is suitable when the space $C'_{0,1}[0, 1]$ is used instead of H_1.)

In the special case of a bilinear form (3.3) we obtain the following statement.

Corollary 3.7b. *If the bilinear form B is positive definite, i.e.,*

$$B[\eta, \eta] > 0 \qquad \text{for all} \quad \eta \neq o \quad \text{in} \quad H, \tag{3.17}$$

then Property IP* *holds.*

Notice that it makes no difference if H is replaced by K in (3.17), since $B[\eta, \eta] = B[\eta^+, \eta^+] + B[\eta^-, \eta^-]$.

Example 3.1. Let $B[u, \eta] = \int_0^1 (u'\eta' + cu\eta)\, dx$ and $\beta = \{0, 1\}$. Clearly, if $c(x) \geqslant 0$ $(0 \leqslant x \leqslant 1)$, then Property IP* holds, since (3.17) is satisfied. □

For B and β in Example 3.1, condition (3.17) even holds if $c(x) \geqslant -\pi^2 + \mu$ $(0 \leqslant x \leqslant 1)$ for some $\mu > 0$. This fact, however, is not so easily deduced from Corollary 3.7b as the statement of Example 3.1. Such weaker sufficient conditions are derived more conveniently from the theory of Section 3.3 (see Example 3.3).

Example 3.2. Let $B[u, \eta] = \int_0^1 [u'(x)\eta'(x) + g(x, u(x))\eta(x)]\, dx$, $\beta = \{0, 1\}$ where $g(x, y)$ is continuously differentiable on $[0, 1] \times \mathbb{R}$. Using (3.15) we calculate for $\eta = (w - v)^-$ with $v, w \in D$

$$B[w + \eta, \eta] - B[w, \eta] = \int_0^1 (\eta'\eta' + c\eta\eta)(x)\, dx$$

with

$$c(x) = \int_0^1 g_y(x, (1 - t)w(x) + tv(x))\, dt.$$

Thus, according to Corollary 3.7a, *Property* IM* *holds for all* $v, w \in D$ *such that the above function c is positive.* □

The sufficient conditions in Theorem 3.7 and its Corollaries 3.7a,b may also be written in terms of the operator A in (3.4). Obviously, if $w + (w - v)^- = \sup\{v, w\}$ belongs to the domain D, then condition (3.16) is equivalent to $\langle A(w + \eta) - Aw, \eta\rangle_1 > 0$ for $\eta = (w - v)^-$ in case $\eta > o$.

The operator A, which maps the subset D of (the Hilbert space) H_1 into $H \subset H_1$ will be called *monotone definite* or, shortly, d-*monotone* if

$$\langle Av - Aw, v - w\rangle_1 > 0 \qquad \text{for arbitrary} \quad v, w \in D \quad \text{with} \quad v \neq w.$$

In particular, a linear d-monotone operator is called *positive definite* or d-*positive*. (Concerning this notation see the Notes.)

With this terminology Corollary 3.7a yields the following result.

Corollary 3.7c. *Suppose that* $\sup\{v, w\} \in D$ *whenever* $v, w \in D$, *and that the operator* $A : D \to H$ *is monotone definite.*

Then Property IM* *holds for arbitrary* $v, w \in D$, *i.e., the operator* $\mathcal{M}$ *in* (3.6) *is inverse-monotone.*

For example, if D is the set of all $u \in H_1$ with $a[u] \in L_2(0, 1)$, $g[u] \in L_1(0, 1)$, then $v, w \in D$ always implies $\sup\{v, w\} \in D$.

Problem

3.1 Many results proved here for the interval (0, 1) can be carried over to unbounded intervals, such as $(-\infty, \infty)$. In this case, consider $B[u, \eta] = \int_{-\infty}^{\infty} [a(x, u, u')\eta' + g(x, u, u')\eta]\, dx$ with a and g satisfying the Caratheodory condition on $(-\infty, \infty) \times \mathbb{R} \times \mathbb{R}$, and redefine the terms H_1, K, D accordingly. For example, let H_1 be the set of all $u \in L_2(-\infty, \infty) \cap C_0(-\infty, \infty)$ which have a derivative $u' \in L_2(-\infty, \infty)$, and D the set of all $u \in H_1$ with $a[u]$, $g[u]$ in $L_2(-\infty, \infty)$. Now $\beta = \emptyset$, and $v \leqslant w \Leftrightarrow v(x) \leqslant w(x)$ $(-\infty < x < \infty)$. Prove that *Theorem* 3.7 *holds for this case also* (with $L_1(0, 1)$ replaced by $L_1(-\infty, \infty)$).

3.3 Further constructive sufficient conditions

For a large class of problems, the sufficient conditions of the preceding sections are easily verified. However, there are certain limits in the immediate applicability of those results (see, for instance, the remarks following Example 3.1). We shall overcome those limits by establishing sufficient conditions which involve a function z—or a family of functions $\mathfrak{z}(\lambda)$—similar to those in Sections 2.1, 2.2. Here, these functions have to satisfy certain weak differential inequalities. In a concrete case, Property IM* can be verified by constructing such functions.

We shall again use the notion of an admissible family $\{\mathfrak{z}(\lambda) : 0 \leqslant \lambda < \infty\}$ and a majorizing family given in Section 2.2. Then, under the general assumptions of Section 3.1.1, we obtain the following result, for fixed $v, w \in D$.

Theorem 3.8. *Suppose that there exist an admissible family* $\{\mathfrak{z}(\lambda): 0 \leqslant \lambda < \infty\} \subset H_1$ *which majorizes* $v - w$, *a function* $h \in C_0(0, \infty)$ *with* $h(\lambda) > 0$ $(0 < \lambda < \infty)$, *and constants* $\varepsilon > 0, k \geqslant 0$, *such that* $w + \mathfrak{z}(\lambda) \in D$ $(0 \leqslant \lambda < \infty)$ *and the following conditions are satisfied.*

For each $\lambda > 0$ *with* $w - v + \mathfrak{z}(\lambda) \ngeq o$, *the inequalities*

$$B[v, \eta] - B[v - \eta, \eta] \geqslant -k \int_0^1 [\eta(x)]^{1+\varepsilon}\, dx \tag{3.18}$$

and

$$B[w + \mathfrak{z}(\lambda), \eta] - B[w, \eta] \geqslant h(\lambda) \int_0^1 \eta(x)\, dx \tag{3.19}$$

hold for $\eta = (w - v + \mathfrak{z}(\lambda))^-$, *if* $\eta \in K$.

Then Property IM* *holds for* v *and* w.

Proof. If v and w satisfy the inequalities in the premise of Property IM*, but $v \nleqslant w$, then there exists a smallest number $\lambda_0 > 0$ such that $v \leqslant w + \mathfrak{z}(\lambda_0)$. Choose $\lambda \in (0, \lambda_0)$ and define $\eta = \eta_\lambda = (w - v + \mathfrak{z}(\lambda))^- > o$.

By arguments similar to those in the proof of Theorem 3.7 we derive from (3.18) that

$$\begin{aligned} B[w + \mathfrak{z}(\lambda), \eta] - B[w, \eta] &\leqslant B[v + (w - v + \mathfrak{z}(\lambda)), \eta] - B[v, \eta] \\ &\leqslant B[v - \eta, \eta] - B[v, \eta] \leqslant k \int_0^1 \eta^{1+\varepsilon}\, dx \\ &\leqslant k\|\eta\|_\infty^\varepsilon \int_0^1 \eta\, dx. \end{aligned}$$

Together with (3.19) this inequality implies $h(\lambda) \leqslant k(\|\eta_\lambda\|_\infty)^\varepsilon$. Since $h(\lambda) \to h(\lambda_0) > 0$ and $\|\eta_\lambda\|_\infty \to 0$ for $\lambda \to \lambda_0 - 0$, we obtain a contradiction. Thus $v \leqslant w$ must be true. □

Of course, it suffices to require that the inequalities (3.18), (3.19) hold for *all* $\eta > o$ in K for which the integrals exist. But, here again, the occurring *side condition* $\eta = (w - v + \mathfrak{z}(\lambda))^-$ can be used to modify those inequalities. As the proof shows, one may further weaken the assumptions. For example, in place of (3.18), (3.19), it suffices to require that

$$B[w + \mathfrak{z}(\lambda), \eta_\lambda] - B[w, \eta_\lambda] > 0 \qquad (0 < \lambda),$$

and

$$\limsup_{\lambda \to \lambda_0 - 0} (B[w + \mathfrak{z}(\lambda), \eta_\lambda] - B[w, \eta_\lambda])^{-1}(B[v - \eta_\lambda, \eta_\lambda] - B[v, \eta_\lambda]) < 1.$$

It will turn out that (3.18) is satisfied under mild assumptions on the functional B. For instance, for the linear case, Theorem 3.8 and Lemma 3.3 together yield the following result, the proof of which is left to the reader.

Corollary 3.8a. *Suppose that the bilinear form B in* (3.3) *satisfies* (3.8) *and that there exist a strictly positive function $z \in H_1$ and a constant $h_0 > 0$ such that*

$$B[z, \eta] \geqslant h_0 \int_0^1 \eta(x)\, dx \qquad \text{for all } \eta \in K. \tag{3.20}$$

Then Property IP* *holds.*

In the following examples we consider only the special case in which

$$\begin{gathered} B[u, \eta] = \int_0^1 (au'\eta' + cu\eta)\, dx, \\ a(x) \geqslant a_0 > 0 \qquad (0 \leqslant x \leqslant 1), \qquad \text{and} \quad \beta = \{0, 1\}. \end{gathered} \tag{3.21}$$

However, the ideas which we explain can be carried over to more general problems.

If the coefficient $a(x)$ has a sufficiently smooth derivative, one may try to find a smooth function z by means of partial integration,

$$B[z, \eta] = \int_0^1 L[z]\eta\, dx \qquad \text{with} \quad L[z] = -(az')' + cz. \tag{3.22}$$

Obviously, inequality (3.20) holds if

$$L[z](x) \geqslant h_0 > 0 \qquad (0 \leqslant x \leqslant 1). \tag{3.23}$$

Thus, z may be constructed, for example, with the methods explained in Section II.3.3.

Example 3.3. *Property* IP* *holds for the case* (3.21) *if* $a(x) = 1$ *and* $c(x) > -\pi^2 + \mu$ $(0 \leqslant x \leqslant 1)$ *with some* $\mu > 0$. To prove this, choose $z = \cos(\pi - \varepsilon)(x - \frac{1}{2})$ with sufficiently small $\varepsilon > 0$, and apply (3.22), (3.23). □

Example 3.4. The inequality $c(x) > -\pi^2 + \mu$ in the preceding example is sufficient, but not necessary. Consider (3.21) with $a(x) \equiv 1$ and

$$c(x) = \begin{cases} -4\pi^2 + \varepsilon & \text{for } 0 \leqslant x \leqslant \xi \\ 0 & \text{for } \xi < x \leqslant 1, \end{cases} \qquad \text{where} \quad \varepsilon > 0 \quad \text{and} \quad \xi \in (0, \tfrac{1}{4}).$$

Then Property IP* holds, since (3.20) is satisfied with

$$z(x) = \begin{cases} \sin \dfrac{\pi}{2}(\delta + (1-\delta)4x) & \text{for} \quad 0 \leqslant x \leqslant \xi \\[2ex] \sin \dfrac{\pi}{2}(\delta + (1-\delta)4\xi) & \text{for} \quad \xi \leqslant x \leqslant 1, \end{cases}$$

$\delta > 0$ sufficiently small. □

The approach in (3.22) may be used even if $a(x)$ is not differentiable. For example, by formally integrating the differential equation $-(az_0')'(x) = q(x)$ for some strictly positive $q \in C_0[0, 1]$, and requiring the boundary conditions $z_0(0) = z_0(1) = 0$, one calculates

$$z_0(x) = \left[\int_0^1 h(s)\,ds\right]^{-1} \int_0^x h(s)\,ds \int_0^1 h(s) \int_0^s q(t)\,dt\,ds - \int_0^x h(s) \int_0^s q(t)\,dt\,ds \tag{3.24}$$

with $h(x) = 1 : a(x)$. For $z(x) = \varepsilon + z_0(x)$ with $\varepsilon > 0$, (3.22) is valid, even if $a(x)$ has no additional smoothness properties.

Example 3.5. *Property* IP* *holds for the case* (3.21), *if there exists a strictly positive function* $q \in C_0[0, 1]$ *such that, for* z_0 *in* (3.24),

$$c(x)z_0(x) > q(x) + \delta \qquad (0 \leqslant x \leqslant 1), \qquad \textit{with} \quad \delta > 0. \tag{3.25}$$

For sufficiently small $\varepsilon > 0$, the function $z = \varepsilon + z_0 \in H_1$ satisfies $L[z](x) = q(x) + c(x)(z_0(x) + \varepsilon) \geqslant h_0 > 0 \ (0 \leqslant x \leqslant 1)$, so that (3.20) follows from (3.22). Through elementary calculations one can show that $z_0 \geqslant o$, which yields that z is strictly positive.

For the special case $a(x) \equiv 1$, $q(x) \equiv 1$, one obtains $z_0 = \frac{1}{2}x(1-x)$, so that (3.25) is satisfied if $c(x) > -8 + \mu \ (0 \leqslant x \leqslant 1)$ with $\mu > 0$. □

Example 3.6. As an application of the previous example we shall compare the Green function $\mathscr{G}(x, \xi)$ corresponding to (3.21) with the Green function $\tilde{\mathscr{G}}(x, \xi)$ corresponding to $\tilde{B}[u, \eta] = \int_0^1 (au'\eta' + \tilde{c}u\eta)\,dx$, $\beta = \{0, 1\}$, $\tilde{c}$ bounded and measurable. We assume that $\tilde{\mathscr{G}}(x, \xi)$ exists and denote by z_0 the function (3.24) with $q(x) \equiv 1$.

If $1 + c(x)z_0(x) \geqslant \mu > 0$, $|c(x) - \tilde{c}(x)| \leqslant \kappa \ (0 \leqslant x \leqslant 1)$, *then the Green function* $\mathscr{G}(x, \xi)$ *exists and*

$$|\mathscr{G}(x, \xi) - \tilde{\mathscr{G}}(x, \xi)| \leqslant d(\xi)z_0(x) \qquad \textit{with} \quad d(\xi) = \mu^{-1}\kappa\gamma(\xi),$$

$$\gamma(\xi) = \max_{0 \leq x \leq 1} |\tilde{\mathscr{G}}(x, \xi)|.$$

According to Example 3.5, Property IP* holds for (3.21). Consequently, $\mathscr{G}(x, \xi)$ exists and is positive (cf. Proposition 3.5). For fixed $\xi \in (0, 1)$ we estimate with $g(x) = \mathscr{G}(x, \xi)$, $\tilde{g}(x) = \tilde{\mathscr{G}}(x, \xi)$ and $\eta \in K$

$$|B[g, \eta] - B[\tilde{g}, \eta]| \leqslant \int_0^1 |c - \tilde{c}|\,|\tilde{g}|\eta\, dx \leqslant \kappa\gamma(\xi) \int_0^1 \eta\, dx$$

$$\leqslant d(\xi) \int_0^1 (1 + cz_0)\eta\, dx = d(\xi)B[z_0, \eta].$$

Due to Property IP*, these inequalities imply the stated estimate. □

Example 3.7. As a special case of Example 3.6, let $a(x) \equiv 1$ and $\tilde{c}(x) \equiv k^2$ with a constant $k > 0$. Then

$$\tilde{\mathscr{G}}(x, \xi) = \mathscr{G}_k(x, \xi) := \frac{1}{k \sinh k} \begin{cases} \sinh kx \sinh k(1 - \xi) & \text{for} \quad 0 \leqslant x \leqslant \xi \\ \sinh k\xi \sinh k(1 - x) & \text{for} \quad \xi \leqslant x \leqslant 1. \end{cases}$$

Using $z_0(x) = \frac{1}{2}x(1 - x)$, one obtains the estimate

$$|\mathscr{G}(x, \xi) - \mathscr{G}_k(x, \xi)| \leqslant \kappa\gamma_k(\xi)\tfrac{1}{2}x(1 - x) \qquad (0 \leqslant x \leqslant 1)$$

with

$$\gamma_k(\xi) = \mathscr{G}_k(\xi, \xi) = \frac{1}{k \sinh k}\left(\sinh^2\frac{k}{2} - \sinh^2\left(\xi - \frac{1}{2}\right)\right) \leqslant \frac{1}{2k},$$

provided $0 \leqslant k^2 - \kappa \leqslant c(x) \leqslant k^2 + \kappa$ $(0 < x < 1)$. □

Now we return to the general nonlinear functional (3.2). For simplicity, however, we require in the following that v' and w' are bounded and that the functions $\mathfrak{z}(\lambda)$ which occur in the theory have a derivative $\partial/\partial x\ \mathfrak{z}(\lambda, x)$ which is bounded on each set $[0, \lambda_0] \times [0, 1]$. Under these conditions the following corollary generalizes Corollary 3.8a.

Corollary 3.8b. *For each number $N > 0$, let there exist constants $a_0 > 0$, $a_1, c_0, c_1, \gamma_0, \gamma_1$ such that the following inequalities hold for $0 \leqslant x \leqslant 1$, $0 \leqslant y \leqslant v(x) - w(x)$, $-N \leqslant p \leqslant N$;*

$$(\alpha(x, v(x), v'(x)) - \alpha(x, v(x), v'(x) - p))p \geqslant a_0 p^2,$$

$$|\alpha(x, v(x), v'(x) - p) - \alpha(x, v(x) - y, v'(x) - p)| \leqslant a_1 y,$$

$$\mathscr{g}(x, v(x), v'(x)) - \mathscr{g}(x, v(x) - y, v'(x)) \geqslant c_0 y,$$

$$|\mathscr{g}(x, v(x) - y, v'(x)) - \mathscr{g}(x, v(x) - y, v'(x) - p)| \leqslant c_1 |p|,$$

$$\mathscr{g}_0(v(0)) - \mathscr{g}_0(v(0) - y) \geqslant \gamma_0 y, \qquad \mathscr{g}_1(v(1)) - \mathscr{g}_1(v(1) - y) \geqslant \gamma_1 y.$$

Then Theorem 3.8 remains true, if the requirement (3.18) is dropped.

Proof. Let $\mathfrak{z}(\lambda)$ denote suitable functions such that condition (3.19) is satisfied. Then there exists a number $N > 0$ such that a.e. $-N \leqslant \eta' \leqslant N$ for all functions $\eta = (w - v + \mathfrak{z}(\lambda))^{-}$. Using the preceding assumptions for this number N and the inequalities in the proof of Lemma 3.3, one shows that condition (3.18) holds with $\varepsilon = 1$. The details are left to the reader. □

Example 3.8. *Let* $B[u, \eta] = \int_0^1 (u'\eta' + u^2\eta)\,dx$, $\beta = \{0, 1\}$. Then Property IM* holds for all $v \in H$ and all $w \in H$ with $w \succ -\pi^2/2$. In particular, it follows that each corresponding boundary value problem (3.7) has at most one solution $u \succ -\pi^2/2$. To prove this, let $\mathfrak{z}(\lambda) = \lambda z$ with z as in Example 3.3. Then we calculate $B[w + \lambda z, \eta] - B[w, \eta] \geqslant \lambda \int_0^1 (z'\eta' + 2wz\eta)\,dx \geqslant h(\lambda) \int_0^1 \eta\,dx$ with $h(\lambda) = \lambda h_0$ and $(2w(x) + (\pi - \varepsilon)^2)z(x) \geqslant h_0 > 0$ $(0 \leqslant x \leqslant 1)$. □

Problems

3.2 Apply Theorem 3.8 and Corollary 3.8b to show that Property IM* holds for $B[u, \eta] = \int_0^1 (u'\eta' + u'^2\eta + u^2\eta)\,dx + u^2(0)\eta(0)$, $\beta = \{1\}$, and $w \geqslant o$, v' and w' bounded.

3.3 Consider the case (3.21) and let $u \in H$. Prove that $B[u, \eta] \geqslant 0$ for all $\eta \in K$ if and only if this inequality holds for all $\eta \in K'$ where $K' \subset K$ denotes the set of all

$$\eta(x) = \eta_{\xi, h}(x) = \begin{cases} h - |x - \xi| & \text{for } |x - \xi| \leqslant h \\ 0 & \text{otherwise} \end{cases}$$

$$\text{with } 0 < h < \tfrac{1}{2}, \quad h < \xi < 1 - h.$$

For $a(x) \equiv 1$ and $\eta = \eta_{\xi, h}$,

$$B[u, \eta] = -u(\xi - h) + 2u(\xi) - u(\xi + h) + \int_{\xi - h}^{\xi + h} cu\eta\,dx.$$

3.4 Consider the same situation as in Example 3.6 except that, now, $\tilde{B}$ is obtained from B when both coefficients a and c are replaced by coefficients $\tilde{a}$ and $\tilde{c}$. Prove that $|\mathscr{G}(x, \xi) - \tilde{\mathscr{G}}(x, \xi)| \leqslant d(\xi)z_0(x)$ with $d(\xi) = \mu^{-1}(\kappa\gamma(\xi) + \alpha)$, if μ, κ, γ, z_0 have the properties explained in Example 3.6 and the function $f = (a - \tilde{a})\,\partial\tilde{\mathscr{G}}/\partial x$ satisfies a Lipschitz condition $|f(x) - f(y)| \leqslant \alpha|x - y|$. (Use the result of Problem 3.3.)

3.4 L_1-differential inequalities

Under appropriate smoothness conditions on the functions a, v, and w, Property IM* can be formulated in terms of differential inequalities in L_1. Under even stronger conditions, the weak differential inequalities are equivalent to pointwise differential inequalities. Thus the results of the previous sections on Property IM* also yield sufficient conditions for the inverse-monotonicity considered in Section 2. Here, we restrict ourselves to a few corresponding remarks.

Whenever the terms are meaningful, we shall use the notation

$$L[u](x) = -\frac{d}{dx}\, \mathscr{a}[u](x) + \mathscr{g}[u](x),$$

$$\mathscr{b}_0[u] = \begin{cases} u(0) & \text{if } 0 \in \beta \\ -\mathscr{a}[u](0) + \mathscr{g}_0(u(0)) & \text{if } 0 \notin \beta, \end{cases}$$

$$\mathscr{b}_1[u] = \begin{cases} u(1) & \text{if } 1 \in \beta \\ \mathscr{a}[u](1) + \mathscr{g}_1(u(1)) & \text{if } 1 \notin \beta. \end{cases}$$

Suppose that v and w are given functions in D such that $\mathscr{a}[v]$ and $\mathscr{a}[w]$ have L_1-derivatives. Then we obtain by partial integration, for $\eta \in K$,

$$B[v, \eta] = \int_0^1 L[v](x)\eta(x)\,dx + \mathscr{b}_0[v]\eta(0) + \mathscr{b}_1[v]\eta(1), \tag{3.26}$$

and a corresponding formula for w. (Observe that $\eta(x) = 0$ for $x \in \beta$.) From this presentation we conclude that *Property* IM* (*for* v, w) *holds if and only if the inequalities*

$$\begin{aligned} L[v](x) &\leqslant L[w](x) \qquad \textit{for a.a. } x \in [0, 1], \\ \mathscr{b}_0[v] &\leqslant \mathscr{b}_0[w], \qquad \mathscr{b}_1[v] \leqslant \mathscr{b}_1[w] \end{aligned} \tag{3.27}$$

together imply $v \leqslant w$.

Obviously, the differential inequalities in (3.27) hold for all $x \in (0, 1)$ if the functions $L[v]$ and $L[w]$ are continuous. In this case, we may write the above implication as $Mv \leqslant Mw \Rightarrow v \leqslant w$ with

$$\begin{aligned} Mu(x) &= L[u](x) \qquad \text{for } 0 < x < 1, \\ Mu(0) &= \mathscr{b}_0[u], \qquad Mu(1) = \mathscr{b}_1[u] \end{aligned} \tag{3.28}$$

and $\leqslant$ denoting the natural order relation. Thus all sufficient conditions for Property IM* are sufficient also for the latter implication. Let us formulate one particular result which follows in this way from Corollary 3.7c, in which we choose D to be the set of all $u \in H_1$ with $\mathscr{a}[u] \in L_2$, $\mathscr{g}[u] \in L_1$.

Theorem 3.9. *Suppose that* $\mathscr{a}(x, y, p)$ *is continuously differentiable and that* $\mathscr{g}(x, y, p)$ *is continuous on* $[0, 1] \times \mathbb{R} \times \mathbb{R}$.

Then the operator $M : C_1[0, 1] \cap C_2(0, 1) \to \mathbb{R}[0, 1]$ *defined by* (3.28) *is inverse-monotone, if the operator* $A : D \to H$ *defined by* (3.4) *is monotone definite.*

Of course, the required smoothness conditions on a and g may be relaxed. Also, one may consider a larger domain of M. Moreover, under appropriate conditions, $A : D \to H$ is monotone definite if

$$\int_0^1 (L[v] - L[w])(v - w)\,dx$$
$$+ (b_0[v] - b_0[w])(v - w)(0) + (b_1[v] - b_1[w])(v - w)(1) \geqslant c > 0$$

for all v, w in the domain of M with $\|v - w\|_1 = 1$. Thus *this type of definiteness of M then implies inverse-monotonicity.* On the other hand, to verify the above inequality one will anyway transform the left-hand side into $\langle Av - Aw, v - w\rangle_1$.

In Sections 2.1 and 2.2 we permitted the functions v, w to have only piecewise continuous first derivatives and to be only piecewise twice differentiable. Similarly, we may consider here functions v, w such that $a[v]$ and $a[w]$ are only piecewise continuous and have only *piecewise L_1-derivatives.* In this way one can also consider jump conditions for u' which differ from those in Conditions S_l and S_r. We shall, however, not carry out this theory here.

3.5 Stronger positivity statements for linear problems

The result of Corollary 3.8a may also be obtained by applying the abstract monotonicity theorem II.1.2 to the operator $\mathcal{M}$ defined in Section 3.1.2 and still using essentially the same ideas as in the above proof. Here, we shall show that even stronger positivity statements can be derived from that theorem. We restrict ourselves to the special case

$$B[u, \eta] = \int_0^1 (au'\eta' + cu\eta)(x)\,dx + \gamma_0 u(0)\eta(0) + \gamma_1 u(1)\eta(1), \quad (3.29)$$

where a and c are bounded and measurable, $a(x) \geqslant a_0 > 0$ $(0 \leqslant x \leqslant 1)$, $\gamma_0 = 0$ if $0 \in \beta$, $\gamma_1 = 0$ if $1 \in \beta$. For this case the following result strengthens Corollary 3.8a. The interval J is defined by $J = [0, 1] \sim \beta$.

Theorem 3.10. (a) *Property* IP* *holds, if there exists a function $z \in H_1$ such that*

$$z \geqslant o, \qquad z(x) > 0 \quad \textit{for} \quad x \in \beta, \qquad B[z, \eta] \geqslant 0 \quad \textit{for all} \quad \eta \in K,$$

and in case $\beta = \varnothing$ also $B[z, \eta] > 0$ for some $\eta \in K$.

(b) *If Property* IP* *holds, then for each $u \in H_1$ even the following is true:*

$$\left.\begin{array}{ll} B[u, \eta] \geqslant 0 & \textit{for all} \quad \eta \in K \\ u(x) \geqslant 0 & \textit{for all} \quad x \in \beta \end{array}\right\} \Rightarrow \left\{\begin{array}{ll} \textit{either} \quad u = o & \\ \textit{or} \quad u(x) > 0 & \textit{for} \quad x \in J. \end{array}\right.$$

Proof. As in Section 3.1.2 we shall rewrite the occurring inequalities by defining a suitable operator $\mathcal{M}$. We shall, however, work with different spaces, so that Riesz' representation theorem is not needed.

Let X denote the space $X = \mathbb{R}^K$ of all vectors $U = (U_\eta) = (U_\eta)_{\eta \in K}$ and define $A : H_1 \to X$ by $Au = ((Au)_\eta)$ with $(Au)_\eta = B[u, \eta]$. Moreover, for the four cases described in (3.1), let, respectively, $\mathcal{S} = X \times \mathbb{R} \times \mathbb{R}$, $\mathcal{S} = X \times \mathbb{R}$, $\mathcal{S} = X \times \mathbb{R}$, $\mathcal{S} = X$. Then we use again notation (3.5) and define an operator $\mathcal{M} : H_1 \to \mathcal{S}$ by (3.6). In the space $\mathcal{S}$, a linear order $\leqslant$ and a further stronger order $\lessdot$ are introduced by defining for $\mathcal{U} \in \mathcal{S}$

$$\mathcal{U} \geqslant o \quad \Leftrightarrow \quad \begin{cases} U_\eta \geqslant 0 & \text{for all} \quad \eta \in K \\ U_0 \geqslant 0 & \text{if} \quad 0 \in \beta \\ U_1 \geqslant 0 & \text{if} \quad 1 \in \beta, \end{cases}$$

$$\mathcal{U} \gtrdot o \quad \Leftrightarrow \quad \begin{cases} \mathcal{U} > o & (\text{i.e. } \mathcal{U} \geqslant o, \mathcal{U} \neq o) \\ U_0 > 0 & \text{if} \quad 0 \in \beta \\ U_1 > 0 & \text{if} \quad 1 \in \beta. \end{cases}$$

Obviously, *Property* IP* *holds if and only if the operator* $\mathcal{M} : (H_1, \leqslant) \to (\mathcal{S}, \leqslant)$ *is inverse-positive* (*on* H_1). Moreover, the above conditions on z can be written as $z \geqslant o$, $\mathcal{M}z \gtrdot o$. Thus, in order to derive statement (a) from Theorem II.1.2 it suffices to show that for all $u \in H_1$ the inequalities $u \geqslant o$ and $\mathcal{M}u \gtrdot o$ together imply $u \succ o$. The proof is carried out only for the third case in (3.1), when $J = [0, 1)$.

Let $u \in H_1$ with $u > o$ and $u \not\succ o$, so that $u(\xi) = 0$ for some $\xi \in [0, 1]$. We shall show that $\mathcal{M}u \not\gtrdot o$ follows. This is clear for $\xi = 1$, since $\mathcal{M}u = (Au, u(1))$ with $Au = (B[u, \eta])$. For each $\xi \in [0, 1)$ we construct a function $h \in K$ with $B[u, h] < 0$.

First we assume that $u(0) = 0$ and $u(x) > 0$ for $0 < x \leqslant 1$. Then let

$$h(x) = \begin{cases} \delta^{-1}(\varepsilon) \displaystyle\int_x^\varepsilon \kappa(t)\,dt + \int_0^x \kappa(t)t\,dt \int_0^\varepsilon \kappa(t)\,dt & \\ \qquad - \displaystyle\int_0^x \kappa(t)\,dt \int_0^\varepsilon \kappa(t)t\,dt & \text{for} \quad 0 \leqslant x \leqslant \varepsilon \\ 0 & \text{for} \quad \varepsilon \leqslant x \leqslant 1 \end{cases} \tag{3.30}$$

with $\delta(\varepsilon) = \int_0^\varepsilon \kappa(t)\,dt$, $\kappa = a^{-1}$ and $\varepsilon > 0$. For small enough ε, a formal calculation yields $h(0) = 1$, $h(\varepsilon) = 0$, $0 < h(x) \leqslant 1$ $(0 \leqslant x < \varepsilon)$.

$$(ah')(x) = -\delta^{-1}(\varepsilon) + \int_0^\varepsilon (x - t)\kappa(t)\,dt \leqslant -\delta^{-1}(\varepsilon) + \varepsilon\,\delta(\varepsilon)$$

$$< 0 \qquad (0 \leqslant x \leqslant \varepsilon),$$

and $(ah')'(x) = \delta(\varepsilon)$ $(0 \leqslant x \leqslant \varepsilon)$. Thus we obtain by partial integration, with $\|u\|_\varepsilon = \max\{|u(x)| : 0 \leqslant x \leqslant \varepsilon\}$,

$$B[u, h] = (ah')(\varepsilon)u(\varepsilon) + \int_0^\varepsilon (-\delta(\varepsilon) + c(x)h(x))u(x)\,dx$$

$$\leqslant (-\delta^{-1}(\varepsilon) + \varepsilon\delta(\varepsilon))u(\varepsilon) + \int_0^\varepsilon |c(x)|\,dx\|u\|_\varepsilon.$$

There exist arbitrarily small $\varepsilon > 0$ such that $\|u\|_\varepsilon = u(\varepsilon)$. Choosing a small enough $\varepsilon > 0$ of this kind, we finally get

$$B[u, h] \leqslant \left[-\delta^{-1}(\varepsilon) + \varepsilon\,\delta(\varepsilon) + \int_0^\varepsilon |c(x)|\,dx\right]u(\varepsilon) < 0.$$

If $u(\xi) = 0$ for some $\xi \in (0, 1)$, we may assume, without loss of generality, that $u(x) > 0$ for $\xi < x \leqslant \xi + \mu$ or $u(x) > 0$ for $\xi - \mu \leqslant x < \xi$, where $\mu > 0$ denotes some sufficiently small constant. Then define

$$h(x) = \begin{cases} 0 & \text{for} \quad 0 \leqslant x \leqslant \xi - \varepsilon \quad \text{and for} \quad \xi + \varepsilon' \leqslant x \leqslant 1 \\ h_1(x) & \text{for} \quad \xi - \varepsilon \leqslant x \leqslant \xi \\ h_2(x) & \text{for} \quad \xi \leqslant x \leqslant \xi + \varepsilon' \end{cases} \tag{3.31}$$

with

$$h_1(x) = \gamma^{-1}(\varepsilon) \int_{\xi-\varepsilon}^x \kappa(t)\,dt - \left\{\int_x^\xi \kappa(t)t\,dt \int_{\xi-\varepsilon}^\xi \kappa(t)\,dt - \int_x^\xi \kappa(t)\,dt \int_{\xi-\varepsilon}^\xi \kappa(t)t\,dt\right\},$$

$$h_2(x) = \delta^{-1}(\varepsilon') \int_x^{\xi+\varepsilon'} \kappa(t)\,dt + \left\{\int_\xi^x \kappa(t)t\,dt \int_\xi^{\xi+\varepsilon'} \kappa(t)\,dt - \int_\xi^x \kappa(t)\,dt \int_\xi^{\xi+\varepsilon'} \kappa(t)\,dt\right\},$$

$$\gamma(\varepsilon) = \int_{\xi-\varepsilon}^\xi \kappa(t)\,dt, \qquad \delta(\varepsilon') = \int_\xi^{\xi+\varepsilon'} \kappa(t)\,dt,$$

and $\varepsilon, \varepsilon' > 0$ small enough.

Formal calculations and simple estimates which are similar to those above again yield $B[u, h] < 0$ for some small enough $\varepsilon, \varepsilon' > 0$. This proves part (a) of the theorem.

To prove part (b), suppose that $\mathcal{M}u \geqslant o$ and $u \neq o$. If $u(\xi) = 0$ for some $\xi \in (0, 1)$, then $B[u, h] < 0$ for a function h as in (3.30). This contradicts

$\mathcal{M}u \geqslant o$, so that $u(x) > 0$ $(0 < x < 1)$ must hold. If $u(0) = 0$, we proceed similarly using h in (3.31). □

The above result together with Proposition 3.4 yields the following statement which improves Proposition 3.5.

Proposition 3.11. *If Property* IP* *holds for the bilinear functional* (3.29) *and given* β, *then the corresponding Green function satisfies* $\mathcal{G}(x, \xi) > 0$ *for all* $x \in J$, $\xi \in J$.

In particular, this inequality holds for $\mathcal{G}$ in Examples 3.6 and 3.7. Moreover, each solution u of the variational problem considered in Proposition 3.6 in case $b = o$ even satisfies $u(x) > 0$ $(x \in J)$, if $F \neq o$.

4 EXISTENCE AND INCLUSION FOR TWO-POINT BOUNDARY VALUE PROBLEMS

In this section our main interest is in boundary value problems of the form

$$Mu(x) = 0 \qquad (0 \leqslant x \leqslant 1) \tag{4.1}$$

with an operator $M : R = C_2[0, 1] \to S = \mathbb{R}[0, 1]$ defined by

$$M = \mathbf{L} + \mathbf{N}, \qquad \mathbf{N}u(x) = F(x, u(x), u'(x)), \tag{4.2}$$

$$\mathbf{L}u(x) = \begin{cases} -u''(x) + b(x)u'(x) & \text{for } 0 < x < 1 \\ -\alpha_0 u'(0) & \text{for } x = 0 \\ \alpha_1 u'(1) & \text{for } x = 1, \end{cases}$$

$$F(x, u(x), u'(x)) = \begin{cases} f(x, u(x), u'(x)) & \text{for } 0 < x < 1 \\ f_0(u(0), u'(0)) & \text{for } x = 0 \\ f_1(u(1), u'(1)) & \text{for } x = 1, \end{cases}$$

where $b : [0, 1] \to \mathbb{R}$, $f : [0, 1] \times \mathbb{R}^2 \to \mathbb{R}$, and $f_0, f_1 : \mathbb{R}^2 \to \mathbb{R}$ are given continuous functions and $\alpha_0, \alpha_1 \in \mathbb{R}$. ($b$, α_0, and α_1 are introduced for notational convenience only.) We assume that

$$\begin{aligned} \alpha_0 \geqslant 0, \quad &\text{and} \quad \alpha_0 \kappa + f_0(y, p - \kappa) - f_0(y, p) \geqslant 0 \\ &\qquad \text{if} \quad \kappa \geqslant 0, \quad y \in \mathbb{R}, \quad p \in \mathbb{R} \\ \alpha_1 \geqslant 0, \quad &\text{and} \quad \alpha_1 \kappa + f_1(y, p + \kappa) - f_1(y, p) \geqslant 0 \\ &\qquad \text{if} \quad \kappa \geqslant 0, \quad y \in \mathbb{R}, \quad p \in \mathbb{R}. \end{aligned}$$

In particular, we shall often consider the following

Special case: $\alpha_0 \in \{0, 1\}$, $\alpha_1 \in \{0, 1\}$, $r_0 \in \mathbb{R}$, $r_1 \in \mathbb{R}$, *and*

$$Mu(0) = \begin{cases} u(0) - r_0 & \text{if } \alpha_0 = 0 \\ -u'(0) + f_0(u(0)) & \text{if } \alpha_0 = 1, \end{cases}$$

$$Mu(1) = \begin{cases} u(1) - r_1 & \text{if } \alpha_1 = 0 \\ u'(1) + f_1(u(1)) & \text{if } \alpha_1 = 1. \end{cases} \tag{4.3}$$

A function $\psi \in R$ with $M\psi \geqslant o$ is called an *upper solution* of $Mu = o$, and a function $\varphi \in R$ with $M\varphi \leqslant o$ is called a *lower solution*. A lower solution φ and an upper solution ψ are said to be (a pair of) *comparison functions* for the equation $Mu = o$ (or for the operator M) if $\varphi \leqslant \psi$. Thus comparison functions are characterized by $\varphi, \psi \in R$ and

$$\varphi \leqslant \psi, \qquad M\varphi \leqslant o \leqslant M\psi. \tag{4.4}$$

We shall derive *sufficient conditions such that the existence of comparison functions $\varphi, \psi \in R$ implies the existence of a function $u^* \in R$ with $Mu^* = o$ and $\varphi \leqslant u^* \leqslant \psi$.*

Section 4.1 contains the basic existence and inclusion theorem (Theorem 4.2). It can be applied to problems for which $F(x, y, p)$ is bounded with respect to p, and it will also be used as a tool for proving the results in Sections 4.2 and 4.3. In these sections the boundedness condition mentioned above is replaced by weaker assumptions. In Section 4.2 additional differential inequalities for functions Φ, Ψ are required which also yield estimates $\Phi \leqslant (u^*)' \leqslant \Psi$. In Section 4.3 more general functions $\Phi(x, y)$, $\Psi(x, y)$ are considered such that estimates $\Phi(x, u^*(x)) \leqslant (u^*)'(x) \leqslant \Psi(x, u^*(x))$ are achieved. Here, special choices of Φ, Ψ lead to results in which the boundedness condition mentioned above is replaced by growth restrictions on $f(x, y, p)$ as a function of p.

Section 4.4 is concerned with "singular" problems, in particular, problems on noncompact domains. Several other methods of proof are briefly discussed in Section 4.5.

For given comparison functions φ, ψ define

$$\omega = \omega[\varphi, \psi] = \{(x, y) : \varphi(x) \leqslant y \leqslant \psi(x), \quad 0 \leqslant x \leqslant 1\}. \tag{4.5}$$

Moreover, for any $u, v, w \in \mathbb{R}[0, 1]$, let $\langle u, v, w\rangle \in \mathbb{R}[0, 1]$ be the function given by

$$\langle v, u, w\rangle(x) = \max\{v(x), \min\{u(x), w(x)\}\} \qquad (0 \leqslant x \leqslant 1). \tag{4.6}$$

The preceding assumptions on the occurring functions may be weakened in many ways. For example, the domain of definition of F may be restricted. *All results below hold also if $f(x, y, p)$, $f_0(y, p)$ and $f_1(y, p)$ are only defined on,* *respectively*, $\omega \times \mathbb{R}$, $[\varphi(0), \psi(0)] \times \mathbb{R}$ *and* $[\varphi(1), \psi(1)] \times \mathbb{R}$. *Moreover, in all results which involve functions* Φ, Ψ, *the domain of definition of* f, f_0, *and* f_1 *may further be restricted to values p such that, respectively,* $p \in [\Phi(x), \Psi(x)]$, $p \in [\Phi(0), \Psi(0)]$, *and* $p \in [\Phi(1), \Psi(1)]$.

Moreover, many of the succeeding results remain true under weaker smoothness assumptions on the comparison functions φ, ψ. (We may use the notion of comparison functions also if φ, ψ do not belong to R, but satisfy (4.4) in a suitable sense.) The existence and inclusion statements below are proved by the method of modification as described in the introduction of Chapter III. Here, in the second step of this method the fixed-point theorem of Schauder is used. This theorem may be applied in the same way, if φ and ψ are only continuous (or piecewise continuous), but do not belong to R. Stronger smoothness assumptions on φ, ψ are needed in the third step, in which Theorem 2.1 or another result on inverse-monotonicity is applied. In Section 2.1 we explained in detail weaker smoothness assumptions such that Theorem 2.1 remains true (cf. Corollary 2.1a). *The assumptions on* φ, ψ *may be weakened in a corresponding way*. For simplicity of presentation, however, we shall not describe these possibilities in detail, but formulate our results for smooth functions φ, ψ.

Similar statements hold for the smoothness of the bounds Φ, Ψ for the derivative of the solution.

4.1 The basic theorem

The main result of this section is Theorem 4.2. It yields the desired inclusion statement for functions $F(x, y, p)$ which are bounded with respect to p. This theorem, however, will also be used to derive the theorems of the following sections, where less restrictive assumptions on F are required.

For given comparison functions $\varphi, \psi \in R$, we introduce a *modified problem* $M^{\#}u = o$ with a *modified operator* $M^{\#}: R \to S$ defined by

$$M^{\#}u = \mathbf{L}u + u - u^{\#} + \mathbf{N}^{\#}u, \qquad \mathbf{N}^{\#}u(x) = F(x, u^{\#}(x), u'(x)),$$
$$u^{\#} = \langle \varphi, u, \psi \rangle.$$

Lemma 4.1. *Each solution* $u^* \in R$ *of the modified boundary value problem* $M^{\#}u = o$ *is also a solution of the given boundary value problem* $Mu = o$ *and it satisfies* $\varphi \leqslant u^* \leqslant \psi$.

Proof. Suppose $u \in C_2[0, 1]$ is a solution of the modified problem. It suffices to show that $\varphi \leqslant u \leqslant \psi$, because then $u^{\#} = u$ and hence $M^{\#}u = Mu$. Since $M^{\#}\varphi = M\varphi$ and $M^{\#}\psi = M\psi$, it follows from (4.4) that $M^{\#}\varphi \leqslant M^{\#}u \leqslant M^{\#}\psi$. We shall only show that $M^{\#}u \leqslant M^{\#}\psi \Rightarrow u \leqslant \psi$. The second inequality $\varphi \leqslant u$ is derived similarly.

In order to prove the above implication we apply Theorem 2.1 to $M^{\#}$, u, ψ instead of M, v, w. Obviously, this is allowed because $M^{\#}$satisfies all assumptions imposed on M in Section 2.1. Also it is easily verified that the corresponding assumption (2.5) holds with $z(x) \equiv 1$, i.e. $M^{\#}\psi(x) < M^{\#}(\psi + \lambda)(x)$ $(0 \leqslant x \leqslant 1,\ \lambda > 0)$. Thus $u \leqslant \psi$ follows. (For proving $\varphi \leqslant u$, inequality (2.6) is used.) □

Theorem 4.2. *Suppose that there exist comparison functions $\varphi, \psi \in R$ for M and that $F(x, y, p)$ is bounded on $\omega \times \mathbb{R}$.*

Then the given problem $Mu = o$ has a solution $u^* \in R$ such that $\varphi \leqslant u^* \leqslant \psi$.

Proof. According to Lemma 4.1 we have only to show that the modified problem $M^{\#}u = o$ has a solution $u^* \in R$.

When we apply Theorem 2.1 to the linear operator $Au = \mathbf{L}u + u$, it follows by choosing $z(x) \equiv 1$ that A is inverse-positive. Therefore, A^{-1} exists and the problem $Au = u^{\#} - \mathbf{N}^{\#}u$ is equivalent to an integro-differential equation $u = Tu$ with T defined on $C_1[0, 1]$ by

$$Tu(x) = \mathcal{G}_0(x)[u^{\#}(0) - f_0(u^{\#}(0), u'(0))] + \mathcal{G}_1(x)[u^{\#}(1) - f_1(u^{\#}(1), u'(1))]$$

$$+ \int_0^1 \mathcal{G}(x, \xi)[u^{\#}(\xi) - f(\xi, u^{\#}, u')]\, d\xi. \tag{4.7}$$

Here, $\mathcal{G}$ is the Green function which belongs to A, and $\mathcal{G}_0$, $\mathcal{G}_1 \in C_2[0, 1]$ are solutions of the differential equation $Au(x) = 0$ $(0 < x < 1)$ which satisfy $A\mathcal{G}_i(k) = \delta_{ik}$ $(i, k = 0, 1)$.

The space $X = C_1[0, 1]$ is a Banach space with respect to the norm $\|u\| = \|u\|_\infty + \|u'\|_\infty$. Since in (4.7) the terms within brackets are uniformly bounded for all $u \in X$, the image set TX is shown to be a relatively compact subset of X by means of the Arzela–Ascoli theorem. As T is a continuous mapping, the equation $u = Tu$ has a solution $u^* \in C_1[0, 1]$, according to the fixed-point theorem of Schauder. □

Example 4.1 (Continuation of Examples 2.3 and 2.5.) The boundary value problem $Mu = o$ with M in Example 2.3 has a solution $u^* \in R$ such that $o \leqslant u^* \leqslant \eta$, since $\varphi = o$ and $\psi = \eta$ constitute a pair of comparison functions. □

Example 4.2. Consider the nonlinear eigenvalue problem

$$L[u](x) - \mu g(x)u(x) + f(x, u(x)) = 0 \qquad (0 < x < 1),$$
$$B_0[u] = B_1[u] = 0,$$

with L and g as in the linear eigenvalue problem (II.3.40) and

$$f(x, 0) \equiv 0, \qquad \lim_{y \to +0} y^{-1} f(x, y) \leqslant 0, \qquad \lim_{y \to +\infty} y^{-1} f(x, y) = +\infty,$$

where both limits hold uniformly for $x \in (0, 1)$. *This problem has a solution* $u^* > o$, *if* $\mu > \lambda_0$, *where* λ_0 *denotes the smallest eigenvalue of the linear problem* (II.3.40).

To prove this one constructs suitable comparison functions. Let φ_0 denote an eigenfunction of the linear problem corresponding to λ_0, as described in Theorem II.3.24, and let $\psi_0 \in C_2[0, 1]$ be any function such that $\psi_0(x) > 0$ $(0 \leqslant x \leqslant 1)$, $B_0[\psi_0] \geqslant 0$, $B_1[\psi_0] \geqslant 0$. Then $\varphi = \alpha\varphi_0$ and $\psi = \beta\psi_0$ are comparison functions for the given problem, if $\alpha > 0$ is sufficiently small and $\beta > 0$ sufficiently large. □

Example 4.3 (Continuation of Example 2.8.) We consider the problem $Mu = o$ in Example 2.8 for $\mu > \pi^2/4$ and extend $f(y, p) = g(y, p) = -\mu y(1 - p^2)^{1/2}$ by defining $f(y, p) = 0$ for $p^2 > 1$, as before. Obviously, $\varphi = \gamma \sin \pi/2x$ and $\psi(x) = x$ constitute a pair of comparison functions for small enough $\gamma > 0$. Thus, according to Theorem 4.2, the "extended problem" has a solution u which satisfies $\varphi \leqslant u \leqslant \psi$.

It remains to be shown that $-1 \leqslant u'(x) \leqslant 1$ $(0 \leqslant x \leqslant 1)$. In Section 4.2 we shall derive these inequalities from general theoretical results. However, in the present simple case, the inequality $u''(x) \leqslant 0$ $(0 < x < 1)$ immediately yields $0 = u'(1) \leqslant u'(x) \leqslant u'(0) \leqslant \psi'(0) = 1$ $(0 \leqslant x \leqslant 1)$. □

Problems

4.1 Apply the statement in Example 4.2 to obtain an inclusion statement $\varphi \leqslant u^* \leqslant \psi$ for a solution u^* of $-u'' - \mu u + u^3 = 0$, $u(0) = u(1) = 0$ with $\mu > \pi^2$, using $\varphi(x) = \alpha \sin \pi x$, $\psi(x) = \beta \cos(\pi - \varepsilon)(x - \frac{1}{2})$ with suitable $\alpha > 0$, $\beta > 0$, $\varepsilon > 0$. □

4.2 Suppose that the function $f(x, y, p)$ which occurs in the definition of M in (4.1) is not necessarily continuous, but satisfies the following assumption: There exist x_i $(i = 0, 1, \ldots, m)$ with $0 = x_0 < x_1 < \cdots < x_m = 1$ such that $f(x, y, p) = \tilde{f}_i(x, y, p)$ holds for $x_{i-1} < x < x_i$, $y \in \mathbb{R}$, $p \in \mathbb{R}$ with a continuous function $\tilde{f}_i$ on $[x_{i-1}, x_i] \times \mathbb{R} \times \mathbb{R}$ $(i = 1, 2, \ldots, m)$. Then redefine R by $R = C_{1,2}[x_0, x_1, \ldots, x_m]$ and interpret the inequalities $M\varphi \leqslant o \leqslant M\psi$ in (4.4) to mean that $M\varphi(x) \leqslant 0 \leqslant M\psi(x)$ for all $x \in [0, 1]$ satisfying $x \neq x_i$ $(i = 1, 2, \ldots, m - 1)$. Prove that *then Lemma* 4.1 *and Theorem* 4.2 *remain true.*

4.3 Show that the assumptions described in Problem 4.2 may be weakened further by allowing φ', ψ' not to be continuous at the points x_i $(i = 1, 2, \ldots, m - 1)$, but requiring instead that $D^+\varphi(x_i) \leqslant D^-\varphi(x_i)$ and $D^+\psi(x_i) \geqslant D^-\psi(x_i)$. (Hint: use Corollary 2.1a.)

4.4 The method of proof used for Lemma 4.1 and Theorem 4.2 can be carried over to initial value problems. Suppose that $M : C_1[0, 1] \to S = \mathbb{R}[0, 1]$ denotes an operator of the form (2.25), where $f(x, y)$ is continuous for

$x \in [0, 1]$, $y \in \mathbb{R}$. Prove the following statement: *If functions* $\varphi, \psi \in C_1[0, 1]$ *exist such that* $\varphi \leqslant \psi$ *and* $M\varphi \leqslant o \leqslant M\psi$, *then the initial value problem* $Mu(x) = 0$ $(0 < x \leqslant 1)$, $u(0) = r_0$ *has a solution* $u^* \in C_1[0, 1]$ *with* $\varphi \leqslant u^* \leqslant \psi$.

4.2 Bounds for the derivative of the solution

In Theorem 4.4, we require additional differential inequalities for a pair of functions Φ, Ψ. These functions will serve as bounds for the derivative of a solution u^*. The existence of such bounds allows us to omit the boundedness conditions on F in Theorem 4.2. Such bounds for $(u^*)'$ are useful even if F does not depend on p. Theorem 4.4 will be derived from the basic Theorem 4.2 with the aid of Theorem 4.3. The latter theorem is of a more general nature; it will also be used in Section 4.5. Without loss of generality, we consider here only the case $b(x) = 0$ $(0 \leqslant x \leqslant 1)$.

Suppose that φ and ψ are comparison functions for M, and let $\Phi, \Psi \in C_0[0, 1]$ denote given functions which satisfy

$$\Phi \leqslant \varphi' \leqslant \Psi, \qquad \Phi \leqslant \psi' \leqslant \Psi. \tag{4.8}$$

We consider a second modified problem $\hat{M}u = o$ with $\hat{M} : R \to S$ given by

$$\hat{M}u = \mathbf{L}u + \hat{\mathbf{N}}u, \qquad \hat{\mathbf{N}}u(x) = F(x, u(x), \hat{u}'(x)) \qquad (0 \leqslant x \leqslant 1) \tag{4.9}$$

where $\hat{v} = \langle \Phi, v, \Psi \rangle$ for $v \in C_0[0, 1]$, $\hat{u}' = \langle \Phi, u', \Psi \rangle$. Observe that $\hat{M}u(0) = Mu(0)$ and $\hat{M}u(1) = Mu(1)$ for the special boundary conditions (4.3).

Theorem 4.3. *Suppose that for all* $u \in R$

$$\left.\begin{array}{r} \hat{M}u = o \\ \varphi \leqslant u \leqslant \psi \end{array}\right\} \Rightarrow \Phi \leqslant u' \leqslant \Psi. \tag{4.10}$$

Then the given problem $Mu = o$ *has a solution* $u^* \in R$ *with* $\varphi \leqslant u^* \leqslant \psi$, $\Phi \leqslant (u^*)' \leqslant \Psi$.

Proof. Applying Theorem 4.2 to the operator $\hat{M}$, we see that there exists a function $u \in R$ with $\hat{M}u = o$, $\varphi \leqslant u \leqslant \psi$. Observe that the function $\hat{F}$ which corresponds to $\hat{M}$ is bounded on $\omega \times \mathbb{R}$; for example, $\hat{f}(x, y, p) = f(x, y, \langle \Phi(x), p, \Psi(x) \rangle)$. Because of (4.10), the function u also satisfies $\Phi \leqslant u' \leqslant \Psi$, so that $\hat{u}' = u'$ and hence $Mu = \hat{M}u = o$. $\square$

The main problem now is to derive sufficient conditions for property (4.10). Theorem 4.4 will provide such conditions in terms of differential inequalities for Φ and Ψ. (Other sufficient conditions for (4.10) are given in Section 4.5.) Theorem 4.4 will be formulated in a rather general way in order to include a series of special cases.

Theorem 4.4. *Suppose that φ, ψ is a pair of comparison functions for M, and that $\Phi, \Psi \in C_0[0, 1]$ are functions satisfying (4.8). For each $u \in R$ with $\varphi \leqslant u \leqslant \psi$, $\hat{M}u = o$, let the following two conditions hold.*

(a) *For each $x_0 \in [0, 1]$, there exists an interval $[\eta, \zeta] \subset [0, 1]$ containing x_0 such that $\Phi \in C_1[\eta, \zeta]$ and either*

$$\Phi'(x) - f(x, u(x), \Phi(x)) \leqslant 0 \qquad (\eta < x \leqslant \zeta), \qquad \Phi(\eta) \leqslant u'(\eta),$$

or

$$\Phi'(x) - f(x, u(x), \Phi(x)) \geqslant 0 \qquad (\eta \leqslant x < \zeta), \qquad \Phi(\zeta) \leqslant u'(\zeta). \quad (4.11)$$

(b) *For each $x_0 \in [0, 1]$, there exists an interval $[\eta, \zeta] \subset [0, 1]$ containing x_0 such that $\Psi \in C_1[\eta, \zeta]$ and either*

$$\Psi'(x) - f(x, u(x), \Psi(x)) \geqslant 0 \qquad (\eta < x \leqslant \zeta), \qquad u'(\eta) \leqslant \Psi(\eta), \quad (4.12)$$

or

$$\Psi'(x) - f(x, u(x), \Psi(x)) \leqslant 0 \qquad (\eta \leqslant x < \zeta), \qquad u'(\zeta) \leqslant \Psi(\zeta). \quad (4.13)$$

Then the given problem $Mu = o$ has a solution $u^* \in R$ with

$$\varphi \leqslant u^* \leqslant \psi, \qquad \Phi \leqslant (u^*)' \leqslant \Psi.$$

Proof. Suppose that $u \in R$ is a fixed function satisfying $\varphi \leqslant u \leqslant \psi$ and $\hat{M}u = o$. In order to derive the above result from Theorem 4.3, we have to show that $\Phi \leqslant u' \leqslant \Psi$. We shall only prove $u' \leqslant \Psi$; the other inequality can be derived in an analogous way.

Given any $x_0 \in [0, 1]$, we choose a corresponding interval $[\eta, \zeta]$ and show that $u'(x) \leqslant \Psi(x)$ $(\eta \leqslant x \leqslant \zeta)$. The ideas involved will become clear when we consider the case (4.12). In this case, define an operator $A : C_0[\eta, \zeta] \cap C_1(\eta, \zeta] \to \mathbb{R}[\eta, \zeta]$ by

$$Aw(x) = w'(x) - f(x, u(x), \hat{w}(x)) \qquad \text{for} \quad \eta < x \leqslant \zeta, \qquad Aw(\eta) = w(\eta). \quad (4.14)$$

With $v := u'$ we obtain from (4.12) and $\hat{M}u = o$ that $Av(x) \leqslant A\Psi(x)$ $(\eta \leqslant x \leqslant \zeta)$. Obviously, the operator A has the form (2.25) on $[\eta, \zeta]$ in place of $[0, 1]$, so that we can apply Theorem 2.1. This theorem yields $v(x) \leqslant \Psi(x)$ $(\eta \leqslant x \leqslant \zeta)$, since the corresponding inequality (2.5) is satisfied for $z(x) = 1 + x$. If (4.13) holds, the inequality $v(x) \leqslant \Psi(x)$ $(\eta \leqslant x \leqslant \zeta)$ is derived using analogous arguments. □

The application of the preceding theorem requires the knowledge of a priori inequalities for the derivative $u'(x)$ at the occurring points η and ζ. For example, one may choose $[\eta, \zeta] = [0, 1]$ in each case. Then a priori inequalities at the boundary points are to be verified. In general, the values $u'(0)$ and $u'(1)$ are not known. However, one can often obtain bounds for these quantities using the boundary conditions $\hat{M}u(0) = \hat{M}u(1) = 0$. (These boundary conditions coincide with $Mu(0) = Mu(1) = 0$ for the case (4.3).)

For example, if $Mu(0) = u(0) - r_0$ and if $M\varphi(0) = 0$, then the relations $\varphi \leqslant u$, $\varphi(0) = u(0)$ imply that $\varphi'(0) \leqslant u'(0)$, so that $\Phi(0) \leqslant \varphi'(0)$ suffices for $\Phi(0) \leqslant u'(0)$. On the other hand, if $\alpha_0 > 0$, $\hat{M}u(0) = 0$, and $\alpha_0 \Phi(0) \leqslant f_0(u(0), \Phi(0))$, then $\Phi(0) \leqslant u'(0)$, because the inequality $\Phi(0) > u'(0)$ would imply $\alpha_0 u'(0) = f_0(u(0), \hat{u}'(0)) = f_0(u(0), \Phi(0)) \geqslant \alpha_0 \Phi(0)$.

By similar means, all statements of the following proposition are proved.

Proposition 4.5. *For each $u \in C_1[0, 1]$ which satisfies $\varphi \leqslant u \leqslant \psi$, the following statements hold:*

(i) $\varphi(0) = u(0) \;\Rightarrow\; \varphi'(0) \leqslant u'(0)$;
$\psi(0) = u(0) \;\Rightarrow\; \psi'(0) \geqslant u'(0)$;
$\varphi(1) = u(1) \;\Rightarrow\; \varphi'(1) \geqslant u'(1)$;
$\psi(1) = u(1) \;\Rightarrow\; \psi'(1) \leqslant u'(1)$.

(ii) *If $\alpha_0 > 0$ and $\hat{M}u(0) = 0$, then*

$$\begin{aligned} \alpha_0 \Phi(0) \leqslant f_0(u(0),\ \Phi(0)) \quad &\Rightarrow \quad \Phi(0) \leqslant u'(0), \\ \alpha_0 \Psi(0) \geqslant f_0(u(0),\ \Psi(0)) \quad &\Rightarrow \quad \Psi(0) \geqslant u'(0). \end{aligned} \tag{4.15}$$

If $\alpha_1 > 0$ and $\hat{M}u(1) = 0$, then

$$\begin{aligned} -\alpha_1 \Phi(1) \geqslant f_1(u(1),\ \Phi(1)) \quad &\Rightarrow \quad \Phi(1) \leqslant u'(1), \\ -\alpha_1 \Psi(1) \leqslant f_1(u(1),\ \Psi(1)) \quad &\Rightarrow \quad \Psi(1) \geqslant u'(1). \end{aligned}$$

By use of this result the occurrence of u' in the conditions (a), (b) of Theorem 4.4 can often be eliminated. In general, the conditions so transformed still contain the unknown function u, so that they have to be satisfied for all $u \in R$ with $\varphi \leqslant u \leqslant \psi$.

For example, the differential inequality in (4.12) may be replaced by

$$\Psi' \geqslant f(x, y, \Psi) \qquad (\eta < x \leqslant \zeta,\ \ \varphi(x) \leqslant y \leqslant \psi(x)).$$

Moreover, if we want to use (4.15), we may require that

$$\alpha_0 \Psi(0) \geqslant f_0(y, \Psi(0)) \qquad (\varphi(0) \leqslant y \leqslant \psi(0)).$$

We point out that, for constant functions Φ, Ψ, the differential inequalities in Theorem 4.4 merely constitute conditions on the sign of f. Thus, if for some constants $N_1 > 0$ and $N_2 > 0$ neither of the two functions $f(x, y, -N_1)$ and $f(x, y, N_2)$ changes sign on ω (the value 0 being allowed), then $\Phi(x) \equiv -N_1$ satisfies one of the differential inequalities in (a) and $\Psi(x) \equiv N_2$ satisfies one of the differential inequalities in (b). (See also Problem 4.6.)

Example 4.3 (continued). Because of the required boundary condition $u'(1) = 0$, the functions $\Phi(x) \equiv 0$ and $\Psi(x) \equiv 1$ satisfy (4.11) and (4.13), respectively, for $[\eta, \zeta] = [0, 1]$. Since (4.8) holds also (for $\gamma \leqslant 2/\pi$), there exists a solution u^* with $\varphi \leqslant u^* \leqslant \psi$, $0 \leqslant (u^*)' \leqslant 1$. □

Example 4.4. When we replace $(1 - (u')^2)^{1/2}$ in Example 4.3 by the first three terms of its Taylor expansion, we obtain the boundary value problem

$$-u'' - \mu u(1 - \tfrac{1}{2}u'^2 - \tfrac{1}{8}u'^4) = 0, \qquad u(0) = u'(1) = 0,$$

where $\mu > \mu_0 = \pi^2/4$ is a given number. A corresponding pair of comparison functions is $\varphi = \gamma \sin(\pi/2)x$, $\psi = \sqrt{2}x$, with a small enough $\gamma > 0$. Because of $u'(1) = 0$, (4.11) holds for $\Phi(x) \equiv 0$ with $\eta = 0$, $\zeta = 1$. For $\Psi(x) \equiv \sqrt{2}$, condition (4.13) is satisfied with $\eta = 0$, $\zeta = 1$. □

Problems

4.5 Suppose that Φ and Ψ above are not continuous, but belong to $C_0[0, \xi, 1]$ for some $\xi \in (0, 1)$. Use Problem 4.2 to prove that then $\hat{M}u = o$ has a solution $u^* \in C_{1,2}[0, \xi, 1]$. Moreover, show that Theorems 4.3 and 4.4 remain true if the assumptions in these theorems are required to hold for all $u \in C_{1,2}[0, \xi, 1]$ (instead of $u \in R$) and if in Theorem 4.4 only intervals $[\eta, \zeta] = [0, \xi]$ or $[\eta, \zeta] = [\xi, 1]$ are used.

4.6 For the case (4.3), derive the following statement from Theorem 4.4 and Proposition 4.5. (Hint: Choose $\Phi(x) \equiv -N_1$, $\Psi(x) \equiv N_2$ with N_1, N_2 sufficiently large.)

Suppose that φ, ψ is a pair of comparison functions for M satisfying $M\varphi(0) = M\psi(0) = 0$ if $\alpha_0 = 0$, $M\varphi(1) = M\psi(1) = 0$ if $\alpha_1 = 0$. Moreover, assume that for all sufficiently large $p > 0$ the functions $f(x, y, -p)$ and $f(x, y, p)$ do not change sign on the set ω defined in (4.5). Then $Mu = o$ has a solution $u^* \in R$ with $\varphi \leqslant u^* \leqslant \psi$.

4.7 It is sometimes convenient to rewrite the differential inequalities in Theorem 4.4 in a "defect form." Suppose that $\alpha_0 = \alpha_1 = 0$, $\varphi(0) = \psi(0)$, $\varphi(1) = \psi(1)$ *and* $f(x, \varphi(x), p) \leqslant f(x, y, p) \leqslant f(x, \psi(x), p)$ for $(x, y) \in \omega$, $p \in \mathbb{R}$. Write

$$\Phi(x) = \begin{cases} \varphi'(x) - v_1(x) & \text{for} \quad 0 \leqslant x \leqslant \xi_0 \\ \psi'(x) - v_2(x) & \text{for} \quad \xi_0 < x \leqslant 1, \end{cases}$$

$$\Psi(x) = \begin{cases} \psi'(x) + w_1(x) & \text{for} \quad 0 \leqslant x \leqslant \xi_1 \\ \varphi'(x) + w_2(x) & \text{for} \quad \xi_1 < x \leqslant 1 \end{cases}$$

with functions $v_i, w_i \geqslant o$ and constants $\xi_0, \xi_1 \in [0, 1]$. Then formulate the assumptions of Theorem 4.4 in terms of these functions v_i, w_i for suitable intervals $[\eta, \zeta]$. For example, show that the differential inequality in (4.12) with $[\eta, \zeta] = [0, \xi_1]$ is equivalent to

$$w_1'(x) - [f(x, \psi(x), \psi'(x) + w_1(x)) - f(x, \psi(x), \psi'(x))] \geqslant M\psi(x) \qquad (0 < x \leqslant \xi_1).$$

4.3 Quadratic growth restrictions, Nagumo conditions

The theory of the preceding section will now be generalized by constructing functions $\Phi(x, y)$, $\Psi(x, y)$ (instead of $\Phi(x)$, $\Psi(x)$ such that $\Phi(x, u^*(x))$ and $\Psi(x, u^*(x))$ constitute two-sided bounds for the derivative of the solution u^*. For the special case in which the functions Φ, Ψ depend only on y results are obtained which generalize Theorem 4.2 such that the boundedness condition on $F(x, y, p)$ is replaced by certain growth restrictions on $f(x, y, p)$ with respect to p. We point out that these results can also be proved differently, as described in Section 4.5.1. As before, we assume $b(x) = 0$ $(0 \leqslant x \leqslant 1)$, without loss of generality.

Suppose again that φ and ψ are given comparison functions for M. In addition, let $\Phi(x, y)$ and $\Psi(x, y)$ be given continuous functions on ω such that $\Phi(x, y) \leqslant \Psi(x, y)$ $(0 \leqslant x \leqslant 1,\ \varphi(x) \leqslant y \leqslant \psi(x))$ and, moreover, for $0 \leqslant x \leqslant 1$

$$\Phi(x, \varphi(x)) \leqslant \varphi'(x) \leqslant \Psi(x, \varphi(x)), \qquad \Phi(x, \psi(x)) \leqslant \psi'(x) \leqslant \Psi(x, \psi(x)). \tag{4.16}$$

Then, for any $u \in C_1[0, 1]$ with $\varphi \leqslant u \leqslant \psi$, define functions $\Phi[u]$, $\Psi[u]$ and "derivatives" $\dot{\Phi}[u]$, $\dot{\Psi}[u]$ by

$$\Phi[u](x) = \Phi(x, u(x)), \qquad \Psi[u](x) = \Psi(x, u(x)),$$

$$\dot{\Phi}[u](x) = (\Phi_x + \Phi_y\Phi)(x, u(x)), \qquad \dot{\Psi}[u](x) = (\Psi_x + \Psi_y\Psi)(x, u(x)).$$

We assume that the general assumptions at the beginning of Section 4 are satisfied. These assumptions, as well as the above conditions on Φ and Ψ, may be weakened in a way similar to that described in Problems 4.2, 4.3, and 4.5. Such weaker assumptions will be used at a few places which will be pointed out explicitly.

To prove the existence of a solution $u^* \in [\varphi, \psi]$, one proceeds in a way similar to that in Section 4.2. While $\Phi \leqslant (u^*)' \leqslant \Psi$ was shown in that section, we shall now prove $\Phi[u^*] \leqslant (u^*)' \leqslant \Psi[u^*]$. For that purpose, a modified operator $\hat{M}$ is defined by (4.9), where now $\hat{u}' = \langle \Phi[u], u', \Psi[u] \rangle$, so that

$$\hat{M}u(x) = \mathbf{L}u(x) + \hat{F}(x, u(x), u'(x)) \quad \text{with} \quad \hat{F}(x, y, p) = F(x, y, \langle \Phi(x, y), p, \Psi(x, y) \rangle).$$

Theorem 4.6. *The statement of Theorem* 4.3 *remains true if the inequalities* $\Phi \leqslant u' \leqslant \Psi$ *and* $\Phi \leqslant (u^*)' \leqslant \Psi$ *are replaced by* $\Phi[u] \leqslant u' \leqslant \Psi[u]$ *and* $\Phi[u^*] \leqslant (u^*)' \leqslant \Psi[u^*]$, *respectively.*

Theorem 4.7. *The statements of Theorem* 4.4 *remain true if the following changes are made.*

In the assumptions, the functions Φ, Ψ, Φ', *and* Ψ' *are to be replaced by* $\Phi[u]$, $\Psi[u]$, $\dot{\Phi}[u]$, *and* $\dot{\Psi}[u]$, *respectively, and the condition* $\Phi \in C_1[\eta, \zeta]$ $\{\Psi \in C_1[\eta, \zeta]\}$ *is to be replaced by the condition that the restriction of* $\Phi\{\Psi\}$ *to* $\{(x, y) : \eta \leqslant x \leqslant \zeta, \varphi(x) \leqslant y \leqslant \psi(x)\}$ *is continuously differentiable.*

The resulting inequalities are to be replaced by $\varphi \leqslant u^* \leqslant \psi$, $\Phi[u^*] \leqslant (u^*)' \leqslant \Psi[u^*]$.

Proposition 4.8. *The statements of Proposition* 4.5 *remain true if* Φ *and* Ψ *are replaced by* $\Phi[u]$ *and* $\Psi[u]$, *respectively.*

These generalized results are proved in essentially the same way as the corresponding results in Section 4.2. Since Φ and Ψ are defined only on ω, we have to use the fact that $f(x, y, p)$ need be defined only on $\omega \times \mathbb{R}$. We also point out the following slight difference in the proof of Theorem 4.7.

In order to show that $u'(x) \leqslant \Psi[u](x)$ $(\eta \leqslant x \leqslant \zeta)$, one uses an operator

$$Aw(x) = \begin{cases} w' + \Psi_y(x, u)(w - u') - f(x, u, \langle \Phi[u], w, \Psi[u] \rangle) & \text{for } \eta < x \leqslant \zeta \\ w(\eta) & \text{for } x = \eta \end{cases}$$

in place of the operator A in (4.14). Moreover, one chooses $z(x) = \exp(Nx)$ with sufficiently large N, instead of $z(x) = 1 + x$. □

For example, the inequalities which generalize (4.12) assume the form

$$\begin{aligned} (\Psi_x + \Psi_y \Psi)(x, u(x)) - f(x, u(x), \Psi(x, u(x))) \geqslant 0 \qquad & (\eta < x \leqslant \zeta), \\ u'(\eta) \leqslant \Psi(\eta, u(\eta)), \quad & \end{aligned} \tag{4.17}$$

for each $u \in R$ satisfying $\varphi \leqslant u \leqslant \psi$, $\hat{M}u = o$. We shall discuss the solution of these inequalities for the special case $\Psi(x, y) = \Psi(y)$, in order to explain an application of Theorem 4.7.

Let $h \in C_0[0, \infty)$ denote a function with $h(p) > 0$ $(0 < p < \infty)$, $\mu > 0$ a constant, and

$$\varphi_0 = \min\{\varphi(x) : 0 \leqslant x \leqslant 1\}, \qquad \psi_1 = \max\{\psi(x) : 0 \leqslant x \leqslant 1\}.$$

(In the following arguments we shall assume that $\varphi_0 < \psi_1$, because otherwise $\varphi = u^* = \psi$, so that the results formulated below become trivial.) The relation

$$\int_{\mu}^{w(t)} \frac{s\,ds}{h(s)} = t - \varphi_0 \tag{4.18}$$

defines a continuously differentiable function $w(t)$ on some interval $[\varphi_0, \varphi_0 + \tau]$. This function is strictly positive and strictly monotone, and is satisfies the differential equation $w\, dw/dt - h(w) = 0$. If

$$\int_\mu^\infty \frac{s\, ds}{h(s)} \geqslant \psi_1 - \varphi_0, \tag{4.19}$$

then $w(t)$ is defined for all $t \in [\varphi_0, \psi_1]$. Using these arguments one proves the following result.

Lemma 4.9. *Let the continuous function h satisfying $h(p) > 0\ (0 \leqslant p < \infty)$ majorize the function $f(x, y, p)$ in the sense that*

$$f(x, y, p) \leqslant h(p) \qquad \text{for} \quad \eta \leqslant x \leqslant \zeta, \quad \varphi(x) \leqslant y \leqslant \psi(x), \quad p > 0. \tag{4.20}$$

Moreover, suppose that (4.19) *holds and $u'(\eta) \leqslant \mu$ with $\mu > 0$.*

Then $\Psi(x, y) := w(y)$ defined by (4.18) *for $\eta \leqslant x \leqslant \zeta$, $\varphi(x) \leqslant y \leqslant \psi(x)$, satisfies the inequalities* (4.17). *If, in addition, $\|\varphi'\|_\infty \leqslant \mu$ and $\|\psi'\|_\infty \leqslant \mu$, then the inequalities $\varphi'(x) \leqslant \Psi(x, \varphi(x))$ and $\psi'(x) \leqslant \Psi(x, \psi(x))$ hold for $\eta \leqslant x \leqslant \zeta$.*

Condition (4.19) ensures that $\Psi(x, y) = w(y)$ has a proper domain of definition. Obviously, this condition is satisfied if

$$\int_\mu^\infty \frac{s\, ds}{h(s)} = +\infty \qquad \text{for some} \quad \mu > 0. \tag{4.21}$$

The latter relation roughly means that $h(s)$ cannot grow very much faster than s^2. For example, (4.21) is satisfied for $h(s) = c(1 + s)^2$, as well as for $h(s) = 1 + c(1 + s)^2 \log(1 + s)$, with an arbitrary constant $c > 0$. However, (4.21) does not hold for $h(s) = c(1 + s)^{2+\varepsilon}$, $c > 0$, $\varepsilon > 0$.

In a similar way, the remaining inequalities required in Theorem 4.7 can be solved, provided the function f satisfies appropriate growth restrictions from above or below and provided a priori bounds for $u'(x)$ can be established at certain points.

For many concrete problems, bounds Φ and Ψ need not actually be constructed, since growth restrictions such as (4.20) ensure the existence of such bounds. For such problems, the existence of comparison functions φ, ψ alone yields the existence of a solution $u^* \in [\varphi, \psi]$. This fact may be considered to be the main advantage of the theory developed in this section. Simple results of this nature are formulated in the next two theorems. (See also Problem 4.10.)

Theorem 4.10. *Suppose that the boundary conditions have the special form* (4.3), *and that comparison functions φ and ψ for M exist. Assume, moreover,* *that the following two conditions* (a), (b) *are satisfied for some constant* $c > 0$.

(a) *Either* $f(x, y, p) \geqslant c(1 + p^2)$ *for* $(x, y) \in \omega$, $p < 0$,
and $\varphi(0) = r_0$ *if* $\alpha_0 = 0$;
or $f(x, y, p) \leqslant -c(1 + p^2)$ *for* $(x, y) \in \omega$, $p < 0$,
and $\psi(1) = r_1$ *if* $\alpha_1 = 0$.

(b) *Either* $f(x, y, p) \leqslant c(1 + p^2)$ *for* $(x, y) \in \omega$, $p > 0$,
and $\psi(0) = r_0$ *if* $\alpha_0 = 0$;
or $f(x, y, p) \geqslant -c(1 + p^2)$ *for* $(x, y) \in \omega$, $p > 0$,
and $\varphi(1) = r_1$ *if* $\alpha_1 = 0$.

Then the given problem $Mu = o$ *has a solution* $u^* \in R$ *such that* $\varphi \leqslant u^* \leqslant \psi$.

Proof. Define w as above and $v(y)$ by $\int_\mu^{v(y)} sh^{-1}(s)\, ds = \psi_1 - y$, where $h(p) = c(1 + p^2)$ and $\mu > 0$. Then, choose $\Phi(x, y) := -w(y)$ or $\Phi(x, y) := -v(y)$ depending on whether the first or the second condition in (a) is satisfied. (If both are satisfied, choose either of these functions.) In a corresponding way, choose $\Psi(x, y) := w(y)$ or $\Psi(x, y) := v(y)$. If μ is large enough, these functions Φ, Ψ satisfy the assumptions required in Theorem 4.7, when $[\eta, \zeta] = [0, 1]$ is chosen in each case. The boundary conditions imposed on φ and ψ ensure that $u'(0)$ and $u'(1)$ can be estimated a priori, as far as needed. □

If the function $|f(x, y, p)|$ satisfies certain growth restrictions for $|p| \to \infty$, then the derivative of the solution can be estimated without using the boundary conditions at all, as seen in the next two theorems.

Theorem 4.11a. *Assume that comparison functions* φ *and* ψ *for* M *exist and that*

$$|f(x, y, p)| \leqslant c(1 + p^2) \quad \text{for all} \quad (x, y) \in \omega, \quad p \in \mathbb{R} \tag{4.22}$$

with a constant $c > 0$.

Then the given problem $Mu = o$ *has a solution* $u^* \in R$ *with* $\varphi \leqslant u^* \leqslant \psi$.

Proof. For each $u \in [\varphi, \psi]$ there exists a value $\xi \in (0, 1)$ such that

$$|u'(\xi)| \leqslant \Lambda := \max(|\psi(1) - \varphi(0)|, \ |\psi(0) - \varphi(1)|). \tag{4.23}$$

(Note that, for example, $\psi(1) - \varphi(0)$ is the slope of the line passing through the points $(0, \varphi(0))$ and $(1, \psi(1))$.) Define

$$\Phi(x, y) = \begin{cases} -w(y) & \text{for } 0 \leqslant x \leqslant \xi, \\ -v(y) & \text{for } \xi < x \leqslant 1, \end{cases}$$

$$\Psi(x, y) = \begin{cases} v(y) & \text{for } 0 \leqslant x \leqslant \xi \\ w(y) & \text{for } \xi < x \leqslant 1, \end{cases}$$

with v, w as in the proof of Theorem 4.10 and $\mu > \Lambda$ large enough.

Now we have an a priori estimate for u' at $x = \xi$. Thus, in the conditions of Theorem 4.7, we may choose $[\eta, \zeta] = [0, \xi]$ or $[\eta, \zeta] = [\xi, 1]$, depending on to which interval the value x_0 belongs.

Using the arguments in proving Lemma 4.9, one can show that all assumptions of Theorem 4.7 are satisfied, with one exception. The considered functions Φ, Ψ may not be continuous at $x = \xi$. However, we already remarked that the required conditions may be relaxed. In particular, the discontinuity which may occur here does not affect the results. □

Example 4.5. The theory of Section 4 can be used to investigate the qualitative behavior of solutions of singular perturbation problems. Here we consider only a simple example: *The problem*

$$-\varepsilon u'' - uu' + u = 0, \qquad u(0) = 1, \qquad u(1) = -1 \qquad \text{with} \quad \varepsilon > 0$$

has a solution $u^* \in C_2[0, 1]$ *such that*

$$\varphi \leqslant u^* \leqslant \psi \qquad \textit{with} \quad \varphi(x) = -\exp \mu(x - 1),$$
$$\psi(x) = \exp(-\mu x), \qquad \mu = \varepsilon^{-1/2}.$$

This statement is derived from Theorem 4.11a by verifying that φ and ψ constitute a pair of comparison functions. Using the technique in Section 2.3.3 one proves, in addition, that *the derivative of the solution satisfies* $(u^*)' \leqslant \Psi$ *with* $\Psi(x) = 1 - (1 + \mu)\exp[\mu(e^{-\mu x} - 1)]$. □

Theorem 4.11b. *Suppose that* φ *and* ψ *are comparison functions for M. Assume, moreover, that a function* $h \in C_0[0, \infty)$ *exists such that* $h(p) > 0$ $(0 \leqslant p < \infty)$,

$$|f(x, y, p)| \leqslant h(|p|) \qquad \textit{for all} \quad (x, y) \in \omega, \qquad p \in \mathbb{R}, \tag{4.24}$$

and (4.19) *holds with* $\mu = \max(\Lambda, \|\varphi'\|_\infty, \|\psi'\|_\infty)$ *and* Λ *in* (4.23).

Then the given problem $Mu = o$ *has a solution* $u^* \in R$ *with* $\varphi \leqslant u^* \leqslant \psi$.

The proof is similar to that of Theorem 4.11a.

In the theory of differential equations $-u'' + f(x, u, u') = 0$, the function $f(x, y, p)$ is said to satisfy the *Nagumo condition on* ω if (4.24) and (4.21) hold. *One-sided Nagumo conditions* are described by requirements such as (4.20) and (4.21). As already mentioned, assumption (4.21) severely restricts the possible behavior of the *growth function h*. Often one also speaks about a Nagumo condition if (4.21) is replaced by a weaker assumption such as (4.19). This latter condition may be satisfied for suitable comparison functions even if $|f(x, y, p)|$ grows "much faster" than p^2. In fact, for any given growth function h, inequality (4.19) holds, if $\psi_1 - \varphi_0$ is small enough. Often, this may be achieved by transforming the given equation $Mu = o$ into an equation for $v = u - u_0$ with u_0 denoting an approximate solution.

The assumptions of Theorems 4.10, 4.11a, and 4.11b also yield $\Phi[u^*] \leqslant (u^*)' \leqslant \Psi[u^*]$, where Φ and Ψ are defined in terms of the functions v, w occurring above. Using $w(u^*(x)) \leqslant w(\psi(x))$ and $v(u^*(x)) \geqslant v(\varphi(x))$ one can obtain bounds for $(u^*)'$ which do not depend on u^*. These bounds may be estimated further. For example, under the assumptions of Theorem 4.11b we obtain

$$-N \leqslant (u^*)' \leqslant N \qquad \text{for} \quad \int_\mu^N sh^{-1}(s)\, ds \geqslant \psi_1 - \varphi_0. \tag{4.25}$$

Problems

4.8 Consider the boundary value problem $-u'' + (u')^{2+\varepsilon} = 0$, $u(0) = 0$, $u(1) = \mu$ with $\varepsilon > 0$, $\mu > 0$. Prove that the existence of comparison functions, such as $\varphi = 0$, $\psi = \mu$, does not necessarily imply the existence of a solution u^* with $\varphi \leqslant u^* \leqslant \psi$. In particular, show that for each $\varepsilon > 0$ there exists a $\mu > 0$ such that the problem has no solution.

4.9 Carry out the details in the proof of Lemma 4.9.

4.10 Show that in Theorem 4.10 the four inequalities for $f(x, y, p)$ which occur in assumptions (a) and (b) may be replaced by

$$f(x, y, p) \geqslant h_1(-p), \qquad f(x, y, p) \leqslant -h_2(-p),$$
$$f(x, y, p) \leqslant h_3(p), \qquad f(x, y, p) \geqslant -h_4(p),$$

in this order. Here, $h_i \in C_0[0, \infty)$, $h_i(t) > 0$ for $t \geqslant 0$, and $\int_\mu^\infty sh_i^{-1}(s)\, ds \geqslant \psi_1 - \varphi_0$ and μ as in Theorem 4.11b.

4.4 Singular problems

Many results of the previous sections can be carried over to problems on noncompact intervals or problems where the occurring functions have "singularities" at the boundary points.

If in the latter case the singularities are "weak enough," only minor changes in the above theory may be necessary. For example, let $R = C_0[0, 1] \cap C_2(0, 1)$ and consider an operator M defined on R by

$$Mu(x) = -u''(x) + b(x)u'(x) + f(x, u(x)) \qquad \text{for} \quad 0 < x < 1,$$
$$Mu(0) = u(0) - r_0, \qquad Mu(1) = u(1) - r_1$$

where $b \in C_0[0, 1]$, $f(x, y)$ is continuous on $(0, 1) \times \mathbb{R}$, and $\int_0^1 |f(x, w(x))|\, dx$ exists for each $w \in C_0[0, 1]$. For this operator the following result is proved in the same way as Theorem 4.2.

Proposition 4.12. *If comparison functions $\varphi, \psi \in R$ for M exist, then the problem $Mu = o$ has a solution $u^* \in R$ such that $\varphi \leqslant u^* \leqslant \psi$.*

Example 4.6. Consider the problem

$$-u'' + x^{-1/2}u^{3/2} = 0, \qquad u(0) = 1, \quad u(1) = 0$$

with the Thomas–Fermi differential equation. (See also Example 4.7.) Here, $\varphi(x) \equiv 0$ and $\psi(x) \equiv 1$ are comparison functions. Since the problem is of the type described above, Proposition 4.12 yields the existence of a solution u^* such that $0 \leqslant u^*(x) \leqslant 1$ $(0 \leqslant x \leqslant 1)$. (Observe that $f(x, y) = x^{-1/2}y^{3/2}$ may be extended to $(0, 1) \times \mathbb{R}$ by defining $f(x, y) = 0$ for $y < 0$.) □

Now let us consider problems $Mu = o$ in which M is defined on a noncompact interval. For example, a differential equation $Mu(x) = 0$ $(0 < x < 1)$ may be given without any boundary conditions. We shall consider only certain problems on $[0, \infty)$; the approach may also be used for problems on other noncompact intervals.

Let $R = C_2[0, \infty)$ and consider an operator M on R defined by

$$Mu(x) = \begin{cases} -u''(x) + b(x)u'(x) + f(x, u(x), u'(x)) & \text{for} \quad 0 < x < \infty \\ -\alpha_0 u'(0) + f_0(u(0), u'(0)) & \text{for} \quad x = 0, \end{cases}$$

where b, f, α_0, and f_0 have the properties described at the beginning of Section 4 with the interval $[0, 1]$ for the variable x replaced by $[0, \infty)$, everywhere.

The equation $Mu = o$ does not include any "boundary condition" at ∞. If in a concrete problem such a condition is posed, one may try to construct suitable comparison functions such that $\varphi \leqslant u^* \leqslant \psi$ guarantees the condition to be satisfied. (See Example 4.7.)

To describe the manner in which the results for compact intervals are employed here, we first prove a result analogous to Theorem 4.2. Comparison functions are again defined by (4.4) where $\leqslant$ now is interpreted to hold pointwise on $[0, \infty)$.

Theorem 4.13. *Suppose that comparison functions $\varphi, \psi \in R$ for M exist, $f(x, y, p)$ is bounded for $0 \leqslant x < \infty$, $\varphi(x) \leqslant y \leqslant \psi(x)$, $p \in \mathbb{R}$ and $f_0(y, p)$ is bounded for $\varphi(0) \leqslant y \leqslant \psi(0)$, $p \in \mathbb{R}$.*

Then the problem $Mu = o$ has a solution $u^* \in R$ such that $\varphi \leqslant u^* \leqslant \psi$.

Proof. First recall that Theorem 4.2 remains true for problems analogous to (4.1) in which $[0, 1]$ is replaced by an arbitrary interval $[0, l]$, $l > 0$. For each $n \in \mathbb{N}$ define an operator M_n on $C_2[0, n]$ by

$$\begin{aligned} M_n u(x) &= Mu(x) \qquad \text{for} \quad 0 \leqslant x < n, \\ M_n u(n) &= u(n) - \beta_n, \qquad \beta_n = \tfrac{1}{2}(\varphi(n) + \psi(n)). \end{aligned} \tag{4.26}$$

The restrictions of φ and ψ to $[0, n]$ are comparison functions for M_n. Thus, according to Theorem 4.2, the equation $M_n u = o$ has a solution $u_n \in C_2[0, n]$ such that $\varphi(x) \leqslant u_n(x) \leqslant \psi(x)$ $(0 \leqslant x \leqslant n)$. This solution also satisfies a fixed-point equation $u_n(x) = T_n u_n(x)$ $(0 \leqslant x \leqslant n)$ with

$$T_n u(x) = \mathscr{G}_0(x)[u(0) - f_0(u(0), u'(0))] + \frac{x}{n} u(n) + \int_0^n \mathscr{G}_n(x, \xi)[u(\xi) - f(\xi, u(\xi), u'(\xi))]\, d\xi,$$

similar to $Tu(x)$ in (4.7). Moreover, for $m \geqslant n$ the function u_m is also defined on $[0, n]$ and satisfies

$$u_m(x) = T_n u_m(x) \qquad (0 \leqslant x \leqslant n). \tag{4.27}$$

Since $T_n C_1[0, n]$ is a relatively compact subset of $C_1[0, n]$ with respect to the norm $\|u\| = \|u\|_\infty + \|u'\|_\infty$ on $[0, n]$, the sequence $\{u_m\}$ contains a subsequence $\{v_m^{(n)}\}$ such that the functions $v_m^{(n)}$ and the derivations $d/dx\, v_m^{(n)}$ converge uniformly on $[0, n]$ for $m \to \infty$. The diagonal sequence $\{w_m\}$ with $w_m = v_m^{(m)}$ then converges uniformly on each interval $[0, n]$ and the same is true for $\{w_m'\}$. Consequently, there exists a function $u^* \in C_1[0, \infty)$ such that $w_m \to u^*$, $w_m' \to (u^*)'$ uniformly on each interval $[0, n]$. Obviously, $\varphi \leqslant u^* \leqslant \psi$. Taking the limit $m \to \infty$ in (4.27), we see that $u^*(x) = T_n u^*(x)$ $(0 \leqslant x \leqslant n)$ for each n, and consequently, $Mu^*(x) = 0$ for $0 \leqslant x < \infty$. □

As we remarked repeatedly, the assumptions which we require may be relaxed in various ways. For example, in the preceding theorem the smoothness conditions on φ, ψ may be weakened as described in Problem 4.3 and the continuity requirement on f may be relaxed as in Proposition 4.12. Both these possibilities will be used in the following example.

Example 4.7. The boundary value problem

$$-u''(x) + x^{-1/2} u^{3/2}(x) = 0 \qquad (0 < x < \infty), \qquad u(0) = 1, \qquad \lim_{x \to \infty} u(x) = 0$$

was considered by Fermi and Thomas. (For references see Kamke [1951], in particular, C 6.100.) Here, φ and ψ defined by

$$\begin{aligned} \varphi(x) \equiv 0, \qquad \psi(x) &= 1 && \text{for} \quad 0 \leqslant x \leqslant x_0 \\ \psi(x) &= 144 x^{-3} && \text{for} \quad x_0 < x < \infty \end{aligned}$$

with $144 x_0^{-3} = 1$ are corresponding comparison functions. (ψ satisfies Condition S_r.) Consequently, the problem has a solution $u^* \in C_0[0, \infty) \cap C_2(0, \infty)$ such that $\varphi(x) \leqslant u^*(x) \leqslant \psi(x)$ $(0 \leqslant x < \infty)$. (Observe that this inequality implies $\lim_{x \to \infty} u^*(x) = 0$.) □

In the following theorem the boundedness conditions on f and f_0 required in Theorem 4.13 are replaced by a different assumption which can be verified in various ways.

Theorem 4.14. *Suppose that $\varphi, \psi \in R$ are comparison functions for M and that the following assumption* (A) *is satisfied.*

(A) *For each $n \in \mathbb{N}$ the existence of the comparison functions φ, ψ implies the existence of a function $u_n \in C_2[0, n]$ such that*

$$\begin{aligned} M_n u_n(x) = 0, \qquad & \varphi(x) \leqslant u_n(x) \leqslant \psi(x), \\ |u_n'(x)| \leqslant N \qquad & \text{for} \quad 0 \leqslant x \leqslant n, \end{aligned} \tag{4.28}$$

where M_n is defined by (4.26) *and N denotes a constant independent of n.*

Then the problem $Mu = o$ has a solution $u^* \in R$ such that $\varphi \leqslant u^* \leqslant \psi$.

The proof is essentially the same as that of Theorem 4.13. Now the set $T_n C_1[0, 1]$ is not relatively compact, but the sequence $\{T_n u_m\}$ still is.

One may try to verify (4.28) by applying the results of Sections 4.2 and 4.3 to the operators M_n $(n = 1, 2, \ldots)$ and observing that (the corresponding restrictions of) φ and ψ are comparison functions for M_n. For example, Theorem 4.4 yields Corollary 4.14a and Theorems 4.11a,b together with (4.25) yield Corollary 4.14b. In these corollaries we assume $b(x) = 0$ $(0 \leqslant x < \infty)$ without loss of generality.

Corollary 4.14a. *Assumption* (A) *may be replaced by the condition that functions $\Phi, \Psi \in C_1[0, \infty)$ exist such that $-N \leqslant \Phi(x) \leqslant \Psi(x) \leqslant N$, $\Phi(x) \leqslant \varphi'(x) \leqslant \Psi(x)$, $\Phi(x) \leqslant \psi'(x) \leqslant \Psi(x)$ $(0 \leqslant x < \infty)$ and for each $u \in C_2[0, \infty)$ satisfying $\hat{M}u(0) = 0$*

$$\begin{aligned} \Phi'(x) - F(x, u(x), \Phi(x)) \leqslant 0 \qquad (0 < x < \infty), \qquad \Phi(0) \leqslant u'(0), \\ \Psi'(x) - F(x, u(x), \Psi(x)) \geqslant 0 \qquad (0 < x < \infty), \qquad u'(0) \leqslant \Psi(0). \end{aligned}$$

Corollary 4.14b. *Assumption* (A) *may be replaced by the condition that the functions $\varphi, \psi, \varphi', \psi'$ are bounded on $[0, \infty)$ and* (4.22) *holds for $\omega = \{(x, y): 0 \leqslant x < \infty, \varphi(x) \leqslant y \leqslant \psi(x)\}$.*

Observe that estimates $\Phi(0) \leqslant u'(0) \leqslant \Psi(0)$ as required in Corollary 4.14a can be obtained from Proposition 4.5.

Problem

4.11 Consider the nonlinear eigenvalue problem

$$-u''(x) + f(x)u^{2i+1}(x) - \mu u(x) = 0 \qquad (0 < x < \infty), \qquad u(0) = 0$$

for $u \in C_2[0, \infty)$ with $u \in L_2(0, 1)$, $u'' \in L_2(0, 1)$, where $i \in \mathbb{N}$. The spectrum of the linearized problem (with f replaced by o) consists of the points $\mu \in [0, \infty)$. *If there exist constants ρ, σ such that $0 < \rho \leqslant f(x)$, $0 \leqslant f'(x)f^{-1}(x) \leqslant \sigma$, $|f''(x)f^{-1}(x)| \leqslant \sigma$ $(0 \leqslant x < \infty)$ and if $\int_0^\infty f^{-1/i}(x)\,dx < \infty$, then the given nonlinear problem has a solution $u_\mu > o$ for each $\mu > 0$.*

To prove this, construct comparison functions φ, ψ of the form $\varphi(x) = \gamma \sin \delta x$ for $0 \leqslant x \leqslant \pi\delta^{-1}$, $\varphi(x) = 0$ for $\pi\delta^{-1} < x$, $\psi(x) = cf^{-1/2i}(x)$ for $x_0 \leqslant x < \infty$, $\psi(x) = \psi(x_0)$ for $0 \leqslant x < x_0$ with γ, δ, c, x_0 depending on μ. Show, in addition, that $f(x) = (1 + x)^\alpha$ *satisfies all assumptions for* $\alpha > i$.

4.5 Different approaches

4.5.1 A priori bounds for the derivative

In the theory of Section 4.3, existence and inclusion statements were derived for problems with f satisfying a Nagumo condition. It will now be shown by a different method that Nagumo conditions yield a priori bounds for the derivative of the solution. Using these a priori bounds one can again derive existence and inclusion statements, by now applying Theorem 4.3 in Section 4.2.

Proposition 4.15. *Let $\varphi, \psi \in C_0[0, 1]$ denote any functions with $\varphi \leqslant \psi$ and define the set ω by* (4.5). *Suppose that the differential equation $-u''(x) + f(x, u(x), u'(x)) = 0$ has a solution $u^* \in C_2[0, 1]$ with its graph in ω and that $f(x, y, p)$ satisfies the growth restriction described in Theorem* 4.11b, *where now $\mu = \Lambda$ as defined in* (4.23).

Then $|(u^*)'| \leqslant \kappa$ with $\int_\Lambda^\kappa sh^{-1}(s)\,ds = \psi_1 - \varphi_0$.

Proof. We write $u = u^*$. Suppose the estimate for u' were false, so that $|u'(\eta)| > \kappa$ for some $\eta \in (0, 1)$. According to (4.23), we have also $|u'(\xi)| \leqslant \Lambda \leqslant \kappa$. To describe the method of proof, it suffices to consider the case $u'(\xi) = \Lambda$, $u'(\eta) > \kappa$, $\xi < \eta$. Without loss of generality, we may also assume that $u'(x) > \Lambda$ for $\xi < x \leqslant \eta$. Since $|u''(x)| \leqslant |f(x, u(x), u'(x))| \leqslant h(u'(x))$ for $\xi \leqslant x \leqslant \eta$, we obtain

$$\left| \int_\xi^\eta \frac{u''(x)u'(x)}{h(u'(x))}\,dx \right| \leqslant \int_\xi^\eta u'(x)\,dx = u(\eta) - u(\xi) \leqslant \psi_1 - \varphi_0.$$

On the other hand, by substituting $s = u'(x)$, the left-hand integral can be transformed into $\int_\Lambda^{u'(\eta)} sh^{-1}(s)\,ds > \int_\Lambda^\kappa sh^{-1}(s)\,ds = \psi_1 - \varphi_0$, so that a contradiction is established. □

The proposition can be used to prove Theorem 4.11b, where now $\mu = \Lambda$ is chosen. Define $\hat{M}$ by (4.9) with $\Phi = -N$, $\Psi = N$, where N denotes a constant which is greater than each of the numbers κ, $\|\varphi'\|_\infty$, and $\|\psi'\|_\infty$. Proposition 4.15 applied to $\hat{M}$ in place of M then yields (4.10), so that a solution $u^* \in [\varphi, \psi]$ exists, according to Theorem 4.3. (Here, the growth function $\hat{h}$ for $\hat{M}$ may be chosen such that $\hat{h}(s) = h(s)$ for $s \leqslant \kappa$.)

In an analogous way, Theorem 4.10 and generalizations involving one-sided Nagumo conditions can be proved.

4.5.2 *The degree of mapping method*

In the preceding theory, existence and inclusion statements are proved with the aid of Schauder's fixed-point theorem and Theorem 2.1. The fixed-point theorem yields the *existence* of a solution u^* of a modified problem; Theorem 2.1 then establishes *inclusion properties* such that u^* is also a solution of the given problem. Such existence and inclusion statements have also been derived by other means. In particular, the degree theory has been applied by several authors (see the Notes).

Here we shall describe the application of Theorem I.3.2 by considering a special case of the boundary value problem (4.1). The object is not only to demonstrate this different method, but also to explain its relation to the above "modification method." Both methods of proof have certain essential elements in common. For instance, in order to verify (I.3.1), a series of arguments will be used which were also used in the proof of Theorem 2.1. However, there are differences, too, in the methods as well as in the corresponding assumptions.

Let a problem $Mu = o$ in $C_2[0, 1]$ be given in which M has the special form $Mu = (-u'' + f(x, u, u'), u(0), u(1))$ with continuous $f\colon [0, 1] \times \mathbb{R} \times \mathbb{R} \to \mathbb{R}$. Suppose that φ and ψ are comparison functions in $C_2[0, 1]$ which satisfy $\varphi(x) < 0 < \psi(x)$ $(0 \leqslant x \leqslant 1)$. Also assume that $f(x, y, p)$ satisfies the Nagumo condition (4.24), (4.21). Under these assumptions, the given problem has a solution u^* with $\varphi \leqslant u^* \leqslant \psi$.

This statement will now be derived from the theorem of Leray–Schauder. Instead of a modified operator $M^\#$, a family of problems

$$-u'' + cu + \alpha(-cu + f(x, u, u')) = 0 \qquad (0 < x < 1), \qquad u(0) = u(1) = 0 \tag{4.29}$$

with a parameter $\alpha \in [0, 1]$ is considered, where $c > 0$ denotes a constant with

$$c\varphi - f(x, \varphi, \varphi') < 0, \qquad c\psi - f(x, \psi, \psi') > 0 \qquad (0 \leqslant x \leqslant 1). \tag{4.30}$$

Since f satisfies the Nagumo condition (4.24), (4.21) with a growth function h, all functions $f_\alpha(x, y, p) = cy + \alpha(-cy + f(x, y, p))$ $(0 \leqslant \alpha \leqslant 1)$ also satisfy such a condition with a function $c_0 + h(p)$ where c_0 does not depend on α.

Thus, it follows from Proposition 4.15 that there exist functions Φ, $\Psi \in C_0[0, 1]$ with

$$\Phi(x) < 0 < \Psi(x), \qquad \Phi(x) < u'(x) < \Psi(x) \qquad (0 \leqslant x \leqslant 1) \quad (4.31)$$

for each solution $u \in [\varphi, \psi]$ of (4.29). Actually, one may choose $-\Phi = \Psi = N$ with a sufficiently large constant N.

For each $\alpha \in [0, 1]$ the problem (4.29) is equivalent to an operator equation

$$u = \alpha T u \qquad (4.32)$$

in the Banach space $C_1[0, 1]$ furnished with the norm $\|u\| = \|u\|_\infty + \|u'\|_\infty$. Here, T is a compact (integro-differential) operator similar to the operator in (4.7).

We define $D = \{u \in C_1[0, 1]: \varphi \leqslant u \leqslant \psi, \Phi \leqslant u' \leqslant \Psi\}$. Then the boundary δD is the set of all $u \in D$ such that for at least one value $\xi \in [0, 1]$ at least one of the following equations holds:

$$\begin{aligned} u(\xi) = \varphi(\xi), \qquad &\text{or} \qquad u(\xi) = \psi(\xi), \\ \text{or} \qquad u'(\xi) = \Phi(\xi), \qquad &\text{or} \qquad u'(\xi) = \Psi(\xi). \end{aligned} \qquad (4.33)$$

Now suppose that assumption (I.3.1) of Theorem I.3.2 were false. Then we have $o = (I - \alpha T)u$ for some $\alpha \in (0, 1)$ and some $u \in \delta D$. In other words, for some $\alpha \in (0, 1)$ the boundary value problem (4.29) has a solution $u \in \delta D$. Because of (4.31), the first or second relation in (4.33) must hold. Suppose that $u(\xi) = \psi(\xi)$ for some $\xi \in [0, 1]$. (The case $u(\xi) = \varphi(\xi)$ is treated analogously.) Since $\varphi \prec o \prec \psi$ and $u(0) = u(1) = 0$, ξ cannot be a boundary point. For $0 < \xi < 1$, however, the relations $u \leqslant \psi$ and $u(\xi) = \psi(\xi)$ imply $u'(\xi) = \psi'(\xi)$ and $u''(\xi) \leqslant \psi''(\xi)$. Since u satisfies (4.29), all the above relations together yield

$$-\psi'' + f(x, \psi, \psi') + (1 - \alpha)(c\psi - f(x, \psi, \psi')) \leqslant 0 \qquad \text{for} \quad x = \xi.$$

This inequality contradicts the fact that φ and ψ are comparison functions and (4.30) holds. This completes the proof. □

In the proof, contradictions to each of the equations in (4.33) had to be derived. The last two cases were excluded using the Nagumo condition. Of course, there are other possibilities to treat these cases. For example, one may require Φ and Ψ to satisfy first order differential inequalities as in Section 4.2 and then derive contradictions.

All results derived in the previous sections essentially can also be obtained by using the theory of Leray–Schauder and, in particular, Theorem I.3.2. Observe, however, that for the simple special case treated above we had to require certain assumptions which were not necessary in using the modification method. For example, we assumed $\varphi \prec \psi$ and $\Phi \prec \Psi$. These inequalities are somewhat inconvenient; it requires additional tedious considerations to remove them. Moreover, when f satisfies a weaker Nagumo condition (4.19), (4.20), then the functions f_α need not satisfy such a condition.

4.5.3 A simple approach for a special case, iterative procedures

We consider problem (4.1), (4.2) under the assumption that (4.3) holds and $f(x, y, p) =: f(x, y)$ does not depend on p. Moreover, we assume that for given comparison functions φ, ψ a constant κ exists such that

$$\mathbf{N}u - \mathbf{N}v \leqslant \kappa(u - v) \qquad \text{for} \quad u, v \in C_0[0, 1] \qquad \text{with} \quad \varphi \leqslant v \leqslant u \leqslant \psi. \tag{4.34}$$

This amounts to requiring one-sided Lipschitz conditions for f, f_0, and f_1. Under this assumption one can easily prove Theorem 4.2 by Schauder's fixed-point theorem without using a modified operator. In addition, one can obtain existence and inclusion statements by iterative procedures.

Theorem 4.16. *Suppose that φ, ψ are comparison functions for the given problem and that* (4.34) *holds for some constant κ.*

Then the given boundary value problem has a solution $u^* \in C_2[0, 1]$ with $\varphi \leqslant u^* \leqslant \psi$.

Moreover, for $\varphi_0 = \varphi$, $\psi_0 = \psi$ and any constant $c \geqslant \kappa$ with $c > 0$ the formulas

$$\mathbf{L}\varphi_{\nu+1} + c\varphi_{\nu+1} = c\varphi_\nu - \mathbf{N}\varphi_\nu,$$
$$\mathbf{L}\psi_{\nu+1} + c\psi_{\nu+1} = c\psi_\nu - \mathbf{N}\psi_\nu \qquad (\nu = 0, 1, 2, \ldots)$$

define sequences $\{\varphi_\nu\}$ and $\{\psi_\nu\}$ in R such that for all ν

$$\varphi_0 \leqslant \varphi_1 \leqslant \cdots \leqslant \varphi_\nu \leqslant \psi_\nu \leqslant \cdots \leqslant \psi_1 \leqslant \psi_0.$$

These sequences converge uniformly toward solutions $\bar{u}$ and $\bar{\bar{u}}$ of the given boundary value problem, so that $\varphi_\nu \leqslant \bar{u} \leqslant \bar{\bar{u}} \leqslant \psi_\nu$ for all ν. Each solution u^* with $\varphi \leqslant u^* \leqslant \psi$ satisfies also $\bar{u} \leqslant u^* \leqslant \bar{\bar{u}}$, that means, $\bar{u}$ is a minimal solution and $\bar{\bar{u}}$ is a maximal solution in $[\varphi, \psi]$.

Proof. For any constant c as described above the operator $A: R \to S$ given by $Au = \mathbf{L}u + cu$ is inverse-positive (see Example II.3.1). On the other hand, the operator B defined on $[\varphi, \psi]_0$ by $Bu = cu - \mathbf{N}u$ is monotone. Consequently $T = A^{-1}B$ is monotone.

Since A is inverse-positive and φ, ψ are comparison functions, the operator T maps the interval $[\varphi, \psi]_0$ into itself. For example, if $u \in [\varphi, \psi]_0$, then $A(Tu) = Bu \leqslant B\psi \leqslant A\psi$ and hence $Tu \leqslant \psi$. Applying Schauder's fixed-point theorem, one sees that the subset $[\varphi, \psi]_0$ of $C_0[0, 1]$ contains a fixed point of T, which then is a solution of the given boundary value problem.

Now, observe that $\varphi \leqslant T\varphi$, $\varphi \leqslant \psi$, and $T\psi \leqslant \psi$, since φ, ψ are comparison functions and A is inverse-positive. Moreover, the iteration formulas are equivalent to $\varphi_{\nu+1} = T\varphi_\nu$, $\psi_{\nu+1} = T\psi_\nu$ $(\nu = 0, 1, 2, \ldots)$. Therefore, the above statements concerning the iterative procedures follow by applying Theorem I.3.4 and the succeeding Remark 2. □

5 EXISTENCE AND INCLUSION BY WEAK COMPARISON FUNCTIONS

In this section we consider (generalized) boundary value problems of the form

$$B[u, \eta] = 0 \qquad \text{for all} \quad \eta \in H \tag{5.1}$$

with B as in (3.2). We use the notation H_1, β, H, K explained in Section 3.1.1 and require that the assumptions on a, g, g_0, g_1 made in that section be satisfied. In particular, a and g are assumed to satisfy the Caratheodory condition. Now let D denote the set of all $u \in H_1$ with $a[u] \in L_2(0, 1)$, $g[u] \in L_1(0, 1)$. The problem is to find a $u \in D \cap H$ such that (5.1) holds. (Nonhomogeneous boundary conditions can be handled by simple transformations.)

We assume that there exist *weak comparison functions* $\varphi, \psi \in D$ for the given problem, i.e., functions $\varphi, \psi \in D$ which satisfy

$$\begin{gathered}\varphi \leqslant \psi, \qquad B[\varphi, \eta] \leqslant 0 \leqslant B[\psi, \eta] \qquad \text{for all} \quad \eta \in K, \\ \varphi(0) \leqslant 0 \leqslant \psi(0) \quad \text{if} \quad 0 \in \beta, \qquad \varphi(1) \leqslant 0 \leqslant \psi(1) \quad \text{if} \quad 1 \in \beta.\end{gathered} \tag{5.2}$$

Then the set ω is defined as in (4.5).

We shall derive sufficient conditions such that the existence of the comparison functions implies the existence of a solution u^* with $\varphi \leqslant u^* \leqslant \psi$. In Section 5.1 Schauder's fixed-point theorem will be used, in Section 5.2 a theorem on monotone (monotone definite) operators will be applied. In both cases, the theorems are applied to a modified problem obtained by a truncation of u. Let $u^\# = \langle \varphi, u, \psi \rangle$ be defined as in (4.6); then

$$(u^\#)'(x) = \begin{cases} \varphi'(x) \\ u'(x) \\ \psi'(x) \end{cases} \quad \text{for a.a.} \quad x \in (0, 1) \quad \text{with} \quad \begin{cases} u(x) \leqslant \varphi(x) \\ \varphi(x) \leqslant u(x) \leqslant \psi(x) \\ \psi(x) \leqslant u(x) \end{cases}$$

and, consequently, $u, \varphi, \psi \in D \Rightarrow u^\# \in D$.

5.1 Application of Schauder's fixed-point theorem

Here we assume that B has the special form

$$B[u, \eta] = B_L[u, \eta] + \langle g[u], \eta \rangle_0 \qquad \text{with} \quad g[u](x) = g(x, u(x)),$$

where B_L is a bilinear form as in (3.3), $a(x) \geqslant a_0 > 0$ $(0 \leqslant x \leqslant 1)$ for some constant a_0 (and $g(x, y)$ satisfies the Caratheodory condition). We introduce a modified problem

$$B^{\#}[u, \eta] = 0 \qquad \text{for all} \quad \eta \in H \tag{5.3}$$

where

$$B^{\#}[u, \eta] = B_L^{\#}[u, \eta] + \langle g[u^{\#}] - c_1 u^{\#}, \eta \rangle_0,$$

$$B_L^{\#}[u, \eta] = B_L[u, \eta] + c_1 \langle u, \eta \rangle_0$$

and $c_1 \geqslant 0$ is a constant such that

$$B_L^{\#}[\eta, \eta] \geqslant c_0 \langle \eta, \eta \rangle_1 \qquad \text{for all} \quad \eta \in H \quad \text{and some} \quad c_0 > 0 \tag{5.4}$$

(cf. Lemma 3.3).

Lemma 5.1. *Each solution $u^* \in D \cap H$ of the modified problem is also a solution of the given problem* (5.1) *and $\varphi \leqslant u^* \leqslant \psi$.*

Proof. We have $B^{\#}[u^*, \eta] = 0 \leqslant B[\psi, \eta] = B^{\#}[\psi, \eta]$ for all $\eta \in K$ and therefore can apply Theorem 3.7 to $B^{\#}$, u^*, ψ in place of B, v, w. Since

$$B^{\#}[\psi + \eta, \eta] - B^{\#}[\psi, \eta] = B_L^{\#}[\eta, \eta] > 0 \tag{5.5}$$

for all η described in the theorem, we conclude that $u^* \leqslant \psi$. The inequality $\varphi \leqslant u^*$ is derived similarly. Thus $(u^*)^{\#} = u^*$ and hence $B[u^*, \eta] = B^{\#}[u^*, \eta] = 0$ for all $\eta \in H$. □

Theorem 5.2. *Suppose that there exist functions $\varphi, \psi \in D$ which satisfy* (5.2), *and that*

$$|g(x, y)| \leqslant k(x) \qquad \textit{for all} \quad (x, y) \in \omega \quad \textit{and some} \quad k \in L_2(0, 1).$$

Then the given boundary value problem (5.1) *has a solution $u^* \in D \cap H$ such that $\varphi \leqslant u^* \leqslant \psi$.*

Proof. According to the preceding lemma, it suffices to show that the modified problem has a solution. To prove the existence we shall define an operator $T: L_2 \to H \subset L_2$ such that the fixed-point equation $u = Tu$ is equivalent to the modified problem.

When we apply the lemma of Lax–Milgram (see Friedman [1969, p. 41], for example) to the bilinear form $B_L^{\#}$ on the Hilbert space H with the inner product $\langle\ ,\ \rangle_1$, we see that for each $f \in L_2(0, 1)$ there exists a unique $w \in H$ such that $B_L^{\#}[w, \eta] = \langle f, \eta \rangle_0$ for all $\eta \in H$. Thus, an operator $A: L_2(0, 1) \to H$ is defined by $Af = w$. It follows from (5.4) that $c_0 \|Af\|_1^2 \leqslant B_L^{\#}[Af, Af] = \langle f, Af \rangle_0 \leqslant \|f\|_0 \|Af\|_1$ and hence $\|Af\|_1 \leqslant c_0^{-1} \|f\|_0$, for all $f \in L_2(0, 1)$.

Since $g^{\#}(x, y) := g(x, \langle \varphi(x), y, \psi(x) \rangle)$ also satisfies the Caratheodory condition and $|g^{\#}(x, y)| \leqslant k(x)$ on $[0, 1] \times \mathbb{R}$, the Nemytsky operator Γ

defined by $\Gamma u(x) = g(x, u^{\#}(x))$ maps $L_2(0, 1)$ into $L_2(0, 1)$. Moreover, this operator is continuous and uniformly bounded, $\|\Gamma u\|_0 \leqslant \|k\|_0$ for all $u \in L_2(0, 1)$. (These statements follow from Theorem 19.1 of Vainberg [1964], for example.)

Because of these facts, $T = -A\Gamma$ is a continuous operator from L_2 into L_2, and $\|Tf\|_1 \leqslant c_0^{-1}\|k\|_0$ for all $f \in L_2$. Thus, T maps $L_2(0, 1)$ into a $\|\ \|_1$-bounded subset, which is relatively compact, according to Rellich's lemma (see Friedman [1969, p. 32], for example). Because of these properties, the fixed-point theorem of Schauder yields the existence of a fixed point $u^* = Tu^* \in H$. Finally, it is easily seen that u^* is a solution of the modified boundary value problem. □

5.2 Existence by the theory on monotone definite operators

Now we treat (5.1) under the condition that $g_0(y) \equiv 0 \equiv g_1(y)$. We consider again a modified problem (5.3), where now

$$B^{\#}[u, \eta] = \int_0^1 [a(x, u^{\#}(x), u'(x))\eta'(x) + g[u^{\#}](x)\eta(x)]\, dx + c\langle u - u^{\#}, \eta\rangle_0$$

with some constant $c > 0$. A solution of this problem is required to be a function $u^* \in H \cap D$ such that also $a(x, u^{\#}, u') \in L_2(0, 1)$.

Lemma 5.3. *Suppose that*

$$\langle a(x, \varphi, \varphi') - a(x, \varphi, \varphi' - \eta'), \eta'\rangle_0 \geqslant 0,$$

$$\langle a(x, \psi, \psi' + \eta') - a(x, \psi, \psi'), \eta'\rangle_0 \geqslant 0$$

for all $\eta \in H$ such that $a(x, \varphi, \varphi' - \eta')\eta' \in L_1(0, 1)$, $a(x, \psi, \psi' + \eta')\eta' \in L_1(0, 1)$.

Then each solution u^* of the modified problem is also a solution of the given problem (5.1) *and $\varphi \leqslant u^* \leqslant \psi$.*

Proof. The proof is analogous to that of Lemma 5.1. Instead of (5.5) we obtain $B^{\#}[\psi + \eta, \eta] - B^{\#}[\psi, \eta] = \langle a(x, \psi, \psi' + \eta') - a(x, \psi, \psi'), \eta'\rangle_0 + c\langle \eta, \eta\rangle_0 > 0$. □

Theorem 5.4. *For given functions $\varphi, \psi \in D$ which satisfy* (5.2), *let there exist constants $\gamma > 0$, $c_1 > 0$ and a function $k \in L_2(0, 1)$ such that for all $(x, y) \in \omega$, $p \in \mathbb{R}$, $p^* \in \mathbb{R}$,*

$$(a(x, y, p) - a(x, y, p^*))(p - p^*) > 0 \qquad \text{if} \quad p \neq p^*,$$

$$|a(x, y, p)| \leqslant k(x) + \gamma|p|, \qquad |g(x, y, p)| \leqslant k(x) + \gamma|p|, \tag{5.6}$$

$$a(x, y, p)p \geqslant c_1|p|^2. \tag{5.7}$$

Then the given boundary value problem (5.1) *has a solution* $u^* \in D \cap H$ *such that* $\varphi \leqslant u^* \leqslant \psi$.

Proof. The modified problem considered above depends on the constant c. According to Lemma 5.3 it suffices to prove that for some constant $c > 0$ the modified problem has a solution. Observe that, due to (5.6), $a(x, u^{\#}, u')$ and $g(x, u^{\#}, (u^{\#})')$ are functions in $L_2(0, 1)$ (compare Krasnosel'skii [1964b], for example).

The existence of such a solution follows from Theorem 2.8 in Chapter 2 of Lions [1969], if we can show that $B^{\#}$ is coercive, i.e.,

$$B^{\#}[\eta, \eta]/\|\eta\|_1 \to +\infty \qquad \text{for} \quad \|\eta\|_1 \to \infty. \tag{5.8}$$

All the remaining assumptions of that theorem are immediately seen to be satisfied. Using (5.6) and (5.7), we estimate

$$\int_0^1 a(x, \eta^{\#}, \eta')\eta' \, dx \geqslant c_1\|\eta'\|_0^2, \qquad c\langle \eta - \eta^{\#}, \eta\rangle_0 \geqslant c(\|\eta\|_0^2 - c_2\|\eta\|_0),$$

$$-\int_0^1 g(x, \eta^{\#}, (\eta^{\#})')\eta \, dx \leqslant \langle k_1, |\eta|\rangle_0 + \beta\langle |\eta|, |\eta'|\rangle_0,$$

$$\langle |\eta|, |\eta'|\rangle_0 \leqslant \delta^2\|\eta'\|_0^2 + \tfrac{1}{4}\delta^{-2}\|\eta\|_0^2$$

with $k_1 = k + \beta(|\varphi'| + |\psi'|)$, suitable $c_2 > 0$, and arbitrary $\delta > 0$. Choosing δ and c such that $\beta\delta^2 < c_1$ and $4\delta^2 c > \beta$ we can then verify (5.8). $\square$

6 ERROR ESTIMATES

The results of the preceding sections can be applied to estimate solutions of nonlinear problems, similarly as the inverse-positivity of linear operators was used in Section II.5.2. The procedures explained there are also helpful in treating nonlinear problems. Here we shall describe certain methods which are based on the theories of Sections 2 and 4. A general approach for constructing error bounds is discussed in Section 6.1. Sections 6.2 and 6.3 contain applications to initial value problems and boundary value problems. Analogous procedures have been applied to vector-valued differential equations. (See Examples V.2.4 and V.4.12.)

6.1 Construction of error bounds by linear approximation

We consider an equation $Mu = o$ with an operator M as in (2.1) or (2.15) (or a more general operator) and suppose that for certain functions φ, ψ the following statement holds.

The inequalities $M\varphi \leqslant o \leqslant M\psi$ *and* $\varphi \leqslant \psi$ *imply that the equation* $Mu = r$ *has a solution* u^* *satisfying* $\varphi \leqslant u^* \leqslant \psi$.

Sufficient conditions for this existence and inclusion statement can be found in Section 4. If, however, the existence of a solution has been established by other means, the preceding implication may also be verified using the theory of inverse-monotonicity in Section 2.

As for linear problems, one may try to compute functions φ, ψ of the form $\varphi = u_0 - \beta\bar{z}$, $\psi = u_0 + \alpha z$, where u_0 is an approximate solution with defect $d[u_0] = -Mu_0$, z and $\bar{z}$ denote suitably chosen positive functions, and α and β are constants to be calculated so that the required inequalities are satisfied. For such functions the above statement becomes a principle of error estimation analogous to that in Section II.5.2.1:

The inequalities $-\beta \leqslant \alpha$ *and*

$$M(u_0 - \beta\bar{z}) - Mu_0 \leqslant d[u_0] \leqslant M(u_0 + \alpha z) - Mu_0 \tag{6.1}$$

imply that $-\beta\bar{z} \leqslant u^* - u_0 \leqslant \alpha z$ *for at least one solution* u^*.

In order to solve these nonlinear inequalities for α and β, one may first approximate the occurring nonlinear functions of α and β by linear functions. In particular, if M has a derivative $M_0 = M'(u_0)$ (in some suitable sense), one may first compute constants $\bar{\alpha}$, $\bar{\beta}$ such that $-\bar{\beta} \leqslant \bar{\alpha}$ and

$$-\bar{\beta}M_0\bar{z} \leqslant d[u_0] \leqslant \bar{\alpha}M_0 z.$$

If these constants are small enough, constants α and β such that (6.1) holds, in general, can also be obtained, for example, by slightly enlarging $\bar{\alpha}$ and $\bar{\beta}$. Of course, if $M(u_0 + v) - Mu_0 - M_0 v$ is positive for $v \geqslant o$ and negative for $v \leqslant o$, then $\alpha = \bar{\alpha}$ and $\beta = \bar{\beta}$ may be chosen.

Moreover, the choice of the bound functions z and $\bar{z}$ is motivated by the nature of the linear operator M_0; that means these functions may be chosen as described in the examples in Section II.5.2.

In the following sections we shall discuss the construction of u_0 and the verification of the inequalities.

6.2 Initial value problems

The procedure described in Section II.5.2.2 can immediately be generalized to problems of the form (II.5.17) with a nonlinear function $f(x, y)$, provided f is sufficiently smooth. We shall not explain all details, but only point out some necessary changes.

For $f(x, y)$ a continuous function on $[0, 1] \times \mathbb{R}$ one proves instead of (II.5.18) the following basic statement (using the result of Problem 4.4):

If for a positive function z

$$|d[u_0](x)| \leqslant z'(x) + f(x, u_0(x) + z(x)) - f(x, u_0(x)) \qquad (0 < x \leqslant 1),$$

$$|r_0 - u_0(0)| \leqslant z(0),$$

then problem (II.5.17) *has a solution* u^* *with* $|u^*(x) - u_0(x)| \leqslant z(x)\,(0 \leqslant x \leqslant 1)$.

The construction of an approximation u_0 and a bound $\Gamma(x)$ for the defect $d[u_0](x)$ (on the entire interval [0, 1]) is carried out as for linear problems. Of course, now $f^{[5]}(x) = d^5/dx^5\, f(x, u_0(x))$ depends nonlinearly on $u_0(x), \ldots, u_0^{(v)}(x)$. (Here f is assumed to be five times continuously differentiable on each subinterval I_k.)

In order to determine the bound function z_k on I_k, first a linear inequality

$$\Gamma_k(x) \leqslant \bar{z}_k'(x) + \bar{c}_k \bar{z}_k(x) \qquad \text{with} \quad \bar{c}_k \leqslant f_y(x, u_0(x)) \quad \text{for} \quad x_{2k-2} \leqslant x \leqslant x_{2k}$$

is solved with the methods explained for linear problems. (For calculating the lower bound $\bar{c}_k$ of f_y one uses the known bounds of u_0, applying a type of interval arithmetic.) In a second step, an inequality of the same form is solved with $\bar{c}_k$ replaced by a constant $c_k \leqslant \bar{c}_k - \tilde{f}_k((1 + \varepsilon)\bar{z}_k(x))$ $(x_{2k-2} \leqslant x \leqslant x_{2k})$, where $\varepsilon > 0$ and $\tilde{f}_k$ is a certain monotone function such that $f(x, u_0 + y) - f(x, u_0) - f_y(x, u_0)y \geqslant -\tilde{f}_k(y)y$ for $x \in I_k$, $y \geqslant 0$. If the corresponding solution z_k satisfies $z_k(x) \leqslant (1 + \varepsilon)\bar{z}_k(x)$ for $x \in I_k$, then z_k is an error bound on I_k. As for linear problems, all estimations are carried out such that rounding errors are taken into account.

In the tables corresponding to the following examples we use the notation $\Delta_k = u_k - u^*(x_k)$, $\beta_k = z_k(x_k)$. Here the approximate values u_k are computed by a Runge–Kutta procedure of order 4 with $h = 0.1$.

Example 6.1 (see Table 8). The problem $u' = u^2 + 3 - (1 + x)^2$, $u(0) = 0$ has the exact solution $u^* = (1 + x) - (1 + x)^{-1}$. Approximate values were calculated for $0 \leqslant x_k \leqslant 2.4$. □

TABLE 8

Errors Δ and error bounds β for Example 6.1

k	x_k	Δ_k	β_k
4	0.4	$2.00E-07$	$1.56E-06$
8	0.8	$1.21E-06$	$6.34E-06$
12	1.2	$8.96E-06$	$2.92E-05$
16	1.6	$5.03E-05$	$1.68E-04$
20	2.0	$3.66E-04$	$1.32E-03$
24	2.4	$3.70E-03$	$1.47E-02$

Example 6.2 (see Table 9). The problem $u' = \cos\frac{1}{2}\pi u + \sin\frac{1}{2}\pi x, u(0) = 0$, which was solved for $0 \leqslant x_k \leqslant 5.6$, shows that the error bounds may decrease for increasing x. Here the approximate values u_k increase from $u_0 = 0$ to $u_{56} = 1.463\ldots$. □

TABLE 9

Error bounds β for Example 6.2

k	x_k	β_k
4	0.4	$1.308E-06$
8	0.8	$2.535E-06$
12	1.2	$2.857E-06$
16	1.6	$2.296E-06$
56	5.6	$1.101E-06$

6.3 Boundary value problems

Now consider a boundary value problem $Mu = o$ as in Section 4, with Dirichlet boundary conditions. According to Theorem 4.2 the existence and inclusion statement required in Section 6.1 holds here, for example, if

$$\begin{aligned} Mu(x) &= -u''(x) + f(x, u(x)) \qquad (0 < x < 1), \\ Mu(0) &= u(0) - r_0, \qquad Mu(1) = u(1) - r_1 \end{aligned} \tag{6.2}$$

and $f(x, y)$ is continuous on $[0, 1] \times \mathbb{R}$.

In order to obtain an approximate solution which we now call $\bar{u}$ instead of u_0, we combine Newton's iterative method with approximation methods for linear problems. Without specifying all necessary assumptions, the procedure can briefly be described as follows.

Starting from an initial approximation $u_0 \in R_0$, we calculate approximations u_ν in finite-dimensional subspaces R_ν such that

$$\rho_\nu := Mu_\nu + M'(u_{\nu-1})(u_\nu - u_{\nu-1})$$

becomes small in some suitable sense. In other words, in each iterative step the linear problem

$$M_\nu u = o \qquad \text{with} \quad M_\nu u = Mu + M'(u_{\nu-1})(u - u_{\nu-1})$$

is solved by an approximation procedure.

For solving these linear problems we applied Method P explained in Section II.5.2.3 as well as other methods. When Method P is used in each step, the whole procedure becomes a *collocation Newton method*. This method is essentially determined by choosing (1) the method for computing the initial approximation u_0, (2) the dimension m_ν of the spaces R_ν, and (3) the criterion for stopping the iteration at a number $\nu = N$, for which the final approximation $\bar{u} = u_N$ is obtained.

Finding an initial approximation u_0 is a possibly difficult problem common to most of the iterative procedures for nonlinear problems. Observe, however, that here the values $u_0(x)$ need only be known for the points x which are used in constructing u_1. In general, it is economical to choose dimensions m_ν increasing in a specific way with ν (see the Notes). The iteration may be stopped at a number N for which the difference of two subsequent approximations is "small enough."

We report here only about the following two rather simple examples, where M has the form (6.2).

Example 6.3 (see Table 10). The problem

$$-u''(x) + \tfrac{3}{2}u^2(x) = 0 \qquad (0 < x < 1), \qquad u(0) = 4, \qquad u(1) = 1$$

has the positive solution $u^*(x) = 4(1 + x)^{-2}$ and a second solution which also assumes negative values. We are here concerned with computing the positive solution. □

TABLE 10

Errors Δ and error bounds β for Example 6.3

m	N	Δ	β
7	4	$3.27E-05$	$1.47E-03$
11	5	$3.41E-08$	$3.63E-06$
15	5	$3.56E-11$	$6.84E-09$
19	5	$5.68E-14$	$9.98E-12$

Example 6.4 (see Table 11). The problem

$$-u''(x) + \kappa \sinh(\kappa u(x)) = 0, \qquad u(0) = 0, \quad u(1) = 1$$

has a unique solution u^* for each $\kappa > 0$. The larger κ is, the more difficult the numerical solution is. We consider here values up to 5. □

To both examples, we applied the collocation Newton method explained above with the initial approximation $u_0(x) = (1 - x)r_0 + xr_1$ satisfying the boundary conditions, using the same dimension $m_\nu = m$ for all steps of the iterative procedure. N denotes the number of iterations carried out to obtain $\bar{u} = u_N$.

Moreover, in both examples, $f(x, \bar{u}(x)) \geqslant 0$ and $f_y(x, y) \geqslant 0$ for $0 < x < 1$ and all relevant values of y. Therefore, since all approximations satisfy the boundary conditions, the function $z(x) = \alpha\frac{1}{2}x(1-x) \leqslant \frac{1}{8}\alpha$ with $\alpha = \|\delta[\bar{u}]\|_\infty$ and $\delta[\bar{u}](x) = \bar{u}''(x) - f(x, \bar{u}(x))$ is a bound for $|u^*(x) - \bar{u}(x)|$. In Tables 10 and 11, Δ and β are defined by $\Delta = \|u^* - \bar{u}\|_G$ and $\beta = 1.1\frac{1}{8}\bar{\alpha}$ with $\bar{\alpha} = \|\delta[\bar{u}]\|_G$ and $\| \;\|_G$ denoting a discrete maximum norm corresponding to 501 equidistant points (as explained in Section II.5.2.3). The error estimate $\|u^* - \bar{u}\|_\infty \leqslant \beta$ holds if $\|\delta[\bar{u}]\|_\infty \leqslant 1.1\|\delta[\bar{u}]\|_G$.

TABLE 11

Error bounds β for Example 6.4

κ	m	N	β
1	10	4	$7.70E-09$
1	13	4	$2.74E-11$
1	16	4	$9.60E-14$
2	20	4	$1.35E-09$
2	23	4	$4.59E-11$
2	26	4	$1.56E-12$
2	29	4	$6.40E-14$
3	41	5	$3.83E-12$
3	44	5	$4.69E-13$
4	63	6	$2.88E-12$
4	66	6	$6.88E-13$
5	87	7	$3.23E-11$
5	90	7	$7.51E-12$

NOTES

Section 2

For sufficiently smooth functions u the main results of Sections 2.1 and 2.2 can also be derived from abstract results such as Theorems IV.2.1a,b. (See Schröder [1963, 1966, 1968b, 1977b].) Very often, statements of the form $Mv \leqslant Mw \Rightarrow v \leqslant w$ for nonlinear operators are reduced to corresponding statements for linear operators by use of the mean value theorem (see, for example, Collatz [1952, 1960]; Protter and Weinberger [1967], Adams and Spreuer [1975], Redheffer [1977]). Problem 2.8 shows how the resulting statements can be obtained directly from the above nonlinear theory.

A series of results on nonlinear differential inequalities was proved by Redheffer without reducing them to linear inequalities (see, for instance, Redheffer [1962a,b] and also Satz 2 in Redheffer [1977]). The results of Redheffer [1962a] implicitly contain side conditions of type (2.18) for the case $\mathfrak{z}(\lambda, x) \equiv \lambda$, where $\mathfrak{z}'(\lambda, x) \equiv 0$. In connection with inverse-monotonicity a family of functions $\mathfrak{z}(\lambda, x)$ has already been used by McNabb [1961].

In the theory of second order differential inequalities many authors have considered functions u with reduced smoothness properties. The conditions S_r and S_l formulated above are not the weakest possible, but seem to be of sufficient generality for many applications. Lemmert [1977] considered smoothness properties as weak as those in Problem 2.3.

Most results of Sections 2.1 and 2.2 can be generalized rather easily to elliptic–parabolic partial differential operators by using essentially the same ideas of proof. Indeed, all papers mentioned above are mainly concerned with partial differential operators. We point out, however, that treating functions with discontinuous derivatives is more complicated for several variables (see Meyn and Werner [1979] and the literature mentioned therein). For further references on partial differential operators see Protter and Weinberger [1967].

Theorem 2.4 in Section 2.3.2 essentially is a special case of a more general result on quasilinear elliptic differential operators of divergence form proved by Douglas *et al.* [1971]. The proof of the general result is more involved. Theorems 2.3 and 2.5 could also be derived from Theorem IV.2.1b. Theorem 2.6 and its corollary constitute an example for the method of absorption described in Section IV.1. Similar results were obtained for partial differential equations by Redheffer [1962a]. Also compare the Phragmèn–Lindelöff principle in Protter and Weinberger [1967]. Problems on unbounded intervals (domains) may also be investigated by first transforming them to problems on compact intervals (domains) (see Collatz [1958]).

Concerning estimates for derivatives see Payne [1976] and Stakgold [1974].

Section 3

In the theory of Browder [1963], Minty [1962], Zarantonello [1960], and others, a monotone operator A in a Hilbert space is characterized by the property that $\langle Av - Aw, v - w\rangle > 0$ for $v \neq w$. On the other hand, in the theory of ordered spaces the notion of a monotone operator A is defined by $v \leqslant w \Rightarrow Av \leqslant Aw$. One may distinguish these two types of monotonicity by using the terms *monotone definite* (or d-*monotone*) and *order monotone* (or o-*monotone*), respectively. The term of a monotone definite operator generalizes the notion of a positive definite linear operator in a natural way. Since in this book d-monotone operators play only a marginal role, but o-monotone operators occur frequently, we use the terms monotone and o-monotone synonymously.

The results of Section 3 were proved by Schröder [1977b]. As already remarked therein, most of these results, in particular those in Sections 3.2, 3.3 can be carried over to weak partial differential inequalities by using the theory of Sobolev spaces. Results which generalize Corollary 3.7c in this way were formulated by Glashoff and Werner [1979]. These authors also prove a corresponding abstract result on L-*operators* in lattice-ordered inner product spaces. The basic idea for deriving such relations between monotone definite operators and inverse-monotone operators goes back to Stieltjes [1887], who investigated positive definite Z-matrices. This idea was applied to differential operators by Bellman [1957]. Similar tools are now frequently being used in the theory of differential equations and variational inequalities. For example, Stampacchia [1966] proved a maximum principle for solutions of weak linear differential inequalities, using a truncation operation as employed above. For earlier results on such maximum principles see Littman [1959, 1963].

Section 3.3 combines the ideas employed in Sections 2.2 and 3.2. As remarked above, these results can also be generalized to partial differential operators.

The proof of Theorem 3.10 relies on the monotonicity theorem in Section II.1. The main difficulty in applying this result is here to prove the pre-inverse-positivity of M, i.e., the fact that $u \geqslant o$ and $Mu \succ o$ imply $u \succ o$. This is done here in an elementary way. For nonlinear weak partial differential inequalities a corresponding property has been derived by Serrin [1970] by means of an inequality proved by Trudinger [1967]. Using this result of Serrin, one can generalize Theorem 3.10 (for the case $\beta = \{0, 1\}$) to certain nonlinear operators by employing an abstract theorem on nonlinear operators, such as Theorem IV.2a or b. Theorem 3.10 was proved in a different way by Clément and Peletier [1979] who made additional statements concerning the behavior of the functions at the boundary points.

Section 4

Nagumo [1937] derived a result like Theorem 4.11b for Dirichlet boundary conditions $u(0) = r_0$, $u(1) = r_1$ under the assumption (4.21), while investigating the continuation of solutions of second order differential equations. In case of constant functions $\varphi = -\psi$ a corresponding result is contained in a paper of Hartman [1960], who in the proof applied a fixed-point theorem to a modified problem (see also Hartman [1964]).

A series of authors then proved a result like Theorem 4.11b in various degrees of generality, under partly different assumptions. The proofs in the following papers are more or less closely related to our approach above: Akô [1965, 1967/70], Jackson and Schrader [1967], Schrader [1967], Schmitt [1970], Erbe [1970]. These authors used various types of modified problems and applied fixed-point theorems for the existence proof; the Nagumo conditions were generally exploited to obtain a priori bounds for u', as described in Section 4.5.1. Schrader [1967] introduced one-sided Nagumo conditions. Schrader [1968] also treated a problem on the unbounded interval $[0, \infty)$. The above approach to the theory was given by Schröder [1972a, 1975b].

The papers of Akô mentioned above differ from those of the other authors in that this author applied the theory of super- and subfunctions. In a different way this theory was used by Fountain and Jackson [1962], Bebernes [1963], Jackson [1968], and others (see the references in the last paper.)

Further related results are due to Peixoto [1949], Epheser [1955], Schmitt [1968], Knobloch [1969], Jackson and Klaasen [1971], Bebernes and Wilhelmsen [1971], Gaines [1972], and Heidel [1974], among others.

Many of these investigations were preceded by the work of Knobloch [1963] on *periodic* solutions of second order boundary value problems. This author derived the existence of such solutions from the existence of corresponding comparison functions, by applying a method of Cesari. He introduced functions $\Phi(x, y)$, $\Psi(x, y)$ with properties very similar to those in Section 4.3 and, in particular, employed the relation between Nagumo conditions and such functions. Schmitt [1968] used such bound functions $\Phi(x, y)$, $\Psi(x, y)$ in case of Dirichlet boundary conditions, relying on a theorem of Jackson and Schrader [1967]. In particular, this author formulated the result of Problem 4.6 for Dirichlet boundary conditions.

For further work on periodic boundary conditions see Knobloch and Schmitt [1977] and the references therein. A unified treatment of periodic boundary conditions similar to the theory presented in Section 4 was given by Stüben [1975].

J. Werner [1969] showed that for periodic boundary conditions no comparison functions φ, ψ exist (except $\varphi = \psi = u^*$), if the differential equation has the form $-u'' - f(u)u' - g(x, u) = 0$, where $g(x, y)$ is monotone (isotone) with respect to y and the relations $g(x, y) = g(x, \bar{y})$, $y \leqslant \bar{y}$ together imply $y = \bar{y}$. Nevertheless, this author derived existence and inclusion statements for such cases by transforming the differential equation into a system of first order differential equations and introducing order relations different from the natural order (see also Werner [1970]).

Newer developments of the theory of comparison functions are contained in papers on vector-valued differential operators. (See the Notes to Sections V.3 and V.6.) In these papers the degree of mapping method, which was explained in Section 4.5.2, is often used.

In Section 4.5.3 a monotone operator T was constructed by the simple "trick" of adding cu on both sides of the given equation $\mathbf{L}u = \mathbf{N}u$ and then inverting $\mathbf{L} + c$. This method has been employed by many authors to problems of the type considered and similar problems (see, for instance, Courant and Hilbert [1962]). The method uses the fact that $\mathbf{L} + c$ is inverse-positive for all sufficiently large constants (or functions) c. This shows that the method can be carried over to more general equations $\mathbf{L}u = \mathbf{N}u$ where $\mathbf{L}$ is a Z-operator and $\mathbf{N}$ satisfies an estimate (4.34). (Obviously, the method can be generalized further by transforming the equation into $\mathbf{L}u + cBu = \mathbf{N}u + cBu$ where $(\mathbf{L}, B)$ is a Z-pair of operators.) While in Theorem 4.16 the function $f(x, u, u')$ must not depend on u', Chandra and Davis [1974] proved analogous results without requiring this restriction. This work was generalized by Bernfeld and Chandra [1977]. For corresponding results on partial differential equations see Amann and Crandall [1978]. Bandle and Marcus [1977] used a quite different approach to prove the existence of a solution by iterative procedures; these authors exploited the fact that a certain nonlinear comparison problem has a solution.

The theory of comparison functions has many interesting applications. In particular, applications of iterative procedures such as those in Section 4.5.3 are so numerous that we cannot give a survey here, but only cite some papers in which further references may be found. For example, these methods have been applied to prove the existence of positive solutions of nonlinear boundary value problems and, in particular, nonlinear eigenvalue problems (see, for example, Cohen [1967], Keller and Cohen [1967], Keller [1969a,b], Sattinger [1973], and the survey given by Amann [1976]). The example in Problem 4.11 is taken from Küpper [1979], who treats more general classes of "singular" bifurcation problems. For the proof of inclusion and existence theorems and their applications see also Voss [1977] and the references in this paper.

The singular perturbation problem treated in Examples 2.3, 2.5, and 4.1 and certain generalizations were investigated by Cohen and Laetsch [1970], Cohen [1971, 1972], Keller [1972], and Sattinger [1972]. A systematical theory of singular perturbation problems based on results such as those in Sections 4.1 and 4.3 has been developed in a series of papers by Howes. For references see Howes [1978]; Example 4.5 is taken from that paper. A first result of this type is due to Briš [1954].

Iterative procedures are also of interest for numerical applications. (See, for instance, Collatz [1964], Parter [1965], Greenspan and Parter [1965], Shampine [1966, 1968], and Bohl [1974].) Adams and Scheu [1975], Spreuer *et al.* [1975], and Adams and Ames [1979] developed computer algorithms for the construction of rigorous numerical upper and lower bounds for solutions of various problems.

References to the work of Čapligin and others can be found in Čapligin [1950]. Many applications of various types of iterative procedures with monotone and antitone sequences are described by Ames and Ginsberg [1975]. There is an immense number of further similar procedures which we cannot describe here (for instance, methods based on quasilinearization investigated by Bellman and Kalaba [1965)]. Also, the subject is related to the theory of interval analysis (see Moore [1975], Nickel [1976]).

Section 5

For Dirichlet boundary conditions Theorem 5.4 is contained in a paper of Deuel and Hess [1975]. These authors prove analogous results for weak solutions $u \in H_{1,q}(\Omega)$ with $1 < q < \infty$ of second order quasilinear partial differential equations on a bounded domain Ω. These results, when applied to the case considered here, require "polynomial" growth restrictions $|a(x, y, p)| \leqslant k(x) + \beta|p|^{q-1}$ for $\varphi(x) \leqslant y \leqslant \psi(x)$ and an analogous estimate for g instead of inequalities (5.6). (Actually, the authors assume $|a(x, y, p)| \leqslant k(x) + \beta(|y|^{q-1} + |p|^{q-1})$ for all $y \in \mathbb{R}$, which seems unnecessary.) On the other hand, $a(x, y, p) \cdot p \geqslant c_1|p|^q$ is assumed instead of (5.7), so that in the important case $a(x, y, p) = a(x)p$, for instance, $q = 2$ is the "optimal" choice. The authors remark that Neumann boundary conditions can be treated in an analogous way. Our proof above is essentially that of Deuel and Hess, except that we apply a monotonicity result (Theorem 3.7) instead of proving $\varphi \leqslant u^* \leqslant \psi$ directly. In particular, Deuel and Hess also use Theorem 2.8 in Chapter 2 of Lions [1969].

Hess published a series of further papers on existence and inclusion statements for differential equations using the method of modification. Here, the papers described in the following are of particular interest. Hess [1976] proved a result as the one above for semilinear equations in $H_{1,2}$ using also Lions' theorem mentioned above. For ordinary differential equations this result essentially is Theorem 5.2 above, which we derived by applying Schauder's fixed-point theorem. Quasilinear equations on unbounded domains were treated by Hess [1975, 1977]. Furthermore, Hess [1978] showed that the linear growth restriction on $g(\cdot, \cdot, p)$ in (5.6) may be replaced by $|g(x, y, p)| \leqslant k(x) + \gamma|p|^{2-\varepsilon}$, where $\varepsilon > 0$ is arbitrarily small. In the latter paper results on variational inequalities were used to prove the existence of a solution of the modified problem.

Section 6

Concerning this section see also the Notes to Section II.5. The numerical results in Section 6.3 were calculated by G. Hübner on the CDC/CYBER 76 of the Computing Center at the University of Cologne. Witsch [1978] developed a convergence theory for projective Newton methods and, in particular, for the collocation Newton method in Section 6.3. He provided criteria for the choice of the number m_ν of parameters to be used in the νth iteration step, such that the quadratic convergence of the Newton method is preserved, but the computational work is minimal.

Chapter IV

An estimation theory for linear and nonlinear operators, range–domain implications

In this chapter the concept of an inverse-monotone operator will be generalized by considering statements of the form

$$Mv \in C \quad \Rightarrow \quad v \in K,$$

where C and K denote certain sets which need not be cones or transposed cones. Such implications allow one to obtain estimates for an element v in cases in which the operator M is not inverse-monotone, or the inverse-monotonicity of M cannot be proved. An implication of the above form will be called a *range–domain implication* (RD-*implication*). We may also speak of an *input–output statement*.

To prove a range–domain implication one may apply quite different methods, depending on the type of the operator. Many of these proofs, however, have certain arguments in common which we present as a *continuity principle* in Section 1.2. The abstract results in this section are very general; they do not immediately yield useful statements in more special cases. However, the continuity principle can serve as a tool to derive such more specific statements by applying simultaneously other tools, which depend on the problem considered. When a concrete problem is to be investigated, one may either incorporate it into the abstract theory, or simply use the general idea of the abstract proof without defining the abstract terms.

Section 2 contains a theory of inverse-monotone operators M, which are characterized by the property $Mv \leqslant Mw \Rightarrow v \leqslant w$. In particular, this theory can be used to obtain two-sided bounds for an unknown function v by use of the implication $M\varphi \leqslant Mv \leqslant M\psi \Rightarrow \varphi \leqslant v \leqslant \psi$. In Section 3 a two-sided estimate is derived from a certain set of inequalities $\Phi \leqslant Mv \leqslant \Psi$. Section 4 treats range–domain implications mainly for linear operators. This theory generalizes parts of the theory on inverse-positive operators in Chapter II. For example, inclusion statements for non-inverse-positive differential operators can be derived (Section 4.2).

The continuity principle will, furthermore, be applied in Sections V.2 and V.4.

1 SUFFICIENT CONDITIONS, GENERAL THEORY

1.1 Notation and introductory remarks

Suppose that M denotes an operator which maps a set D into a set S, and that $K \subset D$ and $C \subset S$ are given subsets. We ask for sufficient conditions such that

$$Mv \in C \quad \Rightarrow \quad v \in K \tag{1.1}$$

holds. In general, v will denote a fixed element in D. If, however, the above RD-implication holds for all $v \in D$, then M is said to be *(C, K)-inverse on D*, or in brief, *(C, K)-inverse*, if the choice of D is clear. Obviously, *if M is invertible and (C, K)-inverse on D, then* (1.1) *is equivalent to $M^{-1}C \subset K$.*

When property (1.1) is applied to a fixed element v, this element is in general unknown, while "sufficient" information on Mv is available. In this case, the set C is characterized by known properties of Mv, while K describes certain desired properties of v. In general, the relation $v \in K$ will be equivalent to certain estimates of v. Property (1.1) then allows one to estimate an element in the inverse set $\{u \in D : Mu \in C\}$ without knowing the inverse operator M^{-1} explicitly or even knowing whether M has an inverse.

When we want to prove (1.1) for a *fixed* element $v \in D$ without proving this property for *all* elements in D, we have to exploit some a priori knowledge on v. That means we have to prove a statement of the form

$$v \in K^{(0)}, \quad Mv \in C \quad \Rightarrow \quad v \in K, \tag{1.2}$$

where $v \in K^{(0)}$ describes the known properties of v. In particular, the relation $v \in K^{(0)}$ may constitute a "rough" a priori estimate, which is improved to a "better" estimate $u \in K$ by applying (1.2). (Sometimes the improvement may be continued by using iterative statements of the form: $u \in K^{(\nu)}$, $Mu \in C \Rightarrow u \in K^{(\nu+1)}$.)

The range–domain implication (1.1) can be used to obtain error estimates. For example, if u^* is a solution and u_0 an approximate solution of a given equation $Mu = r$, then this equation can be rewritten as an equation $\tilde{M}v = \tilde{r}$ for $v = u^* - u_0$. If $\tilde{r}$ belongs to C and (1.1) holds for $\tilde{M}$ instead of M, one obtains the error estimate $u^* - u_0 \in K$. Here again, rough a priori estimates of u^* may have to be known in order to ensure that v has the required properties.

Moreover, this theory may yield the uniqueness of a solution (or the statement that at most one solution with certain prescribed properties exists). One transforms the given problem so that v is the difference of two solutions and tries to prove a suitable statement (1.1) with $K = \{o\}$.

Finally, the theory can often be used as a tool in proving the existence of a solution of an equation $Mu = r$ by the *modification method* (see Sections V.3 and V.5). In this method one defines a modified operator $M^\#$ such that the existence of a solution u^* of the modified problem $M^\# u = r$ can be proved and $Mu = M^\# u$ for $u \in K$. Then $Mu = r$ follows, if (1.1) holds for u^*, $M^\#$ in place of v, M and if $r \in C$.

Problem

1.1 Suppose that D is a linear space, M is a linear operator, K does not contain a line, and $MD \cap C \neq \emptyset$. Prove that M is invertible, if M is (C, K)-inverse.

1.2 A continuity principle

This section provides sufficient conditions of a very general form such that (1.1) holds. We shall use the notation of the preceding section and assume, in particular, that v denotes a fixed element in D. Moreover, let R be a given set which contains D. (In most of the applications below, R and S denote linear spaces, while D need not be linear.)

The results of this section will be formulated as theorems. Actually, what we describe here is better considered as a principle or device for obtaining sufficient conditions in less general cases. The sufficient conditions will rely on a continuity argument which involves certain families of sets.

Suppose that, for each λ which belongs to a given finite or infinite interval $I = [0, \gamma)$, subsets K_λ and $\mathring{K}_\lambda$ are given which satisfy $K = K_0$, $\mathring{K}_\lambda \subset K_\lambda \subset R$. In other words, a family $\mathscr{K} = \{(K_\lambda, \mathring{K}_\lambda) : 0 \leqslant \lambda < \gamma\}$ of pairs of sets with the above properties is given. For convenience, write $\mathring{K} = \mathring{K}_0$, $\Gamma_\lambda = K_\lambda \sim \mathring{K}_\lambda$. (In certain applications $\mathring{K}_\lambda$ will be the interior of K_λ and Γ_λ the boundary.)

We shall say that the family $\mathscr{K}$ is *admissible at u*, with u being a fixed element in R, if the following two conditions are satisfied.

(a) *If* $u \in K_\mu$ *for some* $\mu \in I$, *then there exists a minimal* $\lambda \in I$ *with* $u \in K_\lambda$.
(b) *If* $u \in \mathring{K}_\lambda$ *for some* $\lambda \in (0, \gamma)$, *then* $u \in K_\mu$ *for some* $\mu \in (0, \lambda)$.

If these conditions hold for all $u \in R$, then $\mathscr{K}$ is simply called *admissible* or *admissible on R*.

Theorem 1.1. (i) *Suppose that the family* $\mathscr{K}$ *is admissible at v and that* $v \in K_\lambda$ *for some* $\lambda \in I$.

(ii) *For each* $\lambda \in (0, \gamma)$ *with* $v \in \Gamma_\lambda$, *let there exist a set* $\Omega' \subset R$ *such that* (I) $v \notin \Omega'$, (II) $Mv \in C \Rightarrow v \in \Omega'$.

Then $Mv \in C \Rightarrow v \in K$.

Proof. According to property (i) there exists a minimal $\lambda \geqslant 0$ with $v \in K_\lambda$. We want to prove $\lambda = 0$. Suppose therefore that $\lambda > 0$. In this case v belongs to Γ_λ, since otherwise property (b) would imply $v \in K_\mu$ for some $\mu < \lambda$. Now, if Ω' denotes the set which corresponds to the considered value λ, condition II yields $v \in \Omega'$, which contradicts condition I. □

In the preceding theorem no indication is given on how to choose the occurring sets. In the subsequent sections, however, a series of different and more concrete results will be derived by choosing those sets properly.

First, we shall formulate some modifications of Theorem 1.1. In particular, we remark that in most of the following applications a set Ω' of the form $\Omega' = M^{-1}\Omega = \{u \in R : Mu = U \text{ for some } U \in \Omega\}$ is used. For this case, we obtain

Corollary 1.1a. *Assumption* (ii) *of Theorem* 1.1 *may be replaced by*

(ii′) *For each* $\lambda \in (0, \gamma)$ *with* $v \in \Gamma_\lambda$, *let there exist a set* $\Omega \subset S$ *such that* (I) $Mv \notin \Omega$, (II) $Mv \in C \Rightarrow Mv \in \Omega$.

By strengthening the assumptions a stronger statement on v can be made.

Corollary 1.1b. *If in assumption* (ii) *or* (ii′) *the occurring interval* $(0, \gamma)$ *is replaced by* $[0, \gamma)$, *then statement* (1.1) *can be replaced by*

$$Mv \in C \quad \Rightarrow \quad v \in \mathring{K}. \tag{1.3}$$

One sometimes obtains weaker conditions in a concrete case, if Theorem 1.1 is not immediately applied to the given problem, but the following result is used.

Corollary 1.1c. *Suppose that* $K_m \subset R$ *and* $C_m \subset S$ $(m = 1, 2, \ldots)$ *denote sets such that for each m*: (i) $C \subset C_m$; (ii) $Mv \in C_m \Rightarrow v \in K_m$; *and moreover* (iii) $v \in K$, *if* $v \in K_m$ *for all m*.

Then $Mv \in C \Rightarrow v \in K$.

When this corollary is applied, the sets K_m in general are defined such that $\bigcap K_m \subset K$. To prove $v \in K$, one verifies that $v \notin \tilde{K}_m = R \sim K_m$ for each m. Then, since these sets $\tilde{K}_m$ *absorb* $\tilde{K} = R \sim K$ in the sense that $\bigcup \tilde{K}_m \supset \tilde{K}$, one obtains $v \notin \tilde{K}$, i.e., $v \in K$. A method of this type will be called a *method of absorption.* In particular, we shall speak of the method of absorption, if the corollary is applied.

There are a series of possibilities to modify and generalize the results. For example, one may require that $v \in \mathring{K}_\mu$ for some $\mu \in I$ and then consider $\lambda = \inf\{\mu \in I : v \in \mathring{K}_\mu\}$. In this case, the general assumptions are to be modified such that $v \in \Gamma_\lambda$ if $\lambda > 0$.

Moreover, one may allow λ to be a more general parameter, such as a pair $\lambda = (\lambda_1, \lambda_2)$ of real parameters. For example, suppose that the method of absorption is applied and $Mv \in C_m \Rightarrow v \in K_m$ is derived from Theorem 1.1. Then for each m, families of sets $K_{m,\lambda}$ $(0 \leqslant \lambda < \gamma_m)$ are needed. Obviously, all these sets together constitute a family with a parameter $(m, \lambda) \in \mathbb{R}^2$.

2 INVERSE-MONOTONE OPERATORS

As a first application of the continuity principle described in the previous section, we consider now implications of the form

$$Mv \leqslant Mw \quad \Rightarrow \quad v \leqslant w, \tag{2.1}$$

in ordered linear spaces, and similar statements. Such implications as well as the concept of an inverse-monotone operator were already discussed in Section III.1.

For fixed elements v and w, (2.1) can obviously be written in the form (1.1), so that the methods in Section 1 may be applied. Most of the following results are formulated with the situation in mind, that the element v is unknown, while the element w may be known. Different results can be obtained by first transforming (2.1) and then applying the theorems derived. (See the discussion in Section III.1.)

Section 2.1 provides an abstract theory where the sets $K_\lambda, \mathring{K}_\lambda$ occurring in Section 1.2 are determined by certain elements $\mathfrak{z}(\lambda)$. (The most important result, Theorem 2.1b, is also proved directly, without defining these sets.) The theory is applied to the finite-dimensional case in Section 2.2 and to vector-valued functional differential operators of the second order in Section 2.3. The latter section yields also results for differential operators of the first order related to initial value problems. However, for such operators a different application of the general continuity principle in Section 1 often leads to weaker assumptions. This approach is described in Section 2.4. Here, the sets $K_\lambda, \mathring{K}_\lambda$ are determined by intervals such as $(0, 1 - \lambda]$ or by families $\mathfrak{z}(\lambda)$ and intervals depending on λ.

2.1 Abstract results using majorizing elements

Suppose that M denotes an operator which maps a subset D of an ordered linear space $(R, \leqslant)$ into an ordered space $(S, \leqslant)$. Assume, in addition, that for some pairs u_1, u_2 of elements in R a relation $u_1 \prec u_2$ is defined such that $u_1 \prec u_2 \Rightarrow u_1 < u_2$. Similarly, suppose that for some pairs U_1, U_2 of elements in S a relation $U_1 \prec U_2$ is defined such that the inequalities $U_1 \leqslant U_2$, $U_2 \prec U_3$ imply $U_1 \prec U_3$. (For convenience, we use the same symbols $\leqslant$ and $\prec$ in R and S, since no misunderstanding seems likely.) In most applications the relation $\prec$ in R (S) induces a strong order relation $\preccurlyeq$ in R (S). However, only the properties stated above are needed. The relation $\prec$ in S will only be used in some of the following results.

We shall derive sufficient conditions such that $Mv \leqslant Mw \Rightarrow v \leqslant w$, or $Mv \prec Mw \Rightarrow v \prec w$. The main theorems are contained in Section 2.1.1. Section 2.1.2 treats a special class of abstract operators for which some important assumptions for inverse-monotonicity can be verified rather easily. In Section 2.1.3 this class is related to *pre-inverse-monotone* operators and *quasi-antitone* operators. Section 2.1.4 provides a concrete example for an application of the theory in Section 2.1.1.

2.1.1 The main theorems

The assumptions of this section involve a family $\mathscr{Z} = \{\mathfrak{z}(\lambda) : \lambda \in \mathrm{I} = [0, \gamma)\} \subset R$ with $\mathfrak{z}(0) = o$, $0 < \gamma \leqslant \infty$. Such a family is called *admissible at* $u \in R$, if the following two conditions hold.

(a) *If* $u \leqslant \mathfrak{z}(\mu)$ *for some* $\mu \in I$, *then a minimal* $\lambda \in I$ *with* $u \leqslant \mathfrak{z}(\lambda)$ *exists.*
(b) *If* $u \prec \mathfrak{z}(\lambda)$ *for some* $\lambda \in (0, \gamma)$, *then* $u \leqslant \mathfrak{z}(\mu)$ *for some* $\mu \in (0, \lambda)$.

If $\mathscr{Z}$ is admissible at each $u \in R$, then $\mathscr{Z}$ is simply called an *admissible* family. Moreover, the family $\mathscr{Z}$ is said to *majorize an element* $u \in R$ provided $u \leqslant \mathfrak{z}(\mu)$ for some value $\mu \in I$. If each $u \in R$ is majorized, then $\mathscr{Z}$ is called *a majorizing family*. Of course, all these notions depend on the given relations $\leqslant$ and $\prec$.

Example 2.1. Let the space $(R, \leqslant)$ be Archimedian and $o \prec u$ if and only if $o \prec u$ (i.e., u is strictly positive). Then the elements

$$\mathfrak{z}(\lambda) = \lambda z_0 \qquad \text{for} \quad 0 \leqslant \lambda \leqslant \varepsilon,$$
$$\mathfrak{z}(\lambda) = \varepsilon z_0 + (\lambda - \varepsilon) z \qquad \text{for} \quad \varepsilon < \lambda < \infty$$

constitute an admissible majorizing family if $z_0 \geqslant o$, $z \succ o$, and $\varepsilon \geqslant 0$. □

Example 2.2. Let $(R, \leqslant)$ be an Archimedian space and $p > o$ a fixed element in R. Then define $u \succ o$ to mean that $u \geq \alpha p$ for some $\alpha > 0$. For any $z \succ o$ in R such that $z \leqslant \beta p$ with some $\beta > 0$, the family $\mathscr{Z} = \{\lambda z : 0 \leqslant$

$\lambda < \infty\}$ is admissible. Moreover, if an element $u \in R$ satisfies $u \geqslant -\gamma p$ for some $\gamma \in \mathbb{R}$, then $\mathscr{Z}$ majorizes $-u$. (In particular, one may choose $z = p$.) □

For the following statements we make the *general assumption* that v *and* w *are fixed elements in* D, that $\mathscr{Z} = \{\mathfrak{z}(\lambda): 0 \leqslant \lambda < \gamma\}$ *is an admissible family which majorizes the element* $v - w$, *and that* $w + \mathfrak{z}(\lambda) \in D$ *for* $\lambda \in [0, \gamma)$.

Theorem 2.1a. *Suppose that for each* $\lambda \in (0, \gamma)$

$$\text{I.}\quad \left.\begin{matrix} v \leqslant w + \mathfrak{z}(\lambda) \\ v \nless w + \mathfrak{z}(\lambda) \end{matrix}\right\} \Rightarrow \quad Mv \nless M(w + \mathfrak{z}(\lambda)),$$

$$\text{II.}\quad Mw \prec M(w + \mathfrak{z}(\lambda)).$$

Then $Mv \leqslant Mw \Rightarrow v \leqslant w$.

Theorem 2.1b. *Suppose that for each* $\lambda \in (0, \gamma)$ *with*

$$v \leqslant w + \mathfrak{z}(\lambda) \qquad \text{and} \qquad v \nless w + \mathfrak{z}(\lambda) \tag{2.2}$$

there exists a monotone functional $\not\!\ell$ *on* S *such that*

$$\text{(I)}\quad \not\!\ell Mv \geqslant \not\!\ell M(w + \mathfrak{z}(\lambda)), \qquad \text{(II)}\quad \not\!\ell Mw < \not\!\ell M(w + \mathfrak{z}(\lambda)). \tag{2.3}$$

Then $Mv \leqslant Mw \Rightarrow v \leqslant w$.

Theorem 2.1c. *Suppose that for each* $\lambda \in I$ *assumption* (I) *of Theorem* 2.1a *is satisfied and* (II) $Mw \leqslant M(w + \mathfrak{z}(\lambda))$.
Then $Mv \prec Mw \Rightarrow v \prec w$. (2.4)

Theorem 2.1d. *Suppose that for each* $\lambda \in I$ *satisfying* (2.2) *there exists a monotone functional* $\not\!\ell$ *on* S *such that* $\not\!\ell U < \not\!\ell V$ *for* $U \prec V$ *and*

$$\text{(I)}\quad \not\!\ell Mv \geqslant \not\!\ell M(w + \mathfrak{z}(\lambda)), \qquad \text{(II)}\quad \not\!\ell Mw \leqslant \not\!\ell M(w + \mathfrak{z}(\lambda)).$$

Then $Mv \prec Mw \Rightarrow v \prec w$.

We are going to derive these four theorems from Corollaries 1.1a,b. However we shall also provide a direct proof of Theorem 2.1b, since this is the one most often applied in the following sections.

Proof of the preceding theorems. Define

$$K_\lambda = \{u \in R : u \leqslant w + \mathfrak{z}(\lambda)\}, \qquad \mathring{K}_\lambda = \{u \in R : u \prec w + \mathfrak{z}(\lambda)\}$$

for $\lambda \in I = [0, \gamma)$, $K = K_0$, $\mathring{K} = \mathring{K}_0$. Because of the above general assumption, condition (i) of Theorem 1.1 is satisfied for the family of all pairs $(K_\lambda, \mathring{K}_\lambda)$ and the element v considered. Obviously, $v \in \Gamma_\lambda$ if and only if (2.2) holds. Define moreover,

$$C = \{U \in S : U \leqslant Mw\} \qquad \text{for} \quad a \text{ and } b,$$
$$C = \{U \in S : U \prec Mw\} \qquad \text{for} \quad c \text{ and } d.$$

(Here, a, b, ... refer to the respective theorem.) Then apply Corollary 1.1a to prove a and b, and apply Corollary 1.1b to prove c and d. Obviously, the statements to be proved can be written as (1.1) and (1.3), respectively. The assumptions of the previously mentioned corollaries are satisfied for

$$\Omega = \{U \in S : U \prec M(w + \mathfrak{z}(\lambda))\} \qquad \text{for} \quad a, c;$$

$$\Omega = \{U \in S : \mathscr{f}U < \mathscr{f}M(w + \mathfrak{z}(\lambda))\} \qquad \text{for} \quad b, d.$$

For example, in case of Theorem 2.1a, the relation $Mv \in \Omega$ follows from $Mv \leqslant Mw \prec M(w + \mathfrak{z}(\lambda))$. □

A separate *proof of Theorem* 2.1b. Due to the general assumptions and (a), there exists a minimal $\lambda \in [0, \gamma)$ such that $v \leqslant w + \mathfrak{z}(\lambda)$. For $\lambda = 0$ the statement $v \leqslant w$ is verified. For $\lambda > 0$ we have $v \nless w + \mathfrak{z}(\lambda)$ because of (b). Thus, $Mv \leqslant Mw$ and II together imply $\mathscr{f}Mv \leqslant \mathscr{f}Mw < \mathscr{f}M(w + \mathfrak{z}(\lambda))$, which contradicts I. □

The above assumptions may be modified. In particular, one may apply the preceding theorems to a transformed problem as in (III.1.3), provided S is an ordered linear space. For example, if the operator $\tilde{M}_2$ in (III.1.4) is used, one obtains $M(v - \mathfrak{z}(\lambda)) \prec Mv$ in place of condition II of Theorem 2.1a.

Also, one may formulate sufficient conditions for the above assumptions using functionals, similarly as in Theorem II.1.7, for example. We describe this for Theorem 2.1b. Assume that $\mathscr{L}$ denotes a set of functionals $\ell \neq o$ on R such that for each $u \in R$ (and the fixed element v considered)

$$v \leqslant u, \quad v \nless u \quad \Rightarrow \quad \ell v = \ell u \quad \text{for some} \quad \ell \in \mathscr{L}. \tag{2.5}$$

Corollary 2.1b′. *Suppose that for each $\ell \in \mathscr{L}$ there exists a monotone functional $\mathscr{f} = \mathscr{f}_\ell$ on S such that for each $\lambda \in (0, \gamma)$ the following conditions hold:*

I. *For $u = w + \mathfrak{z}(\lambda)$,*

$$v \leqslant u, \quad \ell v = \ell u \quad \Rightarrow \quad \mathscr{f}Mv \geqslant \mathscr{f}Mu. \tag{2.6}$$

II. $\mathscr{f}Mw < \mathscr{f}M(w + \mathfrak{z}(\lambda))$, *whenever* $v \leqslant w + \mathfrak{z}(\lambda)$, $\ell v = \ell(w + \mathfrak{z}(\lambda))$.

Then $Mv \leqslant Mw \Rightarrow v \leqslant w$.

Observe that the relations $v \leqslant u, \ell v = \ell u$ (for $u = w + \mathfrak{z}(\lambda)$) occur in both assumptions I and II; they play, however, a quite different role in these assumptions. In I these relations are *the* essential tool for proving $\mathscr{f}Mv \geqslant \mathscr{f}Mu$; in II they are merely *side conditions* which weaken the inequalities required for $\mathfrak{z}(\lambda)$.

For the case $(R, \leqslant) \subset (S, \leqslant)$, $\mathscr{L}$ may denote a set of monotone functionals $\neq o$ on S such that (2.5) holds for all $u \in R$, and one may then choose $\mathscr{f}_\ell = \ell$.

2.1.2 A special class of abstract operators, order-quasilinear operators

The preceding theorems and also Corollary 2.1b′ are still of a very general nature. The question arises for which type of operators and in which way can the assumptions be verified. For Corollary 2.1b′ we shall now give a (partial) answer to this question by describing in abstract terms a class of operators to which this corollary can be applied in a specific way. (For certain more concrete operators we shall derive sufficient conditions in Sections 2.2–2.4. A reader who is only interested in these operators may immediately proceed to these sections and skip the following discussion.)

For simplicity, we consider here only the case in which $(R, \leqslant) \subset (S, \leqslant)$ and $D = R$. Suppose that the following assumptions are satisfied, for fixed $v, w \in R$.

(i) There exists a set $\mathscr{L}$ of monotone functionals $\ell \neq o$ on R such that (2.5) holds for each $u \in R$.

(ii) For each $\ell \in \mathscr{L}$ and all $u \in R$,

$$v \leqslant u, \quad \ell v = \ell u \quad \Rightarrow \quad P_\ell v = P_\ell u, \quad Q_\ell v \leqslant Q_\ell u, \tag{2.7}$$

where $P_\ell \colon R \to X_\ell$, $Q_\ell \colon R \to Y_\ell$ denote operators and $(Y_\ell, \leqslant)$ is an ordered space.

(iii) For each $\ell \in \mathscr{L}$, ℓM is a function of P_ℓ and Q_ℓ, i.e., there exists an $F_\ell \colon X_\ell \times Y_\ell \to \mathbb{R}$ such that for all $u \in R$

$$\ell M u = F_\ell(P_\ell u, Q_\ell u). \tag{2.8}$$

In most applications, $\mathscr{L}$ is a set of positive linear functionals, X_ℓ and Y_ℓ are linear spaces, and P_ℓ and Q_ℓ linear operators. Then it suffices to require (2.5) and (2.7) for v replaced by o. We already considered a series of special cases in which $u \geqslant o$, $\ell u = 0$ together imply a set of equations and inequalities. For example, let $u \in C_2[0, 1]$. Then for $\xi \in (0, 1)$, the relations $u \geqslant o$, $u(\xi) = 0$ imply the relations $u(\xi) = u'(\xi) = 0$, $u''(\xi) \geqslant 0$; and the relations $u \geqslant o$, $u(0) = 0$ imply the relations $u(0) = 0$, $u'(0) \geqslant 0$. Assumption (iii) essentially says that Mu must depend only on quantities involved in such equations and inequalities.

The preceding assumptions enable us, in particular, to replace condition I of Theorem 2.1b (or Corollary 2.1b′) by a condition more easily verified. Under the general assumptions required for Theorem 2.1b we obtain here

Corollary 2.1b″. *Suppose that*

I. $F_\ell(P_\ell v, Q_\ell v) \geqslant F_\ell(P_\ell v, q)$ *for each* $\ell \in \mathscr{L}$ *and all* $q \geqslant Q_\ell v$ *in* Y_ℓ,

II. $\ell M w < \ell M(w + \mathfrak{z}(\lambda))$ (2.9)

for each $\ell \in \mathcal{L}$ *and each* $\lambda \in (0, \gamma)$ *satisfying the side conditions*

$$P_\ell v = P_\ell(w + \mathfrak{z}(\lambda)), \qquad Q_\ell v \leqslant Q_\ell(w + \mathfrak{z}(\lambda)). \tag{2.10}$$

Then $Mv \leqslant Mw \Rightarrow v \leqslant w$.

In condition II the form (2.8) of the operator M has not been used explicitly. The construction of a suitable family $\mathfrak{z}(\lambda)$ depends, in general, on the concrete operator M considered. For a certain class of operators, however, we can rewrite (2.9) in a way which is helpful in constructing $\mathfrak{z}(\lambda)$.

Suppose now that all functionals ℓ, the spaces X_ℓ, Y_ℓ, and the operators P_ℓ, Q_ℓ are linear. Then let us say that the operator M is $\mathcal{L}$*-quasilinear*, if assumptions (i)–(iii) are satisfied and for each $\ell \in \mathcal{L}$ the term ℓM depends linearly on Q_ℓ. This means, ℓMu has the form

$$\ell Mu = -a_\ell(P_\ell u) \cdot Q_\ell u + g_\ell(P_\ell u),$$

where $a_\ell(P_\ell u)$ is a linear functional on Y_ℓ. Moreover, we define M to be *order-quasilinear*, if there exists a set $\mathcal{L} \subset R^*$ with the above properties. For such order-quasilinear operators Corollary 2.1b″ can be reformulated as follows.

Corollary 2.1b‴. *Suppose that* M *is* $\mathcal{L}$*-quasilinear,*

I. $a_\ell(P_\ell v) \geqslant o$ *for each* $\ell \in \mathcal{L}$,

II. $-a_\ell(P_\ell v) \cdot Q_\ell \mathfrak{z}(\lambda) + h_\ell(w + \mathfrak{z}(\lambda)) - h_\ell(w) > 0$ *for each* $\ell \in \mathcal{L}$ *and each* $\lambda \in (0, \gamma)$ *satisfying* (2.10), *where* $h_\ell(u) = -a_\ell(P_\ell u) \cdot Q_\ell w + g_\ell(P_\ell u)$.

Then $Mv \leqslant Mw \Rightarrow v \leqslant w$.

Of course, these corollaries describe only one way to apply Theorem 2.1b. For example, the above assumptions may be modified so that ℓMu is replaced by $f_\ell Mu$ with a functional f_ℓ on S. Then the assumption $(R, \leqslant) \subset (S, \leqslant)$ is not needed and the functionals $\ell \in \mathcal{L}$ need only be defined on R.

However, Corollaries 2.1b″ and 2.1b‴ can be applied to a series of concrete operators different from those considered below. For example, these corollaries yield also results on partial differential operators.

2.1.3 Pre-inverse-monotonicity and related concepts

Now we shall discuss certain sufficient conditions for the property

$$v \leqslant u, \quad v \nless u \quad \Rightarrow \quad Mv \nless Mu \tag{2.11}$$

which occurs in the assumptions of Theorems 2.1a and 2.1c (for $u = w + \mathfrak{z}(\lambda)$). Generalizing the term pre-inverse-positive introduced in Section II.1.2, we call M *pre-inverse-monotone on* (D_1, D_2), if the above implication holds for all $v \in D_1$ and $u \in D_2$. (Of course, this property depends also on the order relations involved.) Without describing all assumptions in detail, we shall here briefly discuss some sufficient conditions for pre-inverse-monotonicity which use functionals.

If $\not\ell$ is a monotone functional on S such that $U \prec V \Rightarrow \not\ell U < \not\ell V$, for all $U, V \in S$, then (2.11) is satisfied if the inequalities $v \leqslant u$, $v \prec u$ imply $\not\ell Mv \geqslant \not\ell Mu$. Using also property (2.5), we obtain the sufficient condition

$$v \leqslant u, \quad \ell v = \ell u \quad \Rightarrow \quad \not\ell Mv \geqslant \not\ell Mu. \tag{2.12}$$

If one allows $\not\ell$ to depend on ℓ, v, and u, necessary and sufficient properties for (2.11) can often be written in this form.

One is particularly interested in property (2.12) with $\not\ell = \not\ell_\ell$ depending only on ℓ (see Corollary 2.1b′). Here, the functional $\not\ell_\ell$ may be of a quite different nature than ℓ. (In Section II.3.6, for example, we used point functionals ℓ_x and integrals $\not\ell_x$ which depended on the value of x.) If $(R, \leqslant) \subset (S, \leqslant)$, however, one may formulate (2.12) also for the case that $\not\ell$ is an extension of ℓ. Writing in this case $\not\ell = \ell$, we obtain

$$v \leqslant u, \quad \ell v = \ell u \quad \Rightarrow \quad \ell Mv \geqslant \ell Mu. \tag{2.13}$$

This property of M can be described using the term quasi-antitone. More precisely, M shall be called *$\mathscr{L}$-quasi-antitone on* (D_1, D_2), if (2.13) holds for all $\ell \in \mathscr{L}$, $v \in D_1$, $u \in D_2$. For the case in which $D_1 = D_2 = R$, $(S, \leqslant)$ is an ordered linear space and $\mathscr{L}$ is the set of all linear functionals $\ell > o$ on S, M is simply called *quasi-antitone*. (Moreover, then $-M$ is said to be *quasi-monotone*.)

There are a series of operator classes for which (2.13) can be verified (at least for certain ℓ, v, u). The proofs, in general, apply a method which was described in abstract terms in Section 2.1.2: one shows that ℓM has the form (2.8) and verifies

$$F_\ell(P_\ell v, Q_\ell v) \geqslant F_\ell(P_\ell v, q) \qquad \text{for} \quad q = Q_\ell u \geqslant Q_\ell v. \tag{2.14}$$

Thus the proof of M being quasi-antitone is here reduced to the proof that ℓMu is an antitone function of those quantities $Q_\ell u$ for which only inequality can be established in (2.7). (For precise assumptions and statements see Section 2.1.2.)

As we have seen, property (2.13) and, in particular, (2.8) and (2.14) describe an important special case of pre-inverse-monotonicity. This special case is distinguished by the fact that, if M has property (2.11), then this property can be verified rather easily. In other cases, (2.11) may be very difficult to prove.

Of course, (2.13) constitutes only a sufficient condition. For example, suppose that $R = S = \mathbb{R}^n$ with $\leqslant$ denoting the natural order and write

$Mu = F(u) = (F_i(u))$. Then *M is quasi-antitone if and only if*

$$v \leqslant u, \quad v_k = u_k \quad \Rightarrow \quad F_k(v) \geqslant F_k(u) \tag{2.15}$$

for each index k and all $v, u \in \mathbb{R}^n$. Functions with this property are also called *off-diagonally antitone*. Moreover, we shall use the term *Z-functions*, observing that a matrix M is a Z-matrix if and only if (2.15) holds for $F(u) = Mu$.

Property (2.13) is also closely related to the definition of abstract (linear) Z-operators, since, clearly, property (2.13) holds for M if and only if it holds for $M + \lambda I$ with arbitrary $\lambda \in \mathbb{R}$.

2.1.4 An application

Many of the results in Section III.2 can be derived from the above abstract theory using (linear) point functionals ℓ, $\not\ell$ and defining $u \succ o$ by $u > o$. (See the problems below.) Here we shall explain a somewhat different application, in which M is a differential operator on an unbounded interval. (Such operators were already treated in Theorem III.2.6 by the method of absorption.)

Let $M: R = C_2[0, \infty) \to \mathbb{R}[0, \infty)$ be an operator as in (III.2.30) with $a(x) \geqslant 0$ $(0 \leqslant x < \infty)$ and $\alpha_0 \geqslant 0$, let v and w denote fixed functions in R and $q \in C_0[0, \infty)$ a given function such that $q(x) > 0$ $(0 \leqslant x < \infty)$. The following statement will be derived from Theorem 2.1b using some tools described in Section 2.1.3.

Suppose that there exists a function $z \in R$ satisfying $\alpha q(x) \leqslant z(x) \leqslant \beta q(x)$ $(0 \leqslant x < \infty)$ for some constants $\alpha, \beta > 0$, that $Mz(0) > 0$, and that $Mw(x) < M(w + \lambda z)(x)$ for all $x \in (0, \infty)$ and $\lambda > 0$ with $v(x) = w(x) + \lambda z(x)$, $v'(x) = w'(x) + \lambda z'(x)$.

Then

$$\left.\begin{array}{ll} Mv(x) \leqslant Mw(x) & \text{for} \quad 0 \leqslant x < \infty \\ \liminf\limits_{x \to +\infty} [(w - v)/q](x) \geqslant 0 & \end{array}\right\} \Rightarrow \quad v(x) \leqslant w(x) \quad \text{for } 0 \leqslant x < \infty. \tag{2.16}$$

To prove this statement we define an operator $\tilde{M}: R \to S$ by $\tilde{M}u(x) = M(v + u)(x) - Mv(x)$ for $0 \leqslant x < \infty$, $\tilde{M}u(\infty) = \liminf_{x \to \infty} u(x)/q(x)$, where S is the space of all functions on $[0, \infty]$ which accept real values on $[0, \infty)$ and values in $[-\infty, \infty]$ for $x = \infty$. Both spaces R and S are furnished with the natural (pointwise) order relation $\leqslant$. We shall apply the above theory to $\tilde{M}$, $\tilde{v} = o$, $\tilde{w} = w - v$ in place of M, v, w in order to verify that $\tilde{M}\tilde{w} \geqslant o \Rightarrow \tilde{w} \geqslant o$.

For $u \in R$, let $u \succ o \Leftrightarrow u \succ_q o$, i.e., $u(x) \geqslant \kappa q(x)$ on $[0, \infty)$ for some $\kappa = \kappa(u) > 0$. Moreover, define $\mathscr{L} = \mathscr{L}_\pi \cup \{\ell_\infty\}$ where $\ell_\infty u = \liminf_{x \to \infty} u(x)/q(x)$ and $\mathscr{L}_\pi$ is the set of all point functionals ℓ_x $(0 \leqslant x < \infty)$ (i.e.,

$\ell_x u = u(x)$). Clearly, (2.5) holds (with v replaced by o). Thus, to derive the result from Theorem 2.1b we have to find for each $\ell \in \mathscr{L}$ a suitable functional $\not{\ell} = \not{\ell}_\ell$ on S such that for $u = \tilde{w} + \lambda z, u \geqslant o, \ell u = 0$ the conditions $\not{\ell}\tilde{M}u \leqslant \not{\ell}o$ and $\not{\ell}\tilde{M}\tilde{w} < \not{\ell}\tilde{M}(\tilde{w} + \lambda z)$ are satisfied (cf. also (2.12)). One verifies that these properties hold for $\not{\ell}_\ell = \not{\ell}_x$ if $\ell = \ell_x (0 \leqslant x < \infty)$, $\not{\ell}U = U(\infty)$ if $\ell = \ell_\infty$. □

Example 2.3. The above result can be applied to the boundary value problem in Example III.2.16. Since $f'(y) > 0$, all assumptions are satisfied for $q(x) \equiv z(x) \equiv 1$ and arbitrary v, w with $\lim_{x\to\infty} v(x) = \lim_{x\to\infty} w(x) = 0$. Thus we see again that the problem has at most one solution u^* in $C_2[0, \infty)$ and that the estimate $0 \leqslant u^*(x) \leqslant \gamma\, (0 \leqslant x < \infty)$ holds. Moreover, the above assumptions are also satisfied for $q(x) = \exp\sqrt{2}\, x$ and $z(x) = q(x) + 2 - (1 + x)^{-1}$. □

Problems

2.1 Prove the assertions in Examples 2.1 and 2.2.
2.2 Derive Theorem III.2.1 from Theorem 2.1a.
2.3 Derive Theorem III.2.2 from Theorem 2.1b, under the assumption that $v, w, \mathfrak{z}(\lambda) \in C_1[0, 1] \cap C_2(0, 1)$.
2.4 Apply Corollary 2.1b‴ to the differential operator M in (III.2.15), using a family $\mathfrak{z}(\lambda) = \lambda z$.

2.2 Nonlinear functions in finite-dimensional spaces

For a given function $F: \mathbb{R}^n \to \mathbb{R}^n$ and fixed $v, w \in \mathbb{R}^n$ we are interested in sufficient conditions such that

$$F(v) \leqslant F(w) \quad \Rightarrow \quad v \leqslant w, \tag{2.17}$$

with $\leqslant$ denoting the natural (componentwise) order relation. An application of Theorem 2.1b yields the following result, where an *index* i is defined to be a number $i \in \{1, 2, \ldots, n\}$ and $\preccurlyeq$ denotes the strict natural order.

Theorem 2.2. *Suppose there exists a* $z \succ o$ *in* $\mathbb{R}^n$ *such that for each* $i \in \{1, 2, \ldots, n\}$ *and each* $\lambda > 0$ *satisfying*

$$v \leqslant w + \lambda z \qquad \textit{and} \qquad v_i = w_i + \lambda z_i \tag{2.18}$$

the following two inequalities hold:

$$\text{(I)} \quad F_i(v) \geqslant F_i(w + \lambda z), \qquad \text{(II)} \quad F_i(w) < F_i(w + \lambda z). \tag{2.19}$$

Then $F(v) \leqslant F(w) \Rightarrow v \leqslant w$.

Proof. We define $R = S = \mathbb{R}^n$, $Mu = F(u)$, $u \succ o \Leftrightarrow u \succ o$, $\gamma = +\infty$ and apply Theorem 2.1b. If (2.2) holds for some $\lambda > 0$, then the relations (2.18) are satisfied for this λ and some index i. Because of (2.19) this implies that the inequalities in (2.3) hold with $\not\ell$ defined by $\not\ell U = U_i$ for $U \in S$. $\square$

Now suppose that w is known and that we want to prove (2.17) for all $v \in \mathbb{R}^n$. For this case the next corollary yields sufficient conditions which do not contain v. Here $\partial F_i/\partial u_k$ denotes the derivative of the ith component of F with respect to the kth variable.

Corollary 2.2a. *Suppose that F is continuously differentiable and that there exists a $z \succ o$ in $\mathbb{R}^n$ such that for each index i the following two conditions are satisfied*:

$$\text{(a)} \quad \frac{\partial F_i}{\partial u_k}(w + \lambda z + y) \leqslant 0$$

for all indices $k \neq i$, $\lambda > 0$, $y \in \mathbb{R}^n$ with $y \leqslant o$, $y_i = 0$;

$$\text{(b)} \quad \sum_{k=1}^{n} \frac{\partial F_i}{\partial u_k}(w + \lambda z)z_k > 0 \qquad \textit{for} \quad \lambda > 0.$$

Then for all $v \in \mathbb{R}^n$, $F(v) \leqslant F(w) \Rightarrow v \leqslant w$.

Proof. Inequality I in (2.19) can be written as

$$\sum_{k=1}^{n} \int_0^1 \frac{\partial F_i}{\partial u_k}(w + \lambda z + ty)y_k \, dt \geqslant 0 \qquad \text{with} \quad y = v - (w + \lambda z).$$

Obviously, this inequality holds under the side condition (2.18), if (a) is satisfied. In a similar way, inequality II follows from (b). $\square$

Example 2.4. Let $n = 2$ and $F(u) = Au + B[u, u]$ with a 2×2 matrix A and a bilinear form

$$B[u, u] = \begin{pmatrix} au_1^2 + 2bu_1u_2 + cu_2^2 \\ \alpha u_1^2 + 2\beta u_1u_2 + \gamma u_2^2 \end{pmatrix}.$$

Applying Corollary 2.2a to this function and $w = o$ we obtain the following statement by formal calculations.

Suppose that A is a Z-matrix and

$$c \gtrless 0, \qquad bz_1 + cz_2 \leqslant 0, \qquad \alpha \geqslant 0, \qquad \alpha z_1 + \beta z_2 \leqslant 0,$$
$$Az \geqslant o, \qquad B[z, z] \geqslant o, \qquad Az + B[z, z] \succ o$$

for some $z \succ o$. Then for all $v \in \mathbb{R}^n$, $F(v) \leqslant o \Rightarrow v \leqslant o$. $\square$

Under the smoothness assumptions of the preceding corollary, one may transform the implication (2.17) into

$$\tilde{M}\tilde{w} \geqslant o \quad \Rightarrow \quad \tilde{w} \geqslant o \qquad \text{with} \quad \tilde{w} = w - v,$$

$$\tilde{M}\tilde{w} = \int_0^1 F'((1-t)v + tw)\, dt\, \tilde{w}$$

and then apply the theory of M-matrices to $\tilde{M}$. The resulting assumptions, however, are quite different from those in Corollary 2.2a.

If additional information on v is available (such as $v \geqslant o$, for example), the assumptions in Corollary 2.2a can in general be relaxed. On the other hand, if one wants to prove (2.17) for all $v, w \in \mathbb{R}^n$, stronger assumptions will have to be required.

The function F is called a *Z-function* if (2.15) holds for all $v, w \in \mathbb{R}^n$ and each index k. An *M-function* F is an inverse-monotone Z-function, i.e., a Z-function such that (2.17) holds for all $v, w \in \mathbb{R}^n$. (In an analogous way one can define a *Z-function on D*, an *M-function on D*, etc.) Clearly, these notions generalize the definition of Z-matrices and M-matrices. Often, a Z-function is also called an *off-diagonally antitone function* or a *quasi-antitone function*, as already explained in Section 2.1.3.

Theorem 2.2 immediately yields

Corollary 2.2b. *If F is a Z-function and $F(w) \prec F(w + \lambda z)$ for all $\lambda > 0$, all $w \in \mathbb{R}^n$ and some $z \succ o$ in $\mathbb{R}^n$, then F is an M-function.*

Obviously, the function F in Example 2.4 in general is not a Z-function. However, Z-functions often occur when difference methods are applied to differential equations.

Example 2.5. Consider the boundary value problem $Mu(x) + g(x, u(x)) = 0$ $(0 \leqslant x \leqslant 1)$, where $g: [0, 1] \times \mathbb{R} \to \mathbb{R}$ and M denotes a linear differential operator as in Section II.3.1 with $B_0[u] = u(0)$, $B_1[u] = u(1)$. For this problem consider the ordinary difference method described by

$$M_h\varphi(x_i) + g(x_i, \varphi(x_i)) = 0 \qquad (i = 0, 1, 2, \ldots, n)$$
$$\text{with} \quad x_i = ih, \quad h = n^{-1},$$

and M_h as in Section II.5.1.3, method (1). These equations can be written as $F(\varphi) = o$ with a function $F: \mathbb{R}^{n+1} \to \mathbb{R}^{n+1}$, $\varphi = (\varphi_i)$, $\varphi_i = \varphi(x_i)$ $(i = 0, 1, \ldots, n)$.

Since M_h is a Z-operator and the nonlinear term in the ith equation does not depend on any $\varphi(x_j)$ with $j \neq i$, *the function F is a Z-function, for all sufficiently small h.* Moreover, *if the differential operator M is inverse-positive and if*

$g(x, y) \leqslant g(x, \bar{y})$ for $y \leqslant \bar{y}$ $(0 \leqslant x \leqslant 1, y \in \mathbb{R}, \bar{y} \in \mathbb{R})$, then F is an M-function, for all sufficiently small h.

To prove this latter statement, choose $\zeta \in C_2[0, 1]$ such that $M\zeta(x) = 1$ $(0 \leqslant x \leqslant 1)$. Then $M_h\zeta(x_i) \geqslant \frac{1}{2}$ $(i = 0, 1, \ldots, n)$ for h small enough, because of the consistency of the difference method (1). Consequently, the inequality required in Corollary 2.2b is satisfied for arbitrary $w \in \mathbb{R}^{n+1}$ and $z = (z_i)$, $z_i = \zeta(x_i)$. (Compare also Sections II.5.1.2 and II.5.1.3.) □

2.3 Functional-differential operators of the second order

Here we consider vector-valued differential operators and certain generalizations which, in particular, include integro-differential operators. Let R be the space $C_1^n[0, 1] \cap C_2^n(0, 1)$ (or a larger space as described below) and suppose that $M: R \to S = \mathbb{R}^n[0, 1]$ is defined by

$$Mu(x) = \begin{cases} -a(x)u''(x) + g(x, u(x), u'(x), u) & \text{for} \quad 0 < x < 1 \\ -\alpha^0 u'(0) + g^0(u(0), u) & \text{for} \quad x = 0 \\ \alpha^1 u'(1) + g^1(u(1), u) & \text{for} \quad x = 1, \end{cases} \tag{2.20}$$

where the occurring quantities have the following meaning: $a(x) = (a_i(x)\,\delta_{ik})$, $\alpha^0 = (\alpha_i^0\,\delta_{ik})$, and $\alpha^1 = (\alpha_i^1\,\delta_{ik})$ are diagonal $n \times n$ matrices such that for all indices i $(i = 1, 2, \ldots, n)$

$$a_i(x) \geqslant 0 \qquad (0 < x < 1), \quad \alpha_i^0 \geqslant 0, \quad \alpha_i^1 \geqslant 0;$$

g is a mapping of all $(x, y, p, u) \in (0, 1) \times \mathbb{R}^n \times \mathbb{R}^n \times R$ into $\mathbb{R}^n$; and both g^0 and g^1 are mappings of all $(y, u) \in \mathbb{R}^n \times R$ into $\mathbb{R}^n$.

Instead of defining R as above, we may also define R to be the set of all $u \in C_0^n[0, 1]$ such that all derivatives $u_i'(x)$, $u_i''(x)$ exist which actually occur in $Mu(x)$. (For example, $u_i''(x)$ must exist, if $a_i(x) > 0$.)

Throughout this section we make also the following assumption:

(W) *For each index i, the ith component $g_i(x, y, p, u)$ does not depend on any y_j or p_j with $j \neq i$ and the ith components $g_i^0(y, u)$, $g_i^1(y, u)$ do not depend on any y_j with $j \neq i$.*

For illustration we consider two special cases, where $a(x)$, α^0, and α^1 are defined as above.

Special case 1: *Weakly coupled differential operators.* Let

$$Mu(x) = \begin{cases} -a(x)u''(x) + f(x, u(x), u'(x)) & \text{for} \quad 0 < x < 1 \\ -\alpha^0 u'(0) + f^0(u(0)) & \text{for} \quad x = 0 \\ \alpha^1 u'(1) + f^1(u(1)) & \text{for} \quad x = 1 \end{cases} \tag{2.21}$$

with $f\colon (0,1) \times \mathbb{R}^n \times \mathbb{R}^n \to \mathbb{R}^n, f^0\colon \mathbb{R}^n \to \mathbb{R}^n, f^1\colon \mathbb{R}^n \to \mathbb{R}^n$ and assume that this operator is *weakly coupled*, i.e., *for each index i the component $f_i(x, y, p)$ must not depend on any p_j with $j \neq i$.* Then Mu can be written in the form (2.20) by defining

$$g_i(x, y, p, u) = f_i(x, h, p) \qquad \text{with} \quad h_i = y_i, \qquad h_j = u_j(x) \qquad \text{for} \quad j \neq i$$

and defining g_i^0, g_i^1 in an analogous way. Obviously, here (W) is satisfied, since M is (only) weakly coupled.

Special case 2: *Scalar-valued integro-differential operators.* Let $n = 1$ and

$$Mu(x) = \begin{cases} -a(x)u''(x) + \int_0^1 f(x, t, u(x), u'(x), u(t))\,dt & \text{for} \quad 0 < x < 1 \\ -\alpha^0 u'(0) + \int_0^1 f^0(t, u(0), u(t))\,dt & \text{for} \quad x = 0 \\ \alpha^1 u'(1) + \int_0^1 f^1(t, u(1), u(t))\,dt & \text{for} \quad x = 1 \end{cases}$$

with suitable functions f, f^0, f^1. This operator, too, can be written in the form (2.20), where, for example, $g(x, y, p, u) = \int_0^1 f(x, t, y, p, u(t))\,dt$.

For operators of the above form (2.20) various results on inverse-monotonicity can be derived from the abstract theory in Section 2.1. We restrict ourselves to an application of Corollary 2.1.b″, thus illustrating the use of this corollary. Also, we consider only the case in which $u \succ o \Leftrightarrow u > o$, for $u \in R$, and $\mathfrak{z}(\lambda) = \lambda z$ $(0 \leqslant \lambda < \infty)$ with a $z > o$.

Theorem 2.3. *Suppose that assumption* (W) *holds and assume, moreover*:

I. *For each $u \in R$ satisfying $u \geqslant o$*

$$\begin{aligned} &g(x, v(x), v'(x), v) \geqslant g(x, v(x), v'(x), v + u) \quad (0 < x < 1), \\ &g^0(v(0), v) \geqslant g^0(v(0), v + u), \qquad g^1(v(1), v) \geqslant g^1(v(1), v + u). \end{aligned} \tag{2.22}$$

II. *There exists a $z > o$ in R such that for each index i and all $x \in [0,1]$, $\lambda > 0$ the following condition is satisfied*:

$$(M(w + \lambda z) - Mw)_i(x) > 0, \qquad \textit{if}$$

$$\begin{aligned} &v_i(x) = w_i(x) + \lambda z_i(x), \qquad v_k(x) \leqslant w_k(x) + \lambda z_k(x) \qquad \textit{for} \quad k \neq i, \\ &v_i'(x) = w_i'(x) + \lambda z_i'(x) \quad \textit{in case} \quad x \in (0, 1). \end{aligned}$$

Then $Mv \leqslant Mw \Rightarrow v \leqslant w$.

Proof. As remarked above, we shall apply Corollary 2.1b″. The result could be proved with less formal definitions by using Theorem 2.1b directly. However, in applying Corollary 2.1b″ we explain, at the same time, a general method which can be used in many other cases.

First we have to find a suitable set $\mathscr{L}$ of linear functionals ℓ on $S \supset R$. Obviously, for each $x \in [0, 1]$ and each index i a linear functional $\ell = \ell_{i,x}$ on S is defined by $\ell u = u_i(x)$, and the set $\mathscr{L}$ of all these functionals has property (i) required in Corollary 2.1b″.

Now we suppose that for some $u \in R$ the relations on the left-hand side of (2.7) hold, i.e., $u \geqslant o$ and $\ell u = 0$ for some $\ell = \ell_{i,x}$. We restrict ourselves to considering the case in which $x \in (0, 1)$ and $a_i(x) > 0$. (All other cases are treated analogously.) The function u then also satisfies the following equations and inequalities:

$$u_i(x) = 0, \qquad u_i'(x) = 0, \qquad u_i''(x) \geqslant 0, \qquad u \geqslant o. \tag{2.23}$$

We have to rewrite these relations in the form $P_\ell u = o, Q_\ell u \geqslant o$ as in (2.7), show that (2.8) holds with a suitable F_ℓ, and prove that assumptions I, II of the corollary are satisfied.

It is quite clear that $\ell Mu = (Mu)_i(x)$ depends only on the quantities occurring in (2.23), as required in assumption (iii). Let us, however, carry out the details. The relations (2.23) assume the form $P_\ell u = o, Q_\ell u \geqslant o$, if P_ℓ: $R \to X_\ell = \mathbb{R}^2$ and $Q_\ell: R \to Y_\ell = \mathbb{R} \times R$ are defined by $P_\ell u = (u_i(x), u_i'(x))$, $Q_\ell u = (u_i''(x), u)$ and for $\eta = (\eta_1, \eta_2) \in Y_\ell$ the inequality $\eta \geqslant o$ means that $0 \leqslant \eta_1 \in \mathbb{R}$ and η_2 is a positive element in R. Moreover, (2.8) holds with $F_\ell(\xi, \eta) = -a_i(x)\eta_1 + g_i(x, \xi_1, \xi_2, \eta_2)$, where $\xi = (\xi_1, \xi_2) \in X_\ell$.

One verifies easily that for $\ell = \ell_{i,x}$ as above the assumptions of Corollary 2.1b″ are satisfied, if the assumptions of Theorem 2.3 hold. For example, assumption I of Corollary 2.1b″ requires that $(Mv)_i(x)$ decreases if $v_i''(x)$ and v increase (but $v_i(x) = 0$). □

We point out that for the special case of a weakly coupled differential operator as in (2.21) the condition on g in (2.22) requires that for each index i, all $x \in (0, 1)$ and all $y, p \in \mathbb{R}^n$

$$v(x) \leqslant y, \quad v_i(x) = y_i \quad \Rightarrow \quad f_i(x, v(x), p) \geqslant f_i(x, y, p). \tag{2.24}$$

Problems

2.5 The results of Section 2.3 can be carried over to differential operators in normed linear spaces. For example, let $(Y, \leqslant)$ be an ordered linear space which contains an element $\eta \succ o$, $\|y\| = \|y\|_\eta$ for $y \in Y$, $(Y_0, \leqslant) \subset (Y, \leqslant)$ an Archimedian linear subspace, $\mathscr{L}$ a set of linear functionals $\ell > o$ on Y which describes the positive cone in Y_0 completely, S the set of all functions $u: x \in [0, 1] \to u(x) \in Y$ with $u \geqslant o$ defined pointwise, $R = C_0([0, 1], Y_0) \cap$

$C_2((0,1), Y_0)$, $f\colon (0,1) \times Y_0 \times Y_0 \to Y$, $M\colon R \to S$ given by $Mu(x) = -u''(x) + f(x, u(x), u'(x))(0 < x < 1)$, $Mu(x) = u(x)$ $(x = 0, 1)$. Prove the following statement:

For $v, w \in R$, the inequality $Mv \leqslant Mw$ implies $v \leqslant w$, if (I) *for each $x \in (0,1)$, each $\ell \in \mathscr{L}$ and all $u \in R$ the relations* $(*)$ $v \leqslant u$, $\ell v(x) = \ell u(x)$, $\ell v'(x) = \ell u'(x)$ *imply* $\ell f(x, v(x), v'(x)) \geqslant \ell f(x, u(x), u'(x))$, (II) *there exists a $z \in R$ with $z(x) \succ o$ $(0 \leqslant x \leqslant 1)$ such that $\ell\{-z''(x) + \lambda^{-1}[f(x, u(x), u'(x)) - f(x, w(x), w'(x))]\} > o$ for $u = w + \lambda z$ and all $x \in (0,1)$, $\ell \in \mathscr{L}$, $\lambda > 0$ satisfying* $(*)$.

2.6 Formulate and prove a result analogous to that in Problem 2.5 for the case in which $Mu(0) = \alpha^0 u'(0) + f^0(u(0))$, $Mu(1) = \alpha^1 u'(1) + f^1(u(1))$ with α^0, α^1 denoting suitable linear operators.

2.4 A different approach for initial value problems

The theorems on inverse-monotone operators in Section 2.1 can also be applied to differential operators of the first order related to initial value problems. The results are completely satisfactory if a one-sided Lipschitz condition or a certain weaker estimate holds (cf. Example III.2.12 and Problem III.2.11). However, in the theory on differential inequalities with initial conditions, one usually proceeds differently, requiring still weaker assumptions. We present here some typical results. Theorems 2.7 and 2.6 below are derived with the "usual" methods of proof. Proofs of this nature can also be incorporated into the continuity principle of Section 1. To illustrate this is the main object of Section 2.4.1. For the simple problem treated there the proofs do not become simpler in this way, but lengther. This approach, however, exposes the common element in different proofs and can also lead to generalizations.

Section 2.4.2 is concerned with abstract differential operators. The discussion of an infinite system of first order equations in Section 2.4.3 will show that it can be advantageous to use special methods adapted to the given problem, even if a general theory is available.

2.4.1 Description of the method for a simple special case

To explain an application of the continuity principle which differs from the application in Section 2.1 we derive two typical results for a very simple operator. Let $R = C_0[0,1] \cap C_1(0,1]$, $S = \mathbb{R}(0,1]$, and $M\colon R \to S$ be given by

$$Mu(x) = u'(x) + f(x, u(x)) \qquad \text{with} \quad f\colon (0,1] \times \mathbb{R} \to \mathbb{R} \tag{2.25}$$

and suppose that v, w are fixed functions in R.

Theorem 2.4. *If* $v(0) < w(0)$ *and* $Mv(x) < Mw(x)$ *for* $0 < x \leqslant 1$, *then* $v(x) < w(x)$ *for* $0 \leqslant x \leqslant 1$.

Proof. To derive the statement from Corollary 1.1b we define $C = \{U \in S: U(x) < Mw(x), 0 < x \leqslant 1\}$ and

$$K_\lambda = \{u \in R : u(x) \leqslant w(x), 0 < x \leqslant 1 - \lambda\},$$

$$\mathring{K}_\lambda = \{u \in R : u(x) < w(x), 0 < x \leqslant 1 - \lambda\}$$

for $\lambda \in I = [0, 1)$. Then assumption (i) of Theorem 1.1 is satisfied. Observe, in particular, that $v \in K_\lambda$ for all sufficiently large $\lambda < 1$, since $v(0) < w(0)$.

If $v \in \Gamma_\lambda$ for some $\lambda \in [0, 1)$, then $v(x) \leqslant w(x)$ for all $x \in (0, 1 - \lambda]$ and $v(\xi) = w(\xi)$ for some $\xi \in (0, 1 - \lambda]$; consequently, $v'(\xi) \geqslant w'(\xi)$. Because of these relations, $Mv \notin \Omega := \{U \in S: U(\xi) < Mw(\xi)\}$, while obviously $Mv \in C \Rightarrow Mv \in \Omega$. Thus, assumption (ii′) is satisfied for $[0, 1)$ in place of $(0, \gamma) = (0, 1)$. □

Theorem 2.5. *Suppose there exist functions* $\mathfrak{z}(\lambda) \in R$ *for* $0 \leqslant \lambda \leqslant \varepsilon$ *and some* $\varepsilon > 0$ *such that* $\mathfrak{z}(\lambda, x) := \mathfrak{z}(\lambda)(x)$ *is continuous on* $[0, \varepsilon] \times [0, 1]$, $\mathfrak{z}(0, x) = 0$ *for* $0 \leqslant x \leqslant 1$, $\mathfrak{z}(\lambda, 0) > 0$ *for* $0 < \lambda \leqslant \varepsilon$ *and*

$$\partial/\partial x\, \mathfrak{z}(\lambda, x) + f(x, w(x) + \mathfrak{z}(\lambda, x)) - f(x, w(x)) > 0$$

for all $x \in (0, 1]$ *and* $\lambda \in (0, \varepsilon]$ *with* $v(x) = w(x) + \mathfrak{z}(\lambda, x)$.

Then the inequalities $v(0) \leqslant w(0)$, $Mv(x) \leqslant Mw(x)$ $(0 < x \leqslant 1)$ *together imply* $v(x) \leqslant w(x)$ $(0 \leqslant x \leqslant 1)$. (2.26)

Proof. Now we apply Corollary 1.1a. For $\lambda \in I = [0, 1 + \varepsilon)$ let

$$K_\lambda = \{u \in R : u(x) \leqslant w(x) + \mathfrak{z}(\lambda, x), 0 < x \leqslant 1\} \qquad \text{if} \quad 0 \leqslant \lambda \leqslant \varepsilon,$$

$$K_\lambda = \{u \in R : u(x) \leqslant w(x) + \mathfrak{z}(\varepsilon, x), 0 < x \leqslant 1 + \varepsilon - \lambda\} \qquad \text{if} \quad \varepsilon < \lambda < 1 + \varepsilon,$$

and define $\mathring{K}_\lambda$ by the same formulas with $u(x) \leqslant \cdots$ replaced by $u(x) < \cdots$. Also, let $C = \{U \in S : U(x) \leqslant Mw(x),\ 0 < x \leqslant 1\}$. Obviously, conditions (a) and (b) in Section 1 are satisfied. Also, $v \in K_\lambda$ for all sufficiently large $\lambda < 1 + \varepsilon$, since $v(0) < w(0) + \mathfrak{z}(\varepsilon, 0)$.

If $v \in \Gamma_\lambda$ for some $\lambda \in (0, \varepsilon]$, then

$$\begin{aligned} v(x) &\leqslant w(x) + \mathfrak{z}(\lambda, x) \qquad \text{for all} \quad x \in (0, 1], \\ v(\xi) &= w(\xi) + \mathfrak{z}(\lambda, \xi) \qquad \text{for some} \quad \xi \in (0, 1]. \end{aligned} \tag{2.27}$$

Because of these relations and $v'(\xi) \geqslant w'(\xi) + \partial/\partial x \mathfrak{z}(\lambda, \xi)$, property I holds for $\Omega = \{U \in S : U(\xi) < M(w + \mathfrak{z}(\lambda))(\xi)\}$, while property II follows from $Mv \leqslant Mw$ and the required differential inequality for $\mathfrak{z}(\lambda)$. If $v \in \Gamma_\lambda$ for some

$\lambda \in (\varepsilon, 1 + \varepsilon)$, then the relations (2.27) hold for $\lambda = \varepsilon$ with the interval $(0, 1]$ replaced by $(0, 1 + \varepsilon - \lambda]$. Define then Ω as above with $\mathfrak{z}(\varepsilon)$ in place of $\mathfrak{z}(\lambda)$. □

Example 2.6. Suppose that $f(x, w(x) + y) - f(x, w(x)) \geqslant -d(y)$ for all $x \in (0, 1]$ and $y \in (0, \varepsilon]$ with $v(x) = w(x) + y$, where $d \in C_0[0, \varepsilon]$, $d(0) = 0$, $d(y) > 0$ for $0 < y \leqslant \varepsilon$, and $\int_0^\varepsilon d^{-1}(t)\, dt = \infty$. Then, for any $\delta > 0$, the relations $\mathfrak{z}(0, x) \equiv 0$, $\int_{\mathfrak{z}(\lambda, x)}^{\lambda} d^{-1}(t)\, dt = (1 - \delta)(1 - x)$ for $0 < \lambda \leqslant \varepsilon$ define a family of functions $\mathfrak{z}(\lambda)$ that have all properties required in Theorem 2.5. (Compare these conditions with those in Problem III.2.11.) □

The two preceding theorems can be combined in one theorem and at the same time generalized by considering sets of the form

$$K_\lambda = \{u \in R : u \leqslant_\lambda w + \mathfrak{z}(\lambda)\}, \qquad \mathring{K}_\lambda = \{u \in R : u \prec_\lambda w + \mathfrak{z}(\lambda)\},$$

where $\leqslant_\lambda$, $\prec_\lambda$ denote relations depending on λ. For example, the first relation may be defined by $u \geqslant_\lambda o \Leftrightarrow u(x) \geqslant 0$ for $0 < x \leqslant 1 - h(\lambda)$ with a suitable function h. The proofs assume a simpler form, when the idea of the continuity principle is used in the same way, but the sets K_λ, $\mathring{K}_\lambda$ are not defined explicitly. (See, for example, the proof of Theorem 2.7.)

Instead of applying the continuity principle directly, one may also combine it with the method of absorption, as in the following theorem.

Theorem 2.6. *The implication* (2.26) *holds, if there exists a sequence* $\{z_m\} \subset R$ *such that* $\lim_{m\to\infty} z_m(x) = 0$ *for each* $x \in [0, 1]$, *and for all sufficiently large* m

$$\begin{aligned} &z_m(0) > 0, \\ &z_m'(x) + f(x, w(x) + z_m(x)) - f(x, w(x)) > 0 \qquad (0 < x \leqslant 1). \end{aligned} \tag{2.28}$$

Proof. If $v(0) \leqslant w(0)$ and $Mv(x) \leqslant Mw(x)$ $(0 < x \leqslant 1)$, then the functions $w_m = w + z_m$ satisfy $v(0) < w_m(0)$ and $Mv(x) < Mw_m(x)$ $(0 < x \leqslant 1)$ and hence $v(x) < w_m(x)$ $(0 \leqslant x \leqslant 1)$ for each m, by Theorem 2.4. Consequently, $v(x) \leqslant w(x)$ $(0 \leqslant x \leqslant 1)$. □

Corollary 2.6a. *A sequence* $\{z_m\}$ *as required in Theorem* 2.6 *exists, if there exists a function* $d: [0, 1] \times \mathbb{R} \to \mathbb{R}$ *with the following properties.*

(i) *d is continuous,* $d(x, 0) = 0$ $(0 \leqslant x \leqslant 1)$ *and the initial value problem* $u'(x) = d(x, u(x))$ $(0 < x \leqslant 1)$, $u(0) = 0$ *has the only solution* $u(x) \equiv 0$.

(ii) $f(x, w(x) + t) - f(x, w(x)) \geqslant -d(x, t)$ *for* $0 < x \leqslant 1$, $t \geqslant 0$.

Proof. The functions $z_m \in C_1[0, 1]$ defined by $z_m(0) = m^{-1}$, $z_m'(x) - d(x, \min(1, z_m(x))) = m^{-1}$ $(0 < x \leqslant 1)$ constitute an antitone sequence which converges uniformly on $[0, 1]$ toward the null function. These functions satisfy (2.28) for m so large that $z_m(x) \leqslant 1$ $(0 \leqslant x \leqslant 1)$. □

A function d which has property (i) in the preceding corollary is often called a *uniqueness function*. Obviously, this is a "local" property, which is determined by the behavior of $d(x, t)$ for small t. Consequently, the conditions in the corollary need be required only for all sufficiently small $t \geqslant 0$. For instance, the function $d(x, t) = d(t)$ in Example 2.6 is a uniqueness function. (It can easily be extended to $t \in \mathbb{R}$.)

It is clear from the above proofs that one may require weaker smoothness conditions, as it is often done in connection with differential inequalities of the type considered here. For instance, the functions v and w need not be continuously differentiable on $(0, 1]$; it suffices, for example, that the left-hand derivatives $D^- v(x)$, $D^- w(x)$ exist for each $x \in (0, 1]$. The reader may simply check which conditions have really been used in the proofs.

Various other generalizations are possible. For example, in order to prove (2.26) one may apply the method of absorption somewhat differently than above, deriving $v(x) \leqslant w(x)$ $(0 \leqslant x \leqslant 1)$ from $v(x) \leqslant w(x) + z_m(x)$ $(0 < x_m \leqslant x \leqslant 1)$, where $x_m \to 0$ and the functions z_m satisfy suitable conditions which are weaker than those above.

2.4.2 *Abstract differential equations*

The results of the preceding section can be carried over to much more general problems. Here we shall consider abstract differential operators of the first order and formulate results analogous to those above. For example, these results can be applied to (vector-valued) parabolic partial differential equations.

Let $(Y_0, \leqslant)$ and $(Y, \leqslant) \supset (Y_0, \leqslant)$ denote ordered linear spaces such that Y_0 is Archimedian and contains an element $\eta \succ o$. All topological terms which occur are to be interpreted with respect to the order topology in Y_0, that means, with respect to the norm $\| \ \|_\eta$. Moreover, let a set $\mathscr{L}$ of linear functionals $\ell > o$ on Y be given such that $\mathscr{L}$ describes the positive cone in Y_0 completely. For $y \in Y$ define $y \succ o$ by $\ell y > 0$ for all $\ell \in \mathscr{L}$. Observe in the following that the functionals $\ell \in \mathscr{L}$ are continuous on Y_0, according to Theorem I.1.10(ii).

We consider an operator $M: R \to S$, where $R = C_1([0, 1], Y_0)$, S is the set of all functions $U: x \to U(x)$ from $(0, 1]$ into Y and Mu is given by

$$Mu(x) = u'(x) + f(x, u(x)) \qquad \text{for} \quad 0 < x \leqslant 1 \qquad \text{with} \quad f: (0, 1] \times Y_0 \to Y. \tag{2.29}$$

As before, v and w denote fixed (but possibly unknown) elements in R. We assume that for each $x \in (0, 1]$ and each $\ell \in \mathscr{L}$

$$\ell f(x, v(x)) \geqslant \ell f(x, v(x) + y) \qquad \text{for all} \quad y \in Y_0 \quad \text{satisfying} \quad y \geqslant o, \quad \ell y = 0. \tag{2.30}$$

Under these assumptions the following theorems hold.

Theorem 2.7. *If* $v(0) \prec w(0)$ *and* $Mv(x) \prec' Mw(x)$ *for* $0 < x \leqslant 1$, *then* $v(x) \prec w(x)$ *for* $0 \leqslant x \leqslant 1$.

Theorem 2.8. *Suppose there exist an element* $q \succ o$ *in* Y *and a function* $d\colon [0, 1] \times \mathbb{R} \to \mathbb{R}$ *such that*

$$f(x, w(x) + tq) - f(x, w(x)) \geqslant -d(x, t)q \qquad \textit{for} \quad 0 < x \leqslant 1, \quad t \geqslant 0$$

and d *is a uniqueness function, i.e.,* d *satisfies assumption* (i) *of Corollary* 2.6a.
Then the inequalities $v(0) \leqslant w(0)$, $Mv(x) \leqslant Mw(x)$ $(0 < x \leqslant 1)$ *together imply* $v(x) \leqslant w(x)$ $(0 \leqslant x \leqslant 1)$.

Theorem 2.8 can be derived from Theorem 2.7 similarly as the result of Theorem 2.6 and its corollary was derived from Theorem 2.4. Theorem 2.7 could be proved in essentially the same way as Theorem 2.4. Here we shall give a direct proof which is formally simpler, but essentially uses the same ideas.

Proof of Theorem 2.7. Let $u = w - v$. Since $u(0) \succ o$ and $u(x)$ is continuous, we have $u(x) \succ o$ for $0 \leqslant x < \xi$ and some $\xi \in (0, 1]$. Let ξ denote the maximal number of this kind. If $\xi = 1$ and $u(1) \succ o$, the statement is proved. In all other cases, we have $u(\xi) \geqslant o$, since Y_0 is Archimedian; moreover, $u(\xi) \not\succ o$, because otherwise $\xi < 1$ could not be maximal. Thus, there exists an $\ell \in \mathscr{L}$ such that $(\ell u)(\xi) = \ell u(\xi) = 0$. Since $(\ell u)(x) \geqslant 0$ for $0 \leqslant x < \xi$, we have $\ell u'(\xi) = (\ell u)'(\xi) \leqslant 0$. From this inequality and (2.30), we derive

$$\begin{aligned}\ell Mw(\xi) = \ell w'(\xi) + \ell f(\xi, w(\xi)) &\leqslant \ell v'(\xi) + \ell f(\xi, v(\xi) + u(\xi)) \\ &\leqslant \ell v'(\xi) + \ell f(\xi, v(\xi)) = \ell Mv(\xi),\end{aligned}$$

which contradicts the assumption $Mv(\xi) \prec' Mw(\xi)$. □

The operator M in (2.29) can also be written in the form $Mu(x) = u'(x) + A(x)u(x)$, where for each $x \in (0, 1]$ the operator $A(x)\colon Y_0 \to Y$ is defined by $A(x)y = f(x, y)$. Condition (2.30) then is equivalent to

$$v(x) \leqslant y, \quad \ell v(x) = \ell y \quad \Rightarrow \quad \ell A(x)v(x) \geqslant \ell A(x)y. \tag{2.31}$$

This form is analogous to (2.13). Thus, the required assumptions hold, for example, if $A(x)$ is $\mathscr{L}$-quasi-antitone for each x. (Actually, (2.31) in general, need only be required for certain values of y, such as $y = w(x)$ in Theorem 2.7.)

In order to verify (2.31) one may apply the methods explained in Section 2.1.2 to $A(x)$ instead of M, writing $\ell A(x)y$ for each $x \in (0, 1]$ in a form analogous to (2.8) and then replacing (2.31) by a condition as (2.14). We shall not carry out the details.

Finally, we observe that here the operator M itself has a property like (2.13). One verifies that (2.13) holds for suitable u with ℓ replaced by functionals λ of the form $\lambda u = \ell u(x)$, where $x \in (0, 1]$ and $\ell \in \mathscr{L}$.

2.4.3 *Differential equations for sequences* $(u_i(x))$

In order to apply the theory in the preceding section one has to choose a set $\mathscr{L}$ of linear functionals ℓ which describes K completely and then verify (2.30). For some cases this is easily done; for other cases, however, carrying out the proofs can be difficult, although a suitable set may exist, such as the set $\mathscr{L}$ of all $\ell > o$. Then it may be easier to use a modified approach which takes into account the special properties of the problem considered. This direct approach may also lead to weaker assumptions, as for the following problem.

Now let Y be a linear subspace of the space of all (real) sequences (y_i) such that each sequence with only finitely many $y_i \neq 0$ belongs to Y. Denote by $\leqslant$ the natural (componentwise) order in Y. Depending on the definition of Y the core of the positive cone in Y may be empty or not. Suppose that $q = (q_i)$ with $q_i > 0$ $(i = 1, 2, \ldots)$ is a fixed element in Y and define $y \succ o$ by $y \succ_q o$. Denote by $\mathscr{L}$ the set which consists of the nonlinear functional $\ell_\infty y = \liminf(y_j q_j^{-1})$ and all point functionals $\ell_i y = y_i$ $(i = 1, 2, \ldots)$. Then for $y \in Y: y \geqslant o \Leftrightarrow \ell y \geqslant 0$ $(\ell \in \mathscr{L})$; $y \succ o \Leftrightarrow \ell y > 0$ $(\ell \in \mathscr{L})$.

Suppose moreover that f is a given function on $(0, 1] \times Y$ with values $f(x, y) = (f_i(x, y)) \in Y$, $f(x, o) = o$, and define g: $(0, 1] \times Y \times Y$ by

$$g_i(x, y, z) = f_i(x, z_1, \ldots, z_{i-1}, y_i, z_{i+1}, z_{i+2}, \ldots).$$

We want to derive estimates for a fixed function u: $[0, 1] \to Y$ with components $u_i \in C_0[0, 1] \cap C_1(0, 1]$ and $u'(x) = (u_i'(x)) \in Y$, using inequalities for $Mu(x) = u'(x) + f(x, u(x))$ $(0 < x \leqslant 1)$.

Theorem 2.9. *Suppose that for each* $x \in (0, 1]$ *and each* $y \geqslant o$ *in* Y *the following inequalities are satisfied:*

$$g(x, y, \bar{z}) \leqslant g(x, y, z), \qquad \textit{if} \quad o \leqslant z \leqslant \bar{z}, \tag{2.32}$$

$$\limsup q_i^{-1} g_i(x, y, o) \leqslant 0, \qquad \textit{if} \quad \lim q_i^{-1} y_i = 0. \tag{2.33}$$

Moreover, assume that for each $x \in [0, 1]$ *and each* $\varepsilon > 0$ *there exists a* $\delta > 0$ *such that*

$$\begin{aligned} u(x + h) - u(x) &\geqslant -\varepsilon q, &&\text{if} \quad 0 < h < \delta, \quad x + h \leqslant 1; \\ u'(x) - h^{-1}(u(x) - u(x - h)) &\leqslant \varepsilon q, &&\text{if} \quad 0 < h < \delta, \quad 0 \leqslant x - h. \end{aligned} \tag{2.34}$$

Then

$$u(0) \succ o, Mu(x) \succ o \quad (0 < x \leqslant 1) \quad \Rightarrow \quad u(x) \succ o \quad (0 \leqslant x \leqslant 1).$$

Proof. Due to (2.34), the number $\xi = \sup\{\zeta \in [0, 1] : u(x) \succ o$ for $0 \leqslant x \leqslant \zeta\}$ satisfies $\xi > 0$ and $u(\xi) \not\succ o$ if $\xi < 1$. Moreover $u(\xi) \geqslant o$, because of the continuity of the components u_i. Thus, it suffices to show that the inequalities $u(\xi) \geqslant o, u(\xi) \not\succ o$ cannot hold. If these inequalities are satisfied, then $\ell u(\xi) = 0$ for some $\ell \in \mathscr{L}$. In case this relation holds for a point functional $\ell = \ell_i$ one obtains a contradiction to $Mu(\xi) \succ o$, using $\ell u'(\xi) \leqslant 0$ and $\ell f(\xi, u(\xi)) = g_i(\xi, o, u(\xi)) \leqslant 0$. Suppose, therefore, that $\ell_\infty u(\xi) = 0$.

First, we consider the case where $\ell_\infty u(\xi) = \ell_\infty(-u)(\xi) = 0$, that is, $\lim q_i^{-1} u_i(\xi) = 0$. For $h > 0$ and $\varepsilon > 0$, we estimate

$$\begin{aligned} u'(\xi) &\leqslant [u'(\xi) - h^{-1}(u(\xi) - u(\xi - h))] + h^{-1}u(\xi) \\ &\leqslant \varepsilon q + h^{-1}u(\xi) \qquad \text{if} \quad h < \delta(\varepsilon, \xi) \end{aligned}$$

and thus obtain $\ell_\infty(-u'(\xi)) \geqslant 0$. Assumption (2.32) yields $f(\xi, u(\xi)) = g(\xi, u(\xi), u(\xi)) \leqslant g(\xi, u(\xi), o)$, so that $\ell_\infty(-f(\xi, u(\xi))) \geqslant \ell_\infty(-g(\xi, u(\xi), o)) \geqslant 0$, due to (2.33). Consequently $\ell_\infty(-Mu(\xi)) \geqslant 0$, which contradicts $Mu(\xi) \succ o$.

In general, only a subsequence $q_{i_j}^{-1} u_{i_j}(\xi)$ converges toward 0. Then we prove that $\lim \inf_j (q_{i_j})^{-1}(-g_{i_j}(\xi, u(\xi), o)) \leqslant 0$. Observe that the last inequality holds, since $\ell_\infty(-g(\xi, y, o)) \leqslant 0$, where $y_i = u_i(\xi)$ if i belongs to the above subsequence, $y_i = 0$ otherwise. □

Corollary 2.9a. *Condition* (2.33) *is satisfied if* $g_i(x, y, o) \leqslant d(x, q_i^{-1} y_i) q_i$ *for* $0 < x \leqslant 1$, $y \in Y$ *with* $y \geqslant o$ *and all indices* i, *where it is assumed that* $d(x, 0) = 0$ *and for each* $x \in (0, 1]$ *a number* $\varepsilon > 0$ *exists such that* $d(x, s)$ *is continuous for* $s \in [0, \varepsilon)$.

2.4.4 Maximal and minimal solutions

For initial value problems involving a differential operator M one can often prove the existence of a maximal solution w and, in addition, derive an estimate $v \leqslant w$ from $v(0) \leqslant w(0)$ and the differential inequality $Mv \leqslant o$ without requiring conditions like those in Theorems 2.5 and 2.6, which ensure the inverse-monotonicity. (Analogous statements hold for the minimal solution.) Such a theory can, for example, be carried out using implications such as those in Theorems 2.4 and 2.7, provided a suitable existence theory for the type of initial value problems considered is available.

We explain this method of proof for the simple differential operator M in (2.25) where, for simplicity, f is assumed to be continuous on $[0, \infty) \times \mathbb{R}$. That means we consider an initial value problem of the form

$$u(0) = r_0, \quad Mu(x) := u'(x) + f(x, u(x)) = 0. \qquad (2.35)$$

Each such problem has a (continuously differentiable) solution u which can be extended "up to the boundary" of the domain of definition of f. Let $j[u]$ denote the interval on which such an extended solution u is defined.

Theorem 2.10. *Problem* (2.35) *has a solution* w *such that for each* $l > 0$ *and each* $v \in C_1[0, l]$

$$\left.\begin{array}{l} v(0) \leqslant r_0 \\ Mv(x) \leqslant 0 \quad \text{for} \quad 0 < x \leqslant l \end{array}\right\} \Rightarrow \left\{\begin{array}{l} v(x) \leqslant w(x) \\ \text{for all} \quad x \in j[w] \cap [0, l]. \end{array}\right. \tag{2.36}$$

Proof. Since problem (2.35) has a (local) solution, there exists an $l > 0$ and a $v \in C_1[0, l]$ with the properties in the premise of (2.36). Let v denote a fixed function of this kind. We choose an antitone zero-sequence of real numbers $\varepsilon_k > 0$ and denote by w_k a solution of $w_k(0) = r_0 + \varepsilon_k$, $Mw_k(x) = \varepsilon_k$. Let $j_1 = j[w_1] \cap [0, l]$. Since $v(0) < w_1(0)$ and $Mv(x) < Mw_1(x)$ on each interval $(0, \xi] \subset j_1$, Theorem 2.4 applied to these intervals in place of $(0, 1]$ yields $v(x) < w_1(x)$ on j_1. By analogous arguments one proves that $v(x) < w_k(x) < \cdots < w_1(x)$ for all $x \in [0, \infty)$ for which all functions involved are defined. However, these inequalities also show that all functions w_k $(k \geqslant 2)$ are at least defined on j_1.

The antitone sequence $\{w_k\} \subset C_1[0, \xi]$ is bounded below and hence converges toward a function w. From the usual integral representation of the solutions w_k one concludes that these functions constitute a uniformly bounded and equicontinuous set and thus converge uniformly, so that w is continuous and satisfies $w(x) = r_0 - \int_0^x f(t, w(t))\, dt$ for $x \in [0, \xi]$. Hence w is a solution on j_1 such that (2.36) holds with $j[w]$ replaced by j_1.

Now let ξ denote the largest number such that an extension $w \in C_1[0, \xi)$ exists for which (2.36) holds with $j[w]$ replaced by $J = [0, \xi)$. If this function w, as a solution of (2.35), cannot further be extended to the right, then $J = j[w]$, so that (2.36) is proved. If, however, the solution w can be extended, then it can also be extended to a solution w such that (2.36) holds with $j[w]$ replaced by $[0, \xi + \varepsilon)$ and some $\varepsilon > 0$. To see this, one applies the method used above for constructing $w \in C_1(j_1)$ to the initial value problem (2.35) with $u(0) = r_0$ replaced by $u(\xi) = w(\xi)$. This yields a contradiction, so that $J = j[w]$. □

Obviously, the function w in Theorem 2.10 is the *maximal solution* of (2.35) in the sense that $u(x) \leqslant w(x)$ on $j[u] \cap j[w]$ for each other solution u. An analogous result can be proved for the *minimal solution* which is defined correspondingly. From these results and the fact that each solution can be extended up to the boundary, one derives the following statement.

Corollary 2.10a. *Suppose that for some* $l > 0$ *the initial value problem* (2.35) *has at most one solution in* $C_1[0, l]$ *and that functions* $v, w \in C_1[0, l]$ *exist which satisfy* $v(0) \leqslant r_0 \leqslant w(0)$, $Mv(x) \leqslant 0 \leqslant Mw(x)$ $(0 < x \leqslant l)$.

Then problem (2.35) *has a solution* $u \in C_1[0, l]$ *and* $v(x) \leqslant u(x) \leqslant w(x)$ $(0 \leqslant x \leqslant l)$.

Problems

2.7 Carry out the proof of Corollary 2.6a.
2.8 Show that the assumptions of Theorem 2.5 are satisfied, if the assumptions of Corollary 2.6a hold for a function d which is continuously differentiable for $x > 0$.
2.9 Show that d in Example 2.6 is a uniqueness function.
2.10 Apply Theorems 2.7 and 2.8 to the case $Y_0 = Y = \mathbb{R}^n$ with $\mathscr{L}$ being the set of all functionals ℓ_i $(i = 1, 2, \ldots, n)$ defined by $\ell_i y = y_i$.
2.11 Apply Theorems 2.7 and 2.8 to a differential operator $\mathscr{M}u(x, \xi)$ such that $\mathscr{M}u(0, \xi) = u(0, \xi)$ $(0 \leqslant \xi \leqslant l)$, $\mathscr{M}u(x, \xi) = u_x - u_{\xi\xi} + F(x, \xi, u, u_\xi)$ $(0 < \xi < l, 0 < x \leqslant 1)$, where u satisfies $u_\xi(x, 0) = u_\xi(x, l) = 0$ $(0 \leqslant x \leqslant 1)$.
2.12 Prove Theorem 2.8.
2.13 Apply Theorem 2.9 and Corollary 2.9a to obtain sufficient conditions such that $v(0) \prec w(0)$, $v'(x) + f(x, v(x)) \prec w'(x) + f(x, w(x))$ $(0 < x \leqslant 1)$ imply $v(x) \prec w(x)$ $(0 \leqslant x \leqslant 1)$, for suitable functions v, w.
2.14 Applying the line method to an initial value problem $U_x - U_{\xi\xi} + f(x, \xi, U) = 0$ $(x > 0, \xi > 0)$, $U(x, 0) = 0$ $(x > 0)$, $U(0, \xi) = g(\xi)$ $(\xi > 0)$, one obtains a system of differential equations of the form $(Mu)_i(x) = u_i'(x) - h^{-2}(u_{i-1}(x) - 2u_i(x) + u_{i+1}(x)) + f(x, \xi_i, u_i(x)) = 0$ $(i = 1, 2, \ldots)$ with $\xi_i = ih$, $u_0(x) \equiv 0$. Apply Theorem 2.9 to this operator M.

3 INCLUSION BY UPPER AND LOWER BOUNDS

Vector-valued differential operators as considered in Section 2.3 are inverse-monotone only under restrictive conditions such as (2.22) or (2.24). Consequently, if these conditions are not satisfied, the usual inclusion statement $M\varphi \leqslant Mv \leqslant M\psi \Rightarrow \varphi \leqslant v \leqslant \psi$ for an inverse-monotone operator cannot be used. Nevertheless, one can often obtain inequalities $\varphi \leqslant v \leqslant \psi$, if a certain *set* of conditions of the form $\Phi \leqslant Mv \leqslant \Psi$ is satisfied. This will be described here in an abstract setting, based on the results of Section 1. An application to vector-valued (functional) differential operators will be given in Section V.2.

Suppose that $(R, \leqslant)$ and $(S, \leqslant)$ are ordered linear spaces, $M: R \to S$ is a given operator, and a further relation $u_1 \prec u_2$ is defined in R, as in Section 2.1. Moreover, assume that Mu can be written as $Mu = H(u, u)$ for $u \in R$ where $H: R \times R \to S$ denotes a mapping about which further assumptions are made in the following theorem. We shall use the concept of admissible and majorizing families defined in Section 2.1 and also the interval-arithmetic explained in Section I.1.2, which will yield simpler formulas.

In what follows, let φ, ψ, and v denote fixed elements in R such that $\varphi \leqslant \psi$. We are interested in the property

$$H(\varphi, [\varphi, \psi]) \leqslant Mv \leqslant H(\psi, [\varphi, \psi]) \quad \Rightarrow \quad \varphi \leqslant v \leqslant \psi. \tag{3.1}$$

According to our notation, this property means that $\varphi \leqslant v \leqslant \psi$ if $H(\varphi, u) \leqslant Mv \leqslant H(\psi, u)$ for all $u \in [\varphi, \psi]$.

Theorem 3.1. *Suppose that there exist monotone admissible families* $\{\bar{\mathfrak{z}}(\lambda): 0 \leqslant \lambda < \gamma\} \subset R$, $\{\mathfrak{z}(\lambda): 0 \leqslant \lambda < \gamma\} \subset R$ *such that the first majorizes* $\varphi - v$, *the second majorizes* $v - \psi$, *and the following conditions hold.*

1. *If* $v \in [\varphi - \bar{\mathfrak{z}}(\lambda), \psi + \mathfrak{z}(\lambda)]$ *and* $\varphi - \bar{\mathfrak{z}}(\lambda) \nless v$ *for some* $\lambda > 0$, *then there exists a monotone functional* ℓ *on* S *such that*

   (I) $\ell H(\varphi - \bar{\mathfrak{z}}(\lambda), v) \geqslant \ell H(v, v)$,
   (II) $\ell H(\varphi, \omega) > \ell H(\varphi - \bar{\mathfrak{z}}(\lambda), v)$ *for some* $\omega \in [\varphi, \psi]$.

2. *If* $v \in [\varphi - \bar{\mathfrak{z}}(\lambda), \psi + \mathfrak{z}(\lambda)]$ *and* $v \nless \psi + \mathfrak{z}(\lambda)$ *for some* $\lambda > 0$, *then there exists a monotone functional* ℓ *on* S *such that*

   (I) $\ell H(v, v) \geqslant \ell H(\psi + \mathfrak{z}(\lambda), v)$,
   (II) $\ell H(\psi + \mathfrak{z}(\lambda), v) > \ell H(\psi, \omega)$ *for some* $\omega \in [\varphi, \psi]$.

Under these assumptions $H(\varphi, [\varphi, \psi]) \leqslant Mv \leqslant H(\psi, [\varphi, \psi]) \Rightarrow \varphi \leqslant v \leqslant \psi$.

Proof. To derive the result from Corollary 1.1a we define

$$K_\lambda = [\varphi - \bar{\mathfrak{z}}(\lambda), \psi + \mathfrak{z}(\lambda)], \qquad \mathring{K}_\lambda = \{u \in R : \varphi - \bar{\mathfrak{z}}(\lambda) \prec u \prec \psi + \mathfrak{z}(\lambda)\},$$

for $0 \leqslant \lambda < \gamma$ and $C = \{U \in S : H(\varphi, [\varphi, \psi]) \leqslant U \leqslant H(\psi, [\varphi, \psi])\}$.

There exists a $\lambda \in [0, \gamma)$ with $v \in K_\lambda$. If $v \in \Gamma_\lambda$ for some $\lambda > 0$, then $v \in K_\lambda$ and $\varphi - \bar{\mathfrak{z}}(\lambda) \nless v$ or $v \nless \psi + \mathfrak{z}(\lambda)$. We only treat the second case, for which we choose $\Omega = \{U \in S : \ell U < \ell H(\psi + \mathfrak{z}(\lambda), v)\}$ with ℓ the functional in assumption 2. Obviously, $Mv \notin \Omega$ due to condition 2.I. On the other hand, if $Mv \in C$, then $\ell Mv \leqslant \ell H(\psi, \omega) < \ell H(\psi + \mathfrak{z}(\lambda), v)$ with the element ω in condition 2.II, so that $Mv \in \Omega$. □

Let us make some comments on the preceding assumptions.

With the notation $M_0 u = H(u, v)$, condition 2.I can be written in the form $\ell M_0 v \geqslant \ell M_0(\psi + \mathfrak{z}(\lambda))$, which corresponds to condition I of Theorem 2.1b. This suggests that, for a concrete problem, the more general condition 2.I may be verified with the same methods as condition I of Theorem 2.1b. If, in particular, $H(x, y) = Ax + By$ with operators $A, B: R \to S$, then for linear ℓ condition 2.I is equivalent to $\ell Av \geqslant \ell A(\psi + \mathfrak{z}(\lambda))$. (An analogous remark holds for condition 1.I.)

The inequalities in 1.II and 2.II are the essential conditions imposed on $\mathfrak{z}(\lambda)$ and $\bar{\mathfrak{z}}(\lambda)$. These inequalities contain the (unknown) element v and an arbitrary element $\omega \in [\varphi, \psi]$. Of course, one may replace v by the order interval $[\varphi - \bar{\mathfrak{z}}(\lambda), \psi + \mathfrak{z}(\lambda)]$, so that, for example, the following sufficient condition for 2.II is obtained:

$$\ell H(\psi + \mathfrak{z}(\lambda), [\varphi - \bar{\mathfrak{z}}(\lambda), \psi + \mathfrak{z}(\lambda)]) > \ell H(\psi, \omega) \qquad \text{for some} \quad \omega \in [\varphi, \psi].$$

Here, ω may then be chosen so that $\ell H(\psi, \omega)$ becomes minimal. However, this is not always the best way to proceed (see Example V.2.3).

The element ω in condition 2.II (and, analogously, ω in 1.II) may depend on the functional ℓ, which in turn depends on the (unknown) function v. Therefore, ω in general will be chosen dependent on v (as in Corollary V.2.1a, for example).

By a slight modification of the proof of Theorem 3.1 one can also obtain the following similar result which does not contain a quantity like ω.

Corollary 3.1a. *In Theorem* 3.1 *conditions* 1.II *and* 2.II *may be replaced by, respectively,*

$$\begin{aligned} \sup \ell H(\varphi, [\varphi, \psi]) &> \sup \ell H(\varphi - \bar{\mathfrak{z}}(\lambda), [\varphi - \bar{\mathfrak{z}}(\lambda), \psi + \mathfrak{z}(\lambda)]), \\ \inf \ell H(\psi + \mathfrak{z}(\lambda), [\varphi - \bar{\mathfrak{z}}(\lambda), \psi + \mathfrak{z}(\lambda)]) &> \inf \ell H(\psi, [\varphi, \psi]), \end{aligned} \tag{3.2}$$

provided these terms exist.

Suppose now that R is Archimedian and $\overset{\text{!}}{\leqslant}$ denotes the strict order in this space. Then the elements $\mathfrak{z}(\lambda)$ may be chosen as in Example 2.1 and the elements $\bar{\mathfrak{z}}(\lambda)$ may have a corresponding form. In particular, one may choose $\mathfrak{z}(\lambda) = \bar{\mathfrak{z}}(\lambda) = \lambda z$ $(0 \leqslant \lambda < \infty)$ with a fixed element $z \succ o$ in R. Another possibility is to choose $\varphi \prec z \prec \psi$ and define $\bar{\mathfrak{z}}(\lambda) = \lambda(1 - \lambda)^{-1}(z - \varphi)$ and $\mathfrak{z}(\lambda) = \lambda(1 - \lambda)^{-1}(\psi - z)$ for $\lambda \in [0, \gamma) = [0, 1)$. If $\mathfrak{z}(\lambda)$ and $\bar{\mathfrak{z}}(\lambda)$ have the latter form and $H(x, y) = Ax + By$ with linear operators A, B, then conditions (3.2) are satisfied for a linear $\ell > o$ if $\ell A\varphi + \sup \ell B[\varphi, \psi] < \ell Mz < \ell A\psi + \inf \ell B[\varphi, \psi]$. For this linear case, the results described can also be derived from the theory in the next section.

Problems

3.1 Derive Theorem 2.1b from Theorem 3.1, assuming that an admissible family $\bar{\mathfrak{z}}(\lambda)$ with $\bar{\mathfrak{z}}(\lambda) \succ o$ for $\lambda > 0$ exists.

3.2 Formulate results which generalize Theorems 2.1a,c,d in the same way as Theorem 3.1 generalizes Theorem 2.1b.

4 RANGE–DOMAIN IMPLICATIONS FOR LINEAR AND CONCAVE OPERATORS

In this section, we investigate range–domain implications (1.1) for rather general sets K and C, but restrict the "nonlinearity" of M. Section 4.1 contains the abstract theory, Section 4.2 applications to differential operators of the second order which are not inverse-positive.

4.1 Abstract theory

Suppose that R and S are (real) linear spaces and $M: R \to S$ is a given operator. Moreover, let K, $\mathring{K}$, Γ, C, $\mathring{C}$, E denote nonempty sets such that K is segmentally closed and

$$\mathring{K} \subset K \subset R, \qquad \Gamma = K \sim \mathring{K}, \qquad \mathring{C} \subset C \subset S, \qquad E = C \sim \mathring{C}.$$

The following theorem will contain assumptions I and II similar to the assumptions I and II in the monotonicity theorem and, in addition, assumptions (a)–(c), for which some sufficient conditions will be provided in the succeeding proposition.

Theorem 4.1. *Let v and z denote fixed elements in R such that*

I. *for each $u = (1 - \lambda)v + \lambda z$ with $\lambda \in (0, 1]$,*

$$u \in \Gamma \quad \Rightarrow \quad Mu \notin \mathring{C}; \tag{4.1}$$

II. $z \in K$, $Mz \in \mathring{C}$.

Suppose, in addition, that

(a) *$(1 - \mu)v + \mu z \in K$ for some $\mu \in (0, \lambda)$, if $(1 - \lambda)v + \lambda z \in \mathring{K}$ and $\lambda \in (0, 1)$,*

(b) *$(1 - \lambda)v + \lambda z \in K$ for some $\lambda \in [0, 1)$,*

(c) $$\left.\begin{matrix} Mv \in C \\ Mz \in \mathring{C} \end{matrix}\right\} \Rightarrow M((1 - \lambda)v + \lambda z) \in \mathring{C} \text{ for all } 0 < \lambda \leqslant 1.$$

Then $Mv \in C \Rightarrow v \in K$.

Proof. To derive this result from Theorem 1.1, define

$$\left.\begin{aligned} K_\lambda &= \{u \in R : (1 - \lambda)u + \lambda z \in K\}, \\ \mathring{K}_\lambda &= \{u \in R : (1 - \lambda)u + \lambda z \in \mathring{K}\} \end{aligned}\right. \qquad \text{for} \quad \lambda \in I = [0, 1).$$

Since K is segmentally closed and conditions (a), (b) hold, assumption (i) of Theorem 1.1 is satisfied. To verify (ii), let $v \in \Gamma_\lambda$ for some $\lambda \in (0, 1)$, i.e., $(1 - \lambda)v + \lambda z \in \Gamma$. Then $v \notin \Omega' = \{u \in R : M((1 - \lambda)u + \lambda z) \in \mathring{C}\}$, due to condition I. On the other hand, conditions II and (c) imply that $v \in \Omega'$ if $Mv \in C$. $\square$

Proposition 4.2. (i) *Condition* (c) *in Theorem* 4.1 *is satisfied if C is convex, E is an extremal subset of C, and M is linear.*

(ii) *Condition* (c) *in Theorem* 4.1 *is satisfied if C is a cone, E is an extremal subset of C, and M is concave, that is, $M((1 - \lambda)u + \lambda v) \geqslant (1 - \lambda)Mu + \lambda Mv$ for $u, v \in R$, $0 \leqslant \lambda \leqslant 1$ with $\leqslant$ denoting the order relation induced by C.*

(iii) *Conditions* (a) *and* (b) *are satisfied if $\mathring{K} = K_X^c$ for a convex set X with $K \subset X \subset R$ and $v \in X$.*

Proof. (i) and (ii) are derived using Proposition I.2.4. (iii) If $v_\lambda := (1 - \lambda)v + \lambda z \in K_X^c$ for a $\lambda \in (0, 1)$ and $v \in X$, then $(1 - t)v_\lambda + tv \in K$ for some $t \in (0, 1)$, that is, $(1 - \mu)v + \mu z \in K$ for $\mu = (1 - t)\lambda < \lambda$. Thus (a) holds. Assumptions I and II imply $z \in \mathring{K} = K_X^c$, so that (b) holds by definition of this set. □

By combining the sufficient conditions in the preceding proposition, one can derive a series of more special results from Theorem 4.1. We shall present some of them in the following theorem and corollaries, using (on purpose) different ways to formulate the assumptions. All these results pertain to the following special case.

Case (L): $M: R \to S$ is linear, K is segmentally closed, C is convex, and E an extremal subset of C.

Theorem 4.3. *Suppose that for a convex set X with $K \subset X \subset R$ the following two conditions are satisfied*:

I. $M\partial_X K \cap C \subset E$,
II. $MK \cap C \not\subset E$.

Then $u \in X$, $Mu \in C \Rightarrow u \in K$.

Proof. (See also Problem 4.1.) Let $\mathring{K} = K_X^c$, $\Gamma = \partial_X K$, and $\mathring{C} = C \sim E$. Then assumption I of this theorem implies that (4.1) holds for each $u \in R$, and II guarantees the existence of a suitable z. Moreover, for $v \in X$ assumptions (a)–(c) in Theorem 4.1 are satisfied according to Proposition 4.2(i), (iii). Thus the above statement follows from Theorem 4.1. □

If $X = R$, Theorem 4.3 asserts that M is (C, K)-inverse. For $X = R$ and C and K being cones, the theorem is equivalent to the monotonicity theorem on inverse-positive linear operators.

A series of further results on inverse-positive operators as proved in Section II.1 can also be generalized to Case (L). Some of these generalizations are left as problems (4.2 and 4.3).

Corollary 4.3a. *Suppose that $\Phi: R \to \{\mathbb{R}, -\infty\}$ is a concave functional such that $u \in K$ if and only if $\Phi u \geqslant 0$. Moreover, assume that*

I. $u \in R$, $\Phi u = 0 \Rightarrow Mu \notin C \sim E$,
II. *there exists a* $z \in K$ *with* $Mz \in C \sim E$.

Then for all $u \in R$ the relations $Mu \in C$, $\Phi u > -\infty$ imply $u \in K$.

This statement follows from Theorem 4.3 by using Proposition I.2.7.

In particular, let now $(R, \leqslant)$ be an Archimedian ordered space and K the order interval $K = [\varphi, \psi]$ for given $\varphi, \psi \in R$ satisfying $\varphi \leqslant \psi$. (Since $(R, \leqslant)$ is Archimedian, K is segmentally closed as required here.) Then we obtain the following result, still requiring the assumptions which describe Case (L).

Corollary 4.3b. *Let $p > o$ and $q > o$ denote given elements in R and assume that*

I. $u \in R, \varphi \leqslant u \leqslant \psi, Mu \in C \sim E \Rightarrow \varphi \prec_p u \prec_q \psi$,
II. *there exists a* $z \in R$ *with* $\varphi \leqslant z \leqslant \psi, Mz \in C \sim E$.

Then

$$\left.\begin{array}{ll} u \in R, \quad Mu \in C & \\ \varphi - \gamma p \leqslant u \leqslant \psi + \gamma q & \text{for some} \quad \gamma \in \mathbb{R} \end{array}\right\} \Rightarrow \quad \varphi \leqslant u \leqslant \psi.$$

For the proof define $\Phi u = \sup \Lambda(u)$ with $\Lambda(u) = \{\alpha : \varphi + \alpha p \leqslant u \leqslant \psi - \alpha q\}$ and apply Corollary 4.3a (see also Problem I.2.6).

In an analogous way one obtains the following result, now defining K to be the order cone of the Archimedian ordered linear space $(R, \leqslant)$ and still requiring the assumptions of Case (L).

Corollary 4.3c. *Suppose that for some $p > o$ in R*

I. $u \in R, u \geqslant o, Mu \in C \sim E \Rightarrow u \succ_p o$,
II. $z \geqslant o, Mz \in C \sim E$ *for some* $z \in R$.

Then for each $u \in R$ which is p-bounded below, $Mu \in C \Rightarrow u \geqslant o$.

Problems

4.1 Derive Theorem 4.3 directly, generalizing the proof of Theorem II.1.2 (without using Theorem 4.1 and Proposition 4.2).
4.2 Assumptions I, II of Theorem 4.3 in general are also necessary. Consider the special case $X = R$ and assume that M is (C, K)-inverse. Then prove: (i) $M\,\partial K \cap C \subset \partial C$; (ii) $M\,\partial K \cap C \subset E$, if M is $(C \sim E, K^c)$-inverse; (iii) $MK \cap C \not\subset E$, if $MR \cap C^c \neq \varnothing$ and $E \neq C$. (See the proof of Theorem II.1.6.)
4.3 Suppose that K and C are cones and $E = \{o\}$. For this case, generalize Theorem 4.3 by assuming $M(\partial_X K \sim \{o\}) \cap C = \varnothing, M(K \sim \{o\}) \cap C \neq \varnothing$ instead of I and II. Formulate the result using the notation $u \succ_X o \Leftrightarrow u \in K_X^c$. (Compare Theorem II.1.9.)

4.2 Applications to non-inverse-positive differential operators

As an application of the theory on linear range–domain implications we shall now discuss inclusion statements for (self-adjoint regular) differential operators of the second order which are not inverse-positive.

For $u \in R := C_2[0, 1]$ let

$$Mu(x) = \begin{cases} L[u](x) & \text{for } \ 0 < x < 1 \\ B_0[u] & \text{for } \ x = 0 \\ B_1[u] & \text{for } \ x = 1 \end{cases} \quad \text{with} \quad \begin{aligned} L[u] &= -(pu')' + qu \\ B_0[u] &= -\alpha_0 u'(0) + \gamma_0 u(0) \\ B_1[u] &= \alpha_1 u'(1) + \gamma_1 u(1), \end{aligned}$$

where $p, p', q \in C_0[0, 1]$, $p(x) > 0$ $(0 \leqslant x \leqslant 1)$, $\alpha_0 = 0$ or $\alpha_0 = 1$, $\gamma_0 = 1$ if $\alpha_0 = 0$, $\alpha_1 = 1$ or $\alpha_1 = 0$, $\gamma_1 = 1$ if $\alpha_1 = 0$.

This operator M maps R into $S = \mathbb{R}[0, 1]$. We shall, however, define Mu also for less smooth functions u. Whenever the one-sided limits $L[u](x - 0)$ and $L[u](x + 0)$ exist for some $x \in (0, 1)$, let $L[u](x)$ be the minimum of these two numbers and define then $Mu(x)$ by the above formulas.

We shall derive sufficient conditions such that for $u \in R$ and suitable functions ψ and Ψ

$$|Mu| \leqslant \Psi \quad \Rightarrow \quad |u| \leqslant \psi, \qquad \text{i.e.,} \quad -\Psi \leqslant Mu \leqslant \Psi \quad \Rightarrow \quad -\psi \leqslant u \leqslant \psi. \tag{4.2}$$

(More general implications $\Phi \leqslant Mu \leqslant \Psi \Rightarrow \varphi \leqslant u \leqslant \psi$ can be treated analogously.) There are various possibilities to proceed. Therefore, we shall not only present some results (in Theorems 4.8 and 4.9 and some corollaries), but also emphasize the methods for obtaining these results.

If u^* is a solution of a given equation $Mu = r$ and u_0 an approximate solution with defect $d[u_0] = r - Mu_0$, then the above implication applied to $u = u^* - u_0$ yields

$$|d[u_0]| \leqslant \Psi \quad \Rightarrow \quad |u^* - u_0| \leqslant \psi.$$

For inverse-positive M we can in general rather easily obtain suitable $\psi \in C_2[0, 1]$ and $\Psi = M\psi$ such that $|d[u_0]| \leqslant \Psi$ holds, provided the defect $d[u_0]$ has been estimated on the whole interval with sufficient accuracy (see Section II.5.2). For non-inverse-positive M we shall also try to obtain suitable ψ and Ψ in as simple a way as possible.

Here we cannot use $\Psi = M\psi$ for a $\psi \in C_2[0, 1]$, since no $\psi \in C_2[0, 1]$ satisfying $\psi \geqslant o$, $M\psi > o$ exists, if M is not inverse-positive (cf. Theorem II.3.17(i)). However, there exist less smooth functions $\psi \geqslant o$ with $M\psi > o$, due to the fact that the inverse-positivity of M in some way depends on the length of the interval considered (see Theorem II.3.8, for example). More precisely, one can show that there always exist $x_1, x_2, \ldots, x_k$ with $0 < x_1 < \cdots < x_k < 1$ and a $\psi \in C_{0,2}[0, x_1, \ldots, x_k, 1]$ such that $\psi \geqslant o$ and $M\psi(x) > 0$ $(0 \leqslant x \leqslant 1)$.

To explain certain methods for treating non-inverse-positive M, we shall here only try to obtain estimates with functions $\psi \in C_{0,2}[0, x_1, 1]$; this means we consider only the case $k = 1$. First, let us discuss some of the mathematical background.

For a given $x_1 \in (0, 1)$ define operators M_1: $C_2[0, x_1] \to \mathbb{R}[0, x_1]$ and M_2: $C_2[x_1, 1] \to \mathbb{R}[x_1, 1]$ by

$$M_1 u(x) = \begin{cases} Mu(x) & \text{for} \quad 0 \leqslant x < x_1 \\ u(x_1) & \text{for} \quad x = x_1, \end{cases}$$
$$M_2 u(x) = \begin{cases} Mu(x) & \text{for} \quad x_1 < x \leqslant 1 \\ u(x_1) & \text{for} \quad x = x_1. \end{cases} \tag{4.3}$$

We shall need the inverse-positivity of both these operators. This property of M_1 and M_2 may be verified by the methods developed in Section II.3.

Example 4.1. Let $Mu = (-u'' + qu,\ u(0),\ u(1))$ with $q(x) > -4\pi^2$ $(0 \leqslant x \leqslant 1)$, and $x_1 = \frac{1}{2}$. Then both M_1 and M_2 are inverse-positive, since $M_1 z_1 > o$, $M_2 z_2 > o$ for $z_1 = \sin 2\pi x$, $z_2 = \sin 2\pi(1 - x)$. We could also use functions $z_1 = \cos(2\pi - \varepsilon)(x - \frac{1}{4})$ and $z_2 = \cos(2\pi - \varepsilon)(x - \frac{3}{4})$ with small enough $\varepsilon > 0$, for which z_1, z_2, Mz_1, and Mz_2 are pointwise >0 on the respective intervals. Observe that in the latter case $z(x) > 0$, $Mz(x) > 0$ $(0 \leqslant x \leqslant 1)$ for $z \in C_{0,2}[0, x_1, 1]$ defined by $z(x) = z_1(x)$ $(0 \leqslant x \leqslant x_1)$, $z(x) = z_2(x)$ $(x_1 < x \leqslant 1)$. □

More generally, one derives the following statement from the theory in Section II.3 (use, in particular, Theorem II.3.17(i)).

Proposition 4.4. (i) *Both operators M_1 and M_2 are inverse-positive if and only if there exists a function $z \in C_{0,2}[0, x_1, 1]$ such that $z \geqslant o$, $z(x_1) > 0$, and $Mz \geqslant o$.*

(ii) *If both M_1 and M_2 are inverse-positive, then for all $u \in R$, all $\psi \in C_{0,2}[0, x_1, 1]$ and each constant $\mu \geqslant 1$:*

$$|Mu| \leqslant M\psi, \quad |u(x_1)| \leqslant \mu\psi(x_1) \quad \Rightarrow \quad |u| \leqslant \mu\psi. \tag{4.4}$$

Of course, we cannot immediately apply the latter implication, since $u(x_1)$ is supposed to be unknown. Our goal is to derive sufficient conditions on ψ and μ such that the inequality for $u(x_1)$ in (4.4) need not be required. Then the resulting implication is equivalent to (4.2) with $\Psi = \mu^{-1}M\psi$. The parameter μ has been introduced to obtain weaker assumptions on ψ.

Our result in Theorem 4.8 will be derived from the abstract Theorem 4.3. In order to motivate the assumptions to be required, we shall first present some results related to the possible choice of ψ and x_1 and the behavior of the Green function corresponding to M.

Proposition 4.5. *Let $\alpha, \beta \in C_2[0, 1]$ denote nontrivial solutions of $L[\alpha](x) = 0$ $(0 \leqslant x \leqslant 1)$, $B_0[\alpha] = 0$, $L[\beta](x) = 0$ $(0 \leqslant x \leqslant 1)$, $B_1[\beta] = 0$.*

Then for a given $x_1 \in (0, 1)$ the operator M_1 in (4.3) is inverse-positive if and only if $\alpha(x) \neq 0$ for $0 < x \leqslant x_1$; and M_2 is inverse-positive if and only if $\beta(x) \neq 0$ for $x_1 \leqslant x < 1$.

Proof. If $\alpha(x) \neq 0$ on $(0, x_1]$, we may assume $\alpha(x) > 0$ $(0 < x \leqslant x_1)$. Then the inverse-positivity of M_1 follows from Theorem II.3.17(i) applied to M_1, α instead of M, z. On the other hand, if M_1 is inverse-positive and hence invertible, we have $\alpha(x_1) \neq 0$ and may assume $\alpha(x_1) > 0$. Then $M_1\alpha > 0$ implies $\alpha(x) > 0$ $(0 < x \leqslant x_1)$, according to the theorem mentioned above. The result concerning β and M_2 is proved analogously. □

According to this proposition the operators M_1 and M_2 can be inverse-positive only if the smallest $\xi \neq 0$ with $\alpha(\xi) = 0$ and the largest $\eta \neq 1$ with $\beta(\eta) = 0$ satisfy $\eta < \xi$, provided these functions vanish somewhere in $(0, 1)$. We then have to choose $x_1 \in (\eta, \xi)$. However, practically we intend to prove the inverse-positivity of M_1 and M_2 by constructing a suitable function $z \in C_{0,2}[0, x_1, 1]$ (see Example 4.1). Sometimes such a function z may show us that M itself is inverse-positive as explained now.

For a function $u \in \mathbb{R}[0, 1]$ such that the one-sided limits $u(x_1 + 0)$ and $u(x_1 - 0)$ exist, we define $\Delta u(x_1)$ to be the "jump" $\Delta u(x_1) = u(x_1 + 0) - u(x_1 - 0)$, so that, for example,

$$\Delta u'(x_1) = u'(x_1 + 0) - u'(x_1 - 0) \qquad \text{for} \quad u \in C_{0,2}[0, x_1, 1].$$

Proposition 4.6. *If a* $z \in C_{0,2}[0, x_1, 1]$ *exists such that* $Mz \geqslant o$ *and* $\Delta z'(x_1) < 0$, *then* M *is inverse-positive on* R.

Proof. First observe that $z(x_1) > 0$ due to $\Delta z'(x_1) < 0$; consequently, both M_1 and M_2 are inverse-positive, according to Proposition 4.4. Now let $u \in R$ satisfy $Mu \geqslant o$ and $u \ngeqslant o$. Then a smallest $\lambda > 0$ with $v := (1 - \lambda)u + \lambda z \geqslant o$ exists. Again, $v(x_1) > 0$ follows from $\Delta v'(x_1) < 0$. When we now apply Theorem II.3.16B to M_1 and M_2, we see that $v(x) > 0$ $(0 < x < 1)$, $v'(0) > 0$ if $v(0) = 0$, $v'(1) < 0$ if $v(1) = 0$. In each case, we can then conclude that λ cannot be minimal. □

Proposition 4.7. (i) *If* M *is one-to-one on* R *and the Green function* $\mathcal{G}$ *corresponding to* M *satisfies* $\mathcal{G}(x_1, x) \geqslant 0$ $(0 \leqslant x \leqslant 1)$ *for the fixed value* $x_1 \in (0, 1)$, *then* M *is inverse-positive and hence* $\mathcal{G}(y, x) \geqslant 0$ $(0 \leqslant x, y \leqslant 1)$.

(ii) *Suppose that both* M_1 *and* M_2 *are inverse-positive, and that* M *is one-to-one on* R *but not inverse-positive. Then the Green function* $\mathcal{G}$ *corresponding to* M *satisfies*

$$\mathcal{G}(x_1, x) = \mathcal{G}(x, x_1) < 0 \qquad (0 < x < 1). \tag{4.5}$$

Proof. (i) follows from Proposition 4.6 with $z(x) = \mathcal{G}(x_1, x)$. (ii) As a consequence of Proposition 4.5, $\mathcal{G}(x_1, x)$ does not vanish for any $x \in (0, 1)$ (use the representation of $\mathcal{G}$ in terms of α and β). Moreover, $\mathcal{G}(x_1, x)$ is not positive according to (i). □

Using this Green function $\mathscr{G}(x_1, x)$, we can obtain a *theoretical* estimate for the value $u(x_1)$ occurring in (4.4). Suppose that $\mathscr{G}$ exists and satisfies (4.5), $u \in R$ satisfies the boundary conditions $B_0[u] = B_1[u] = 0$, and $|Mu| \leqslant M\psi$ for some $\psi \in C_{0,2}[0, x_1, 1]$. Then multiplying the inequality $L[u](x) \leqslant L[\psi](x)$ $(0 \leqslant x \leqslant 1)$ by $-g(x)$ with $g(x) := \mathscr{G}(x_1, x)$, integrating over x and using partial integration, one obtains

$$\begin{aligned}-u(x_1) \leqslant -\psi(x_1) + p(x_1)\,\Delta\psi'(x_1)|g(x_1)| \\ + p(0)W[\psi, g](0) - p(1)W[\psi, g](1),\end{aligned}$$

where $W[\psi, g]$ denotes the Wronskian determinant and

$$\begin{aligned} W[\psi, g](0) &= \begin{cases}\psi(0)g'(0) & \text{for} \quad \alpha_0 = 0\\ B_0[\psi]g(0) & \text{for} \quad \alpha_0 = 1,\end{cases}\\ -W[\psi, g](1) &= \begin{cases}-\psi(1)g'(1) & \text{for} \quad \alpha_1 = 0\\ B_1[\psi]g(1) & \text{for} \quad \alpha_1 = 1.\end{cases}\end{aligned} \tag{4.6}$$

The same estimate holds for u in place of $-u$.

Thus, under the conditions imposed *the inequality* $|u(x_1)| \leqslant \mu\psi(x_1)$ *in* (4.4) *may be replaced by the inequality*

$$\begin{aligned}(1 + \mu)\psi(x_1) \geqslant p(x_1)\,\Delta\psi'(x_1)|\mathscr{G}(x_1, x_1)| \\ + \{p(0)W[\psi, g](0) - p(1)W[\psi, g](1)\}.\end{aligned}$$

Observe that the term within braces could be dropped, since it is a number $\leqslant 0$. The application of this conditions requires that one know the numerical value of $|\mathscr{G}(x_1, x_1)|$ or an upper bound of it.

Now we shall derive a sufficient condition similar to the one above in which $\mathscr{G}(x_1, x)$ is replaced by a function $g(x)$ which may be an approximation of the Green function. By applying Theorem 4.3 on linear range–domain implications we obtain the following result.

Theorem 4.8. *Suppose that for a given* $x_1 \in (0, 1)$ *there exist functions* $\psi \in C_{0,2}[0, x_1, 1]$, $g \in C_{0,2}[0, x_1, 1]$ *and a constant* $\mu \geqslant 1$ *with the following properties*:

$$\begin{gathered}\psi(x) > 0 \quad (0 < x < 1), \qquad M\psi > o;\\ \psi'(0) > 0 \quad \text{if} \quad \psi(0) = 0, \qquad \psi'(1) < 0 \quad \text{if} \quad \psi(1) = 0,\end{gathered} \tag{4.7}$$

$$\begin{gathered}g(x) < 0 \quad (0 < x < 1), \qquad B_0[g] = B_1[g] = 0, \qquad \Delta g'(x_1) < 0,\\ g'(0) < 0 \quad \text{if} \quad g(0) = 0, \qquad g'(1) > 0 \quad \text{if} \quad g(1) = 0,\end{gathered} \tag{4.8}$$

$$(\mu + 1)\{(p\psi|\Delta g'|)(x_1) - \int_0^1 |L[g](x)|\psi(x)\,dx\} \geqslant (p|g|\,\Delta\psi')(x_1). \tag{4.9}$$

Then, for each $u \in R$ which satisfies the boundary conditions $B_0[u] = B_1[u] = 0$, the inequality $|Mu| \leqslant M\psi$ implies $|u| \leqslant \mu\psi$.

Proof. We want to prove that $Mu \in C \Rightarrow u \in K$ with

$$K = \{u \in R : B_0[u] = B_1[u] = 0, |u| \leqslant \mu\psi\}, \qquad C = \{U \in S : |U| \leqslant M\psi\}.$$

In order to apply Theorem 4.3 we define an extremal set E of C by $C \sim E = \{U \in S : |U| < M\psi\}$. (Recall that $|U| < \Psi$ if and only if $|U| \leqslant \Psi$ and $|U(\xi)| < \Psi(\xi)$ for at least one value $\xi \in [0, 1]$.)

To prove the above statement one has to treat several cases distinguished by the behavior of ψ at the boundary points. We shall restrict ourselves to the case in which $\psi(0) > 0$ and $\psi(1) = 0$, since here all essential ideas can be explained. In this case define $X = \{u \in R : u(1) = 0\}$; then K_X^c is the set of all $u \in K$ such that $|u(x)| < \mu\psi(x)$ $(0 \leqslant x < 1)$ and $|u'(1)| < -\mu\psi'(1)$.

Condition II of Theorem 4.3 holds, since $z = o$ satisfies $z \in K$ and $Mz \in C \sim E$. To verify condition I, suppose that $u \in K$ satisfies $Mu \in C \sim E$, but $|u(\xi)| = \mu\psi(\xi)$ for some $\xi \in [0, 1)$ or $|u'(1)| = -\mu\psi'(1)$. It suffices to consider the cases $u(\xi) = \mu\psi(\xi)$ and $u'(1) = \mu\psi'(1)$.

For $u(x_1) = \mu\psi(x_1)$ we derive from $-M\psi < Mu$ that

$$-\int_0^1 g(x)L[u + \psi](x)\,dx - p(0)W[u + \psi, g](0) + p(1)W[u + \psi, g](1) > 0. \tag{4.10}$$

(In proving this inequality one has to use the conditions imposed on u, g on the boundary and consider a series of cases. For example, let $\alpha_0 = 1$ and $-L[\psi](x) \equiv L[u](x)$, $-B_1[\psi] = B_1[u] = 0$, so that necessarily $-B_0[\psi] < B_0[u] = 0$. Then $W[u + \psi, g](0) = B_0[\psi]g(0) < 0$, since $g(0)$ cannot vanish. Observe that $g(0) = 0$ and $B_0[g] = 0$ imply $g'(0) = 0$.) Applying partial integration and using the relations $u(x_1) = \mu\psi(x_1)$ and $|u| \leqslant \mu\psi$ we obtain from the above inequality

$$\begin{aligned} -(\mu + 1)(p\psi|\Delta g'|)(x_1) + (p|g|\,\Delta\psi')(x_1) & \\ + \int_0^1 (\mu|L[g](x)| - L[g](x))\psi(x)\,dx & \\ \geqslant (p(u + \psi)\,\Delta g')(x_1) - (pg\,\Delta\psi')(x_1) - \int_0^1 L[g](x)(u + \psi)(x)\,dx > 0. & \end{aligned}$$

Clearly, this result contradicts (4.9), so that $u(x_1) < \mu\psi(x_1)$.

The remaining cases can be treated using arguments of inverse-positivity. For example, we obtain $v(\xi) > 0$ $(x_1 < \xi < 1)$ and $v'(1) < 0$ for $v = \mu\psi - u$, if we apply property (II.3.37) to the interval $[x_1, 1]$ instead of $[0, 1]$ and v instead of u. ☐

The assumptions in the preceding theorem can be modified. Here we state only

Corollary 4.8a. *Theorem* 4.8 *remains true if conditions* (4.8), (4.9) *are replaced by the following two conditions*:

$$L[\psi](\xi) > 0 \qquad \textit{for some} \quad \xi \in (0, 1),$$

$$(\mu + 1)\left\{(p\,|\Delta g'|\,\psi)(x_1) - \int_0^1 |L[g](x)|\,\psi(x)\,dx\right\}$$
$$\geqslant (p\,\Delta\psi'\,|g|)(x_1) + p(0)W[\psi, g](0) - p(1)W[\psi, g](1)$$
$$- \int_0^1 (|L[g](x)| + L[g](x))\psi(x)\,dx,$$

with $W[\psi, g](0)$ *and* $W[\psi, g](1)$ *as in* (4.6).

To prove this statement one proceeds as in the proof of Theorem 4.8, but defines E by $C \sim E = \{U \in C : |U(x)| < M\psi(x)$ for some $x \in (0, 1)\}$ and replaces (4.10) by $-\int_0^1 g(x)L[u + \psi](x)\,dx > 0$. □

The conditions imposed on ψ and g suggest $g = -\psi$ as a possible choice, which yields the following statement.

Corollary 4.8b. *Suppose that for a given* $x_1 \in (0, 1)$ *there exists a function* $\psi \in C_{0,2}[0, x_1, 1]$ *such that* (4.7) holds *and, in addition,*

$$B_0[\psi] = B_1[\psi] = 0, \qquad B[\psi] := -(p\psi\,\Delta\psi')(x_1) + \int_0^1 L[\psi](x)\psi(x)\,dx < 0.$$

Then, for each $u \in C_2[0, 1]$ *with* $B_0[u] = B_1[u] = 0$,

$$|Mu| \leqslant M\psi \quad \Rightarrow \quad |u| \leqslant \mu\psi \tag{4.11}$$

with $\mu \geqslant 1$ *satisfying* $\mu \geqslant -(B[\psi])^{-1} \int_0^1 L[\psi](x)\psi(x)\,dx$.

Observe that

$$B[\psi] = \int_0^1 [p(x)(\psi'(x))^2 + q(x)\psi^2(x)]\,dx + \gamma_0\alpha_0 p(0)\psi^2(0) + \alpha_1\gamma_1 p(1)\psi^2(1).$$

Example 4.2. Let $Mu = (-u'' + qu, u(0), u(1))$ and $x_1 = \frac{1}{2}$. For ψ such that $\psi(x) = \psi(1 - x)$ and $\psi(x) = \sin\gamma\pi x$ $(0 \leqslant x \leqslant \frac{1}{2})$, all conditions in the preceding corollary except $B[\psi] < 0$ are satisfied, if $1 < \gamma < 2$ and $q(x) > -\gamma^2\pi^2$. If $q(x) \leqslant (-\gamma^2 + \delta)\pi^2$ $(0 \leqslant x \leqslant 1)$ with a number δ such that $\delta < \delta_0 = (|\sin\gamma\pi| + \gamma\pi)^{-1}2\gamma^2|\sin\gamma\pi|$, then also $B[\psi] < 0$, and the implication (4.11) holds with $\mu = 1$ if $\delta \leqslant \frac{1}{2}\delta_0$, $\mu = (\delta_0 - \delta)^{-1}\delta$ otherwise. □

In applying the preceding results, inequalities involving integrals have to be verified. In the following we shall present sufficient conditions which contain only pointwise inequalities. This method, too, will be illustrated only for the case $k = 1$. Moreover, we restrict ourselves to considering Dirichlet boundary operators. For $U \in S$ define $U \succ o \Leftrightarrow U(x) > 0$ $(0 \leqslant x \leqslant 1)$.

Theorem 4.9. *Suppose that* $B_0[u] = u(0)$, $B_1[u] = u(1)$, *and that for a given* $x_1 \in (0, 1)$ *there exist functions* φ, ψ *in* $C_{0,2}[0, x_1, 1]$ *which satisfy*

$$\psi \succ o, \qquad M\psi \succ o, \qquad L[\varphi - \psi](x) \geqslant 0 \qquad (0 < x < 1),$$

$$(\varphi - \psi)(0) \geqslant 0, \qquad (\varphi - \psi)(1) \geqslant 0, \qquad (\varphi + \psi)(x_1) = 0, \qquad \Delta\varphi'(x_1) \leqslant 0.$$

Then for $u \in C_2[0, 1]$, $|Mu| \leqslant M\psi \Rightarrow |u| \leqslant \psi$.

Proof. Since here the strong inequality $M\psi \succ o$ is assumed, we can apply Theorem 4.3 in a simpler way than in the proof of Theorem 4.8. Now define $K = \{u \in R : |u| \leqslant \psi\}$, $C = \{U \in S : |U| \leqslant M\psi\}$, $C \sim E = \{U \in S : |U| \prec M\psi\}$. Then $o \in MK \cap (C \sim E)$, so that condition II of Theorem 4.3 is satisfied. In order to verify condition I, consider a $u \in \partial K$. We have $|u(\xi)| = \psi(\xi)$ for some $\xi \in [0, 1]$ and may assume $u(\xi) = \psi(\xi)$ without loss of generality. If $\xi \neq x_1$, arguments often used above lead to the inequality $Mu(\xi) \geqslant M\psi(\xi)$, which implies $Mu \notin C \sim E$.

Thus, let now $u(x_1) = \psi(x_1)$. Since $\psi(x) \geqslant 0$ $(0 \leqslant x \leqslant x_1)$ and $M_1\psi(x) > 0$ $(0 \leqslant x \leqslant x_1)$ the operator M_1 defined in (4.3) is inverse-positive, according to Theorem II.3.2. Using the assumptions on φ, we obtain $M_1(u + \varphi)(x) \geqslant M_1(u + \psi)(x) \geqslant 0$ $(0 \leqslant x \leqslant x_1)$ and hence $(u + \varphi)(x) \geqslant 0$ $(0 \leqslant x \leqslant x_1)$. In an analogous way the inequality $(u + \varphi)(x) \geqslant 0$ $(x_1 \leqslant x \leqslant 1)$ is derived using the operator M_2 defined in (4.3). These inequalities and $(u + \varphi)(x_1) = 0$ imply $(u + \varphi)'(x_1 + 0) \geqslant 0 \geqslant (u + \varphi)'(x_1 - 0)$ and hence $\Delta\varphi'(x_1) \geqslant 0$. Since the opposite inequality is assumed above, we have $\Delta\varphi'(x_1) = 0$. This is only possible, if $(u + \varphi)'(x_1 + 0) = 0 = (u + \varphi)'(x_1 - 0)$ and, therefore, also $(u + \varphi)''(x_1 + 0) \geqslant 0$, $(u + \varphi)''(x_1 - 0) \geqslant 0$. The relations thus obtained imply $L[u](x_1) \leqslant -L[\varphi](x_1 + 0) \leqslant -L[\psi](x_1 + 0)$ and analogous inequalities for $x_1 - 0$. Therefore, $Mu \notin C \sim E$. □

In the above theorem one may, for example, use a function $\varphi \in C_2[0, 1]$ and, in addition, replace ψ by an upper bound, as in the following corollary, where we use $x_1 = \frac{1}{2}$, for simplicity.

Corollary 4.9a. *Suppose that* $B_0[u] = u(0)$, $B_1[u] = u(1)$, *and that* $\varphi \in C_2[0, 1]$ *and* $\psi \in C_{0,2}[0, \frac{1}{2}, 1]$ *are functions which satisfy*

$$\psi \geqslant o, \qquad M\psi \geqslant M\varphi > o, \qquad \psi(\tfrac{1}{2}) \geqslant -\varphi(\tfrac{1}{2}) > 0.$$

Then for each $u \in C_2[0, 1]$, $|Mu| \leqslant M\varphi \Rightarrow |u| \leqslant \psi$.

Proof. The inequalities required of ψ yield the inverse-positivity of the operators M_1 and M_2, as in the preceding proof. Therefore, a function $\tilde{\psi} \succ o$ in $C_{0,2}[0, \frac{1}{2}, 1]$ is defined by $M_1\tilde{\psi}(x) = M_1\varphi(x)$ $(0 \leqslant x < \frac{1}{2})$, $\tilde{\psi}(\frac{1}{2}) = -\varphi(\frac{1}{2})$, $M_2\tilde{\psi}(x) = M_2\varphi(x)$ $(\frac{1}{2} < x \leqslant 1)$. When we apply Theorem 4.9 to $\tilde{\psi}$ instead of ψ, we see that $|Mu| \leqslant M\varphi \Rightarrow |u| \leqslant \tilde{\psi}$. Moreover, the inequalities $M_1\tilde{\psi}(x) \leqslant M_1\psi(x)$ $(0 \leqslant x \leqslant \frac{1}{2})$ and $M_2\tilde{\psi}(x) \leqslant M_2\psi(x)$ $(\frac{1}{2} \leqslant x \leqslant 1)$ yield $\tilde{\psi} \leqslant \psi$. □

When this result is applied, one first determines a suitable function φ on $[0, 1]$ and then calculates a bound ψ on each of the intervals $[0, \frac{1}{2}]$, $[\frac{1}{2}, 1]$ separately, observing the continuity condition $\psi(\frac{1}{2} - 0) = \psi(\frac{1}{2} + 0)$. (As the preceding proof shows, this condition is not even necessary.) For calculating the bound ψ numerically, one may apply the methods explained in Section II.5.2.3 to each of the two subintervals. Thus the additional work to be done for a non-inverse-positive operator M (in case $k = 1$) is the computation of a function φ as above.

To illustrate this corollary let us again consider a very simple example.

Example 4.3. For $Mu = (-u'' - \gamma^2 u,\ u(0),\ u(1))$ with $\gamma = \sqrt{2\pi}$ the functions $\varphi(x) = -\gamma^{-2}(1 + \alpha \cos \gamma(x - \frac{1}{2}))$ and $\psi(x) = 2\varphi(0)\ (\frac{1}{2} - x) + 2|\varphi(\frac{1}{2})|x + \beta x(\frac{1}{2} - x) = \psi(1 - x)$ $(0 \leqslant x \leqslant \frac{1}{2})$ satisfy the assumptions of Corollary 4.9a for suitably chosen $\alpha > 0$, $\beta > 0$; in particular, $L[\varphi](x) \equiv 1$. For example, if α is determined such that $\varphi(0) = \varepsilon > 0$, then ψ depends also on ε, $\psi = \psi_\varepsilon$. Of course, if ψ_ε is a bound of $|u|$ for all sufficiently small $\varepsilon > 0$, then the function ψ_0 obtained by taking the limit $\varepsilon \to 0$ is also a bound. Observe also that the functions φ, ψ may be multiplied by a constant $\delta > 0$. Carrying out the details, one sees that the relations $u(0) = u(1) = 0$, $|Mu(x)| \leqslant \delta$ $(0 < x < 1)$ imply $|u| \leqslant \delta\psi_0$ with $\psi_0(x) = \psi_0(1 - x) = (1 + \mu)2\gamma^{-2}x + (\frac{3}{2} + \mu)(1 - \frac{1}{32}\gamma^2)^{-1}x(\frac{1}{2} - x)$ $(0 \leqslant x \leqslant \frac{1}{2})$, $\mu = |\cos \frac{1}{2}\gamma|^{-1}$. □

NOTES

Section 1

The text was published by Schröder [1977a]. The method of absorption described here essentially is the abstract formulation of a method which has been used by many authors in concrete cases (see, for instance, Protter and Weinberger [1967, Theorem 19], Redheffer [1962b, Definition 3], and for initial value problems Walter [1970, p. 84]).

Section 2

For the abstract Theorems 2.1a–d see Schröder [1962, 1963, 1966, 1968b, 1977a]. In this theory the description of certain assumptions and properties by linear (or nonlinear) functionals as discussed in Section 2.1.3 is a standard tool. An approach similar to that in Section 2.1.2, for example, was described in the third and fourth of the above papers. The term quasi-monotone for abstract operators M as defined above was first used by Volkmann [1972] in considering differential operators $du/dx - Mu$. This term is a natural generalization of a corresponding concept in finite dimensions introduced by Walter [1970] (compare (2.15)). The importance of this concept for initial value problems was already recognized by Muller [1927a], Kamke [1932], Szarski [1947], and Wazewski [1950]. There exist a series of other generalizations of the term quasi-monotone (see Volkmann [1972]). For linear operators, the notion of a Z-operator also is such a generalization (see Section II.4.2). Furthermore, see the "distance condition" mentioned later in these Notes.

A theory of M-functions as defined in Section 2.2 was developed by Rheinboldt [1970]. In particular, this author investigated iterative procedures involving M-functions, generalizing results for linear equations as described in Section II.2.5. See also Ortéga and Rheinboldt [1970], Moré and Rheinboldt [1973], and Moré [1972].

There exists an extensive literature on (scalar-valued or vector-valued) inequalities related to initial value problems; see the books of Lakshmikantham and Leela [1969], Szarski [1965], and Walter [1970]. Important early contributions are, for example, due to Nagumo and Wazewski and their co-workers. We shall not attempt to give a survey, but refer the reader to the bibliographies in the above books.

Differential inequalities in Banach spaces of the type considered in Section 2.4.2 were investigated, for example, by Walter [1971], Volkmann [1972, 1974], Redheffer [1973], and Martin [1975]. For further work in this area and references see also Deimling [1977]. Moreover, this theory was investigated in a series of papers by Lakshmikantham and his co-workers. For references we mention here the recent papers of Deimling and Lakshmikantham [1978a,b] and Lakshmikantham [1979].

The property of inverse-monotonicity of operators (2.29) can be incorporated into the theory of flow invariance. A set A is called *flow-invariant* (with respect to M) if each solution of $u'(x) + f(x, u(x)) = o\,(0 < x \leqslant 1)$ with $u(0) \in A$ satisfies also $u(x) \in A\ (0 \leqslant x \leqslant 1)$. By transforming an inequality $Mv \leqslant Mw$ into $u' + \tilde{f}(x, u) = o$, $u = w - v$, $\tilde{f}(x, u) = f(x, v + u) - f(x, v) - f_0(x)$ with $f_0(x) \geqslant o$, inverse-monotonicity can be interpreted as flow invariance of the cone $A = \{y : y \geqslant o\}$. A series of authors has contributed to the theory of flow invariance. We mention here only some recent papers on flow invariance in Banach spaces by Volkmann [1973, 1976] and Redheffer and Walter [1975], in which further references can be found. Moreover, see Deimling [1977] for a survey. In particular, Redheffer and Walter [1975] discuss the relations between various properties and conditions which occur in this theory. One important assumption used in many papers is the *distance condition* $\operatorname{dist}(y - hf(x, y), A) = O(h)$ for $h \to +0$. For the special case of inverse-monotonicity this property is equivalent to $-f(x, y)$ being quasi-monotone. Many results in this field do not assume that $A^c \neq \emptyset$, so that they include statements on inverse-monotonicity for the case in which the space Y_0 does not contain a strictly positive element. The estimates in Section V.4 contain some results on flow invariance with A being a sphere in an inner-product space. Thompson [1977] investigated invariance properties of second order differential operators in Banach spaces.

Section 2.4.3 incorporates results of Walter [1969] into our theory. This author considered operators in a Banach space of generalized sequences $\{u_\alpha\}$ with norm $\sup|u_\alpha|p_\alpha^{-1}$. For applications to the line method for parabolic differential equations, as in Problem 2.14, see the references in Walter [1970].

Section 3

For this theory see Schröder [1977a]. Compare also the Notes to Section V.2.

Section 4

The abstract theory in Section 4.1 was developed by Schröder [1970a, 1971].

This theory was applied to *non*-inverse-positive linear ordinary differential operators by Küpper [1976a, 1978a]. In the first paper, this author developed methods for finding points x_i and constructing a function $\psi \in C_{0,2}[0, x_1, \ldots, x_k, 1]$ such that $|Mu| \leqslant \Psi$ implies $|u| \leqslant \psi$, for a given second order differential operator M and given Ψ. Here, certain integral inequalities are to be satisfied at the "jump points" x_i. The second paper provides a modified approach which, in particular, contains Theorem 4.9 and its corollary. Similar estimation methods for differential operators of the fourth order (and higher order) were investigated by Küpper [1978b]. Moreover, Küpper [1976b] used his techniques to provide pointwise bounds for eigenfunctions of second order eigenvalue problems. The estimates in that paper are related to a method of Bazley and Fox [1966]. These authors used the boundary maximum principle to estimate eigenfunctions outside a compact interval of the given unbounded interval.

Chapter V

Estimation and existence theory for vector-valued differential operators

1 DESCRIPTION OF THE PROBLEM, ASSUMPTIONS, NOTATION

The theory of Chapter III will here be generalized to vector-valued differential operators and systems of (ordinary) differential equations. For the vector-valued functions $v = (v_i)$ considered, we shall derive pointwise estimates, that is, estimates of $v(x)$ by bounds depending on x. Our main interest is in operators M which map $R = C_1^n[0, 1] \cap C_2^n(0, 1)$ (or another suitable subspace R of $C_0^n[0, 1]$) into $S = \mathbb{R}^n[0, 1]$ and have the form

$$M = \mathbf{L} + \mathbf{N} \tag{1.1}$$

with

$$\mathbf{L}u(x) = \begin{cases} -a(x)u''(x) + b(x)u'(x) & \text{for } \ 0 < x < 1 \\ -\alpha^0 u'(0) & \text{for } \ x = 0 \\ \alpha^1 u'(1) & \text{for } \ x = 1 \end{cases} \tag{1.2}$$

and

$$\mathbf{N}u(x) = F(x, u(x), u'(x)) = \begin{cases} f(x, u(x), u'(x)) & \text{for} \quad 0 < x < 1 \\ f^0(u(0)) & \text{for} \quad x = 0 \\ f^1(u(1)) & \text{for} \quad x = 1. \end{cases} \tag{1.3}$$

Now, $u = (u_i)$ and $Mu = ((Mu)_i)$ denote vector-valued functions with indices $i \in \{1, 2, \ldots, n\}$. (In the following an *index* shall always mean such a number.) Moreover, $a(x) = (a_i(x)\,\delta_{ik})$, $b(x) = (b_i(x)\,\delta_{ik})$ $(0 < x < 1)$, and $\alpha^0 = (\alpha_i^0\,\delta_{ik})$, $\alpha^1 = (\alpha_i^1\,\delta_{ik})$ are diagonal matrices satisfying

$$a_i(x) \geqslant 0 \qquad (0 < x < 1), \qquad \alpha_i^0 \geqslant 0, \qquad \alpha_i^1 \geqslant 0$$

for all indices i, and $f = (f_i)$, $f^0 = (f_i^0)$, $f^1 = (f_i^1)$ denote functions with values in $\mathbb{R}^n$ such that $f(x, y, p)$ is defined for all $x \in (0, 1)$, $y \in \mathbb{R}^n$, $p \in \mathbb{R}^n$ and $f^0(y)$, $f^1(y)$ are defined for all $y \in \mathbb{R}^n$.

We shall, however, also consider a more general class of *functional differential operators* M which, in particular, includes certain integro-differential operators. Differential operators of the first order will be treated as a special case.

The pointwise estimates which we shall derive can be written in the form $v(x) \in D(x)$ $(0 \leqslant x \leqslant 1)$ or $(x, v(x)) \in D$ $(0 \leqslant x \leqslant 1)$ with suitably defined sets $D(x) \in \mathbb{R}^n$ and $D \subset \mathbb{R}^{n+1}$. This formulation already indicates possibilities to generalize the succeeding theory. Our purpose in this chapter, however, is not to provide a theory as general as possible, but to consider some useful cases and discuss these in more detail. The methods of proof then can also be used for other cases.

Sections 2 and 3 will be concerned with estimates by two-sided bounds

$$\varphi \leqslant v \leqslant \psi, \qquad \text{i.e.,} \qquad \begin{array}{l} \varphi_i(x) \leqslant v_i(x) \leqslant \psi_i(x) \\ (0 \leqslant x \leqslant 1; \qquad i = 1, 2, \ldots, n); \end{array} \tag{1.4}$$

Sections 4 and 5 with pointwise norm estimates

$$\|v\| \leqslant \psi, \qquad \text{i.e.} \qquad \|v(x)\| \leqslant \psi(x) \qquad (0 \leqslant x \leqslant 1), \tag{1.5}$$

where $\| \;\|$ denotes a certain vector norm in $\mathbb{R}^n$. In Sections 2 and 4 these estimates will be derived from properties of Mv (which means we consider range–domain implications here); in Sections 3 and 5 we shall prove that a given equation $Mu = r$ has a solution u^* such that $v = u^*$ satisfies estimates as above. The existence theory in Section 5 will be based on the range–domain implications in Section 4. Similarly, the results of Section 2 could be used in Section 3. However, in the case of two-sided bounds it is simpler to derive the results on existence and inclusion from the theory of inverse-monotone scalar-valued differential operators in Section III.2.1.

Besides the preceding estimates of v, simultaneous estimates of v and v' will also be considered in Sections 2–5. Moreover, various other generalizations and modifications will be discussed in Section 6.

The implications considered in Sections 2 and 4 can be written in the form $Mv \in C \Rightarrow v \in K$, and the results could be derived from the theory in Section IV.1. Indeed, the results in Section 2 will be obtained from the abstract Theorem IV.3.1, which in turn was based on Corollary IV.1.1a. We thus provide an example of the immediate application of the abstract theory. It may be simpler, however, to use the idea of the continuity principle described in Section IV.1 without defining all the abstract terms which occur there. Such a "direct" approach will be used in Sections 4 and 6.

For vector-valued problems, a series of additional difficulties arise which are related to the coupling of the components.

We shall say that a given function $h: \mathbb{R}^n \to \mathbb{R}^n$ is *not coupled*, if for each index i the ith component $h_i(y)$ does not depend on any variable y_k with $k \neq i$; this means $h_i(y)$ is a function of y_i alone. Otherwise the function is said to be *coupled*. As already remarked, the function is called *quasi-antitone* if for each index i

$$h_i(y) \geqslant h_i(\bar{y}) \qquad \text{for all} \quad y, \bar{y} \in \mathbb{R}^n \quad \text{with} \quad y \leqslant \bar{y}, \quad y_i = \bar{y}_i .$$

Clearly, this property describes a very special type of coupling.

For functions $h(y, p, \ldots)$ with values in $\mathbb{R}^n$ which also depend on variables $p, \ldots$ other than y, the analogous terms *not coupled with respect to* y, *coupled with respect to* y, *quasi-antitone with respect to* y are defined in an obvious way. Moreover, we shall also speak of *equations not coupled with respect to* y, etc.

Finally, the operator M is said to be *weakly coupled*, if $f(x, y, p)$ is not coupled with respect to p, this means that the ith component $(Mu)_i(x)$ does not depend on any derivative $u_k'(x)$ with $k \neq i$. Otherwise M is called *strongly coupled*. Also, by definition, a weakly coupled operator M is *quasi-antitone* if all functions $f(x, y, p)$, $f^0(y)$, $f^1(y)$ are quasi-antitone with respect to y.

In Section IV.2.3 we derived sufficient conditions for the inverse-monotonicity of the operator M, essentially requiring (among other assumptions) that M be weakly coupled and quasi-antitone. In the theory of this chapter, the operator M need not be quasi-antitone, although some results on two-sided bounds assume a simpler form if M has this property. In the main results on two-sided bounds essentially we still have to require that M be weakly coupled, while this assumption is not necessary for obtaining pointwise norm bounds. However, also for two-sided bounds this restriction can partly be overcome, for example, by considering estimates of both v and v', simultaneously.

Due to the coupling of such vector-valued problems, estimates for v or u^* require a *set* of inequalities to be satisfied for the functions φ, ψ in (1.4), or ψ in (1.5), respectively.

The range–domain implications in Sections 2 and 4 can be used to obtain uniqueness results and error estimates for solutions of problems $Mu = r$. *If M is linear and* (1.4) *or* (1.5) *is proved for each $v \in R$, then M is invertible and hence each problem $Mu = r$ has at most one solution* (cf. Problem IV.1.1). For nonlinear problems, the uniqueness can often be established by proving that (1.4) holds with $\varphi = \psi = o$ or (1.5) holds with $\psi = o$ and v denoting the difference of two solutions. Here the theory has to be applied to a transformed problem which has this function v as a solution. In the same way one may prove that at most one solution with certain given properties, such as positivity, exists (restricted uniqueness).

Error estimates based on an estimation (1.4) can be obtained in a way similar to that in Section III.6. The numerical methods in Sections II.5.2 and III.6 can be generalized. We shall give only some examples and references here. (If (1.5) is applied for an error estimation, one has first to transform the problem so that $v = u^* - u_0$ is the error of an approximation u_0.)

The preceding assumptions and also those which will be formulated in the succeeding sections can be relaxed and modified in many ways. For instance, this is applicable to the domain of definition and the smoothness of many of the functions which occur. We shall not explain this in detail, but refer the reader to analogous discussions for the case $n = 1$ in Chapter III. Moreover, some generalizations will be explained in examples. Also, other methods of proof may be used. For example, the degree of mapping method may be employed in Sections 3 and 5, instead of the method of modification.

In most of this chapter, in particular Sections 2 and 3, it is necessary to distinguish carefully between a function u and its value $u(x)$. Thus, if $u \in C_0^n[0, 1]$, then $u(x) \in \mathbb{R}^n$, $u_i \in C_0[0, 1]$, and $u_i(x) \in \mathbb{R}$. The inequality sign $\leqslant$ always denotes the natural order relation for the functions or vectors considered. Thus, if $u, v \in \mathbb{R}^n[0, 1]$, then

$$\begin{aligned} u \leqslant v \quad &\Leftrightarrow \quad u_i(x) \leqslant v_i(x) \quad \text{for all} \quad x \in [0, 1] \quad \text{and all} \quad i \in \{1, 2, \ldots, n\} \\ &\Leftrightarrow \quad u(x) \leqslant v(x) \quad \text{for all} \quad x \in [0, 1] \\ &\Leftrightarrow \quad u_i \leqslant v_i \quad \text{for all} \quad i \in \{1, 2, \ldots, n\}. \end{aligned}$$

Also $\preccurlyeq$ denotes the corresponding strict order. Furthermore, we shall use the notation for matrices in Section II.2, such as B^+, B^-, etc.

2 TWO-SIDED BOUNDS BY RANGE–DOMAIN IMPLICATIONS

In the theory of two-sided bounds presented in Sections 2 and 3 the resulting conditions assume a simpler form, if the operator M is rewritten such that in the expression for $(Mu)_i$ the ith component u_i is distinguished from the components u_k with $k \neq i$. Due to this it is scarcely more complicated to develope a more general theory by considering functional differential operators as described in the following.

Suppose now that $M: R \to S$ is an operator of the form (1.1) with $\mathbf{L}$ as above, but $\mathbf{N}$ given by

$$\mathbf{N}u(x) = G(x, u(x), u'(x), u, u') = \begin{cases} g(x, u(x), u'(x), u, u') & \text{for} \quad 0 < x < 1 \\ g^0(u(0), u) & \text{for} \quad x = 0 \\ g^1(u(1), u) & \text{for} \quad x = 1, \end{cases} \tag{2.1}$$

where g, g^0, g^1 denote functions with values in $\mathbb{R}^n$ such that $g(x, y, p, u, U)$ is defined for all $x \in (0, 1)$, $y \in \mathbb{R}^n$, $p \in \mathbb{R}^n$, $u \in C_0^n[0, 1]$, $U \in C_0^n[0, 1]$, and $g^0(y, u)$, $g^1(y, u)$ are defined for all $y \in \mathbb{R}^n$, $u \in C_0^n[0, 1]$. *We assume here that all three functions g, g^0, g^1 are not coupled with respect to y, and, moreover, that g is not coupled with respect to p.*

Clearly, an operator $\mathbf{N}$ as in (1.3) can always be written in the form (2.1) by defining

$$\begin{aligned} g_i(x, y, p, u, U) &= f_i(x; u_1(x), \ldots, u_{i-1}(x), y_i, u_{i+1}(x), \ldots, u_n(x); \\ &\qquad U_1(x), \ldots, U_{i-1}(x), p_i, U_{i+1}(x), \ldots, U_n(x)), \\ g_i^0(y, u) &= f_i^0(u_1(0), \ldots, u_{i-1}(0), y_i, u_{i+1}(0), \ldots, u_n(0)) \end{aligned} \tag{2.2}$$

and choosing g_i^1 analogously. In the following, whenever an operator $\mathbf{N}$ of the form (1.3) occurs, the functions g, g^0, g^1 are supposed to be defined as explained here.

In this section, we are interested in sufficient conditions such that

$$H(\varphi, [\varphi, \psi]) \leqslant Mv \leqslant H(\psi, [\varphi, \psi]) \quad \Rightarrow \quad \varphi \leqslant v \leqslant \psi \tag{2.3}$$

where

$$H(u, w)(x) = \mathbf{L}u(x) + G(x, u(x), u'(x), w, w') \qquad (0 \leqslant x \leqslant 1).$$

Here $[\varphi, \psi] = [\varphi, \psi]_R$ denotes an order interval in R, and we use again the interval-analytical terminology explained in Section I.1.2. Thus the inequalities in the premise of (2.3) are equivalent to the condition that

$$H(\varphi, w) \leqslant Mv \leqslant H(\psi, w) \quad \text{for all} \quad w \in R \quad \text{satisfying} \quad \varphi \leqslant w \leqslant \psi. \tag{2.4}$$

Due to our notation, the required inequalities can be written in a very short form. However, it is certainly useful to rewrite these conditions as componentwise and pointwise inequalities. For example, let $a(x) \equiv I$, $b(x) \equiv O$, and g be given by (2.2). Then for a fixed index i and a fixed $x \in (0, 1)$, condition (2.4) contains the requirement that

$$-\varphi_i''(x) + f_i(x, h, k) \leqslant (Mv)_i(x) \leqslant -\psi_i''(x) + f_i(x, \bar{h}, \bar{k}) \qquad (2.5)$$

for all $h, k, \bar{h}, \bar{k} \in \mathbb{R}^n$ such that $h, \bar{h} \in [\varphi(x), \psi(x)]$, $h_i = \varphi_i(x)$, $\bar{h}_i = \psi_i(x)$, $k_i = \varphi_i'(x)$, $\bar{k}_i = \psi_i'(x)$.

Example 2.1. Consider an operator M where

$$\begin{aligned} Mu(0) &= u(0), \qquad Mu(1) = u(1), \\ Mu(x) &= -u''(x) + C(x)u(x) \qquad (0 < x < 1) \end{aligned} \qquad (2.6)$$

with $C(x) = (c_{ik}(x))$ an $n \times n$ matrix $(0 < x < 1)$, and let $C(x) = D(x) - B(x)$ with diagonal $D(x) = (c_{ii}(x)\,\delta_{ik})$. Defining $g(x, y, p, u, U) = D(x)y - B(x)u(x)$ we obtain here $H(u, w)(x) = -u''(x) + D(x)u(x) - B(x)w(x)$ $(0 < x < 1)$, $H(u, w)(0) = u(0)$, $H(u, w)(1) = u(1)$, where $(Bw)(x) = (B^+ - B^-)(x)w(x) \leqslant (B^+\psi - B^-\varphi)(x)$ for $\varphi \leqslant w \leqslant \psi$, and $(Bw)(x)$ can be estimated from below in an analogous way. Thus, if (2.3) holds, then the inequalities

$$\begin{aligned} &-\varphi''(x) + (D - B^+)(x)\varphi(x) + B^-(x)\psi(x) \\ &\qquad \leqslant Mv(x) \leqslant -\psi''(x) + (D - B^+)(x)\psi(x) + B^-(x)\varphi(x) \qquad (0 < x < 1) \end{aligned}$$

and $\varphi(0) \leqslant v(0) \leqslant \psi(0)$, $\varphi(1) \leqslant v(1) \leqslant \psi(1)$ together imply $\varphi \leqslant v \leqslant \psi$.

For the special choice $\varphi = -\psi$, this is equivalent to

$$\left.\begin{aligned} &|Mv(x)| \leqslant -\psi''(x) + (D - |B|)(x)\psi(x) \qquad (0 < x < 1) \\ &|v(0)| \leqslant \psi(0), \qquad |v(1)| \leqslant \psi(1) \end{aligned}\right\} \Rightarrow \quad |v| \leqslant \psi. \qquad (2.7)$$

For quasi-antitone M, i.e., for the case $B(x) \geqslant O$ $(0 \leqslant x \leqslant 1)$, the latter implication is equivalent to $|Mv| \leqslant M\psi \Rightarrow |v| \leqslant \psi$. If $B^-(x) \not\equiv O$, however, the lower bound $D - |B| \neq C$ of C occurs in the estimate of $|Mv|$. The question is what effect this has. Some remarks related to this problem will be made in Example 2.2. □

The following theorem, too, is written in a concentrated form, which will be explained by the succeeding examples.

Theorem 2.1. *Suppose there exist functions $z \succ o$ and $\bar{z} \succ o$ in R with the following properties.*

1. *For each index i, each $x \in [0, 1]$, and each $\lambda > 0$ with*

$$\varphi - \lambda\bar{z} \leqslant v \leqslant \psi + \lambda z, \qquad (v - \varphi + \lambda\bar{z})_i(x) = 0, \qquad (2.8)$$

there exists an $\omega \in [\varphi, \psi]$ *such that* $H_i(\varphi - \lambda\bar{z}, v)(x) < H_i(\varphi, \omega)(x)$.

2. *For each index i, each* $x \in [0, 1]$, *and each* $\lambda > 0$ *with*

$$\varphi - \lambda\bar{z} \leqslant v \leqslant \psi + \lambda z, \qquad (\psi + \lambda z - v)_i(x) = 0, \tag{2.9}$$

there exists an $\omega \in [\varphi, \psi]$ *such that* $H_i(\varphi, \omega)(x) < H_i(\psi + \lambda z, v)(x)$.

Then $H(\varphi, [\varphi, \psi]) \leqslant Mv \leqslant H(\psi, [\varphi, \psi]) \Rightarrow \varphi \leqslant v \leqslant \psi$.

Proof. We shall derive this result from Theorem IV.3.1 using $\mathfrak{z}(\lambda) = \lambda z$, $\bar{\mathfrak{z}}(\lambda) = \lambda\bar{z}$ and defining $u > o \Leftrightarrow u \succ o$ for $u \in R$.

If $v \in [\varphi - \lambda\bar{z}, \psi + \lambda z]$ and $v \not< \psi + \lambda z$, then $\ell_{i,x}(\psi + \lambda z - v) = 0$ for some index i and some $x \in [0, 1]$, where $\ell_{i,x}U = U_i(x)$ (for $U \in S \supset R$). One shows by arguments often employed above that then condition 2.I of Theorem IV.3.1 holds with $\ell = \ell_{i,x}$. Notice, for example, that $u_i(x) = u_i'(x) = 0$, $u_i''(x) \geqslant 0$ whenever $u \geqslant o$ and $\ell_{i,x}u = 0$ for some $x \in (0, 1)$. Condition 1.I is verified analogously.

Due to the above assumptions, conditions 1.II and 2.II are also satisfied for the considered functionals $\ell_{i,x}$. Therefore, the statement follows from Theorem IV.3.1. □

The functions ω which occur in the preceding assumptions may be chosen such that they depend on the value x occurring in (2.8) or (2.9); that is, for each $x \in [0, 1]$ one may choose a different function $\omega = \omega_{(x)}$. (The index x must not be confused with the independent variable of the function.) In particular, one may choose the function $\omega = (\omega_k)$ corresponding to $x \in (0, 1)$ such that $\omega_k = \varphi_k$ if $v_k(x) \leqslant \varphi_k(x)$; $\omega_k = \psi_k$ if $\psi_k(x) \leqslant v_k(x)$; and ω_k equals some function $u_k \in [\varphi_k, \psi_k]$ with $u_k(x) = v_k(x)$ and $u_k'(x) = v_k'(x)$, if $\varphi_k(x) < v_k(x) < \psi_k(x)$. If this function is evaluated at the point x, one obtains

$$\omega_{(x)}(x) = v^{\#}(x), \qquad \omega_{(x)}'(x) = (v^{\#})'(x) \tag{2.10}$$

where $v^{\#} = (v_k^{\#}) \in C_0^n[0, 1]$ with $v_k^{\#} = \langle \varphi_k, v_k, \psi_k \rangle$ (see (III.4.6)) and the discontinuous function $(v^{\#})' = ((v_k^{\#})')$ is defined by

$$(v_k^{\#})'(x) = \begin{cases} \varphi_k'(x) & \text{for} \quad v_k(x) \leqslant \varphi_k(x) \\ v_k'(x) & \text{for} \quad \varphi_k(x) < v_k(x) < \psi_k(x) \\ \psi_k'(x) & \text{for} \quad \psi_k(x) \leqslant v_k(x) \end{cases} \qquad (0 \leqslant x \leqslant 1).$$

For $x = 0$ or $x = 1$ the function ω may be chosen correspondingly.

Now, if **N** has the special form (1.3) and g, g^0, g^1 are defined by (2.2) and an analogous formula, only values as in (2.10) occur. Thus, defining $H(u, v^{\#})(x)$ for $v^{\#} \notin R$ in an obvious way, we obtain here the following result.

Corollary 2.1a. *Suppose that* **N** *has the special form* (1.3) *and that functions* $z \succ o$ *and* $\bar{z} \succ o$ *in R with the following properties exist.*

1. *The inequality*

$$H_i(\varphi - \lambda \bar{z}, v)(x) < H_i(\varphi, v^*)(x) \tag{2.11}$$

holds for each index i, each $x \in [0, 1]$, *and each* $\lambda > 0$ *such that* (2.8) *is satisfied.*

2. *The inequality*

$$H_i(\psi, v^*)(x) < H_i(\psi + \lambda z, v)(x) \tag{2.12}$$

holds for each index i, each $x \in [0, 1]$, *and each* $\lambda > 0$ *such that* (2.9) *is satisfied.*

Then $H(\varphi, [\varphi, \psi]) \leqslant Mv \leqslant H(\psi, [\varphi, \psi]) \Rightarrow \varphi \leqslant v \leqslant \psi$.

Observe that the relations (2.8) and (2.9) imply certain side conditions which can be used in verifying (2.11) and (2.12), respectively. For example, if (2.8) holds and $x \in (0, 1)$, then $v_i(x) = \varphi_i(x) - \lambda \bar{z}_i(x)$ and $v_i'(x) = \varphi_i'(x) - \lambda \bar{z}_i'(x)$. Such conditions can be exploited similarly as explained for inverse-monotone operators in Section III.2.2 (see Example 2.4).

Again it seems worthwhile to formulate the conditions in the above corollary more explicitly. For example, *assumption* 2 *is equivalent to the requirement that for each index i and all* $\lambda > 0$ *with* $\varphi - \lambda \bar{z} \leqslant v \leqslant \psi + \lambda z$ *the following conditions are satisfied*:

$$(\mathbf{L}z)_i(x) + \lambda^{-1}[f_i(x, v(x), v'(x)) - f_i(x, v^*(x), v^{*\prime}(x))] > 0, \tag{2.13}$$

if $x \in (0, 1)$, $v_i(x) = \psi_i(x) + \lambda z_i(x)$, *and* $v_i'(x) = \psi_i'(x) + \lambda z_i'(x)$;

$$-\alpha_i^0 z_i'(0) + \lambda^{-1}[f_i^0(v(0)) - f_i^0(v^*(0))] > 0, \qquad \text{if} \quad v_i(0) = \psi_i(0) + \lambda z_i(0);$$

$$\alpha_i^1 z_i'(1) + \lambda^{-1}[f_i^1(v(1)) - f_i^1(v^*(1))] > 0, \qquad \text{if} \quad v_i(1) = \psi_i(1) + \lambda z_i(1).$$

In verifying these inequalities the estimate $|v(x) - v^*(x)| \leqslant \lambda z(x)$ may be used, if $\bar{z} = z$.

Example 2.2. Let M denote an operator as in (2.6) and suppose that the coefficients $c_{ik}(x)$ of $C(x)$ are continuous on $[0, 1]$. Then the assumptions of Corollary 2.1a are satisfied for arbitrary $v \in R$, if

$$z(x) \succ o, \qquad \hat{M}z(x) \succ o \qquad (0 \leqslant x \leqslant 1) \tag{2.14}$$

for some $z(=\bar{z})$ in R, where $\hat{M}: R \to S$ denotes the quasi-monotone operator defined by $\hat{M}u(0) = u(0)$, $\hat{M}u(1) = u(1)$, $\hat{M}u(x) = -u''(x) + (D - |B|)(x)u(x)$ for $0 < x < 1$. Such a function $z \in R$ exists if and only if $\hat{M}$ is inverse-positive. On the other hand, $\hat{M}$ has this property if and only if all (real) eigenvalues of the problem $(\hat{M}\Phi + \Lambda\Phi)(x) = 0$ $(0 < x < 1)$, $\hat{M}\Phi(0) = \hat{M}\Phi(1) = 0$ satisfy $\Lambda < 0$ (cf. Theorem II.4.15).

Now suppose that C does not depend on x and is irreducible. Then the matrix $D - |B|$ has an eigenvector $\Phi \succ o$ corresponding to its smallest eigenvalue $\lambda_{\min}(D - |B|)$ (see Section II.2.4). Here $z(x) = \cos(\pi - \varepsilon)(x - \frac{1}{2}) \cdot \Phi$

satisfies (2.14) for small enough $\varepsilon > 0$, if

$$\lambda_{\min}(D - |B|) > -\pi^2. \tag{2.15}$$

Let us consider three special cases in which C equals one of the matrices

$$C_1 = \begin{pmatrix} \mu & -\nu \\ -\nu & \mu \end{pmatrix}, \quad C_2 = \begin{pmatrix} \mu & \nu \\ \nu & \mu \end{pmatrix}, \quad C_3 = \begin{pmatrix} \mu & \nu \\ -\nu & \mu \end{pmatrix} \quad \text{with} \quad \nu > 0. \tag{2.16}$$

In each case condition (2.15) is equivalent to $\mu - \nu > -\pi^2$. This condition is "natural" for C_1 and C_2, since the corresponding operators M_1 and M_2 are not one-to-one for $\mu - \nu = -\pi^2$. The operator M_3 corresponding to C_3, however, is one-to-one for arbitrary $\mu > -\pi^2$, so that here the condition (2.15) cannot be explained in the same way. This operator M_3 is an example for more general operators for which pointwise norm bounds as considered in Sections 4 and 5 have certain advantages to two-sided bounds. □

Theorem 2.1 and Corollary 2.1a can be modified in many ways. For example, instead of λz and $\lambda \bar{z}$ one may use nonlinear $\mathfrak{z}(\lambda)$ and $\bar{\mathfrak{z}}(\lambda)$, as for scalar-valued operators in Section III.2.2.

Example 2.3. We consider the boundary value problem

$$L[u_1](x) - u_1(x)u_2(x) - 8\mu x^2 = 0, \qquad u_1(0) = u_1(1) = 0$$

$$L[u_2](x) + \tfrac{1}{2}(u_1(x))^2 = 0, \qquad u_2(0) = u_2'(1) - \nu u_2(1) = 0$$

with $\mu > 0$, $0 < \nu < 1$, and

$$L[y](x) = -x(x^{-1}(xy(x))')' = -xy''(x) - y'(x) + x^{-1}y(x),$$

for functions $u_1, u_2 \in C_2[0, 1]$. This problem describes the bending of circular plates. (See Bromberg [1956], Keller and Reiss [1958].) Here, we shall derive (rough) a priori bounds for a solution v. (In Section 3.4.3 the existence of a solution will be proved. Also see Example 4.12.)

The given problem can be written as $Mu = o$ with M defined on $R = C_0^2[0, 1] \cap C_2^2(0, 1)$ by

$$Mu(x) = L[u](x) + \bar{g}(x, u(x), u) \quad \text{for} \quad 0 < x < 1,$$

$$Mu(0) = u(0), \qquad Mu(1) = \begin{pmatrix} u_1(1) \\ u_2'(1) - \nu u_2(1) \end{pmatrix}, \tag{2.17}$$

$$\bar{g}(x, y, u) = \begin{pmatrix} -y_1 u_2(x) - 8\mu x^2 \\ \tfrac{1}{2}(u_1(x))^2 \end{pmatrix}$$

where $L[u]$ has the components $L[u_i]$ $(i = 1, 2)$. Obviously, M also has the form (1.1), (1.2), (2.1) with $g(x, y, u) = \bar{g}(x, y, u) + x^{-1}y$ (the function

$\bar{g}$ is introduced here for later use in Section 3). The corresponding term $H(u, w)$ is obtained from Mu when $\bar{g}(x, u(x), u)$ is replaced by $\bar{g}(x, u(x), w)$. Accordingly, for φ, ψ, v satisfying the boundary conditions, the inequalities in the premise of (2.3) are satisfied if for $0 < x < 1$

$$L[\varphi_1](x) - \varphi_1(x)[\varphi_2, \psi_2](x) \leqslant 8\mu x^2 \leqslant L[\psi_1](x) - \psi_1(x)[\varphi_2, \psi_2](x),$$
$$L[\varphi_2](x) + \tfrac{1}{2}([\varphi_1, \psi_1](x))^2 \leqslant 0 \leqslant L[\psi_2](x) + \tfrac{1}{2}([\varphi_1, \psi_1](x))^2.$$

For example, these conditions hold for

$$\varphi_1(x) \equiv 0, \qquad \psi_2(x) \equiv 0,$$
$$\psi_1(x) = \mu(x - x^3), \qquad \varphi_2(x) = \frac{\mu^2}{16}(-(3 - \nu)(1 - \nu)^{-1}x + x^3). \tag{2.18}$$

We shall prove that each solution $v \in R$ satisfies $\varphi \leqslant v \leqslant \psi$. Applying Corollary 2.1a we have to construct functions $z, \bar{z} \in R$ which satisfy (2.11), (2.12). For $x \in (0, 1)$ the following inequalities constitute sufficient conditions:

$$L[\bar{z}_1](x) - |v_1(x)| z_2(x) > 0, \qquad L[z_1](x) - |v_1(x)| z_2(x) > 0,$$
$$L[z_2](x) > 0, \qquad L[\bar{z}_2](x) - \tfrac{1}{2}(\psi_1(x) + |v_1(x)|) z_1(x) > 0, \tag{2.19}$$
$$L[\bar{z}_2](x) - \tfrac{1}{2}|v_1(x)| \bar{z}_1(x) > 0.$$

To show this, first consider inequality (2.11) for $i = 1$;

$$(\lambda L[\bar{z}_1] - \varphi_1 v_2^\# + (\varphi_1 - \lambda\bar{z}_1)v_2)(x) > 0. \tag{2.20}$$

Using the side condition $(\varphi_1 - \lambda\bar{z}_1)(x) = v_1(x)$, this relation can be transformed into $(\lambda L[\bar{z}_1] + v_1(v_2 - v_2^\#) - \lambda\bar{z}_1 v_2^\#)(x) > 0$. Since here $v_1(x) = (\varphi_1 - \lambda\bar{z}_1)(x) \leqslant 0$, $v_2^\#(x) \leqslant 0$, and $(v_2 - v_2^\#)(x) \leqslant \lambda z_2(x)$, we obtain the sufficient condition $L[\bar{z}_1](x) - |v_1(x)| z_2(x) > 0$.

For $i = 2$, inequality (2.11) assumes the form $\lambda L[\bar{z}_2](x) + \frac{1}{2}[(v_1^\#(x))^2 - v_1^2(x)] > 0$. The term within brackets vanishes for the case $\varphi_1(x) \leqslant v_1(x) \leqslant \psi_1(x)$. If $\varphi_1(x) - \lambda\bar{z}_1(x) \leqslant v_1(x) < 0 = \varphi_1(x) = v_1^\#(x)$, we calculate $-v_1^2(x) \geqslant \lambda v_1(x)\bar{z}_1(x)$ and obtain the sufficient condition $L[\bar{z}_2](x) - \frac{1}{2}|v_1(x)|\bar{z}_1(x) > 0$. If $\psi_1(x) < v_1(x) \leqslant \psi_1(x) + \lambda z_1(x)$, similar arguments yield $L[\bar{z}_2](x) - \frac{1}{2}(\psi_1(x) + |v_1(x)|) z_1(x) > 0$.

The inequalities (2.12) can be treated in an analogous (but simpler) way.

One verifies by formal calculations that for small enough $\varepsilon > 0$, $\bar{z}_2 = 1 + (1 + \nu)(1 - \nu)^{-1}x$, $z_1 = \bar{z}_1 = \varepsilon\bar{z}_2$, $z_2 = \varepsilon z_1$ satisfy all conditions required above.

Observe that by proceeding differently, conditions different from (2.19) can be obtained. For example, since $\varphi_1(x) = 0$, the inequality $L[\bar{z}_1](x) - v_2(x)\bar{z}_1(x) > 0$ certainly is sufficient for (2.20). However, for unknown v_2, we are not able to solve this inequality. On the other hand, one can show by arguments of inverse-positivity that $v_2(x) \leqslant 0$ for each solution v, so that $L[\bar{z}_1](x) > 0$ is sufficient. □

Differential operators of the first order can be incorporated into the above theory. Now suppose that $M = \mathbf{L} + \mathbf{N}$ with

$$\mathbf{L}u(x) = \begin{cases} u'(x) & \text{for } 0 < x \leqslant 1 \\ o & \text{for } x = 0, \end{cases}$$

$$\mathbf{N}u(x) = \begin{cases} f(x, u(x)) & \text{for } 0 < x \leqslant 1 \\ u(0) - r^0 & \text{for } x = 0, \end{cases}$$

and define $R = C_0^n[0, 1] \cap C_1^n(0, 1]$. Here $H(u, w)(x)$ has the components $H_i(u, w)(0) = u_i(0) - r_i^0$ for $x = 0$ and

$$H_i(u, w)(x) = u_i'(x) + f_i(x, w_1(x), \ldots, w_{i-1}(x), u_i(x), w_{i+1}(x), \ldots, w_n(x)) \quad \text{for } 0 < x \leqslant 1.$$

We leave it to the reader to carry out the details and prove the following result (Problems 2.4 and 2.5).

Corollary 2.1b (a) *Suppose that $f(x, y)$ is a continuous function on $[0, 1] \times \mathbb{R}^n$ which satisfies a Lipschitz condition with respect to y on each bounded subset of its domain. Then for given $\varphi, \psi, v \in R$*

$$H(\varphi, [\varphi, \psi]) \leqslant Mv \leqslant H(\psi, [\varphi, \psi]) \quad \Rightarrow \quad \varphi \leqslant v \leqslant \psi.$$

(b) *Moreover, if $H(\varphi, [\varphi, \psi]) \leqslant o \leqslant H(\psi, [\varphi, \psi])$, the initial value problem $Mu = o$ has a solution $u^* \in R$ with $\varphi \leqslant u^* \leqslant \psi$.*

Based on this result a procedure of error estimation analogous to those in Sections II.5.2.2 and III.6.2 was developed by Marcowitz [1973] for vector-valued initial value problems. Here the coupling of $f(x, y)$ causes additional difficulties. Also the procedure is more general in the sense that it yields upper and lower bounds separately. We shall not explain the details, but only provide an example.

Example 2.4. The reentry problem (for a space vehicle) treated by Stoer and Bulirsch [1973] is a control problem which is described by six differential equations of the first order. After the control variable has been calculated, the velocity v, the "direction" γ, and the (normalized) height ξ of the capsule are determined as solutions of an initial value problem for three first order differential equations. To this system the method of error estimation mentioned above was applied.

For the case that the control variable is determined such that the heat of the capsule becomes minimal, the initial value problem is to be solved on the interval $[0, l]$ with $l = 224.9007\ldots$. Up to $x = 224$ the approximate solution was calculated with a Runge–Kutta method of order 4 and step size $h = 1:32$. For the case considered one has $v(0) = 0.36$, $\gamma(0) = -8.1°\pi: 180°$, $\xi(0) = 4R^{-1}$, $v(l) = 0.27$, $\gamma(l) = 0.0$, $\xi(l) = 2.5R^{-1}$, $R = 209$. Table 12 contains lower bounds β_l and upper bounds β_u for the errors $\Delta_v, \Delta_\gamma, \Delta_\xi$. The CPU time of 5.9 sec for the error estimation, naturally, was larger than the time of 0.34 sec for the solution of the initial value problem. Observe, however, that the solution of the original control problem required about 50 to 60 sec. □

TABLE 12

Lower bounds β_l and upper bounds β_u for the errors $\Delta_v, \Delta_\gamma, \Delta_\xi$ in Example 2.4

x	Quantity estimated	β_l	β_u
50	Δ_v	$-0.5151E-09$	$-0.4970E-09$
100	Δ_v	$0.7833E-10$	$0.7237E-09$
160	Δ_v	$0.3475E-08$	$0.6113E-08$
224.9	Δ_v	$0.1308E-07$	$0.2181E-07$
50	Δ_γ	$0.1096E-08$	$0.1128E-08$
100	Δ_γ	$0.3511E-08$	$0.5341E-08$
160	Δ_γ	$0.1411E-07$	$0.2190E-07$
224.9	Δ_γ	$0.4530E-07$	$0.7087E-07$
50	Δ_ξ	$-0.2443E-11$	$-0.1930E-11$
100	Δ_ξ	$0.1398E-09$	$0.1992E-09$
160	Δ_ξ	$0.7940E-09$	$0.1222E-08$
224.9	Δ_ξ	$0.3080E-08$	$0.4782E-08$

Problems

2.1 For the linear operator M in Example 2.1, the statement (2.7) holds, if $\pi^2 e + (D(x) - |B(x)|)e \geqslant \delta e$ $(0 < x < 1)$ for some $\delta > 0$, $e^T = (1, 1, \ldots, 1)$ (see Example 2.2).

2.2 Sufficient conditions for the assumptions of Corollary 2.1a may be obtained by using certain *quasi-antitone majorants* for f, f^0, and f^1. For example, let $f(x, y, p)$ be independent of p and suppose that for each index i the inequality $f_i(x, v(x)) - f_i(x, v^\#(x)) \geqslant \tilde{f}_i(x, |y|)$ holds for $0 < x < 1$, $y = v(x) - v^\#(x)$, and $y_i \geqslant 0$ where $\tilde{f} = (\tilde{f}_i)$ is a quasi-antitone function of y. Show that for $\bar{z} = z$ (2.13) may be replaced by the sufficient condition $(\mathbf{L}z)_i(x) + \lambda^{-1}\tilde{f}_i(x, \lambda z(x)) > 0$. Derive similar sufficient conditions for all assumptions. Discuss also the case in which f does depend on p. Use, in particular, majorants which are linear in y and p.

2.3 Quasi-antitone majorants as in the preceding problem can also be used in verifying the inequalities in the premise of (2.3). For example, let again $f(x, y, p) = f(x, y)$, and $\varphi = u_0 - z$, $\psi = u_0 + z$. Moreover, suppose that $\tilde{f}(x, y)$ denotes a function which is quasi-antitone in y such that for each index i the inequality $f_i(x, u_0(x) + y) - f_i(x, u_0(x)) \geqslant \tilde{f}_i(x, y)$ holds for $0 < x < 1$, $y \in \mathbb{R}^n$, $y_i = z_i(x)$, $|y_k| \leqslant z_k(x)$. Then for $0 < x < 1$

$$(\mathbf{L}u_0)_i(x) + f_i(x, u_0(x)) \leqslant (\mathbf{L}z)_i(x) + \tilde{f}_i(x, z(x)) \leqslant H_i(\psi, [\varphi, \psi])(x).$$

Such estimates may be used, when error bounds for an approximate solution u_0 of a boundary value problem $Mu = o$ are to be computed. Derive similar estimates for $x = 0$ and $x = 1$, and also estimates of $H(\varphi, [\varphi, \psi])$ from above.

2.4 Derive part (a) of Corollary 2.1b from Corollary 2.1a using $\bar{z} = z$. For verifying (2.11) and (2.12) write these conditions in a form analogous to (2.13). Show also that the required Lipschitz conditions may be replaced by inequalities of the form $f_i(x, v(x)) - f_i(x, v^{\#}(x)) \geqslant d_i(x)y_i - \sum_k b_{ik}(x)|y_k|$ for $0 < x \leqslant 1$, $y = v(x) - v^{\#}(x)$, and $y_i \geqslant 0$, where $b_{ii}(x) \equiv 0$, $b_{ik}(x) \geqslant 0$, and $d_i(x) - \sum_k b_{ik}(x) > -N$ $(0 < x \leqslant 1)$.

2.5 Prove part (b) of Corollary 2.1b by applying the usual theory on the existence and continuation of solutions of initial value problems and applying part (a) to intervals $[0, l]$ on which the solution is defined.

3 EXISTENCE AND INCLUSION BY TWO-SIDED BOUNDS

Suppose that

$$Mu(x) = o \qquad (0 \leqslant x \leqslant 1) \tag{3.1}$$

is a given boundary value problem with an operator $M: R \to S$ as described in (1.1), (1.2), and (2.1). All the preceding general assumptions are supposed to be satisfied, so that, in particular, the function G is not coupled with respect to y and p. Using the method of modification, we shall derive conditions on M such that a certain set of differential inequalities for functions φ and ψ implies the existence of a solution u^* with $\varphi \leqslant u^* \leqslant \psi$. In the following, the letter v will *not* denote a fixed function to be estimated, as in Section 2. Thus, we may now denote the independent variables by x, y, p, v, w (instead of x, y, p, u, U).

Our main interest is in problems of the second order, which will be treated in Sections 3.1–3.3. In these sections we define $R = C_2^n[0, 1]$ and assume that

(i) $a_i, b_i \in C_0[0, 1]$, $a_i(x) > 0$ for $0 \leqslant x \leqslant 1$ $(i = 1, 2, \ldots, n)$;

(ii) $g(x, y, p, v, w)$ is a continuous function on $[0, 1] \times \mathbb{R}^n \times \mathbb{R}^n \times C_0^n[0, 1] \times C_0^n[0, 1]$ and $g^0(y, v)$, $g^1(y, v)$ are continuous functions on $\mathbb{R}^n \times C_0^n[0, 1]$;

(iii) the function g is bounded on each bounded subset of its domain (with respect to the norm $\| \ \|_\infty$ in $C_0^n[0, 1]$).

(The latter assumption will be applied in Sections 3.2 and 3.3. Clearly, it is satisfied if g is given by (2.2) with $f(x, y, p)$ continuous on $[0, 1] \times \mathbb{R}^n \times \mathbb{R}^n$.)

The basic theorem is formulated in Section 3.1. It requires rather strong assumptions concerning the dependence of Mu on u'. These assumptions are weakened in Section 3.2 by constructing bounds Φ, Ψ for $(u^*)'$ also, and in Section 3.3 by providing a priori bounds for $(u^*)'$. The theory is similar to the one for a single differential equation in Section III.4, although here additional difficulties arise. Moreover, many proofs use essentially the same tools as those in Section III.4. Therefore, a series of proofs will only be sketched. Section 3.4 is concerned with problems of other forms, such as initial value problems of the first order and singular boundary value problems. Here the assumptions and some parts of the proofs have to be modified appropriately. The general approach, however, is the same.

For $u, v, w \in C_0^n[0, 1]$ let $\langle v, u, w \rangle$ be defined componentwise, i.e., $\langle v, u, w \rangle = (h_i)$ with $h_i = \langle v_i, u_i, w_i \rangle$ and the latter term defined in (III.4.6).

3.1 The basic theorem

The following theorem can immediately be applied to problems in which the differential operator M does not depend on u', except the linear term $b(x)u'(x)$in $\mathbf{L}u(x)$. In addition, the theorem will serve as a tool in treating more general problems.

Theorem 3.1. *Suppose that functions $\varphi, \psi \in C_2^n[0, 1]$ exist which satisfy $\varphi \leqslant \psi$ and*

$$\mathbf{L}\varphi(x) + G(x, \varphi(x), \varphi'(x), v, w) \leqslant o \leqslant \mathbf{L}\psi(x) + G(x, \psi(x), \psi'(x), v, w) \quad (3.2)$$

for $0 \leqslant x \leqslant 1$ and all $v \in [\varphi, \psi]_0$, $w \in C_0^n[0, 1]$. Moreover, let the function g be bounded on $\omega \times \mathbb{R}^n \times [\varphi, \psi]_0 \times C_0^n[0, 1]$, $\omega = \{(x, y): 0 \leqslant x \leqslant 1, \varphi(x) \leqslant y \leqslant \psi(x)\}$.

Then the given problem $Mu = o$ has a solution $u^* \in C_2^n[0, 1]$ such that $\varphi \leqslant u^* \leqslant \psi$.

Proof. We shall only sketch the proof, since it is very similar to the proof of Theorem III.4.2. First, a modified operator $M^\#$ as in Section III.4.1 is defined, where now $\mathbf{N}^\# u(x) = G(x, u^\#(x), u'(x), u^\#, u')$, $u^\# = \langle \varphi, u, \psi \rangle$. Then the modified problem $M^\# u = o$ is transformed into a fixed-point equation $u = Tu$ in $C_1^n[0, 1]$ where T has a form similar to (III.4.7). Since T is continuous and $TC_1^n[0, 1]$ is relatively compact, Schauder's fixed-point theorem yields the existence of a fixed point u^*, which is also a solution of the modified problem.

To prove that u^* is a solution of (3.1), one shows for each index i that $\varphi_i(x) \leqslant u_i^*(x) \leqslant \psi_i(x)$ $(0 \leqslant x \leqslant 1)$. For example, the inequality $u_i^*(x) \leqslant \psi_i(x)$ $(0 \leqslant x \leqslant 1)$ is established by applying Theorem III.2.1 with $z(x) \equiv 1$ to A_i, u_i^*, ψ_i instead of M, v, w, where the operator $A_i: C_2[0, 1] \to \mathbb{R}[0, 1]$ is given by

$$A_i u_i = (M^{\#}u)_i, \qquad M^{\#}u = \mathbf{L}u + u - u^{\#} + \mathbf{N}^*u, \tag{3.3}$$
$$\mathbf{N}^*u(x) = G(x, u^{\#}(x), u'(x), (u^*)^{\#}, (u^*)').$$

(Observe that $(M^{\#}u)_i$ does not depend on any component u_j with $j \neq i$.) The procedure is analogous to that in the proof of Lemma III.4.1. $\square$

We point out that the inequalities (3.2) assume the simple form $M\varphi \leqslant o \leqslant M\psi$ if $G(x, y, p, v, w)$ does not depend on w and if this function is antitone in v.

Example 3.1. We ask for nontrivial positive solutions $u \in C_4[0, 1]$ of the nonlinear eigenvalue problem

$$u^{\text{iv}} + ((u'')^3)'' - \lambda u = 0 \qquad (0 < x < 1),$$
$$u(0) = u''(0) = u(1) = u''(1) = 0,$$

with $\lambda > \pi^4$. This problem will be rewritten as a system of differential equations in Example 3.3. Here we transform it into an integro-differential equation (for the case $n = 1$). By applying the integral operator $\int_0^1 \mathscr{G}(x, \xi) \cdot d\xi$ with $\mathscr{G}(x, \xi) = \mathscr{G}(\xi, x) = x(1 - \xi)$ for $0 \leqslant x \leqslant \xi \leqslant 1$ to the differential equation and observing the boundary conditions for u'', we get

$$h(-u''(x)) - \lambda \int_0^1 \mathscr{G}(x, \xi)u(\xi)\, d\xi = 0 \qquad \text{with} \quad h(s) = s + s^3.$$

Solving this equation for $-u''(x)$, we finally obtain an equivalent problem $Mu = o$ for $u \in C_2[0, 1]$, where M is an operator of the form (1.1), (1.2), (2.1) given by $Mu(0) = u(0)$, $Mu(1) = u(1)$, and $Mu(x) = -u''(x) + g(u)$ for $0 < x < 1$ with $g(v) = -h^{-1}[\lambda \int_0^1 \mathscr{G}(x, \xi)v(\xi)\, d\xi]$ and h^{-1} denoting the (monotone) inverse function of h. The functions $\varphi(x) = c \sin \pi x$ and $\psi(x) = dx(1 - x)$ satisfy $M\varphi \leqslant o \leqslant M\psi$, if $c > 0$ is small enough and d is sufficiently large. Since $g(\varphi) \geqslant g(v) \geqslant g(\psi)$ for $\varphi \leqslant v \leqslant \psi$, the inequalities (3.2) hold also. Thus Theorem 3.1 yields the existence of a solution u^* with $\varphi \leqslant u^* \leqslant \psi$, and hence *the given eigenvalue problem has a nontrivial solution* $u^* \geqslant o$. $\square$

Problems

3.1 Let $Mu = (-u'' + Bu' + Cu, u(0), u(1))$ with $B, C \in \mathbb{R}^{n,n}$. Show that Theorem 3.1 can be applied only if B is diagonal.
3.2 Prove in detail that the operator T in the proof of Theorem 3.1 is continuous.

3.2 Simultaneous estimation of the solution and its derivative

The boundedness assumptions made in Theorem 3.1 can be omitted, when additional differential inequalities for bounds Φ, Ψ of the derivative of the solution are required. The results are similar to those in Section III.4.2. Here, however, the differential inequalities for φ and ψ, in general, depend also on Φ and Ψ. For simplicity, we restrict ourselves to the case of Dirichlet boundary conditions, i.e., we assume in this section that

$$Mu(0) = u(0) - r^0, \qquad Mu(1) = u(1) - r^1, \qquad a(x) \equiv I.$$

Moreover, let $b(x) \equiv O$, without loss of generality.

Theorem 3.2. *Suppose that functions $\varphi, \psi \in C_2^n[0, 1]$ and $\Phi, \Psi \in C_0^n[0, 1]$ exist such that $\varphi \leqslant \psi$, $\Phi \leqslant \Psi$ hold and the following three conditions are satisfied.*

$$\text{(i)} \quad -\varphi''(x) + g(x, \varphi(x), \langle \Phi, \varphi', \Psi \rangle(x), [\varphi, \psi]_0, [\Phi, \Psi]_0) \leqslant o \qquad (0 < x < 1),$$

$$-\psi''(x) + g(x, \psi(x), \langle \Phi, \psi', \Psi \rangle(x), [\varphi, \psi]_0, [\Phi, \Psi]_0) \geqslant o \qquad (0 < x < 1),$$

$$\varphi(0) \leqslant r^0 \leqslant \psi(0), \qquad \varphi(1) \leqslant r^1 \leqslant \psi(1).$$

(ii) *For each index $i \in \{1, 2, \ldots, n\}$ there exists a $\xi_i \in [0, 1]$ with the following properties:*

if $\xi_i > 0$, then $\Phi_i \in C_1[0, \xi_i]$ and

$$\Phi_i'(x) - g_i(x, [\varphi, \psi](x), \Phi(x), [\varphi, \psi]_0, [\Phi, \Psi]_0) \leqslant 0 \quad (0 < x \leqslant \xi_i),$$

$$\varphi_i(0) = r_i^0, \qquad \Phi_i(0) \leqslant \varphi_i'(0);$$

if $\xi_i < 1$, then $\Phi_i \in C_1[\xi_i, 1]$ and

$$\Phi_i'(x) - g_i(x, [\varphi, \psi](x), \Phi(x), [\varphi, \psi]_0, [\Phi, \Psi]_0) \geqslant 0 \quad (\xi_i \leqslant x < 1),$$

$$\psi_i(1) = r_i^1, \qquad \Phi_i(1) \leqslant \psi_i'(1).$$

(iii) *For each index* $i \in \{1, 2, \ldots, n\}$ *there exists an* $\eta_i \in [0, 1]$ *with the following properties*:

if $\eta_i > 0$, *then* $\Psi_i \in C_1[0, \eta_i]$ *and*

$$\Psi_i'(x) - g_i(x, [\varphi, \psi](x), \Psi(x), [\varphi, \psi]_0, [\Phi, \Psi]_0) \geqslant 0 \quad (0 < x \leqslant \eta_i),$$
$$\psi_i(0) = r_i^0, \quad \psi_i'(0) \leqslant \Psi_i(0);$$

if $\eta_i < 1$, *then* $\Psi_i \in C_1[\eta_i, 1]$ *and*

$$\Psi_i'(x) - g_i(x, [\varphi, \psi](x), \Psi(x), [\varphi, \psi]_0, [\Phi, \Psi]_0) \leqslant 0 \quad (\eta_i \leqslant x < 1),$$
$$\varphi_i(1) = r_i^1, \qquad \varphi_i'(1) \leqslant \Psi_i(1).$$

Under these assumptions, the given problem (3.1) *has a solution* $u^* \in C_2^n[0, 1]$ *with* $\varphi \leqslant u^* \leqslant \psi$, $\Phi \leqslant (u^*)' \leqslant \Psi$.

Proof. For the operator $\hat{M}$: $C_2^n[0, 1] \to \mathbb{R}^n[0, 1]$ defined by

$$\begin{aligned} \hat{M}u(x) &= -u''(x) + \hat{g}(x, u(x), u'(x), u, u') \qquad \text{for} \quad 0 < x < 1 \\ \hat{M}u(0) &= u(0) - r^0, \qquad \hat{M}u(1) = u(1) - r^1 \end{aligned} \tag{3.4}$$

with

$$\hat{g}(x, y, p, v, w) = g(x, y, \langle\Phi(x), p, \Psi(x)\rangle, v, \langle\Phi, w, \Psi\rangle),$$

Theorem 3.1 yields the existence of a function $u^* \in C_2^n[0, 1]$ such that $\hat{M}u^* = o$, $\varphi \leqslant u^* \leqslant \psi$. It suffices to show that $\Phi_i \leqslant (u_i^*)' \leqslant \Psi_i$ for each index i.

We consider only the inequality $(u_i^*)'(x) \leqslant \Psi_i(x)$ $(0 \leqslant x \leqslant \eta_i)$ assuming $\eta_i > 0$. To prove this inequality one proceeds in a way similar to that in the proof of Theorem III.4.4. Now use $A: C_1[0, \eta_i] \to \mathbb{R}[0, \eta_i]$ defined by $Ah(0) = h(0)$ and $Ah(x) = h'(x) - g_i(x, u^*(x), \hat{h}(x), u^*, \langle\Phi, (u^*)', \Psi\rangle)$ for $0 < x \leqslant \eta_i$, where the ith component of $\hat{h}$ is $\hat{h}_i = \langle\Phi_i, h, \Psi_i\rangle$. $\square$

Analogous results can be derived for other boundary conditions. The essential point is that "initial inequalities" such as $\Phi_i(0) \leqslant (u_i^*)'(0)$ can be established by using properties of φ and ψ. (Compare Theorem III.4.4 and Proposition III.4.5 for the methods of proof.)

Example 3.2. For illustrating the preceding result let us consider the simple problem

$$-u''(x) + tBu'(x) = r(x) \qquad (0 < x < 1),$$

$$u(0) = u(1) = 0 \qquad \text{with} \quad B = \begin{pmatrix} 0 & 1 \\ 1 & 0 \end{pmatrix}$$

where $g(x, y, p, v, w) = tBw(x) - r(x)$. We shall try to satisfy the assumptions of Theorem 3.2 with $\xi_1 = \eta_1 = \xi_2 = \eta_2 = 1$. For the special case $\varphi = -\psi$, $\Phi = -\Psi$, we then obtain the conditions $\psi \geqslant o$, $\Psi \geqslant o$, $\psi(0) = \psi(1) = o$, $\Psi(0) \geqslant \psi'(0)$, and

$$-\psi'' - |t|\begin{pmatrix} 0 & 1 \\ 1 & 0 \end{pmatrix}\Psi \geqslant |r|, \qquad \Psi' - |t|\begin{pmatrix} 0 & 1 \\ 1 & 0 \end{pmatrix}\Psi \geqslant |r|,$$

Because of the symmetry of these relations, we may choose $\psi_1 = \psi_2 =: h$ and $\Psi_1 = \Psi_2 =: k$. Then, assuming $|r| \leqslant 1$, the above relations hold if

$$-h'' - |t|k = 1, \qquad h(0) = h(1) = 0, \qquad k' - |t|k = 1, \qquad k(0) = h'(0).$$

A formal calculation shows that a solution $h \geqslant o$, $k \geqslant o$ exists if and only if $t = 0$ or $2|t| + 1 > \exp|t|$, i.e., $|t| < 1.256\ldots$.

Obviously, the assumptions of Theorem 3.2 implicitly restrict the way in which the differential expression $Mu(x)$ may depend on $u'(x)$. These restrictions may be weakened by first transforming the problem and then applying the theorem. In this example, one obtains two separate boundary value problems for v_1 and v_2 by an appropriate rotation $v = Pu$. For the problem so transformed suitable functions ψ, Ψ exist if $t > -\pi^2$; the estimate now obtained has the form $\varphi \leqslant Pu^* \leqslant \psi$. Observe that these inequalities can also be written as $P\hat{\varphi} \leqslant Pu^* \leqslant P\hat{\psi}$ with suitable $\hat{\varphi}$, $\hat{\psi}$. Thus, transforming the problem amounts here to using another order relation. □

Example 3.3. The eigenvalue problem in Example 3.1 can be transformed into an equivalent system

$$-u_1''(x) - u_2(x) = 0 \qquad (0 < x < 1), \qquad u_1(0) = u_1(1) = 0,$$

$$-(1 + 3(u_2(x))^2)u_2''(x) - 6u_2(x)(u_2'(x))^2 - \lambda u_1(x) = 0 \qquad (0 < x < 1),$$

$$u_2(0) = u_2(1) = 0$$

(by introducing $u_1 = u$, $u_2 = -u''$). This system can then be written as an equation (3.1) where

$$g_1(x, y, p, v, w) = -v_2(x),$$

$$g_2(x, y, p, v, w) = -(1 + 3(y_2)^2)^{-1}[6y_2(p_2)^2 + \lambda v_1(x)].$$

Let φ and ψ be given by $\varphi_1(x) = c\sin\pi x$, $\varphi_2 = -\varphi_1''$, $\psi_1(x) = 12dx(1-x)$ and $\psi_2(x) + (\psi_2(x))^3 = \lambda dx(1-x)[1 + x(1-x)]$. Moreover, choose $-\Phi_1 = -\Phi_2 = \Psi_1 = \Psi_2 = N$ with a constant $N > 0$. Then $\varphi \leqslant \psi$ and $M\varphi \leqslant o \leqslant M\psi$, if $c > 0$ is small enough and d is sufficiently large. Since g_1 decreases with increasing v_2 and g_2 decreases with increasing v_1, the differential inequalities in assumption (i) of Theorem 3.2 hold also, provided N is so large that $\varphi' = \langle \Phi, \varphi', \Psi \rangle$, $\psi' = \langle \Phi, \psi', \Psi \rangle$.

For $v \in [\varphi, \psi]$, g_1 assumes only values $\leqslant 0$. Thus, for the index $i = 1$, conditions (ii) and (iii) in Theorem 3.2 are satisfied for $\xi_1 = 0, \eta_1 = 1$, and N sufficiently large. In a similar way, one sees that for the index $i = 2$ conditions (ii), (iii) hold with $\xi_2 = 0, \eta_2 = 1$, and N sufficiently large.

Consequently, *the given eigenvalue problem has a solution u^* such that $\varphi_1 \leqslant u^* \leqslant \psi_1$, $-N \leqslant (u^*)' \leqslant N$, $\varphi_2 \leqslant (-u^*)'' \leqslant \psi_2$, $-N \leqslant (-u^*)''' \leqslant N$,* where the constants c, d, N involved could be computed. □

Problems

3.3 Carry out the transformation $v = Pu$ in Example 3.2, and then calculate bounds.

3.4 Reformulate the result of Theorem 3.2 for the special case in which g has the form (2.2) and Φ, Ψ denote constant functions. In particular, use conditions on the sign of the functions f_i and use $\xi_i = 0, \eta_i = 1$ or $\xi_i = 1, \eta_i = 0$. Also observe that in the assumptions only values $f(x, y, p)$ occur with p lying on the surface of the cube $\{p : \Phi \leqslant p \leqslant \Psi\}$.

3.3 Growth restrictions concerning u'

For the case of a single differential equation ($n = 1$) we were concerned in Sections III.4.3 and III.4.5.1 with growth properties of $f(x, y, p)$ with respect to p such that the existence of comparison functions φ, ψ implies the existence of a solution and its inclusion by φ and ψ. This theory cannot immediately be generalized to systems of equations as considered here, except in a certain "trivial" case which we shall discuss. This case is essentially characterized by the property that each function $g_i(x, y, p, v, w)$ is bounded (above or below) with respect to $w \in C_0^n[0, 1]$ while the dependence on p_i is only restricted by a (one-sided) Nagumo condition. For more general problems, the difficulties in developing a corresponding theory are mainly due to the fact that, in general, the comparison functions φ, ψ depend also on the bounds Φ, Ψ.

In the following theorem let $\nu_i > 0$, $\mu_i > 0$ denote fixed constants and h_i, k_i fixed functions with

$$h_i \in C_0[0, \infty), \qquad k_i \in C_0[0, \infty),$$
$$h_i(s) > 0, \qquad k_i(s) > 0 \qquad \text{for} \quad 0 \leqslant s < \infty.$$

This theorem generalizes Theorem III.4.10 in several respects. A corresponding generalization of Theorem III.4.11a,b is also possible (see Problem 3.5).

Theorem 3.3. *Suppose that functions $\varphi, \psi \in C_2^n[0, 1]$ exist which satisfy the inequalities (3.2). Moreover, for each $u \in [\varphi, \psi]_2$ with $Mu(0) = Mu(1) = o$ and for each index $i \in \{1, 2, \ldots, n\}$, let one of the conditions* (a$_1$), (a$_2$) *and* *one of the conditions* (b_1), (b_2) *below hold for all* $x \in (0, 1)$, *all* $w \in C_0^n[0, 1]$, *some* N *satisfying* $N > \max\{\|\varphi_i'\|_\infty, \|\psi_i'\|_\infty\}$ $(i = 1, 2, \ldots, n)$, *and* $\Delta_i = \max\{\psi_i(x) : 0 \leqslant x \leqslant 1\} - \min\{\varphi_i(x) : 0 \leqslant x \leqslant 1\}$.

(a_1) $g_i(x, u(x), p, u, w) \geqslant -h_i(-p_i)$ *for* $p_i < 0$,
$u_i'(0) \geqslant -\nu_i$, $\int_{\nu_i}^N s h_i^{-1}(s)\, ds > \Delta_i$;

(a_2) $g_i(x, u(x), p, u, w) \leqslant h_i(-p_i)$ *for* $p_i < 0$,
$u_i'(1) \geqslant -\nu_i$, $\int_{\nu_i}^N s h_i^{-1}(s)\, ds > \Delta_i$.

(b_1) $g_i(x, u(x), p, u, w) \leqslant k_i(p_i)$ *for* $p_i > 0$,
$u_i'(0) \leqslant \mu_i$, $\int_{\mu_i}^N s k_i^{-1}(s)\, ds > \Delta_i$;

(b_2) $g_i(x, u(x), p, u, w) \geqslant -k_i(p_i)$ *for* $p_i > 0$,
$u_i'(1) \leqslant \mu_i$, $\int_{\mu_i}^N s k_i^{-1}(s)\, ds > \Delta_i$.

Then the given problem (3.1) *has a solution* $u^* \in C_2^n[0, 1]$ *such that*

$$\varphi \leqslant u^* \leqslant \psi, \quad -Ne \leqslant (u^*)' \leqslant Ne.$$

Remark. The preceding conditions contain a function $u \in [\varphi, \psi]_2$ not determined, so that these conditions have to be verified for all such functions. Constants ν_i and μ_i may be calculated similarly as in Proposition III.4.5, using $\varphi_i \leqslant u_i \leqslant \psi_i$ and $(Mu)_i(0) = (Mu)_i(1) = 0$.

To prove the theorem, we could proceed as in Section III.4.3, using functions $\Phi_i(x, y_i)$ and $\Psi_i(x, y_i)$ to obtain bounds for $(u_i^*)'$. Instead, we shall briefly explain how the method of a priori bounds can be applied (compare Section III.4.5.1). We introduce an operator $\hat{M}$ of the form (3.4) with $\Phi = -Ne$ and $\Psi = Ne$. According to Theorem 3.1, the modified problem $\hat{M}u = o$ has a solution $u \in [\varphi, \psi]_2$, for which then $-N \leqslant u_i' \leqslant N$ has to be shown for each index i. This is achieved by proving that each solution of the equation $-u''(x) + \hat{g}(x, u(x), u'(x), u, u') = o$ satisfies the a priori inequalities $u_i'(x) < N$ and $u_i'(x) > -N$, if appropriate "initial estimates" of $u_i'(x)$ at fixed points x are known and $u \in [\varphi, \psi]$. The procedure is essentially the same as in the proof of Proposition III.4.15. Now the arguments are applied to the ith equation $-u_i''(x) + \hat{g}_i(x, u(x), u'(x), u, u') = 0$ and estimates of $u_i'(0)$ or $u_i'(1)$ are used instead of the relation $|u'(\xi)| \leqslant \Lambda$. The details are left to the reader. □

Example 3.4. We consider again the problem in Example 3.3. Since the functions φ, ψ defined there satisfy the given Dirichlet boundary conditions, the values $u'(0)$ and $u'(1)$ can be estimated for $u \in [\varphi, \psi]_2$ (cf. Proposition III.4.5). Moreover, all conditions on g_1, g_2 occurring in Theorem 3.3 are satisfied, respectively, with constant functions h_i, k_i. □

Problem

3.5 Show that Theorem 3.1 remains true if the boundedness condition on g is replaced by the following assumption: There exists a constant $c > 0$ such that $|g_i(x, u(x), p, u, w)| \leqslant c(1 + p_i^2)$ for $0 < x < 1$, all indices i, each $u \in [\varphi, \psi]_2$, $p \in \mathbb{R}^n$, $w \in C_0^n[0, 1]$. Hint: Combine the methods used in Section III.4.5 and in the proof of Theorem 3.3. Generalize this result further by requiring Nagumo conditions instead of the quadratic growth restrictions above.

3.4 Application of the methods to other problems

For regular boundary value problems of the second order we explained methods for obtaining existence and inclusion statements in the preceding sections. Now we shall show that these methods can also be adapted to other problems. Section 3.4.1 contains a brief description of first order initial value problems. In Sections 3.4.2 and 3.4.3 we consider some typical examples of different forms in order to explain some ideas which can also be applied to more general cases. (See also Problems 3.6–3.11.)

3.4.1 First order initial value problems

Let M be an operator of the form

$$Mu(x) = \begin{cases} u'(x) + g(x, u(x), u) & \text{for} \quad 0 < x \leqslant 1 \\ u(0) - r^0 & \text{for} \quad x = 0, \end{cases}$$

which maps $R = C_1^n[0, 1]$ into $S = \mathbb{R}^n[0, 1]$. Here g: $[0, 1] \times \mathbb{R}^n \times C_0^n[0, 1] \to \mathbb{R}^n$ denotes a continuous mapping which is bounded on bounded subsets of its domain. Assume, moreover, that $g(x, y, u)$ is not coupled with respect to y. Define then

$$H(u, w)(x) = u'(x) + g(x, u(x), w) \qquad (0 < x \leqslant 1),$$

$$H(u, w)(0) = u(0) - r^0.$$

We ask for solutions of the problem $Mu = o$. An initial value problem for a system of first order differential equations (in the usual sense) can be written in this form by using a definition analogous to (2.2).

Theorem 3.4. *If there exist functions $\varphi, \psi \in R$ such that*

$$\varphi \leqslant \psi, \qquad H(\varphi, [\varphi, \psi]) \leqslant o \leqslant H(\psi, [\varphi, \psi]),$$

then the problem $Mu = o$ has a solution $u^* \in R$ which satisfies $\varphi \leqslant u^* \leqslant \psi$.

The proof uses essentially the same tools as the proof of Theorem 3.1. The details are left to the reader.

3.4.2 *Mixed problems, a differential equation of the third order on* $[0, \infty)$

Problems consisting of first and second order equations often can also be treated by the methods which lead to Theorems 3.1 and 3.4. Here we consider a third order problem which can be transformed into such a "mixed system."

The equations

$$w''' + ww'' + \lambda(1 - (w')^2) = 0, \qquad w(0) = w'(0) = 0, \qquad w'(\infty) = 1 \quad (3.5)$$

describe a two-dimensional incompressible boundary layer (Falkner and Skan [1931]; for an exposition see Hartman [1964]). They are equivalent to a problem on $[0, \infty)$ which consists of the conditions $u_2(\infty) = 1$ and

$$\begin{gathered} u_1' - u_2 = 0, \qquad -u_2'' - u_1 u_2' - \lambda(1 - (u_2)^2) = 0, \\ u_1(0) = 0, \qquad u_2(0) = 0. \end{gathered} \quad (3.6)$$

To solve this problem we first discuss a problem in which the boundary condition at ∞ is replaced by

$$u_2(l) = c \quad (3.7)$$

with constants $l > 0$ and c. This problem (3.6), (3.7) can be treated with the method of modification used for proving Theorems 3.1 and 3.4, if suitable comparison functions φ, ψ are known (Problem 3.8). For example, the conditions on $\varphi = (\varphi_i)$ assume the form

$$\begin{gathered} \varphi_1' - \varphi_2 \leqslant 0, \qquad -\varphi_2'' - [\varphi_1, \psi_1]\varphi_2' - \lambda(1 - \varphi_2^2) \leqslant 0 \qquad \text{on} \quad (0, l), \\ \varphi_1(0) \leqslant 0, \qquad \varphi_2(0) \leqslant 0, \qquad \varphi_2(l) \leqslant c. \end{gathered}$$

In particular, if we choose $\varphi_2 \equiv \varphi_1'$ and $\psi_2 \equiv \psi_1'$ and write $\underline{w} = \varphi_1$, $\overline{w} = \varphi_2$, then the preceding inequalities and the corresponding inequalities for ψ can be written as

$$\underline{w}''' + [\underline{w}, \overline{w}]\underline{w}'' + \lambda(1 - (\underline{w}')^2) \geqslant 0, \quad (3.8)$$

$$\overline{w}''' + [\underline{w}, \overline{w}]\overline{w}'' + \lambda(1 - (\overline{w}')^2) \leqslant 0, \quad (3.9)$$

$$\begin{aligned} &\underline{w}(0) \leqslant 0, \qquad \underline{w}'(0) \leqslant 0, \qquad \overline{w}(0) \geqslant 0, \qquad \overline{w}'(0) \geqslant 0, \\ &\underline{w}'(l) \leqslant c, \qquad\qquad\qquad\quad \overline{w}'(l) \geqslant c. \end{aligned} \quad (3.10)$$

If there exist sufficiently smooth functions $\underline{w}, \overline{w}$ (e.g., $\underline{w}, \overline{w} \in C_3[0, l]$) which satisfy these inequalities, then the problem (3.6), (3.7) has a solution u^* such that $\varphi(x) \leqslant u^*(x) \leqslant \psi(x)$ $(0 \leqslant x \leqslant l)$.

Using arguments similar to those in the proof of Theorem III.4.13 one can extend this result to the problem (3.6) with $u_2(\infty) = 1$ (see also Example III.4.7), so that we obtain the following statement for the original problem (3.5).

Proposition 3.5. *If there exist functions* $\underline{w}, \overline{w} \in C_3[0, \infty)$ *which satisfy inequalities* (3.8)–(3.10) *and* $\underline{w}'(\infty) = \overline{w}'(\infty) = 1$, *then problem* (3.5) *has a solution* $w \in C_3[0, \infty)$ *such that*

$$\underline{w}(x) \leqslant w(x) \leqslant \overline{w}(x), \qquad \underline{w}'(x) \leqslant w'(x) \leqslant \overline{w}'(x) \qquad (0 \leqslant x < \infty). \quad (3.11)$$

We observe that *this statement remains true if* $\overline{w}$ *and* $\underline{w}$ *have less strong smoothness properties*. For instance, certain discontinuities are allowed, as in Example III.4.7. In particular, φ_2 need only satisfy Condition S_l (defined in Section III.2.1), that means $\underline{w}''$ may have certain jumps. For $\lambda > 0$, the following functions have all properties which then are to be required: $\overline{w}(x) = \alpha + x$ with $\alpha \geqslant 0$, $\underline{w}(x) = 0$ for $0 \leqslant x \leqslant \gamma$, $\underline{w}(x) = x - \gamma - \gamma \log \gamma^{-1} x$ for $\gamma < x < \infty$ with $\gamma = (2/\lambda)^{1/2}$. Thus *the given problem* (3.5) *has a solution* w *which satisfies* (3.11) *with these functions* $\underline{w}, \overline{w}$.

Since $\underline{w}(0) = 0 \leqslant \alpha = \overline{w}(0)$, $\underline{w}'(0) = 0$, and $\overline{w}'(0) = 1$, the same statement is true if the conditions at $x = 0$ are replaced by $w(0) = \alpha$ and $w'(0) = \beta$ with $\alpha \geqslant 0$, $0 \leqslant \beta \leqslant 1$. Furthermore, we remark that suitable functions $\underline{w}$ and $\overline{w}$ can also be constructed for $\lambda = 0$ and values $\lambda \in (\lambda_0, 0)$ with a certain $\lambda_0 < 0$.

3.4.3 A singular problem

We consider again the problem in Example 2.3, which can be written as $Mu = o$ with M defined on $R = C_0^2[0, 1] \cap C_2^2(0, 1)$ by (2.17). We shall prove by the method of modification that this problem has a solution $u^* \in R$ which satisfies $\varphi \leqslant u^* \leqslant \psi$ with φ, ψ given by (2.18).

Due to the singularity of the differential operator at $x = 0$, we have to proceed somewhat differently than in the proof of Theorem 3.1. Now we define a modified operator $M^\#$ on R by

$$M^\# u(x) = L[u](x) + \bar{g}(x, u^\#(x), u^\#) \qquad \text{for} \quad 0 < x < 1,$$

$$M^\# u(x) = Mu(x) \qquad \text{for} \quad x \in \{0, 1\}.$$

(Observe that here a term $u - u^\#$ is not added.) The modified problem $M^\# u = o$ is again transformed into a fixed-point equation $u = Tu$ with $Tu(x) = -\int_0^1 \mathscr{G}(x, \xi)\bar{g}(\xi, u^\#(\xi), u^\#)\, d\xi$, where the 2×2 matrix $\mathscr{G}(x, \xi) = (\mathscr{G}_i(x, \xi)\, \delta_{ik})$ is given by $\mathscr{G}(x, \xi) = \mathscr{G}(\xi, x)$ and

$$\left.\begin{aligned} \mathscr{G}_1(x, \xi) &= \tfrac{1}{2}(1 - \xi^2)\xi^{-1} x \\ \mathscr{G}_2(x, \xi) &= \tfrac{1}{2}[(1 + \nu)(1 - \nu)^{-1}\xi + \xi^{-1}]x \end{aligned}\right\} \quad \text{for} \quad 0 \leqslant x \leqslant \xi \leqslant 1, \quad x \neq 0.$$

One can prove that this operator T maps the space $C_0^2[0, 1]$ (furnished with the norm $\|u\|_\infty$) into a uniformly bounded and equicontinuous subset of itself. Thus by Schauder's fixed-point theorem, T has a fixed point u^*, which then is also a solution of the modified problem.

It remains to be shown that $\varphi \leqslant u^* \leqslant \psi$. This is achieved essentially as in the proof of Theorem 3.1. For example, in order to prove $u_i^*(x) \leqslant \psi_i(x)$ $(0 \leqslant x \leqslant 1)$ for $i = 1$ or $i = 2$, one applies Theorem III.2.1 to A_i, u_i^*, ψ_i instead of M, v, w, where the operator $A_i : C_0[0, 1] \cap C_2(0, 1) \to \mathbb{R}[0, 1]$ is given by

$$A_i u_i(x) = L[u_i](x) + \bar{g}_i(x, u^\#(x), (u^*)^\#) \qquad (0 < x < 1),$$

$$A_i u_i(x) = (Mu)_i(x) \qquad (x = 0, 1),$$

analogously to (3.3). Now, we cannot choose the simple function $z(x) \equiv 1$ in each case. However, all required assumptions are satisfied for $z(x) = 1 + (1 + \nu)(1 - \nu)^{-1}x$. $\square$

Problems

3.6 Prove Theorem 3.4 using a modified differential equation (a) $u'(x) + u(x) - u^\#(x) + g(x, u^\#(x), u^\#) = o$ or (b) $u'(x) + g(x, u^\#(x), u^\#) = o$.

3.7 Consider a problem

$$\begin{aligned} u_1'(x) + f_1(x, u_1(x), u_2(x)) &= 0 \qquad (0 < x \leqslant 1), \qquad u_1(0) = r_1^0, \\ -u_2''(x) + f_2(x, u_1(x), u_2(x)) &= 0 \qquad (0 < x < 1), \qquad u_2(0) = r_2^0, \\ & \qquad\qquad\qquad\qquad\qquad\qquad u_2(1) = r_2^1 \end{aligned}$$

for $u_1 \in C_1[0, 1]$, $u_2 \in C_2[0, 1]$ with sufficiently smooth f_1, f_2. Define a function $H(u, w) = H(u_1, u_2, w_1, w_2)$ such that the given problem has a solution $u^* \in [\varphi, \psi]$ if $H(\varphi, [\varphi, \psi]) \leqslant o \leqslant H(\psi, [\varphi, \psi])$. For the existence proof use the modification method, replacing, for example, the second order equation by $-u_2'' + u_2 - u_2^\# + f_2(x, u_1^\#, u_2^\#) = 0$.

3.8 Generalize the preceding problem, replacing the second order differential equation by $-u_2''(x) + f_2(x, u_1(x), u_2(x), u_2'(x)) = 0$. Derive an existence and inclusion statement under the condition that $f_2(x, y_1, y_2, p_2)$ grows at most quadratically in p_2 (or satisfies a Nagumo condition) for x, y_1, y_2 bounded.

3.9 Generalize Problem 3.7 by assuming that all functions u_1, u_2, f_1, f_2 are vector-valued.

3.10 Consider the problem

$$-u''(x) + g(x, u(x), u'(x), u) = 0 \qquad (0 < x < l),$$

$$u(0) = 0, \qquad u(l) = c$$

with $g(x, y, p, u) = -\int_0^x u(t)\,dt\,p - \lambda(1 - y^2)$ and $\lambda > 0$. Define $\gamma = (2/\lambda)^{1/2}$, $\varphi(x) = 0$ $(0 \leqslant x \leqslant \gamma)$, $\varphi(x) = 1 - x^{-1}\gamma$ $(\gamma < x < \infty)$, $\psi(x) \equiv 1$. Show that for each constant $c \in [\varphi(l), 1]$, the problem has a solution $u^* \in C_2[0, l]$.

Hint: Apply the result of Problem 3.5 and observe that φ satisfies Condition S_l.

3.11 Define g, φ, ψ as in the preceding problem. The third order boundary value problem in Section 3.4.2 is equivalent to

$$-u''(x) + g(x, u(x), u'(x), u) = 0 \qquad (0 < x < \infty),$$
$$u(0) = 0, \qquad u(\infty) = 1.$$

Prove that this problem has a solution $u^* \in C_2[0, \infty)$ with $\varphi(x) \leqslant u^*(x) \leqslant \psi(x)$ $(0 \leqslant x < \infty)$. Hint: use the result of the preceding problem and apply an absorption method as used in Section 3.4.2 and in the proof of Theorem III.4.13.

4 POINTWISE NORM BOUNDS BY RANGE–DOMAIN IMPLICATIONS

This section is concerned with pointwise norm estimates, such as (1.5), derived from properties of Mv. The continuity principle described in Section IV.1 will be used to obtain sufficient conditions. To explain the approach and the nature of the results obtained, sufficient conditions are derived in Section 4.2 for a simple class of operators (with Dirichlet boundary terms), and these conditions are discussed in more detail in Section 4.3. As Sections 4.4 and 4.5 (and some problems) will show, the approach can also be used for more general and other operators. For initial value problems of the first order (Section 4.5) the estimation can easily be combined with existence statements for solutions. The results can then be used to investigate the stability of (stationary) solutions.

4.1 Notation and auxiliary means

In this section and the succeeding Section 5 we denote by $\langle\ ,\ \rangle$ an inner product in $\mathbb{R}^n$ and $\|y\| = \langle y, y\rangle^{1/2}$. All theorems (and corollaries) hold for $\langle y, \eta\rangle = y^{\mathrm{T}}H\eta$ with an arbitrary positive definite symmetric matrix H, except when the contrary is indicated, as in Theorem 4.6. In all examples, however, we shall assume that $\langle y, \eta\rangle = y^{\mathrm{T}}\eta$ and hence $\|y\| = \|y\|_2$, unless otherwise stated.

For a given inner product and a given $n \times n$ matrix C, we define $C_{\mathrm{H}} = \frac{1}{2}(C + C^*)$ to be the symmetric (hermitian) part of C, where C^* is characterized by $\langle y, C\eta\rangle = \langle C^*y, \eta\rangle$ $(y, \eta \in \mathbb{R}^n)$. Of course, if $H = I$, then $C^* = C^{\mathrm{T}}$. For symmetric C, $\tau(C)$ and $\sigma(C)$ denote the minimal and maximal eigenvalue of C, respectively.

If $u, v \in \mathbb{R}^n[0, 1]$, then $\langle u, v\rangle$ and $\|u\|$ are functions in $\mathbb{R}[0, 1]$ defined by $\langle u, v\rangle(x) = \langle u(x), v(x)\rangle$, $\|u\|(x) = \|u(x)\|$ for $0 \leqslant x \leqslant 1$. Thus, an inequality $\|u\| \leqslant \psi$ is equivalent to $\|u(x)\| \leqslant \psi(x)$ $(0 \leqslant x \leqslant 1)$, since the inequality sign here always denotes the natural order relation.

In some applications of our theory we shall use an inner product

$$\langle y, \eta\rangle = (\Phi^{-1}y)^{\mathrm{T}}\Phi^{-1}\eta \qquad \text{with} \quad \Phi^{-1}C\Phi = J \tag{4.1}$$

where C is a given matrix and J is a certain normal form. To understand the significance of this transformation, it may suffice to consider the following simple example.

If $n = 2$ and C has an eigenvalue $\lambda = \mu + i\nu$ with real μ and $\nu > 0$ corresponding to an eigenvector $h = \alpha + i\beta$ with real (column) vectors α and β then

$$\Phi^{-1}C\Phi = J = \begin{pmatrix} \mu & \nu \\ -\nu & \mu \end{pmatrix} \qquad \text{for} \quad \Phi = (\alpha, \beta). \tag{4.2}$$

For these matrices and all $y \in \mathbb{R}^2$

$$y^{\mathrm{T}}Jy = \mu y^{\mathrm{T}}y \qquad \text{and} \quad \langle y, Cy\rangle = \mu\langle y, y\rangle \tag{4.3}$$

with the inner product defined as in (4.1). The important point is that the imaginary part ν does not occur in this relation. Such a relation may be used for theoretical purposes, but also to obtain numerical estimates.

In order to obtain analogous results for a general (real) $n \times n$ matrix C we make use of the real Jordan form $J = J_\varepsilon$ of C. The details are not necessary for understanding the theory in the succeeding sections; the reader may skip the rest of this section until he needs it.

We shall briefly explain the structure of J_ε and the transformation matrix $\Phi = \Phi(\varepsilon)$ using the complex eigenvalues, eigenvectors, and principal vectors (generalized eigenvectors) for C. According to matrix theory, there exist n linear independent principal vectors of C such that the set of these n vectors is the union of subsets with the following properties. Let $\{h^1, h^2, \ldots, h^m\}$ be such a subset (or chain), then $(C - \lambda I)h^j = \varepsilon h^{j-1}$ $(j = 1, 2, \ldots, m)$ with $m \geqslant 1$, $h^0 = o$, $\varepsilon > 0$, and $\lambda = \mu + i\nu$ an eigenvalue of C $(\mu, \nu$ real). The vector h^1 is an eigenvector; for $j > 1$ the vectors h^j are principal vectors of higher order. Observe that to each eigenvalue (at least) one such chain exists. Moreover, if $\lambda = \mu + i\nu$ with $\nu \neq 0$ is an eigenvalue, then $\bar{\lambda} = \mu - i\nu$ is also an eigenvalue and the corresponding chains can be obtained from that of λ. For this reason it suffices to consider values $\nu \geqslant 0$.

Suppose that $\nu > 0$ and $h^j = \alpha^j + i\beta^j$ with $\alpha^j, \beta^j \in \mathbb{R}^n$. Then the latter (column) vectors are used to define a matrix $\tilde{\Phi} = (\alpha^1, \alpha^2, \ldots, \alpha^m, \beta^1, \beta^2, \ldots, \beta^m)$, which has the property that $C\tilde{\Phi} = \tilde{\Phi}\tilde{J}$ for

$$\tilde{J} = \begin{pmatrix} J_1 & J_2 \\ -J_2 & J_1 \end{pmatrix} \quad \text{with } m \times m \text{ matrices}$$

$$J_1 = \begin{pmatrix} \mu & \varepsilon & & & 0 \\ & \mu & \varepsilon & & \\ & & \ddots & \ddots & \\ & & & \mu & \varepsilon \\ 0 & & & & \mu \end{pmatrix}, \qquad J_2 = \begin{pmatrix} \nu & & & & 0 \\ & \nu & & & \\ & & \ddots & & \\ & & & \nu & \\ 0 & & & & \nu \end{pmatrix}.$$

If $\nu = 0$, one defines $\tilde{\Phi} = (\alpha^1, \alpha^1, \ldots, \alpha^m)$ and obtains $C\tilde{\Phi} = \tilde{\Phi}\tilde{J}$ with $\tilde{J} = J_1$ and J_1 as before.

Now, for each chain which belongs to an eigenvalue with imaginary part $\geqslant 0$ we take the corresponding matrix $\tilde{\Phi}$ and combine all these matrices as column blocks to an $n \times n$ matrix Φ. Then $C\Phi = \Phi J_\varepsilon$ where J_ε is a block-diagonal matrix and the diagonal blocks are the matrices $\tilde{J}$ described above.

This normal form J_ε has the property that $\frac{1}{2}(J_\varepsilon + J_\varepsilon^{\mathrm{T}})$ does not depend on the imaginary parts of the eigenvalues. Let μ_0 denote the smallest of the real parts of the eigenvalues of C. In the special case that to each eigenvalue with real part μ_0 no principal vector of an order >1 exists (i.e., all corresponding chains have length $m = 1$), one estimates $y^{\mathrm{T}} J_\varepsilon y \geqslant \mu_0 y^{\mathrm{T}} y$ for $\varepsilon > 0$ sufficiently small. Thus, *for all $y \in \mathbb{R}^n$ and all sufficiently small $\varepsilon > 0$:*

$$y^{\mathrm{T}} J_\varepsilon y \geqslant \mu_\varepsilon y^{\mathrm{T}} y \qquad \textit{with} \quad \begin{cases} \mu_\varepsilon = \mu_0 - \varepsilon \\ \mu_\varepsilon = \mu_0 \quad \textit{in the above special case} \end{cases} \tag{4.4}$$

and, consequently,

$$\langle y, Cy \rangle \geqslant \mu_\varepsilon \langle y, y \rangle \qquad \textit{with} \quad \langle\ ,\ \rangle \textit{ defined in } (4.1). \tag{4.5}$$

4.2 General estimation theorems for Dirichlet boundary conditions

Let $R_0 = C_0[0, 1] \cap C_2(0, 1)$, $R = C_0^n[0, 1] \cap C_2^n(0, 1)$, and $M: R \to \mathbb{R}^n[0, 1]$ be an operator of the form

$$Mu(x) = \begin{cases} -u''(x) + f(x, u(x), u'(x)) & \text{for} \quad 0 < x < 1 \\ u(0) & \text{for} \quad x = 0 \\ u(1) & \text{for} \quad x = 1 \end{cases} \tag{4.6}$$

with $f: [0, 1] \times \mathbb{R}^n \times \mathbb{R}^n \to \mathbb{R}^n$. We shall estimate the norm $\|v\|$ of an unknown (vector-valued) function $v \in R$ by a function $\psi \in R_0$ using properties of Mv. Both functions v and ψ shall be considered fixed, unless otherwise indicated. For convenience we introduce the sets

$$\begin{aligned} P &= \{(x, \eta) : 0 < x < 1, \eta \in \mathbb{R}^n, \|\eta\| = 1\}, \\ Q &= \{(x, \eta, q) : (x, \eta) \in P, q \in \mathbb{R}^n, \langle \eta, q \rangle = 0\}. \end{aligned} \tag{4.7}$$

Theorem 4.1. *Suppose that functions $\psi_\lambda \in R_0$ $(0 \leqslant \lambda < \infty)$ exist such that $\psi_0 = \psi$, $\psi_\lambda \succ o$ for $\lambda > 0$, $\psi(\lambda, x) := \psi_\lambda(x)$ is continuous on $[0, \infty) \times [0, 1]$, $\|v\| \leqslant \psi_\lambda$ for some $\lambda \geqslant 0$ and the following two conditions hold.*

(i)
$$\begin{aligned} \|Mv(x)\| < & -\psi''_\lambda(x) + \langle q, q \rangle \psi_\lambda(x) \\ & + \langle \eta, f(x, \psi_\lambda(x)\eta, \psi'_\lambda(x)\eta + \psi_\lambda(x)q) \rangle \end{aligned} \tag{4.8}$$

for all $(x, \eta, q) \in Q$ and $\lambda > 0$ which satisfy

$$v(x) = \psi_\lambda(x)\eta, \qquad v'(x) = \psi'_\lambda(x)\eta + \psi_\lambda(x)q. \tag{4.9}$$

(ii) $\|Mv(0)\| < \psi_\lambda(0)$, $\|Mv(1)\| < \psi_\lambda(1)$ *for* $\lambda > 0$.

Then

$$\|v(x)\| \leqslant \psi(x) \qquad (0 \leqslant x \leqslant 1). \tag{4.10}$$

Proof. For reasons of continuity there exists a *smallest* $\lambda \geqslant 0$ such that $\|v\| \leqslant \psi_\lambda$. For $\lambda = 0$ the statement (4.10) is proved. Suppose therefore $\lambda > 0$. Because λ is minimal, we have $\rho(\xi) = \psi^2_\lambda(\xi)$ for $\rho = \langle v, v \rangle \in R_0$ and some $\xi \in [0, 1]$. Due to assumption (ii), ξ cannot be a boundary point. Consequently, $h := \psi^2_\lambda - \|v\|^2$ satisfies $h(\xi) = h'(\xi) = 0$, $h''(\xi) \geqslant 0$, that is,

$$\langle v, v \rangle(\xi) = \psi^2_\lambda(\xi), \qquad \langle v, v' \rangle(\xi) = \psi_\lambda(\xi)\psi'_\lambda(\xi), \tag{4.11}$$

$$\langle v, v'' \rangle(\xi) + \langle v', v' \rangle(\xi) \leqslant \psi_\lambda(\xi)\psi''_\lambda(\xi) + \psi'_\lambda(\xi)\psi'_\lambda(\xi). \tag{4.12}$$

Now define $\eta, q \in \mathbb{R}^n$ by (4.9) with ξ in place of x. Because of (4.11), these quantities satisfy $\|\eta\| = 1$ and $\langle \eta, q \rangle = 0$. Moreover, (4.12) yields the inequality $\langle \eta, v''(\xi) \rangle + \psi_\lambda(\xi)\langle q, q \rangle \leqslant \psi''_\lambda(\xi)$, which contradicts (4.8) for $x = \xi$, when $v''(\xi)$ is eliminated using $Mv(\xi)$. □

As the proof shows, the assumptions of Theorem 4.1 can be relaxed as follows.

Corollary 4.1a. *The statements of Theorem* 4.1 *remain true, if for each $x \in (0, 1)$ as well as for $x = 0$ and $x = 1$ the term $\|Mv(x)\|$ is replaced by $\langle \eta, Mv(x) \rangle$ with $\|\eta\| = 1$, $v(x) = \psi_\lambda(x)\eta$.*

Due to this corollary, there is no loss of generality in assuming $f(x, o, o) = o$ $(0 < x < 1)$ in our theory, because adding a term $s(x)$ to $Mv(x)$ for $0 < x < 1$ results only in adding the term $\langle \eta, s(x) \rangle$ on both sides of the inequality $\langle \eta, Mv(x) \rangle \leqslant \cdots$ which is required of $Mv(x)$. On the other hand, when considering a differential equation $Mu(x) = s(x)$ $(0 < x < 1)$, we may also assume that $s(x) = o$ $(0 < x < 1)$, incorporating s into Mu.

Example 4.1. Suppose that all components f_i $(i = 1, 2, \ldots, n)$ have the form $f_i(x, y, p) = y_i h_i(x, y, p)$ with $h_i(x, y, p) \geqslant 0$ for all x, y, p under consideration. Then the last summand on the right-hand side of (4.8) is positive and, therefore, all assumptions of the above theorem are satisfied for $\psi_\lambda = \delta(2 + x(1 - x)) + \lambda$ with $\|Mv(x)\| \leqslant 2\delta$ $(0 \leqslant x \leqslant 1)$. Consequently,

$$\|v(x)\| \leqslant \sup\{\|Mv(\xi)\| : 0 \leqslant \xi \leqslant 1\} \cdot [1 + \tfrac{1}{2}x(1 - x)] \qquad \text{for each} \quad v \in R.$$

Applying Corollary 4.1a we establish, in addition, that

$$v = o \qquad \text{if} \quad \langle v(x), Mv(x)\rangle \leqslant 0 \qquad (0 \leqslant x \leqslant 1). \quad \square$$

Under the special assumptions of the preceding example, the last summand in the differential inequality (4.8) need not be considered. In other cases, however, one may not be able to solve (4.8) without observing the side conditions (4.9). These side conditions can be used similarly as the side conditions (III.2.18) in the theory of inverse-monotone operators. Several special cases and examples will be considered in the subsequent section.

The functions ψ_λ may be written as $\psi_\lambda = \psi + \mathfrak{z}_\lambda$ and each inequality required for ψ_λ may then be replaced by two (sufficient) inequalities, one for $\mathfrak{z}_\lambda$ and one for ψ. In this way the following theorem and corollary are obtained. Here the idea is to verify the conditions on $\mathfrak{z}_\lambda$ a priori for certain *classes* of problems, so that for a concrete problem belonging to such a class only the inequalities for ψ have to be solved. (We point out that in some results of Section 5 conditions on ψ are required which are similar to those below, while assumptions involving $\mathfrak{z}_\lambda$ do *not* occur.)

Theorem 4.2. *Suppose functions $\mathfrak{z}_\lambda \in R_0$ $(0 \leqslant \lambda < \infty)$ exist such that $\mathfrak{z}_0 = o$, $\mathfrak{z}(\lambda, x) := \mathfrak{z}_\lambda(x)$ is continuous on $[0, \infty) \times [0, 1]$, $\psi + \mathfrak{z}_\lambda \succ o$ for $\lambda > 0$, $\|v\| \leqslant \psi_\lambda := \psi + \mathfrak{z}_\lambda$ for some $\lambda \geqslant 0$, and the following conditions hold.*

$$\text{(i)} \quad 0 < -\mathfrak{z}_\lambda''(x) + \mathfrak{z}_\lambda(x)\langle q, q\rangle + \langle \eta, f(x, \psi_\lambda(x)\eta, \psi_\lambda'(x)\eta + \psi_\lambda(x)q)\rangle - \langle \eta, f(x, \psi(x)\eta, \psi'(x)\eta + \psi(x)q)\rangle$$

for all $(x, \eta, q) \in Q$ and $\lambda > 0$ satisfying (4.9).

(ii) $0 < \mathfrak{z}_\lambda(0)$, $0 < \mathfrak{z}_\lambda(1)$ *for all $\lambda > 0$.*

Then the inequalities

$$\|Mv(x)\| \leqslant -\psi''(x) + \langle q, q\rangle\psi(x) + \langle \eta, f(x, \psi(x)\eta, \psi'(x)\eta + \psi(x)q)\rangle \qquad \text{for} \quad (x, \eta, q) \in Q, \tag{4.13}$$

$$\|Mv(0)\| \leqslant \psi(0), \qquad \|Mv(1)\| \leqslant \psi(1) \tag{4.14}$$

together imply $\|v\| \leqslant \psi$.

Corollary 4.2a. *In* (4.13) *the term* $\|Mv(x)\|$ *may be replaced by* $\langle\eta, Mv(x)\rangle$, *and the resulting inequality need only be required for all* $(x, \eta, q) \in Q$ *such that* (4.9) *holds for some* $\lambda > 0$.

The inequalities in (4.14) *may be replaced by* $\langle\eta, Mv(x)\rangle \leqslant \psi(x)$ *for* $x \in \{0, 1\}$, $\|\eta\| = 1$, $v(x) = \psi_\lambda(x)\eta$.

If side conditions such as $v(x) = \psi_\lambda(x)\eta = (\psi(x) + \mathfrak{z}_\lambda(x))\eta$ are used in the inequalities for ψ, as described in the corollary, the resulting assumptions on ψ depend also on $\mathfrak{z}_\lambda$, which we wanted to avoid. However, it is still possible to employ such a side condition without using the special form of $\mathfrak{z}_\lambda$. Observe, in particular, that the side condition implies $o \neq v(x) = \|v(x)\|\eta$.

We point out that the preceding assumptions assume a much simpler form, if $f(x, y, p)$ does not depend on p. In particular, the term $\langle q, q\rangle$ may be replaced by zero wherever it occurs. For example, inequality (4.13) is then equivalent to

$$\|Mv(x)\| \leqslant -\psi''(x) + \langle\eta, f(x, \psi(x)\eta)\rangle \qquad \text{for} \quad (x, \eta) \in P. \tag{4.15}$$

According to the preceding discussion, the set P may here be replaced by

$$P(v) = \{(x, \eta) : 0 < x < 1, \|\eta\| = 1, o \neq v(x) = \|v(x)\|\eta\}. \tag{4.16}$$

We shall discuss these results in more detail in the succeeding section. Here let us briefly compare the preceding differential inequalities for ψ with the corresponding assumption (2.5) for two-sided bounds. (Of course, the letter ψ is defined differently there.)

First we assume that f does not depend on p. Then, in (4.15) exactly those values of $f(x, y)$ occur for which y lies on the surface of the ball given by $\|y\| \leqslant \psi(x)$. In the set of inequalities (2.5), however, exactly those values of $f(x, y)$ are used for which y lies on the surface of the cube given by $\varphi_i(x) \leqslant y_i \leqslant \psi_i(x)$ $(i = 1, 2, \ldots, n)$. Thus, in both cases values $y \in \partial K(x)$ occur where $K(x)$ is a set such that $v(x) \in K(x)$ describes the estimate to be proved. (Naturally, this observation may be used to generalize the theory, at least in a formal way.) What consequences the occurrence of different sets $\partial K(x)$ has, may be explained by the examples of the following sections.

If f depends also on p, additional parameters occur, named q in (4.13) and $k, \bar{k}$ in (2.5). The values p for which $f(x, y, p)$ is evaluated in (2.5) lie on certain hyperplanes which contain the point $\varphi'(x)$ or $\psi'(x)$ and have a direction dependent on y. The values p used in (4.13) for $f(x, y, p)$ lie on a hyperplane described by $\langle y, p\rangle = \psi(x)\psi'(x)$, for $\|y\| = \psi(x)$. For a linear operator M, the occurrence of the parameters $k, \bar{k}$ in (2.5) requires f not be coupled with respect to p. For nonlinear f, the component f_i may depend on u_j with $j \neq i$ only in a very restricted way. If f is not coupled with respect to p, then the parameters $k, \bar{k}$ can be eliminated from (2.5), so that these conditions assume a simpler form. A corresponding elimination of q and hence simplification of (4.13) is not possible. On the other hand, even for linear M, the occurrence of q does not require that f be only weakly coupled.

Problems

4.1 Show that for the case $n = 1$ the result of Theorem 4.2 is contained in Theorem III.2.2.
4.2 Derive results analogous to Theorems 4.1 and 4.2 for a functional differential operator M given by (4.6) with $f(x, u(x), u'(x))$ replaced by $f(x, u(x), u'(x), u)$. In these results sets of differential inequalities are to be required similarly as in Theorem 2.1. For instance, the summand involving f in (4.13) is to be replaced by $\langle \eta, f(x, \psi(x)\eta, \psi'(x)\eta + \psi(x)q, u)\rangle$ with $\|u\| \leqslant \psi$.

4.3 Special cases and examples

In this section we shall derive a series of results for special types of operators M in (4.6). These results are to be considered as examples for applying the preceding theorems. There are many other ways to proceed. In all these results ψ is supposed to be a function in R_0 satisfying $\psi \geqslant o$ and v denotes a fixed function in R. The set $P(v)$ then is defined by (4.16).

First suppose that

$$f(x, y, p) = f(x, y) \qquad \text{and} \qquad f(x, o) = o. \tag{4.17}$$

where the second relation produces no loss of generality. The next result is derived from Theorem 4.1 by using a family $\psi_\lambda = \psi + \lambda z$ and observing the side condition $v(x) = \psi_\lambda(x)\eta$.

Theorem 4.3. *Let M be defined by* (4.6) *with f satisfying* (4.17). *Suppose that*

$$\langle v(x), f(x, v(x))\rangle \geqslant \kappa(x)\langle v(x), v(x)\rangle \qquad \textit{for} \quad 0 < x < 1 \tag{4.18}$$

with some $\kappa \in \mathbb{R}(0, 1)$ and that a $z \succ o$ in R_0 exists which satisfies

$$0 < -z''(x) + \kappa(x)z(x) \qquad \textit{for} \quad 0 < x < 1. \tag{4.19}$$

Then

$$\left.\begin{array}{ll} \|Mv(x)\| \leqslant -\psi''(x) + \kappa(x)\psi(x) & \textit{for} \quad 0 < x < 1 \\ \|Mv(0)\| \leqslant \psi(0), \quad \|Mv(1)\| \leqslant \psi(1) & \end{array}\right\} \Rightarrow \quad \|v\| \leqslant \psi. \tag{4.20}$$

Observe that for $z = \cos(\pi - \varepsilon)(x - \frac{1}{2})$ with $\varepsilon > 0$ sufficiently small *inequality* (4.19) *is satisfied if*

$$\kappa(x) > -\pi^2 + \delta \qquad \textit{for} \quad 0 < x < 1 \quad \textit{and some} \quad \delta > 0. \tag{4.21}$$

For a linear operator M, where

$$f(x, y, p) = f(x, y) = C(x)y \qquad \text{with an} \quad n \times n \text{ matrix } C(x) \tag{4.22}$$

we obtain $\langle y, f(x, y)\rangle = \langle y, C(x)y\rangle = \langle y, C_H(x)y\rangle$. Since the matrix $C_H(x)$ is hermitian with respect to the inner product $\langle\ ,\ \rangle$, we estimate $\langle y, C_H(x)y\rangle \geqslant \tau(C_H(x))\langle y, y\rangle$.

Corollary 4.3a. *Let M be defined by* (4.6) *and* (4.22), *and suppose that*

$$\tau(C_H(x)) > -\pi^2 + \delta \qquad \text{for} \quad 0 < x < 1 \quad \text{and some} \quad \delta > 0. \tag{4.23}$$

Then (4.20) *holds for* $\kappa(x) = \tau(C_H(x))$.

For two-sided bounds we obtained condition (2.15) instead of (4.23). These two conditions (2.15) and (4.23) will be compared in the following examples.

Example 4.2. Let $n = 2$ and $Mu = (-u'' + Cu, u(0), u(1))$ with C denoting one of the three constant matrices in (2.16). For all of these matrices, (2.15) was equivalent to $\mu - \nu > -\pi^2$ (see Example 2.2). For C_1 and C_2 the latter inequality is also equivalent to (4.23). For $C = C_3$, however, (4.23) yields the weaker requirement $\mu > -\pi^2$. Observe that the operator M corresponding to C_3 is invertible for $\mu > -\pi^2$ and all $\nu > 0$, but not invertible for $\mu = -\pi^2$, $\nu = 0$. □

Example 4.3. Generalizing the preceding example let $n = 2$ and $Mu = (-u'' + Cu, u(0), u(1))$, where C denotes a constant matrix which has complex eigenvalues $\mu \pm i\nu$ with $\mu \in \mathbb{R}$, $\nu > 0$. According to (4.3) one has $\langle v, Cv\rangle = \mu\langle v, v\rangle$ for the inner product (4.1) with Φ as in (4.2). Thus, *for an operator as considered here,* (4.20) *holds with* $\|y\| = \|\Phi^{-1}y\|_2$ *and* $\kappa = \mu$, *if* $\mu > -\pi^2$. □

Example 4.4. The statement at the end of the last example can be generalized further. Suppose that $Mu = (-u'' + Cu, u(0), u(1))$, where n is arbitrary and C is a constant $n \times n$ matrix. Denote by μ_0 the minimum of the real parts of the eigenvalues of C. Then Corollary 4.3a and the estimates in Section 4.1, in particular (4.5), yield the following result. *If* $\mu_0 > -\pi^2$, *then to each sufficiently small* $\varepsilon > 0$ *there exists an inner product* $\langle\ ,\ \rangle$ *in* $\mathbb{R}^n$ *such that* (4.20) *holds with the corresponding norm* $\|\ \|$ *and* $\kappa = \mu_\varepsilon$ *as defined in* (4.4). □

For the nonlinear case (4.17) results similar to Corollary 4.3a can be obtained, for example, if for each $x \in (0, 1)$ the derivative $f_y(x, y)$ exists as a continuous function of y. Then $\langle v(x), f(x, v(x))\rangle = \langle v(x), C(x)v(x)\rangle$ with $C(x) = \int_0^1 f_y(x, tv(x))\, dt$. Of course now $C(x)$ may depend on v and the estimate (4.18) with $\kappa(x) = \tau(C_H(x))$, in general, is not the best possible. Thus, one may also try to choose $\kappa(x)$ differently.

For nonlinear operators M one may not even be able to determine $\kappa(x)$ in (4.18) without some knowledge of v. For example, consider a symmetric bilinear form

$$f(x, y) = \mathscr{B}[x, y, y] \qquad \text{with components} \quad f_i(x, y) = y^T B_i y,$$

where $B_i = B_i^T$ is an $n \times n$ matrix. Here inequality (4.18) assumes the form

$$\langle v(x), \mathscr{B}[x, v(x), v(x)]\rangle \geqslant \kappa(x)\langle v(x), v(x)\rangle \qquad (0 < x < 1). \quad (4.24)$$

Under suitable assumptions on $\mathscr{B}$ one may, for instance, obtain statements of the form (4.20) for all positive v.

Example 4.5. *Let $\mathscr{B}[x, y, y] \geqslant 0$ for all $x \in (0, 1)$ and $y \geqslant o$ in $\mathbb{R}^n$. Then for each $v \geqslant o$ in R with $\|Mv(x)\| \leqslant \delta$ $(0 < x < 1)$, $\|Mv(0)\| \leqslant \delta_0$, $\|Mv(1)\| \leqslant \delta_1$, we have*

$$\|v(x)\| \leqslant \tfrac{1}{2}\delta x(1 - x) + \delta_0(1 - x) + \delta_1 x \qquad (0 \leqslant x \leqslant 1). \quad (4.25)$$

For deriving this statement from Theorem 4.3, one shows that $\kappa(x) \equiv 0$ satisfies (4.18) and applies (4.20) with ψ denoting the bound of $\|v(x)\|$ in (4.25). □

Example 4.6. Suppose that $v \in C_2^2[0, 1]$ is a solution of the boundary value problem

$$-u_1''(x) + (u_2(x))^2 = s_1(x), \qquad -u_2''(x) + (u_1(x))^2 = s_2(x) \qquad (0 < x < 1),$$
$$u(0) = r^0, \qquad u(1) = r^1$$

with $s_1, s_2 \in C_0[0, 1]$. Obviously, this problem can be written as $Mu = r$ with $f(x, y)$ a bilinear form in y. According to the statement in Example 4.5, (4.25) *holds for each $v \geqslant o$.*

More generally, for $\kappa(x) \leqslant 0$ inequality (4.24) is equivalent to $[\|v\|(\cos\varphi + \sin\varphi)\cos\varphi\sin\varphi](x) \geqslant \kappa(x)$ $(0 < x < 1)$, where $v_1(x) = \|v(x)\|\cos\varphi(x)$, $v_2(x) = \|v(x)\|\sin\varphi(x)$. For example, this inequality is satisfied if $\|v(x)\| \leqslant -\sqrt{2}\kappa(x)$ $(0 < x < 1)$. Thus, if $\|v(x)\| < \sqrt{2}\pi^2$ $(0 \leqslant x \leqslant 1)$, one can find a κ such that (4.21) and (4.24) hold.

If $\|v(x)\| \leqslant \sqrt{2}|\kappa|$ with a constant $\kappa \in (-8, 0]$, $v(0) = v(1) = 0$, and $\|Mv(x)\| \leqslant \delta$ $(0 < x < 1)$, then (4.20) yields $\|v(x)\| \leqslant \psi(x) = \delta(8 + \kappa)^{-1}4x(1 - x) \leqslant \delta(8 + \kappa)^{-1}$ $(0 \leqslant x \leqslant 1)$. For sufficiently small δ, this estimate constitutes an improvement of the a priori bound $\sqrt{2}|\kappa|$. □

Since, in general, $\kappa(x)$ in (4.18) depends on v, the bound ψ determined from (4.20) also depends on this function. Thus, if only a rough a priori estimate of v is known, ψ may be a rough bound too. Applying Theorem 4.2 and Corollary 4.2a, we shall now derive somewhat different results where ψ is calculated independently of v.

Theorem 4.4. *Let M be defined by* (4.6), (4.17), *and suppose that for each $x \in (0, 1)$ the derivative $f_y(x, y)$ exists as a continuous function of $y \in \mathbb{R}^n$. Assume, moreover, that there exists a function $z \succ o$ in R such that*

$$0 < -z''(x) + \langle \eta, C(x)\eta \rangle z(x) \qquad \text{with} \quad C(x) = \int_0^1 f_y(x, tv(x) + (1-t)\psi(x)\eta)\, dt \tag{4.26}$$

for all $(x, \eta) \in P$ satisfying $v(x) = (\psi(x) + \lambda z(x))\eta$ for some $\lambda > 0$.
Then

$$\left.\begin{aligned} &\|Mv(x)\| \leqslant -\psi''(x) + \langle \eta, f(x, \psi(x)\eta)\rangle \quad \text{for} \quad (x, \eta) \in P(v) \\ &\|Mv(0)\| \leqslant \psi(0), \qquad \|Mv(1)\| \leqslant \psi(1) \end{aligned}\right\} \Rightarrow \|v\| \leqslant \psi.$$

Corollary 4.4a. *Let*

$$f(x, y) = C(x)y + \mathscr{B}[x, y, y] \tag{4.27}$$

with a matrix $C(x)$ and a symmetric bilinear form $\mathscr{B}[x, \cdot, \cdot]$. Suppose that (4.24) *holds and*

$$\langle v(x), \mathscr{B}[x, v(x), v(x)]\rangle \geqslant \gamma(x)\|v(x)\|^3 \qquad (0 < x < 1) \tag{4.28}$$

with some $\gamma \in \mathbb{R}(0, 1)$. Assume, moreover, that there exists a $z \succ o$ in R_0 such that

$$0 < -z''(x) + [\tau(C_{\mathrm{H}}(x)) + \kappa(x) + \gamma(x)\psi(x)]z(x) \qquad \text{for} \quad 0 < x < 1. \tag{4.29}$$

Then

$$\left.\begin{aligned} \|Mv(x)\| \leqslant &-\psi''(x) + \tau(C_{\mathrm{H}}(x))\psi(x) \\ &+ \gamma(x)\psi^2(x) \qquad (0 < x < 1) \\ \|Mv(0)\| \leqslant \psi(0),& \qquad \|Mv(1)\| \leqslant \psi(1) \end{aligned}\right\} \Rightarrow \|v\| \leqslant \psi. \tag{4.30}$$

The term γ occurring in the conditions for ψ again is defined by a formula which involves v. Notice, however, that there always exists a function $\gamma \in \mathbb{R}(0, 1)$ which satisfies (4.28) for *arbitrary* $v \in R$, condition (4.28) is independent of the length $\|v(x)\|$ of $v(x)$.

By applying Theorem 4.3 instead of Theorem 4.4 to the special case (4.27), (4.24), one obtains a differential inequality for z which has the form (4.29) with $\gamma(x)\psi(x)$ omitted. Thus (4.29) constitutes a stronger condition. On the other hand, here the conditions on ψ, in general, are weaker, so that smaller bounds ψ may be obtained. Observe also that the term $\gamma(x)\psi(x)$ is "small," if ψ is "small." In general, this is the case, if the results are used for an error estimation, where Mv is the "small" defect of the error v.

Example 4.7. Consider again the boundary value problem in Example 4.6 and assume that $v(0) = v(1) = 0$ and $\|Mv(x)\| \leqslant \delta$ $(0 < x < 1)$. Here (4.24) and (4.28) hold with $\kappa(x) = -\frac{1}{2}\sqrt{2}\,\|v(x)\|$ and $\gamma(x) = -\frac{1}{2}\sqrt{2}$, respectively, so that (4.29) and the differential inequality in (4.30) may be replaced by the sufficient conditions $0 < -z''(x) - \frac{1}{2}\sqrt{2}(\|v(x)\| + \psi(x))z(x)$, $\delta \leqslant -\psi''(x) - \psi^2(x)$ $(0 < x < 1)$. Using $z(x) = \cos(\pi - \varepsilon)(x - \frac{1}{2})$ with $\varepsilon > 0$ sufficiently small, one obtains the following statement. *If for a given ψ the function v satisfies the a priori inequality $\frac{1}{2}\sqrt{2}(\|v(x)\| + \psi(x)) < \pi^2$ $(0 \leqslant x \leqslant 1)$, then the implication* (4.30) *holds.*

For small δ the nonlinear term $\psi^2(x)$ in the above differential inequality for ψ has little influence. First neglecting this term, we calculate $\psi_0(x) = \frac{1}{2}\delta x(1 - x)$ as a good approximation for ψ. For sufficiently small δ we obtain a suitable bound ψ by slightly enlarging ψ, say by 1%. For example, if $\delta = 0.01$ then $\psi(x) = 1.01\psi_0(x) = 0.00505x(1 - x)$ satisfies the inequality required for ψ. Verifying the above a priori inequality for v, we see that $\|v(x)\| \leqslant 0.00505x(1 - x) \leqslant 0.0013$, *if* $\|v(x)\| \leqslant 13.9$ *and* $\|Mv(x)\| \leqslant 0.01$ $(0 \leqslant x \leqslant 1)$.

These estimates can be improved, if one has suitable information on the direction of the vector $v(x)$. For example, (4.28) holds for $\gamma(x) = 0$, if $v(x) \geqslant o$. We remark also, that the existence of a solution v satisfying $\|v\| \leqslant \psi$ can be proved without constructing κ and z, by applying the theory in Section 5. □

Let us now turn to operators (4.6) which depend also on u' and, in particular, consider the linear case.

Proposition 4.5a. *Let M denote an operator* (4.6) *with*

$$f(x, y, p) = B(x)p + C(x)y \tag{4.31}$$

and $n \times n$ matrices $B(x)$, $C(x)$. Suppose there exists a $z \succ o$ in R_0 which satisfies $0 < L[z](x, \eta)$ for $(x, \eta) \in P$, where

$$L[\varphi](x, \eta) = -\varphi''(x) + \langle \eta, B(x)\eta \rangle \varphi'(x) + \langle \eta, C(x)\eta \rangle \varphi(x) + c_B(x, \eta)\varphi(x)$$

with

$$c_B(x, \eta) = -\tfrac{1}{4}(\|B^*(x)\eta\|^2 - \langle \eta, B(x)\eta \rangle^2). \tag{4.32}$$

Then for each $v \in R$

$$\left.\begin{array}{ll} \|Mv(x)\| \leqslant L[\psi](x, \eta) & \text{for} \quad (x, \eta) \in P(v) \\ \|Mv(0)\| \leqslant \psi(0), & \|Mv(1)\| \leqslant \psi(1) \end{array}\right\} \Rightarrow \|v\| \leqslant \psi. \tag{4.33}$$

Proof. When we apply Theorem 4.2 to the linear case presently considered, we obtain inequalities for z and ψ which contain the term $J(x, \eta, q) = \langle q, q \rangle + \langle \eta, B(x)q \rangle$. For fixed $(x, \eta) \in P$ this function of q under the side condition $\langle \eta, q\rangle = 0$ attains the minimum $c_B(x, \eta)$. Replacing $J(x, \eta, q)$ by this minimal value and also observing Corollary 4.2a, we arrive at the above result. □

In applying the above result the terms involving η have to be estimated. As in Corollary 4.3a, the inner product $\langle \eta, C(x)\eta\rangle$ may be replaced by $\tau(C_{\mathrm{H}}(x))$. Similarly, one may use the inequality $\tau(B_{\mathrm{H}}(x)) \leqslant \langle \eta, B(x)\eta\rangle \leqslant \sigma(B_{\mathrm{H}}(x))$ for estimating the factor of $z'(x)$ and $\psi'(x)$. The term $c_B(x, \eta)$ is of a more complicated form. We shall consider some special cases and examples which show that $B(x)$ need not be diagonal, in other words, f in (4.31) may be strongly coupled.

Proposition 4.5b. *If* $B(x) = B^*(x)$ *for some* $x \in (0, 1)$, *then*

$$c_B(x, \eta) \geqslant -\tfrac{1}{16}[\sigma(B(x)) - \tau(B(x))]^2.$$

If $B(x) = -B^*(x)$ *for some* $x \in (0, 1)$, *then* $c_B(x, \eta) \geqslant \tfrac{1}{4}\tau(B^2(x))$.

Proof. The second statement is clear. To prove the first, one verifies

$$\|By\|^2\|y\|^2 - \langle y, By\rangle^2 = \frac{1}{2}\sum_{i,j} (\beta_i - \beta_j)^2 c_i^2 c_j^2 \qquad \text{for} \quad y = \sum_i c_i\varphi_i$$

with $\beta_i = \beta_i(x)$ the eigenvalues of $B = B(x)$ and $\varphi_i = \varphi_i(x)$ corresponding eigenvectors, $\langle \varphi_i, \varphi_j\rangle = \delta_{ij}$. Then one shows that this quadratic form in the variables c_i^2 attains its maximum under the side condition $\sum_i c_i^2 = 1$ for $c_1^2 = c_{n-1}^2 = \tfrac{1}{2}$, if $\beta_1 \leqslant \beta_2 \leqslant \cdots \leqslant \beta_n$ (Problem 4.7). □

The following three examples concern the special case $n = 2$.

Example 4.8. Let $n = 2$ and $B(x) = B^*(x)$ for each $x \in (0, 1)$. Denote by $\mu(x) \pm \nu(x)$ with $\nu(x) \geqslant 0$ the two eigenvalues of $B(x)$ and let $\gamma(x) = \tau(C_{\mathrm{H}}(x))$. Then for $\varphi \geqslant o$ in R_0 and $\|\eta\| = 1$

$$\begin{aligned} L[\varphi](x, \eta) \geqslant \hat{L}[\varphi](x) := -\varphi''(x) + \mu(x)\varphi'(x) - \nu(x)|\varphi'(x)| \\ + [\gamma(x) - \tfrac{1}{4}\nu^2(x)]\varphi(x). \end{aligned}$$

For constructing a function $z \succ o$ satisfying $\hat{L}[z](x) > 0$ one may apply the methods in Section II.3.3. □

Example 4.9. The implication (4.33) holds for

$$n = 2, \qquad C(x) \equiv O, \qquad B = t\begin{pmatrix} 0 & 1 \\ 1 & 0 \end{pmatrix} \qquad \text{with} \quad |t| < 4.$$

This is proved by verifying $\hat{L}[z](x) = -z''(x) - |t|z'(x) - \tfrac{1}{4}t^2 z(x) > 0$ $(0 < x < 1)$ for $z(x) = x\exp(-2x) + \varepsilon$ $(0 < x < 1)$ with $\varepsilon > 0$ sufficiently small. □

Example 4.10. The implication (4.33) holds for

$$n = 2, \qquad C(x) \equiv O, \qquad B(x) = t\begin{pmatrix} 0 & 1 \\ -1 & 0 \end{pmatrix} \qquad \text{with} \quad |t| < 2\pi.$$

One verifies $L[z](x, \eta) = -z''(x) - \frac{1}{4}t^2 z(x) > 0$ $(0 < x < 1)$ for $z = \cos(\pi - \varepsilon)(x - \frac{1}{2})$ and $\varepsilon > 0$ sufficiently small. □

For operators M which depend on u' in a nonlinear way, the assumptions of Theorems 4.1 and 4.2 are, in general, quite restrictive, in particular, if no additional properties of v' are known. However, if a sufficient knowledge on v' is available, one may use the side conditions (4.9) and apply the techniques used in the linear case. Consider, for instance, a bilinear form $f(x, y, p) = \mathscr{B}[x, p, p]$ in p. Here

$$\mathscr{B}[x, \psi_\lambda'(x)\eta + \psi_\lambda(x)q, \psi_\lambda'(x)\eta + \psi_\lambda(x)q] = \psi_\lambda'(x)B(x)\eta + \psi_\lambda(x)C(x)q$$

with $n \times n$ matrices $B(x)$, $C(x)$ such that $\mathscr{B}[x, v'(x), \eta] = B(x)\eta$, $\mathscr{B}[x, v'(x), q] = C(x)q$.

More generally, if $f(x, y, p)$ is sufficiently smooth and $f(x, o, o) = o$, one can reduce the differential inequality (4.8) for ψ_λ to a linear inequality which depends on v and v', by using the mean value theorem.

Problems

4.3 For a (real) 2×2 matrix C with eigenvalues $\lambda = \mu \pm i\nu$, $\mu \in \mathbb{R}$, $\nu > 0$, construct a matrix Φ with $\Phi^{-1}C\Phi = \begin{pmatrix} \mu & \nu \\ -\nu & \mu \end{pmatrix}$ in the following way. $\Phi^{-1} = DB$, D diagonal, B orthogonal, $B^{-1}CB = \begin{pmatrix} a & b \\ c & a \end{pmatrix}$. Apply this method to $C = \begin{pmatrix} 5 & 10 \\ -10 & 1 \end{pmatrix}$.

4.4 As an application of Example 4.2 consider the fourth order problem

$$\begin{aligned} u^{\text{iv}}(x) + cu(x) &= s(x) \qquad (0 < x < 1), \\ u(0) = u''(0) &= u(1) = u''(1) = 0 \end{aligned} \qquad \text{with} \quad c = \alpha^2, \quad \alpha > 0$$

for $u \in C_4[0, 1]$. Transform this problem into an equation $Mu = r$ with M as in Example 4.2, $\mu = 0$, $\nu = \alpha$. Derive then from Theorem 4.3 that *for each $u \in C_4[0, 1]$ satisfying the above boundary conditions the inequality*

$$(c|u(x)|^2 + |u''(x)|^2)^{1/2} \leqslant \int_0^1 \mathscr{G}(x, \xi)|u^{\text{iv}}(\xi) + cu(\xi)|\, d\xi \qquad (0 \leqslant x \leqslant 1)$$

holds, where $\mathscr{G}(x, \xi) = \mathscr{G}(\xi, x) = x(1 - \xi)$ $(0 \leqslant x \leqslant \xi \leqslant 1)$.

4.5 Apply Theorem 4.3 to the operator (4.6) with $n = 2$, $f_1(x, y, p) = y_1^2 + \gamma y_2^2$, $f_2(x, y, p) = \gamma y_1^2 + y_2^2$.

4.6 Generalize Theorem 4.4 requiring one-sided *local* Lipschitz conditions of the form $\langle \eta, f(x, (\psi(x) + t)\eta) - f(x, \psi(x)\eta)\rangle \geqslant C(x)\langle \eta, \eta\rangle$.

4.7 Complete the proof of Proposition 4.5b, proceeding as follows. (a) Reduce the problem of calculating the maximum of the quadratic form in the variables c_i^2 to a problem of the following type: Find $\sup\{p^T Ap : p \in E\}$, where $A = (a_{ik})$ is an $n \times n$ matrix with $a_{ii} = 0$, $a_{ki} = a_{ik} = (\sum_{j=1}^{k-1} \gamma_j^2)$ for $i < k$, $\gamma_j \geqslant 0$, and $E = \{p \in \mathbb{R}^n : p \geqslant o, \sum_{i=1}^n p_i = 1\}$. (b) Prove that $p^T Ap \leqslant q^T Aq$ for all $p \in E$ and the vector q defined by $q_1 = q_n = \frac{1}{2}$, $q_i = 0$ for $1 < i < n$. Hint: show that $\varphi(p) = p^T Ap$ is concave on E by verifying that $x^T Ax \geqslant 0$ for $x \in \mathbb{R}^n$ with $\sum_{i=1}^n x_i = 0$.

4.4 More general operators and other boundary conditions

Most of the results of Sections 4.2 and 4.3 can be carried over to a series of other cases, as described in the following.

4.4.1 A more general differential operator

Now define M by (4.6) with $-u''$ replaced by

$$L[u](x) = -a(x)u''(x) + b(x)u'(x),$$

where $a, b \in \mathbb{R}(0, 1)$ and $a(x) \geqslant 0$ $(0 < x < 1)$. We state without proof:

For this case all theorems and corollaries in Sections 4.2 *and* 4.3 *remain true, if* $-\psi_\lambda''$, $-\mathfrak{z}_\lambda''$, $-z''$, *and* $-\psi''$ *are replaced by* $L[\psi]$, $L[\mathfrak{z}_\lambda]$, $L[z]$, *and* $L[\psi]$, *respectively, and if* $\langle q, q\rangle$ *is replaced by* $a(x)\langle q, q\rangle$, *whenever the former term occurs.*

This statement means, for instance, that the particularly simple results obtained for the case (4.17) also hold, when $Mu(x)$ has a summand $b(x)u'(x)$ with scalar-valued $b(x)$ for $0 < x < 1$.

4.4.2 Sturm–Liouville boundary operators

Now suppose that the operator M has the general form (1.1)–(1.3) and assume, in addition, that for each index i:

$$\begin{array}{cccc} & \alpha_i^0 \in \{0, 1\}, & & \alpha_i^1 \in \{0, 1\}, \\ f_i^0(y) = y_i & \text{if } \alpha_i^0 = 0, & f_i^1(y) = y_i & \text{if } \alpha_i^1 = 0. \end{array} \tag{4.34}$$

Also define $R_0 = C_1[0, 1] \cap C_2(0, 1)$ and choose, for simplicity, $R = C_1^n[0, 1] \cap C_2^n(0, 1)$. (We may also weaken the differentiability conditions, assuming only that "all occurring derivatives exist." This fact will be exploited in Section 4.5.)

Finally, we shall require that the element $v \in R$ considered satisfies

$$\begin{aligned} v_i(0) &= 0 \quad \text{if} \quad \alpha_i^0 = 0 \quad \text{and} \quad \alpha^0 \neq O, \\ v_i(1) &= 0 \quad \text{if} \quad \alpha_i^1 = 0 \quad \text{and} \quad \alpha^1 \neq O. \end{aligned} \tag{4.35}$$

For such a function v the following result is derived in a way similar to the proof of Theorem 4.1. *In Theorem* 4.6 *we assume that the inner product has the special form* $\langle y, \eta \rangle = y^{\mathrm{T}}\eta$, i.e., the statements of this theorem are not proved for an arbitrary inner product as described in Section 4.1. We point out, however, that for certain special cases, Theorem 4.6 indeed is true for $\langle y, \eta \rangle = y^{\mathrm{T}}H\eta$ with an arbitrary positive definite matrix $H = H^{\mathrm{T}}$. Such a special case is described by $\alpha^0 = O$, $\alpha^1 = I$. (The latter case will occur in Section 4.5.)

Theorem 4.6. *For the operator M described above and v satisfying* (4.35) *the statements of Theorem* 4.1 *remain true if* $-\psi_\lambda''(x) + \langle q, q \rangle \psi_\lambda(x)$ *is replaced by* $L[\psi_\lambda](x) + a(x)\langle q, q \rangle \psi_\lambda(x)$ *and, furthermore, the following changes are made.*

If $\alpha^0 \neq O$, *the first inequality in* (ii) *is replaced by*

$$\begin{aligned} \|Mv(0)\| &< -\psi_\lambda'(0) + \langle \eta, f^0(\psi_\lambda(0)\eta) \rangle \\ &\quad \textit{for} \quad \eta \in \mathbb{R}^n \quad \textit{with} \quad \|\eta\| = 1, \quad v(0) = \psi_\lambda(0)\eta. \end{aligned} \tag{4.36}$$

If $\alpha^1 \neq O$, *the second inequality in* (ii) *is replaced by*

$$\begin{aligned} \|Mv(1)\| &< \psi_\lambda'(1) + \langle \eta, f^1(\psi_\lambda(1)\eta) \rangle \\ &\quad \textit{for} \quad \eta \in \mathbb{R}^n \quad \textit{with} \quad \|\eta\| = 1, \quad v(1) = \psi_\lambda(1)\eta. \end{aligned}$$

Proof. The proof of Theorem 4.1 has to be modified slightly; now the cases $\xi = 0$ and $\xi = 1$ require a more detailed analysis. Suppose, for example, that $\langle v, v \rangle(x) \leqslant \psi_\lambda^2(x)$ $(0 \leqslant x \leqslant 1)$, $\langle v, v \rangle(0) = \psi_\lambda^2(0)$, and $\alpha^0 \neq O$. These relations together imply $\langle v(0), v'(0) \rangle \leqslant \psi_\lambda(0)\psi_\lambda'(0)$, where the inner product is interpreted to be the sum of all $v_i(0)v_i'(0)$ with $\alpha_i^0 = 1$. The latter inequality contradicts (4.36). □

From this theorem a result such as Theorem 4.2 and results similar to those in Section 4.3 can be derived for the more general operator considered here in essentially the same way as for the operator (4.6). We shall not carry out the details.

Example 4.11. Let $n = 3$ and suppose that $Mu(0)$ has the three components $u_1(0)$, $-u_2'(0) + 2u_3(0)$, and $-u_3'(0) - u_2(0)$. Then (4.35) requires $v_1(0) = 0$, and (4.36) holds, if $w = Mv(0)$ satisfies $[(w_2)^2 + (w_3)^2]^{1/2} < -\psi_\lambda'(0) - \frac{1}{2}\psi_\lambda(0)$. (Observe that $\langle \eta, f^0(\psi_\lambda(0)\eta) \rangle = \eta_2\eta_3\psi_\lambda(0) \geqslant \frac{1}{2}\psi_\lambda(0)$, with equality holding for suitable η_2, η_3.) □

Example 4.12. We consider again the boundary value problem in Example 2.3, which describes the bending of circular plates. Now we transform the problem before applying our theory. Suppose that $u \in C_2^2[0, 1]$ is a solution and $\omega \in C_2^2[0, 1]$ an approximate solution satisfying the boundary conditions. Then $v = u - \omega$ also satisfies the boundary conditions and, moreover,

$$L[v](x) + f(x, v(x)) = \delta[\omega](x) \qquad \text{for} \quad 0 < x < 1$$

where $L[v]$ has the components $L[v_i]$ $(i = 1, 2)$ and

$$f(x, y) = \begin{pmatrix} -\omega_2(x)y_1 - \omega_1(x)y_2 - y_1y_2 \\ \omega_1(x)y_1 + \frac{1}{2}y_1^2 \end{pmatrix},$$

$$\delta[\omega](x) = \delta(x) = -L[\omega](x) + \begin{pmatrix} \omega_1(x)\omega_2(x) + 8\mu x^2 \\ -\frac{1}{2}(\omega_1(x))^2 \end{pmatrix}.$$

We shall apply Theorem 4.6 to this transformed problem for $\langle y, \eta \rangle = y^{\mathrm{T}}\eta$. Using the side condition $v(x) = \psi_\lambda(x)\eta$ and $u = v + \omega$, we calculate

$$\langle \eta, f(x, \psi_\lambda(x)\eta) \rangle = -\tfrac{1}{2}(u_2(x) + \omega_2(x))\eta_1^2\psi_\lambda(x).$$

We shall see that this term is positive for suitable ω_2.

First, observe that there exists a function $z \in C_2[0, 1]$ satisfying $z \succ o$, $L[z](x) > 0$ $(0 < x < 1)$, and $z'(1) - vz(1) > 0$. For example, $z(x) = 1 + \delta x - x^3$ has this property for $\delta > 3(1 - v)^{-1}$. Consequently, the inequalities $L[u_2](x) \leqslant 0$ $(0 < x < 1)$, $u_2(0) = u_2'(1) - vu_2(1) = 0$ together imply $u_2(x) \leqslant 0$ $(0 \leqslant x \leqslant 1)$ (cf. Theorem II.3.2).

Next we choose a very simple function ω with $\omega_2(x) \leqslant 0$ $(0 \leqslant x \leqslant 1)$. A formal calculation shows that

$$\omega_1(x) = \alpha(x - x^3), \qquad \omega_2(x) = -\frac{1}{16}\alpha^2\left(\frac{3 - v}{1 - v}x - x^3\right)$$

$$\text{with} \quad \alpha = \mu - \frac{\alpha^3}{128}\frac{3 - v}{1 - v}$$

satisfy the boundary conditions. The defect $\delta[\omega]$ has the components

$$\delta_1[\omega] = \frac{1}{16}\alpha^3 x^4((4 - 2v)(1 - v)^{-1} - x^2), \qquad \delta_2[\omega] = \alpha^2 x^4(1 - \tfrac{1}{2}x^2).$$

Because of $u_2 \leqslant o$, $\omega_2 \leqslant o$, and the properties of z stated above, all assumptions of Theorem 4.6 are satisfied for $\psi_\lambda = \psi + \lambda z$, if

$$\begin{aligned} \psi \geqslant o, \qquad \|\delta[\omega](x)\| \leqslant L[\psi](x) \qquad \text{for} \quad 0 < x < 1, \\ \psi'(1) - v\psi(1) \geqslant 0. \end{aligned} \tag{4.37}$$

By verifying these inequalities, we see that *each solution u of the given boundary value problem satisfies* $\|u - \omega\| \leqslant \psi$ *for* $\psi(x) = \gamma[(5 - \nu)(1 - \nu)^{-1}x - x^5]$ *and* $24\gamma = \sup_x x^{-4}\|\delta[\omega](x)\|$.

For example, if $8\mu = 1$ and $\nu = 0.3$, we obtain $\alpha = 0.125$, $\max\|\omega(x)\| = 0.0481$, $\gamma \leqslant 0.000652$, $\max \psi(x) \leqslant 0.00373$. Of course, for large μ and ν close to 1 we cannot expect the simple approximation ω to be very accurate.

However, using a collocation Newton method similar to that in Section III.6.3 we also calculated approximations ω with very small defect $\delta[\omega]$ and corresponding bounds $\psi(x) = \gamma[1 + \nu(1 - \nu)^{-1}x]$ for values μ up to 1000 and various ν. For m (Tschebychev) collocation points, the approximations ω had the form $\sum_{i=1}^{2m} \alpha_i \varphi_i(x)$ with $\varphi_i \in C_2^2[0, 1]$ denoting suitable functions such that one of the two components of each φ_i is identically zero and the other is a twice integrated Legendre polynomial. The above constant γ is the maximum over $x \in [0, 1]$ of $\|\delta(x)\|_2(L[z](x))^{-1}$, where $z(x) = 1 + \nu(1 - \nu)^{-1}x$, $L[z](x) = x^{-1}$. As in Section III.6.3, this constant was not determined exactly, but replaced by $1.1\gamma_0$, where γ_0 is the corresponding maximum over 501 equidistant points. Table 13 provides some numerical results. Here N denotes the number of iterations (of the Newton method) and $\beta = 1.1\gamma_0[1 + \nu(1 - \nu)^{-1}]$ is a bound of the error $\|u^*(x) - \omega(x)\|_2$ (under the assumption that $\gamma \leqslant 1.1\gamma_0$). For example, $\max\{\|u^*(x)\|_2 : 0 \leqslant x \leqslant 1\}$ approximately assumes the values 0.384, 3.66, and 172 for μ equal to 1, 10, and 1000, respectively.

Using these tools, one can also show that *the given boundary value problem has at most one solution.* For the proof one assumes that now ω denotes a

TABLE 13

Error bounds β *for Example* 4.12 *with* $\nu = 0.3$

μ	m	N	β
1	6	4	$1.79E - 05$
1	8	4	$2.51E - 06$
1	10	4	$5.91E - 10$
10	12	6	$1.32E - 06$
10	15	6	$2.45E - 09$
10	18	6	$1.38E - 11$
10^2	16	8	$1.29E - 05$
10^2	20	10	$2.89E - 08$
10^2	24	10	$2.99E - 10$
10^3	30	13	$3.50E - 08$
10^4	44	15	$5.42E - 08$

solution, so that $\delta[\omega] = o$. Then (4.37) is satisfied for $\psi = o$ and, consequently, $\|u(x) - \omega(x)\| \leqslant 0$ $(0 \leqslant x \leqslant 1)$. □

4.4.3 *Periodic boundary conditions*

Let $R_0 = C_2[0, 1]$, $R = C_2^n[0, 1]$, and

$$Mu(x) = L[u](x) + f(x, u(x), u'(x)) \qquad \text{for} \quad 0 < x < 1 \tag{4.38}$$

with L and f as before. Here we are interested in periodic boundary conditions $u(0) = u(1)$, $u'(0) = u'(1)$. These conditions will not be incorporated into the operator M.

Theorem 4.7. *For M in* (4.38) *and $v \in R$ satisfying $v(0) = v(1)$, the statements of Theorem* 4.1 *remain true, if $-\psi''_\lambda(x) + \langle q, q\rangle\psi_\lambda(x)$ is replaced by $L[\psi_\lambda](x) + a(x)\langle q, q\rangle\psi_\lambda(x)$ and if, instead of condition* (ii), *it is required that $\psi_\lambda(0) = \psi_\lambda(1)$ for $\lambda \geqslant 0$ and*

$$\|-v'(0) + v'(1)\| < -\psi'_\lambda(0) + \psi'_\lambda(1) \qquad \textit{for} \quad \lambda > 0. \tag{4.39}$$

Proof. We remark that the relations $\langle v, v\rangle(0) = \psi^2_\lambda(0)$ and $\langle v, v\rangle(1) = \psi^2_\lambda(1)$ can hold only simultaneously. If these relations are true and $\langle v, v\rangle \leqslant \psi^2_\lambda$, then $\langle v(0), v'(0)\rangle \leqslant \psi_\lambda(0)\psi'_\lambda(0)$ and $\langle v(1), v'(1)\rangle \geqslant \psi_\lambda(1)\psi'_\lambda(1)$. From these inequalities a contradiction to (4.39) is derived. □

4.5 First order initial value problems

Differential operators of the first order can be treated as a special case of the operators considered in Theorem 4.6. Section 4.5.1 contains some simple results which immediately follow from this theorem. In Section 4.5.2 we demonstrate that these results can be used to obtain quantitative statements on the stability of solutions.

4.5.1 *A priori estimates*

Let M denote an operator mapping $R = C_0^n[0, l] \cap C_1^n(0, l]$ into $S = \mathbb{R}^n[0, l]$ such that

$$Mu(x) = \begin{cases} u'(x) + f(x, u(x)) & \text{for} \quad 0 < x \leqslant l \\ u(0) & \text{for} \quad x = 0, \end{cases} \tag{4.40}$$

where $0 < l < \infty$ and $f(x, y) \in \mathbb{R}^n$ is defined for all $x \in (0, l]$ and $y \in \mathbb{R}^n$. Suppose that $v \in R$ and $\psi \in R_0 := C_0[0, l] \cap C_1(0, l]$ with $\psi \geqslant o$ are fixed functions and define

$$P(v) = \{(x, \eta) : 0 < x \leqslant l, \|\eta\| = 1, o \neq v(x) = \|v(x)\|\eta\}.$$

Theorem 4.8. *Suppose that a function $z \in R_0$ exists such that $z(x) > 0$ $(0 \leqslant x \leqslant l)$ and*

$$0 < z'(x) + \lambda^{-1}\langle \eta, f(x, (\psi(x) + \lambda z(x))\eta) - f(x, \psi(x)\eta)\rangle$$

for all $x \in (0, l]$, $\lambda > 0$, $\eta \in \mathbb{R}^n$ satisfying $\|\eta\| = 1$, $v(x) = (\psi(x) + \lambda z(x))\eta$.

Then

$$\left.\begin{array}{l} \|Mv(x)\| \leqslant \psi'(x) + \langle \eta, f(x, \psi(x)\eta)\rangle \\ \qquad\qquad \text{for} \quad (x, \eta) \in P(v) \\ \|v(0)\| \leqslant \psi(0) \end{array}\right\} \Rightarrow \|v(x)\| \leqslant \psi(x) \quad \text{for} \quad 0 \leqslant x \leqslant l. \tag{4.41}$$

Proof. First suppose that $l = 1$. Then M has the form considered in Section 4.4.2 with $L[u] = u'$, $\alpha^0 = O$, $\alpha^1 = I$, and $f^1(y) = f(1, y)$. By combining the inequalities required for ψ and z, one proves that the assumptions of Theorem 4.6 are satisfied for this operator and $\psi_\lambda = \psi + \lambda z$. Thus, the above statement follows from this theorem. For $l \neq 1$ the statement can be proved, for example, by a simple transformation. □

This result can be generalized in a manner similar to that explained in the preceding section. For example, a nonlinear family $\mathfrak{z}_\lambda$ may be used and $\|Mv(x)\|$ may be replaced by $\langle Mv(x), \eta\rangle$. (See also Problem 4.8.) Here, we are more interested in particularly simple conditions, as those in the following corollary.

Corollary 4.8a. *The implication* (4.41) *holds, if for each constant $c > 0$ there exists a constant $\kappa = \kappa(c) > 0$ such that*

$$\langle \eta, f(x, t_2\eta) - f(x, t_1\eta)\rangle \geqslant -\kappa(t_2 - t_1)$$
$$\text{for} \quad 0 < x \leqslant l, \quad \|\eta\| = 1, \quad 0 \leqslant t_1 \leqslant t_2 \leqslant c.$$

Proof. Suppose that $\|v(x)\| \leqslant c$ and $\psi(x) \leqslant c$ for $x \in [0, l]$. Then $z(x) = \exp 2\kappa x$ with $\kappa = \kappa(c)$ satisfies the assumptions of Theorem 4.8. (Observe the side condition for $v(x)$.) □

4.5.2 *Estimation of solutions, stability*

Now we consider initial value problems of the form

$$u'(x) + f(x, u(x)) = o, \qquad u(x_0) = r^0 \tag{4.42}$$

where $x_0 \geqslant 0$ and $r^0 \in \mathbb{R}^n$. We assume that $f(x, y)$ is a continuous function defined for $0 \leqslant x < \infty$, $y \in \mathbb{R}^n$, and that, in addition, f satisfies a *local Lipschitz condition* with respect to y, i.e., to each $l > 0$ and each $c > 0$ there exists a constant κ such that

$$\|f(x, y) - f(x, \bar{y})\| \leqslant \kappa\|y - \bar{y}\| \quad \text{for} \quad x \in [0, l], \quad \|y\| \leqslant c, \quad \|\bar{y}\| \leqslant c. \tag{4.43}$$

Observe that such a condition holds for any norm $\| \ \|$ on $\mathbb{R}^n$ if and only if it holds for $\| \ \|_2$.

Under these assumptions each problem (4.42) has a unique solution $u^* \in C_1^n[x_0, l)$ for some $l \in (x_0, \infty]$, where the largest interval $[x_0, l)$ to which this solution may be extended, in general, is not known a priori. This solution will also be denoted by $u(x, x_0, r^0)$.

We point out that the succeeding results can also be applied to problems (4.42) where $f(x, y)$ is a continuous function defined for $0 \leqslant x < \infty$, $\|y\| \leqslant \rho < \infty$ and $f_y(x, y)$ is also continuous. Such a function can be extended to a function with the properties required above by defining $f(x, y) = f(x, \rho\|y\|^{-1}y)$ for $\|y\| > \rho$. Moreover, the results could be generalized to the case in which the Lipschitz condition is not required and hence the solution may not be unique.

For simplicity, the following basic estimation theorem is formulated for $x_0 = 0$ only.

Theorem 4.9. *Suppose a function* $\psi \in C_1[0, \infty)$ *exists such that* $\psi(x) \geqslant 0$ $(0 \leqslant x < \infty)$, $\|r^0\| \leqslant \psi(0)$, *and*

$$0 \leqslant \psi'(x) + \langle \eta, f(x, \psi(x)\eta)\rangle \quad \textit{for} \quad 0 < x < \infty, \quad \|\eta\| = 1. \tag{4.44}$$

Then the initial value problem

$$u'(x) + f(x, u(x)) = o \qquad (0 < x < \infty), \qquad u(0) = r^0 \tag{4.45}$$

has a unique solution $u^* \in C_1^n[0, \infty)$, *which satisfies* $\|u^*(x)\| \leqslant \psi(x)$ $(0 \leqslant x < \infty)$.

Proof. Let u^* be defined on an interval $[0, l]$ with finite $l > 0$. Due to (4.43) the assumptions of Corollary 4.8a are satisfied, so that $\|u^*(x)\| \leqslant \psi(x)$ $(0 \leqslant x \leqslant l)$ follows by applying (4.41) to $v = u^*$. Since this is true for each interval $[0, l]$ on which u^* is defined, one concludes by arguments used in the theory of differential equations that the solution u^* can be extended to $[0, \infty)$ and the stated estimate holds on this inverval. □

Let us first consider some examples for linear equations. For the special case $f(x, y) = C(x)y$ with $C(x) \in \mathbb{R}^{n,n}$ condition (4.44) is equivalent to $0 \leqslant \psi'(x) + \tau(C_H(x))\psi(x)$ for $0 < x < \infty$ (see the remarks following (4.22)).

Example 4.13. The solution u^* of the problem

$$u_1' + \mu u_1 + \nu u_2 = 0, \qquad u_2' - \nu u_1 + \mu u_2 = 0,$$
$$u_1(0) = r_1^0, \qquad u_2(0) = r_2^0$$

satisfies $\|u^*(x)\| \leqslant \|r^0\| \exp(-\mu x)$ $(0 \leqslant x < \infty)$. Observe that here $\tau(C_H) = \mu$. □

Example 4.14. Let $f(x, y) = Cy$ with $C \in \mathbb{R}^{n,n}$. For a suitable inner product $\langle\ ,\ \rangle$ the estimate (4.5) holds with μ_ε defined in (4.4). Consequently, (4.44) is satisfied for $\psi(x) = \|r^0\| \exp(-\mu_\varepsilon x)$, so that $\|u^*(x)\| \leqslant \psi(x)$ $(0 \leqslant x < \infty)$ for the solution u^* of (4.45). Estimating the norm $\|\ \|$ by the norm $\|\ \|_2$ one also obtains $\|u^*(x)\|_2 \leqslant \kappa(\varepsilon)\|r^0\|_2 \exp(-\mu_\varepsilon x)$ where $\kappa(\varepsilon)$ denotes the spectral condition number $\|\Phi\|_s \|\Phi^{-1}\|_s$ of the matrix Φ (which depends on ε). □

The next example essentially generalizes the estimates of the preceding example, though in a somewhat different form.

Example 4.15. Let $f(x, y) = Cy$ with $C \in \mathbb{R}^{n,n}$ and suppose there exist positive definite symmetric $n \times n$ matrices H and B satisfying

$$\tfrac{1}{2}(C^T H + HC) = B. \tag{4.46}$$

Then define

$$\begin{aligned} \nu_0 &= \{\inf(\eta^T B \eta) \cdot (\eta^T H \eta)^{-1} : \eta \in \mathbb{R}^n, \eta \neq o\}, \\ \langle y, \eta \rangle &= y^T H \eta, \qquad \|y\| = \langle y, y \rangle^{1/2}. \end{aligned} \tag{4.47}$$

By applying Theorem 4.9 one sees that here the solution u^* of (4.45) satisfies $\|u^*(x)\| \leqslant \|r^0\| \exp(-\nu_0 x)$ $(0 \leqslant x < \infty)$ and, consequently, also $\|u^*(x)\|_2 \leqslant \|r^0\|_2 (\kappa(H))^{1/2} \exp(-\nu_0 x)$ $(0 \leqslant x < \infty)$ with $\kappa(H)$ the spectral condition number $\kappa(H) = \sigma(H)(\tau(H))^{-1}$. (Observe that $\langle \eta, C\eta \rangle : \langle \eta, \eta \rangle = \eta^T B \eta : \eta^T H \eta$.)

For instance, this result may be applied to the simple case $H = I$, for which $B = C_H$. On the other hand, one may choose $H = (\Phi^T)^{-1}\Phi^{-1}$ with $\Phi^{-1} C \Phi = J_\varepsilon$ and sufficiently small $\varepsilon > 0$. In the latter case we have $(\eta^T B \eta) \cdot (\eta^T H \eta)^{-1} = \zeta^T[\tfrac{1}{2}(J_\varepsilon + J_\varepsilon^T)]\zeta(\zeta^T \zeta)^{-1} \geqslant \mu_\varepsilon$ for $\zeta = \Phi^{-1}\eta$ and obtain the same estimate of u^* as in Example 4.14. □

Obviously, the preceding examples yield results on the stability and asymptotic stability of solutions of linear problems. In the following we shall derive stability results for somewhat more general nonlinear problems and, in particular, furnish quantitative estimates.

Suppose now that

$$f(x, o) = o \qquad (0 \leqslant x < \infty),$$

so that the differential equation in (4.42) has the trivial solution $\bar{u}(x) \equiv o$. This trivial solution is called *stable*, if for each $\varepsilon > 0$ and each $x_0 \in [0, \infty)$ a $\delta > 0$ exists such that for all r^0 satisfying $\|r^0\|_2 < \delta$ the solution $u(x, x_0, r^0)$ of (4.42) is defined on $[x_0, \infty)$ and satisfies $\|u(x, x_0, r^0)\|_2 \leqslant \varepsilon (x_0 \leqslant x < \infty)$. Moreover, the trivial solution is called *asymptotically stable*, if it is stable and if for each $\varepsilon > 0$ and each $x_0 \in [0, \infty)$ a $\delta_0 \in (0, \delta)$ exists such that $\lim_{x\to\infty} u(x, x_0, r^0) = o$ for $\|r^0\|_2 < \delta_0$. One speaks of *uniform stability* and *uniform asymptotic stability*, if the numbers δ and δ_0 do not depend on x_0.

Obviously, sufficient conditions for stability properties may be obtained by applying Theorem 4.9 and analogous results for the interval $[x_0, \infty)$. This theorem often yields not only the existence of quantities such as δ and δ_0 above, but in addition numerical estimates of these quantities and estimates of the solution. We explain this for the special case where

$$f(x, y) = Cy + h(x, y) \quad \text{with} \quad C \in \mathbb{R}^{n,n}, \quad h(x, o) = o \quad (0 \leqslant x < \infty), \tag{4.48}$$

and f satisfies the general assumptions above.

Theorem 4.10. *Suppose there exist positive definite symmetric matrices B, H such that* (4.46) *holds; define then v_0, $\langle\ ,\ \rangle$, $\|\ \|$ by* (4.47). *Moreover, assume that a continuously differentiable monotone function $\chi : [0, \rho) \to [0, \infty)$ with $0 < \rho \leqslant \infty$ exists such that $\chi(0) = 0$ and*

$$\langle y, h(x, y)\rangle \geqslant -\chi(\|y\|)\|y\|^2 \quad \text{for} \quad 0 \leqslant x < \infty, \quad \|y\| < \rho. \tag{4.49}$$

Then, for any number $\delta \in (0, \rho)$ with $\chi(\delta) \leqslant v_0$, each $x_0 \geqslant 0$ and each r^0 satisfying $\|r^0\| \leqslant \delta$, the solution $u(x, x_0, r^0)$ of (4.42) *is defined for all $x \geqslant x_0$ and satisfies*

$$\|u(x, x_0, r^0)\| \leqslant \psi(x, x_0, \|r^0\|) \quad (x_0 \leqslant x < \infty), \tag{4.50}$$

where $\psi(x, x_0, \|r^0\|)$ denotes the solution of the initial value problem

$$0 = \psi'(x) + v_0\psi(x) - \chi(\psi(x))\psi(x) \quad (x_0 < x < \infty), \quad \psi(x_0) = \|r^0\|. \tag{4.51}$$

If $\delta \in (0, \rho)$ satisfies $\chi(\delta) \leqslant v_0 - v_1$ for some $v_1 \in (0, v_0]$, then for $\|r^0\| \leqslant \delta$

$$\|u(x, x_0, r^0)\| \leqslant \|r^0\| \exp(-v_1(x - x_0)) \quad (x_0 \leqslant x < \infty). \tag{4.52}$$

Proof. We need only consider the case $x_0 = 0$. Observe first that $\alpha(x) \equiv 0$ and $\beta(x) \equiv \delta$ are comparison functions for problem (4.51), so that $\psi(x, 0, \|r^0\|)$ is defined for all $x \geqslant 0$ and $0 \leqslant \psi(x, 0, \|r^0\|) \leqslant \delta$ (see Theorem 3.4).

The statement concerning $u(x, 0, r^0)$ follows from Theorem 4.9 by verifying (4.44) for $\psi(x) = \psi(x, 0, \|r^0\|)$. We have $\langle \eta, C\eta\rangle \geqslant v_0$ for $\|\eta\| = 1$, as in Example 4.15. Moreover, due to (4.49), the term $\langle \eta, h(x, \psi(x)\eta)\rangle$ can be estimated from below by $-\chi(\psi(x))\psi(x)$. Thus the inequality $\psi' + v_0\psi - \chi(\psi)\psi \geqslant 0$ is sufficient. Note, finally, that under the conditions stated the function $\|r^0\| \exp(-v_1 x)$ also satisfies this inequality. □

Corollary 4.10a. *Suppose that all eigenvalues of C have a real part >0 and that*

$$\|h(x, y)\|_2 \leqslant \gamma(\|y\|_2)\|y\|_2 \quad \textit{for} \quad 0 \leqslant x < \infty, \qquad \|y\|_2 < \rho \quad \textit{with} \quad 0 < \rho \leqslant \infty \tag{4.53}$$

and some continuously differentiable monotone function $\gamma: [0, \rho) \to [0, \infty)$ satisfying $\gamma(0) = 0$.

Then the trivial solution of $u' + f(x, u) = o$ with f as in (4.48) *is uniformly asymptotically stable.*

Proof. For $H = (\Phi^{\mathrm{T}})^{-1}\Phi^{-1}$, $\Phi^{-1}C\Phi = J_\varepsilon$ the matrix B in (4.46) is positive definite (cf. Example 4.15). (The existence of positive definite B, H satisfying (4.46) is also asserted by the matrix theorem of Lyapunov.) Moreover, due to (4.53), condition (4.49) holds for a suitable function χ. Therefore, the above statement follows from the estimate (4.52). (Choose, for example, $v_1 = \frac{1}{2}v_0$ and δ satisfying $\chi(\delta) = \frac{1}{2}v_1$.) □

Example 4.16. We consider a system

$$u_i'(x) - u_i(x)\left(b_i + \sum_{k=1}^{n} a_{ik} u_k(x)\right) = 0 \qquad (i = 1, 2, \ldots, n), \tag{4.54}$$

where $b = (b_i) \in \mathbb{R}^n$ and $A = (a_{ik}) \in \mathbb{R}^{n,n}$ are given. Equations of this form often occur in biological problems. One is interested in solutions u^* which are pointwise strictly positive ($u^*(x) \succ o$ for all x considered) and, in particular, in the stability of stationary solutions (i.e., constant solutions). According to the following lemma, the strict positivity is guaranteed by the initial value. □

Lemma 4.11. *If a solution $u \in C_1^n[x_0, l]$ of* (4.54) *with $0 \leqslant x_0 < l < \infty$ satisfies $u(x_0) \succ o$, then $u(x) \succ o$ for all $x \in [x_0, l]$.*

Proof. Suppose the latter inequality were false and denote by ξ the smallest value in $[x_0, l]$ such that $u(\xi) \not\succ o$ and hence $u_j(\xi) = 0$ for some index j; assume $j = n$, without loss of generality. The differential equations for $u_1, \ldots, u_{n-1}$ which are obtained from (4.54) when n is replaced by $n - 1$ have a solution $\omega_1, \ldots, \omega_{n-1}$ on some interval $[\xi - \varepsilon, \xi]$ such that $\omega_i(\xi) = u_i(\xi)$ $(i = 1, 2, \ldots, n - 1)$. The function $\bar{u} \in C_1^n[\xi - \varepsilon, \xi]$ with $\bar{u}_i = \omega_i$ $(i = 1, 2, \ldots, n - 1)$ and $\bar{u}_n(x) \equiv 0$ then is a solution of the given system (4.54) and $\bar{u}(\xi) = u(\xi)$. Because this value can be assumed only by one solution, it follows $\bar{u}(x) = u(x)$ $(\xi - \varepsilon \leqslant x \leqslant \xi)$, which contradicts the definition of ξ. □

Now suppose that $b + A\bar{y} = o$ for some $\bar{y} \in \mathbb{R}^n$ with $\bar{y} \succ o$; then $\bar{u}(x) \equiv \bar{y}$ is a stationary solution of the above differential equations. Introducing $v = u - \bar{u}$, one obtains the transformed system

$$v'(x) + Cv(x) + h(v(x)) = o \tag{4.55}$$

where $C = -YA$, $Y = (\bar{y}_i\,\delta_{ik})$, $h_i(y) = -y_i(Ay)_i$. This system has the trivial solution $\bar{v}(x) \equiv 0$, which corresponds to $\bar{u}$. The stability properties of $\bar{u}$ are defined by those of $\bar{v}$. Clearly, h satisfies an estimate of the form (4.53). Thus, by Corollary 4.10a, *the stationary solution* $\bar{u}$ *of* (4.54) *is* (*uniformly*) *asymptotically stable, if all eigenvalues of the matrix* YA *have a real part* <0. □

Example 4.17. As a special case of the preceding example we consider (for $n = 2$) the equations

$$u_1' - 15u_1 + 0.5u_1^2 + u_1u_2 = 0, \qquad u_2' + 9u_2 + 0.1u_2^2 - u_1u_2 = 0, \tag{4.56}$$

which describe a preditor–prey problem. Here we have a stationary solution $\bar{u}(x) \equiv \bar{y}$ with $\bar{y}_1 = \bar{y}_2 = 10$, and we obtain

$$C = -YA = \begin{pmatrix} 5 & 10 \\ -10 & 1 \end{pmatrix}, \qquad h(y) = \begin{pmatrix} 0.5y_1^2 + y_1y_2 \\ 0.1y_2^2 - y_1y_2 \end{pmatrix}$$

in the transformed problem (4.55) for $v = u - \bar{u}$. Since C has eigenvalues $\mu \pm i\nu$ with $\mu = 3$ and $\nu = \sqrt{96}$, the stationary solution $\bar{u}$ is asymptotically stable.

Using Theorem 4.10 we shall also derive quantitative estimates for the case $x_0 = 0$. There are several possibilities depending on the choice of H. For example, if $H = I$, then $B = \begin{pmatrix} 5 & 0 \\ 0 & 1 \end{pmatrix}$ and hence $\nu_0 = 1$, so that the bound cannot decrease faster than $\exp(-x)$.

Let us here choose a matrix Φ such that $\Phi^{-1}A\Phi = \begin{pmatrix} \mu & \nu \\ -\nu & \mu \end{pmatrix}$ and define $H = (\Phi^{-1})^{\mathrm{T}}\Phi^{-1}$; then $\mu = \nu_0 = 3$. For

$$\Phi = \frac{1}{2}\sqrt{2}\begin{pmatrix} 1 & -1 \\ 1 & 1 \end{pmatrix}\begin{pmatrix} \gamma^{-1} & 0 \\ 0 & \gamma \end{pmatrix} \quad \text{with} \quad \gamma = \left(\frac{3}{2}\right)^{1/4}$$

and

$$\langle y, \eta \rangle = (\Phi^{-1}y)^{\mathrm{T}}(\Phi^{-1}\eta) = \frac{1}{4}\sqrt{\frac{2}{3}}\, y^{\mathrm{T}}\begin{pmatrix} 5 & 1 \\ 1 & 5 \end{pmatrix}\eta,$$

inequality (4.49) holds for $\chi(t) = 1.23t$. Therefore, Theorem 4.10 yields the following result.

If $\|r^0 - \bar{y}\| \leqslant \delta \leqslant 3(1.23)^{-1} =: \beta < 2.5$, *then the given differential equations* (4.56) *have a solution* $u^* \in C_1^2[0, \infty)$ *such that* $u(0) = r^0$ *and*

$$\|u(x) - \bar{y}\| \leqslant \psi(x) := \delta \exp(-3x)[1 + \beta\delta + \beta\delta \exp(-3x)]^{-1}.$$

Here $\|y\|$ denotes the norm corresponding to $\langle\ ,\ \rangle$. By using $\gamma^{-1}\|u\|_2 \leqslant \|u\| \leqslant \gamma\|u\|_2$, one can also obtain estimates in terms of the norm $\|\ \|_2$. □

Problems

4.8 The set of parameters η for which the differential inequality in the premise of (4.41) has to be satisfied can be further restricted, if suitable a priori estimates of v are known. For example, *if* $v(x) \succ -\bar{y}$ $(0 \leqslant x \leqslant 1)$ *for some* $\bar{y} \succ o$ (as in Examples 4.16 and 4.17), *then this differential inequality need only hold for* $(x, \eta) \in P(v)$ *with* $\eta\psi(x) \geqslant -\bar{y}$. (To prove this use the side condition $\psi_\lambda(x)\eta = v(x)$.)
4.9 Apply Theorem 4.10 to the problem $u_1' - u_2 = 0$, $u_2' + \sin u_1 = 0$.
4.10 Apply Theorem 4.10 to the Lienard equations $u_1' - u_2 = 0$, $u_2' + g(u_1) + u_2 = 0$, where $g \in C_1(-\infty, \infty)$, $g(0) = 0$, $g'(0) > 0$, $tg(t) > 0$ for $t \neq 0$.

5 EXISTENCE AND ESTIMATION BY NORM BOUNDS

Now we shall derive existence statements for solutions u^* of boundary value problems and norm estimates of these solutions, simultaneously. Up to a certain extent the procedure is analogous to the one in Section 3, where two-sided bounds were considered. First a basic theorem is derived which requires certain boundedness conditions (Section 5.1). These conditions can be replaced by differential inequalities leading to estimates of the derivative of the solution (Section 5.2). Finally, growth restrictions concerning u' are considered (Section 5.3).

For simplicity, we restrict ourselves here to second order problems with Dirichlet boundary conditions. This is sufficient to explain the general approach and the type of the resulting statements. Similar results can be obtained, for example, for Sturm–Liouville boundary conditions by essentially the same methods.

In this section define $R = C_2^n[0, 1]$ and let $M : R \to \mathbb{R}^n[0, 1]$ be an operator of the form

$$Mu(x) = \begin{cases} L[u](x) + f(x, u(x), u'(x)) & \text{for } 0 < x < 1 \\ u(0) & \text{for } x = 0 \\ u(1) & \text{for } x = 1, \end{cases} \tag{5.1}$$

where $L[u](x) = -u''(x) + b(x)u'(x)$ with (*scalar-valued*) $b \in C_0[0, 1]$ and $f: [0, 1] \times \mathbb{R}^n \times \mathbb{R}^n \to \mathbb{R}^n$ is continuous. We consider a boundary value problem

$$Mu = r$$

for a given $r \in \mathbb{R}^n[0, 1]$ such that $r(x) = s(x)$ $(0 < x < 1)$ with some $s \in C_0^n[0, 1]$ and $r(0) = r^0 \in \mathbb{R}^n$, $r(1) = r^1 \in \mathbb{R}^n$.

We shall again use the notation in Section 4.1 and define the sets P and Q as in (4.7.).

5.1 The basic theorem

Theorem 5.1. *Suppose that the following two assumptions are satisfied.*

(i) *There exists a function* $\psi \geqslant o$ *in* $C_2[0, 1]$ *such that*

$$\|r^0\| \leqslant \psi(0), \qquad \|r^1\| \leqslant \psi(1), \tag{5.2}$$

$$L[\psi](x) + c\psi(x) \geqslant 0 \qquad (0 < x < 1) \qquad \textit{for some constant} \quad c > 0, \tag{5.3}$$

$$\|s(x)\| \leqslant L[\psi](x) + \langle q, q\rangle\psi(x) + \langle \eta, f(x, \psi(x)\eta, \psi'(x)\eta + \psi(x)q)\rangle \quad \textit{for} \quad (x, \eta, q) \in Q. \tag{5.4}$$

(ii) *The function* $\|f(x, y, p)\|$ *is bounded on*

$$Q' := \{(x, y, p): 0 \leqslant x \leqslant 1, \|y\| \leqslant \psi(x), \ p \in \mathbb{R}^n\}.$$

Then the given problem $Mu = r$ *has a solution* $u^* \in C_2^n[0, 1]$ *such that* $\|u^*\| \leqslant \psi$.

Proof. Without loss of generality, we shall here assume that $s = o$. (The right-hand side s can be incorporated into f.) Then we define a *modified operator* $M^\#$ on R by

$$M^\# u(x) = \begin{cases} L[u](x) + c(u(x) - u^\#(x)) + \Delta(x, u(x)) f(x, u^\#(x), u'(x)) \\ \qquad\qquad \text{for} \quad 0 < x < 1 \\ u(x) \qquad \text{for} \quad x = 0 \quad \text{and} \quad x = 1, \end{cases} \tag{5.5}$$

where

$$\Delta(x, y) = \begin{cases} 1 & \text{if} \quad \|y\| \leqslant \psi(x) \\ \|y\|^{-1}\psi(x) & \text{if} \quad \|y\| > \psi(x), \end{cases} \qquad u^\#(x) = \Delta(x, u(x))u(x).$$

The *modified problem* $M^\# u = r$ can be written in the form

$$L[u](x) + cu(x) = \mathscr{H}(x, u(x), u'(x)) \qquad (0 < x < 1),$$

$$u(0) = r^0, \qquad u(1) = r^1$$

with a certain well-defined continuous function $\mathscr{H}(x, y, p)$ on $[0, 1] \times \mathbb{R}^n \times \mathbb{R}^n$. Using the (matrix-valued) Green function $\mathscr{G}(x, \xi)$ corresponding to $L[u] + cu$ and the Dirichlet boundary conditions, one can transform this problem into an equivalent fixed-point equation $u = Tu$ in the Banach space $C_1^n[0, 1]$. Since $\|u^\#\| \leqslant \psi$ holds and f is bounded on Q', the function $\mathscr{H}$ is bounded on $[0, 1] \times \mathbb{R}^n \times \mathbb{R}^n$ and, consequently, the (continuous) operator T maps $C_1^n[0, 1]$ into a relatively compact subset of this space. Schauder's fixed-point theorem then yields the existence of a fixed point u^* of T, which is also a solution of the modified problem. The procedure is analogous to the one in the proofs of Theorems 3.1 and III.4.2. The proof is completed by use of the following lemma. □

Lemma 5.2. *If $s = o$ and assumption* (i) *of Theorem* 5.1 *holds, then each solution $u^* \in R$ of the modified problem $M^\# u = r$ (with $M^\#$ in* (5.5)) *satisfies $\|u^*\| \leqslant \psi$ and hence is also a solution of the given problem $Mu = r$.*

Proof. The statement will be proved for the special case $L[u] = -u''$ by applying Theorem 4.1 to the modified operator $M^\#$ (in place of M) and $v = u^*$, choosing $\psi_\lambda = \psi + \lambda$. In the general case one proceeds analogously observing the statement in Section 4.4.1.

For fixed $\lambda > 0$ we insert $\psi q = (\psi + \lambda)p$ into the differential inequality (5.4), multiply the resulting relation by ψ, add the term $\lambda(L[\psi](x) + c\psi(x)) + \lambda^2 c > 0$ on the right-hand side, and finally divide by $\psi + \lambda$. Then we arrive at the inequality

$$0 < L[\psi](x) + \langle p, p\rangle(\psi(x) + \lambda) + c\lambda \\ + \langle \eta, \psi(x)(\psi(x) + \lambda)^{-1} f(x, \psi(x)\eta, (\psi(x) + \lambda)p + \psi'(x)\eta)\rangle.$$

Due to the definition of p, this inequality holds for all $(x, \eta, p) \in Q$ and $\lambda > 0$. Moreover, since $\psi_\lambda'' = \psi''$, $\psi_\lambda' = \psi'$, $\Delta(x, \psi_\lambda(x)\eta) = \psi(x)(\psi(x) + \lambda)^{-1}$, and $(\psi_\lambda(x)\eta)^\# = \psi(x)\eta$, this inequality is equivalent to inequality (4.8) (formulated for $v = u^*$, $\psi_\lambda = \psi + \lambda$, $M^\#$ in place of M, p in place of q). Consequently, $\|u^*\| \leqslant \psi$ follows from Theorem 4.1. This inequality then implies $\Delta(x, u^*(x)) = 1$ and $(u^*)^\# = u^*$, so that $Mu^* = M^\# u^* = r$. □

Assumption (ii) of Theorem 5.1 is satisfied, for example, if

$$Mu(x) = -u''(x) + b(x)u'(x) + f(x, u(x)) \qquad (0 < x < 1),$$

where f does not depend on $u'(x)$. Here inequality (5.4) is equivalent to

$$\|s(x)\| \leqslant L[\psi](x) + \langle \eta, f(x, \psi(x)\eta)\rangle \quad \text{for} \quad (x, \eta) \in P. \tag{5.6}$$

Example 5.1. The boundary value problem in Example 4.6 with $r^0 = r^1 = o$ has a solution $u^* \in R$ with $\|u^*(x)\| \leqslant \psi(x) := \frac{1}{2}1.01\, \delta x(1 - x)$ $(0 \leqslant x \leqslant 1)$, if $\|s(x)\| \leqslant \delta < 0.8$ $(0 \leqslant x \leqslant 1)$. To prove this, one verifies

(5.6) observing that f has the form (4.27) with (4.28) holding for $\gamma = -\frac{1}{2}\sqrt{2}$ and all $v(x)$, as shown in Example 4.7. Here a function z need not be constructed and hence no a priori bound such as 13.9 in Example 4.7 occurs. □

Problem

5.1 Formulate Theorem 5.1 for the case $n = 1$ and compare the result with that of Theorem 3.1.

5.2 Simultaneous estimation of u^* and $(u^*)'$

Our next aim is to remove the boundedness condition in assumption (ii) of Theorem 5.1. This will be achieved here by requiring a further differential inequality for a function Ψ which bounds the derivative of the solution. In Section 5.3 a different method will be applied.

For a fixed $\Psi \in C_1[0, 1]$ we define a *second modified operator* $\hat{M}$ on R by

$$\hat{M}u(x) = \begin{cases} L[u](x) + \hat{f}(x, u(x), u'(x)) & \text{for} \quad 0 < x < 1 \\ Mu(x) & \text{for} \quad x = 0 \quad \text{and} \quad x = 1, \end{cases} \tag{5.7}$$

where

$$\left.\begin{aligned} &\hat{f}(x, y, p) = f(x, y, \hat{p}), \\ &\hat{p} = \delta(x, p)p, \qquad \delta(x, p) = \begin{cases} 1 & \text{for} \quad \|p\| \leqslant \Psi(x) \\ \Psi(x)\|p\|^{-1} & \text{for} \quad \|p\| > \Psi(x). \end{cases} \end{aligned}\right\} \tag{5.8}$$

Lemma 5.3. *Suppose that*

(i) *assumption* (i) *of Theorem* 5.1 *holds with f replaced by $\hat{f}$,*
(ii) *for each $u \in C_2^n[0, 1]$*

$$\hat{M}u = r, \quad \|u\| \leqslant \psi \quad \Rightarrow \quad \|u'\| \leqslant \Psi. \tag{5.9}$$

Then the given problem $Mu = r$ has a solution $u^* \in R$ such that $\|u^*\| \leqslant \psi$ and $\|(u^*)'\| \leqslant \Psi$.

Proof. Theorem 5.1 applied to the operator $\hat{M}$ (instead of M) yields the existence of a function u^* which satisfies $\hat{M}u^* = r$ and $\|u^*\| \leqslant \psi$. According to assumption (ii) we have also $\|(u^*)'\| \leqslant \Psi$ and hence $Mu^* = \hat{M}u^* = r$. □

In the following theorem we formulate sufficient conditions on ψ and Ψ for assumption (ii) to hold. In particular, we shall need here an "initial estimate" for $\|u'(0)\|$. Such an estimate can be obtained if $\psi(0) = 0$. We point out that analogous conditions can be formulated under the assumption $\psi(1) = 0$, which allows us to estimate $\|u'(1)\|$.

Theorem 5.4. *Suppose that positive functions $\psi \in C_2[0, 1]$ and $\Psi \in C_1[0, 1]$ exist such that*

(i) *inequalities* (5.2), (5.3) *hold and*

$$\|s(x)\| \leqslant L[\psi](x) + \langle q, q \rangle \psi(x) + \langle \eta, \hat{f}(x, \psi(x)\eta, \psi'(x)\eta + \psi(x)q) \rangle \quad \text{for} \quad (x, \eta, q) \in Q \quad \text{with } \hat{f} \text{ defined in (5.8)}; \tag{5.10}$$

(ii) $\psi(0) = 0$, $\psi'(0) \leqslant \Psi(0)$, *and*

$$\|s(x)\| \leqslant \Psi'(x) - b(x)\Psi(x) - \langle p, f(x, \psi(x)y, \Psi(x)p) \rangle$$

for all $(x, y, p) \in (0, 1] \times \mathbb{R}^n \times \mathbb{R}^n$ *with* $\|y\| \leqslant 1$, $\|p\| = 1$.

Then the given problem $Mu = r$ *has a solution* $u^* \in C_2^n[0, 1]$ *such that* $\|u^*\| \leqslant \psi$, $\|(u^*)'\| \leqslant \Psi$.

Proof. We verify (5.9), that is, we derive $\|u'\| \leqslant \Psi$ for an arbitrary fixed function $u \in R$ satisfying $\hat{M}u = r$ and $\|u\| \leqslant \psi$. Let an operator $M_0 : C_1^n[0, 1] \to \mathbb{R}^n[0, 1]$ be defined by

$$M_0 w(x) = w'(x) - b(x)w(x) - f(x, u(x), \hat{w}(x)) \quad \text{for} \quad 0 < x \leqslant 1, \quad M_0 w(0) = w(0).$$

Obviously, $M_0 u'(x) = -s(x)$ $(0 < x \leqslant 1)$ and, since $\psi(0) = 0$, $\|u'(0)\| = \lim_{x \to 0} x^{-1}\|u(x)\| \leqslant \lim_{x \to 0} x^{-1}\psi(x) = \psi'(0)$. These relations enable us to apply Theorem 4.8 (to M_0, u', Ψ, ... in place of M, v, ψ, ...). One verifies that $z(x) = \exp(1 + \gamma)x$ with γ an upper bound of $b(x)$ has all required properties. Consequently, $\|u'\| \leqslant \Psi$. $\square$

The differential inequalities required of ψ and Ψ have a rather complicated form, and it may often be difficult to solve them. However, when the above results are applied (to a transformed problem) to obtain an error estimate, the situation arises that s is "small" and $f(x, o, o) = o$. In this case, expecting ψ and Ψ to be small, one may try to approximate $f(x, y, p)$ by a function which is linear in y and p and estimate the nonlinear remainder roughly.

Thus, let us now write $Mu(x)$ in the form

$$Mu(x) = L[u](x) + B(x)u'(x) + C(x)u(x) + g(x, u(x), u'(x)) \quad (0 < x < 1) \tag{5.11}$$

with $n \times n$ matrices $B(x)$ and $C(x)$ (so that here $f(x, y, p) = B(x)p + C(x)y + g(x, y, p)$). The following corollary is formulated with the case in mind that $g(x, y, p)$ may be small for small y, p.

Corollary 5.4a. *Suppose that there exist a function $\psi \geqslant o$ in $C_2[0, 1]$, a function $\Psi \geqslant o$ in $C_1[0, 1]$, and a monotone function $h \in C_0[0, \infty)$ such that the following conditions are satisfied.*

(1) $\|g(x, y, p)\| \leqslant h(\|p\|)$ *for* $0 \leqslant x \leqslant 1$, $\|y\| \leqslant \psi(x)$, $p \in \mathbb{R}^n$.

(2) *Inequalities* (5.2), (5.3) *hold and*

$$\|s(x)\| \leqslant L[\psi](x) + \kappa\langle \eta, B(x)\eta\rangle\psi'(x) + \langle \eta, C(x)\eta\rangle\psi(x) + c_B(x, \eta)\psi(x) - h(\Psi(x)) \tag{5.12}$$

for all $(x, \eta) \in P$ *and* $\kappa \in [0, 1]$, *where* c_B *is defined as in* (4.32).

(3) $\psi(0) = 0$, $\psi'(0) \leqslant \Psi(0)$, *and for* $0 < x \leqslant 1$, $\|p\| = 1$

$$\|s(x)\| \leqslant \Psi'(x) - b(x)\Psi(x) - \langle p, B(x)p\rangle\Psi(x) - \|C(x)\|\psi(x) - h(\Psi(x)).$$

Then the given problem $Mu = r$ has a solution $u^* \in R$ such that $\|u^*\| \leqslant \psi$ and $\|(u^*)'\| \leqslant \Psi$.

Proof. This statement follows from Theorem 5.4. We only verify the differential inequality (5.10) for the case presently considered. Since $w(x) = \psi'(x)\eta + \psi(x)q$ satisfies $\hat{w}(x) = \kappa(x)w(x)$ with $0 \leqslant \kappa(x) \leqslant 1$, we obtain

$$\begin{aligned}\langle q, q\rangle\psi &+ \langle \eta, \hat{f}(x, \psi\eta, w)\rangle \\ &= \kappa\langle \eta, B\eta\rangle\psi' + \langle \eta, C\eta\rangle\psi + \{\langle q, q\rangle + \kappa\langle B^*\eta, q\rangle\}\psi \\ &\quad + \langle \eta, g(x, \psi\eta, \hat{w})\rangle.\end{aligned} \tag{5.13}$$

The term within braces can be estimated from below by $\kappa(\langle q, q\rangle + \langle B^*\eta, q\rangle) \geqslant \kappa c_B(x, \eta) \geqslant c_B(x, \eta)$ (cf. Proposition 4.5a). Moreover, since $\|\hat{w}\| \leqslant \Psi$, the last summand in (5.13) is bounded below by $-h(\Psi(x))$. Consequently, inequality (5.12) is sufficient for (5.10) to hold. □

Example 5.2. Consider the boundary value problem

$$\begin{aligned}-u_1''(x) + u_1(x) - 2u_2(x) + (u_2'(x))^3 &= s_1(x) \quad (0 < x < 1), \\ u_1(0) &= u_1(1) = 0,\end{aligned}$$

$$\begin{aligned}-u_2''(x) + u_1(x) + u_2(x) + (u_1'(x))^3 &= s_2(x) \quad (0 < x < 1), \\ u_2(0) &= u_2(1) = 0,\end{aligned}$$

which obviously can be written as $Mu = r$ with M in (5.11), $b = o$, $B = O$, $C = \begin{pmatrix} 1 & -2 \\ 1 & 1 \end{pmatrix}$. We assume that $\|s(x)\| \leqslant \delta$ $(0 < x < 1)$.

Here the inequality in assumption (1) of Corollary 5.4a holds for $h(t) = t^3$ (and all $y \in \mathbb{R}^2$). Since $\tau(C_H) = \frac{1}{2}$ and $\sigma(C^T C) = \|C\|^2 \leqslant \gamma^2$ with $\gamma = 2.31$, the following conditions are sufficient for assumptions (2) and (3):

$$\begin{aligned}&\delta \leqslant -\psi''(x) + \tfrac{1}{2}\psi(x) - \Psi^3(x) &&(0 < x < 1), &&\psi(0) = 0, \\ &\delta \leqslant \Psi'(x) - \gamma\psi(x) - \Psi^3(x) &&(0 < x \leqslant 1), &&\psi'(0) \leqslant \Psi(0).\end{aligned}$$

First we neglect the nonlinear terms. The linear inequalities so obtained have the solution $\psi_0(x) = \frac{1}{2}\delta x(1-x)$, $\Psi_0(x) = \delta[\frac{1}{2} + x + \frac{1}{6}\gamma x^2(3-2x)]$. For small enough δ, the functions $\psi = 1.01\psi_0$, $\Psi = 1.01\,\Psi_0$ then satisfy the nonlinear inequalities. Here $\delta \leqslant 0.02$ is sufficient. □

5.3 Quadratic growth restrictions

The boundedness assumption on $f(x, y, p)$ with respect to p in Theorem 5.1 can be replaced by a quadratic growth restriction, as we shall prove in Theorem 5.6. First, however, we shall formulate a more general assumption, which requires an a priori estimate for the derivatives of solutions of a certain modified problem. Let M be defined as in (5.1), but now assume $b(x) \equiv 0$, without loss of generality.

Again, we introduce an operator $\hat{M}$ of the form (5.7), where now

$$\hat{f}(x, y, p) = cy + \delta(\|p\|)(f(x, y, p) - cy) \tag{5.14}$$

with a continuous function $\delta : [0, \infty) \to [0, 1]$ such that $\delta(t) = 1$ for $0 \leqslant t \leqslant N$, $\delta(t) = 0$ for $2N \leqslant t$, and $c > 0$, $N > 0$ are constants on which further conditions will be imposed later.

Lemma 5.5. *Let $s = o$ and suppose that assumption* (i) *of Theorem* 5.1 *is satisfied with c the constant in* (5.14). *Assume, moreover, that* (5.9) *holds for $\Psi(x) \equiv N$ and each $u \in R$.*

Then there exists a $u^* \in R$ such that $Mu^* = r$ and $\|u^*\| \leqslant \psi$.

Proof. By multiplying the inequality (5.4) by $\delta = \delta(\|\psi'\eta + \psi q\|)$ and adding the term $(1-\delta)(-\psi'' + c\psi + \|q\|^2\psi)$ on the right-hand side, we see that ψ satisfies also (5.10) with $\hat{f}$ defined by (5.14). Therefore, we can proceed as in the proof of Lemma 5.3. □

Theorem 5.6. *Theorem* 5.1 *remains true, if instead of the boundedness condition* (ii) *it is required that constants $\beta > 0$ and $\tau > 0$ exist such that*

$$\begin{aligned} &\|f(x, y, p)\| \leqslant \beta\|p\|^2 + \tau && \textit{for} \quad 0 < x < 1, \quad \|y\| \leqslant \psi(x), \quad p \in \mathbb{R}^n \\ &\beta\psi(x) < 1 && \textit{for} \quad 0 \leqslant x \leqslant 1. \end{aligned} \tag{5.15}$$

Proof. We shall show that (5.9) holds, provided the number N, which also appears in the definition of $\hat{M}$, is sufficiently large. Without loss of generality, we assume that $s(x) \equiv o$. In the following let u denote a fixed function in R satisfying $\hat{M}u = r$ and $\|u\| \leqslant \psi$. To verify $\|u'\| \leqslant N$, we proceed in three steps. First, an estimate

$$\|u''\| \leqslant k'' \qquad \text{for} \quad k(x) = \alpha\langle u, u\rangle(x) + \kappa x^2 \tag{5.16}$$

with suitable constants $\alpha \geqslant 0$, $\kappa \geqslant 0$ is derived. Observe that this function k is bounded due to $\|u\| \leqslant \psi$. In a second step, $\|u'(x)\|$ is estimated using $k'(x)$. Finally, the estimate $\|u'\| \leqslant N$ is established by an indirect proof which employs a further integration.

(1) Condition (5.15) yields

$$\|u''\| \leqslant \|\hat{f}(x, u, u')\| \leqslant \beta\|u'\|^2 + \tau_1$$

with

$$\tau_1 = c\rho + \tau, \quad \rho = \max\{\psi(x) : 0 \leqslant x \leqslant 1\}.$$

Thus, for k as above with $\alpha = \frac{1}{2}\beta(1 - \beta\rho)^{-1}, \kappa = \frac{1}{2}\tau_1(1 - \beta\rho)^{-1}$ we calculate

$$\begin{aligned} k'' = 2\alpha\{\langle u, u''\rangle + \langle u', u'\rangle\} + 2\kappa &\geqslant 2\alpha\{-\rho\|u''\| + \|u'\|^2\} + 2\kappa \\ &\geqslant 2\alpha\{-\rho\|u''\| + \beta^{-1}\|u''\| - \beta^{-1}\tau_1\} + 2\kappa \\ &= \|u''\|, \end{aligned}$$

so that (5.16) is verified.

(2) *From* $\|u''\| \leqslant k''$, $\|u\| \leqslant \rho$, $0 \leqslant k \leqslant \gamma$ $(=\alpha\rho^2 + \kappa)$ *it follows that* $\|u'(x)\| \leqslant \mu + k'(x)\,\mathrm{sgn}(x - \frac{1}{2})$ *with* $\mu = 4(\rho + \gamma)$. To prove this, first let $0 \leqslant x \leqslant \frac{1}{2}$. Then we have

$$u(x + \tfrac{1}{2}) - u(x) - \tfrac{1}{2}u'(x) = \int_x^{x+1/2} (x + \tfrac{1}{2} - t)u''(t)\, dt$$

and estimate

$$\begin{aligned} \tfrac{1}{2}\|u'(x)\| &\leqslant \|u(x + \tfrac{1}{2})\| + \|u(x)\| + \int_x^{x+1/2} (x + \tfrac{1}{2} - t)k''(t)\, dt \\ &\leqslant 2\rho + k(x + \tfrac{1}{2}) - k(x) - \tfrac{1}{2}k'(x) \leqslant \tfrac{1}{2}(\mu - k'(x)). \end{aligned}$$

In an analogous way, the inequality $\|u'(x)\| \leqslant \mu + k'(x)$ $(\frac{1}{2} \leqslant x \leqslant 1)$ is established, and both estimates together yield $\|u'(\frac{1}{2})\| \leqslant \mu$.

(3) Assume now that N has been chosen so large that

$$\int_\mu^N sh^{-1}(s)\, ds > \tfrac{1}{2}\mu + 2\gamma, \qquad \text{where} \quad h(s) = \tau_1 + \beta s^2. \tag{5.17}$$

Then, in particular, $\|u'(\frac{1}{2})\| \leqslant \mu < N$. We shall show that $\|u'(x)\| < N$ for all $x \in [0, 1]$.

Suppose that $\|u'(x_0)\| \geqslant N$ for some value $x_0 \in (\frac{1}{2}, 1]$. Let ζ denote the smallest number in $(\frac{1}{2}, 1]$ for which $\|u'(\zeta)\| = N$ and η the largest number in

$(\frac{1}{2}, \zeta)$ for which $\|u'(\eta)\| = \mu$. Then

$$\begin{aligned}\int_{\mu}^{N} sh^{-1}(s)\,ds &= \int_{\eta}^{\zeta} h^{-1}(\|u'(t)\|)\langle u'(t), u''(t)\rangle\,dt \\ &\leqslant \int_{\eta}^{\zeta} h^{-1}(\|u'(t)\|)\|u''(t)\|\,\|u'(t)\|\,dt \leqslant \int_{\eta}^{\zeta} \|u'(t)\|\,dt \\ &\leqslant \int_{\eta}^{\zeta} (\mu + k'(t))\,dt = \mu(\zeta - \eta) + k(\zeta) - k(\eta) \leqslant \tfrac{1}{2}\mu + 2\gamma,\end{aligned}$$

which contradicts (5.17). The case $x_0 \in [0, \frac{1}{2})$ is treated analogously. □

Let us again consider the special case (5.11), where now $b(x) \equiv 0$, so that f in (5.1) has the form

$$f(x, y, p) = B(x)p + C(x)y + g(x, y, p). \tag{5.18}$$

Corollary 5.6a. *Suppose that there exist a positive function ψ in $C_2[0, 1]$ and a constant $\beta > 0$ such that*

$$\|g(x, y, p)\| \leqslant \beta\|p\|^2 \qquad \text{for} \quad 0 < x < 1, \quad \|y\| \leqslant \psi(x), \quad p \in \mathbb{R}^n, \tag{5.19}$$

$\beta\psi(x) < 1$ $(0 \leqslant x \leqslant 1)$, the inequalities (5.2), (5.3) hold, and

$$\begin{aligned}\|s(x)\| \leqslant L[\psi](x) &+ \langle \eta, B(x)\eta\rangle\psi'(x) + \langle \eta, C(x)\eta\rangle\psi(x) \\ &+ c_B(x, \eta)(1 - \beta\psi(x))^{-1}\psi(x) - \beta(\psi'(x))^2 \qquad \text{for} \quad (x, \eta) \in P\end{aligned} \tag{5.20}$$

with c_B defined in (4.32).

Then there exists a $u^* \in R$ such that $Mu^* = r$ and $\|u^*\| \leqslant \psi$.

Proof. Due to (5.19) the function f in (5.18) satisfies the inequality (5.15) with β replaced by $\beta + \varepsilon$ and some $\tau = \tau(\varepsilon)$, where $\varepsilon > 0$ is arbitrarily small.

To verify (5.4), observe that the term $\langle \eta, f\rangle$ on the right-hand side of this inequality contains the summand $\langle \eta, g(x, \psi(x)\eta, \psi'(x)\eta + \psi(x)q)\rangle$, which may be replaced by the lower bound $-\beta[(\psi'(x))^2 + \psi^2(x)\|q\|^2]$. In the inequality so obtained the term $(\|q\|^2 + \langle \zeta, q\rangle)\psi(x)(1 - \beta\psi(x))$ with $\zeta = (1 - \beta\psi(x))^{-1}B^*\eta$ occurs. When this term is replaced by its minimum

$$-\tfrac{1}{4}(\|\zeta\|^2 - \langle \zeta, \eta\rangle^2)\psi(x)(1 - \beta\psi(x)) = c_B(x, \eta)\psi(x)(1 - \beta\psi(x))^{-1},$$

the differential inequality (5.20) is obtained. □

The question arises whether the growth restriction (5.15) may be weakened. Checking the proof of Theorem 5.6, one sees that the following result holds, which is formulated for the case $s(x) \equiv 0$.

Theorem 5.7. *Suppose that the following three assumptions are satisfied.*

(i) *Condition* (i) *of Theorem* 5.1 *holds.*

(ii) $$\|f(x, y, p)\| \leqslant 2\alpha(\langle y, f(x, y, p)\rangle + \langle p, p\rangle) + 2\kappa \tag{5.21}$$

for $0 < x < 1$, $\|y\| \leqslant \psi(x)$, $p \in \mathbb{R}^n$ *and some constants* $\alpha \geqslant 0$, $\kappa \geqslant 0$.

(iii) *There exists a function* $h \in C_0[0, \infty)$ *such that* $h(s) > 0$ $(0 \leqslant s < \infty)$, $\int^{\infty} sh^{-1}(s)\, ds = \infty$ *and*

$$\|f(x, y, p)\| \leqslant h(\|p\|) \quad \text{for} \quad 0 < x < 1, \quad \|y\| \leqslant \psi(x), \quad p \in \mathbb{R}^n.$$

Then there exists a $u^* \in R$ *such that* $Mu^* = r$ *and* $\|u^*\| \leqslant \psi$.

The growth restriction in assumption (iii) of this theorem is certainly weaker than (5.15). However, inequality (5.21), in general, also constitutes a growth restriction. If $\|y\| \leqslant \sigma$ with $4\alpha\sigma < 1$, for example, (5.21) yields $\|f(x, y, p)\| \leqslant 4(\alpha\|p\|^2 + \kappa)$.

Problems

5.2 Show that for $s = o$, $\psi(x) \equiv \rho = \text{const}$, condition (5.4), which is used in Theorems 5.1, 5.6, and 5.7, is equivalent to

$$0 \leqslant \langle p, p\rangle + \langle y, f(x, y, p)\rangle \quad \text{for} \quad 0 < x < 1, \quad \|y\| = \rho, \quad \langle y, p\rangle = 0.$$

5.3 If $f(x, y, p) = h(x, y) + g(x, p)$ with continuous functions h, g on $[0, 1] \times \mathbb{R}^n$ satisfies (5.21), then $\|f(x, y, p)\| \leqslant \beta\|p\|^2 + \tau$ with constants β, τ such that $\beta\rho < 1$. Prove this.

6 MORE GENERAL ESTIMATES

The results of the preceding sections can be generalized by also proving estimates of other forms. In Section 6.1 we shall consider range–domain implications $Mv \in C \Rightarrow v \in K$ for differential operators of the second order, using simultaneously two-sided bounds and norm bounds. These estimates could also be combined with existence statements for solutions by either using the method of modification or the degree of mapping method. This, however, will not be carried out here. In Section 6.2 estimates by Lyapunov functions will be described for first order operators.

Certainly, one can prove still more general estimates which can be written as $v \in K$. However, the more general a class of sets K we consider, the less handy the results will be. Generalizing "as far as possible" one finally arrives at an abstract setting like that in Section IV.1.

6.1 Estimation by two-sided bounds and norm bounds

Let M denote a vector-valued differential operator of the second order as considered above. (As seen before, operators of the first order related to initial value problems can often be treated as a special case.) We describe here certain range–domain implications $Mv \in C \Rightarrow v \in K$ which generalize those treated in Sections 2 and 4. It is not our purpose to present detailed results, but rather to discuss various possible estimates and methods for obtaining sufficient conditions. The implications discussed in the following are such that the methods of proof used in Sections 2 and 4 can rather easily be carried over to these problems, so that results of similar type are obtained.

Since two-sided bounds and norm bounds each have certain advantages, one may try to combine both types of estimates in defining the set K of functions u by relations such as

$$u = \begin{pmatrix} u^1 \\ u^2 \end{pmatrix}, \qquad \varphi^1 \leqslant u^1 \leqslant \psi^1, \qquad \|u^2\| \leqslant \psi^2, \tag{6.1}$$

where $u \in \mathbb{R}^n$ is split into two subvectors $u^1 \in \mathbb{R}^{n_1}$, $u^2 \in \mathbb{R}^{n_2}$ with $n_1 + n_2 = n$. Moreover, one may describe K by several sets of such inequalities. For example, if one is interested in norm estimates of positive functions, K may be defined by

$$\|u\| \leqslant \psi, \qquad u \geqslant o. \tag{6.2}$$

Of course, this would also be possible for differential operators of the first order.

In defining K for second order operators, one may add some properties of the derivative u'. In this way, one obtains, in general, less strong assumptions concerning the dependence of Mu on u'. In particular, if two-sided bounds as in Section 2 are to be derived, strongly coupled operators can be treated by simultaneously estimating the derivative of the function considered.

The properties to be required of u' for $u \in K$ may consist of two-sided estimates or norm estimates. Of course, combined conditions such as (6.1) with u replaced by u' may also be used, etc. Here the splitting of u' need not be the same as that of u. As a special example, K may be described by

$$\|u\| \leqslant \psi, \qquad \Phi \leqslant u' \leqslant \Psi. \tag{6.3}$$

If K is chosen, range–domain implications may be derived using the idea of the continuity principle in Section IV.1. That means, suitable properties involving a parameter λ are formulated which describe families of sets K_λ, like in Sections 2 and 4. Most often, these sets will be defined by certain families of functions. If K is given by (6.3), for example, K_λ may consist of all elements u which satisfy inequalities of the form

$$\|u\| \leqslant \psi + \mathfrak{z}(\lambda), \qquad \Phi - \bar{\mathfrak{y}}(\lambda) \leqslant u' \leqslant \Psi + \mathfrak{y}(\lambda). \tag{6.4}$$

For first order initial value problems one may also choose K_λ differently, generalizing the approach in Section IV.2.4.1. For example, when norm estimates are to be proved, one may define K_λ to be the set of all u such that $\|u(x)\| \leqslant \psi(x) + \mathfrak{z}(\lambda, x)$ for $0 \leqslant x \leqslant h(\lambda)$ with suitable functions $\mathfrak{z}(\lambda, x)$ and $h(\lambda)$. Similarly, if (6.2) is considered, an inequality of the form $u(x) \geqslant -\bar{\mathfrak{z}}(\lambda, x)$ for $0 \leqslant x \leqslant \bar{h}(\lambda)$ may be added as a condition for $u \in K_\lambda$.

In each case, for a minimal λ with $v \in K_\lambda$ certain relations, consisting of equations and inequalities, are derived. For example, in generalizing the proofs in Sections 2 and 4, one may obtain relations of the form $h(\xi) = h'(\xi) = 0$, $h''(\xi) \geqslant 0$ or relations of the form $h(0) = 0$, $h'(0) \geqslant 0$, where the nature of the function h depends on the case considered. (Of course relations involving functionals more general than point functionals may also be used. Compare the proofs in Section II.3.6, for example.)

Finally, properties of Mv are to be formulated which contradict the relations obtained. These properties then describe a set C such that $Mv \in C \Rightarrow v \in K$. They may consist of two-sided estimates or norm estimates of Mv or combinations. In order to write these properties in a manageable form, one may use the tools applied in Sections 2 and 4.

Of course, the set C is not uniquely determined and one may want to prescribe the general form of this set. As an example, let us finally consider the fourth order differential operator M in (II.4.5). We had seen in Section II.4.1.3 that a positive load r always yields a positive solution u of $Mu = r$ if and only if $c \in (-\pi^4, c_0]$ with c_0 in (II.4.6). Now one may not be interested in load functions r of an arbitrary form, but only in those which satisfy certain conditions, such as $|r'(x)| \leqslant \alpha r(x)$, to give just one example. These conditions for $r = Mu$ then describe a set C, and one may ask for which constants c the implication $Mu \in C \Rightarrow u \geqslant o$ holds. This example shows that there are many interesting properties to which the methods described may be applied.

Problems

All possible generalizations which we mentioned above can be carried out as problems. The following two problems are typical examples.

6.1 Let M denote an operator as in (4.40). Find a set C and formulate sufficient conditions on M and ψ such that $Mv \in C \Rightarrow \|v\| \leqslant \psi$, $v \geqslant o$. For this purpose apply the continuity principle using sets K_λ described by (6.4) and corresponding $\mathring{K}_\lambda$.

6.2 For the operator M in the preceding problem derive statements of the form $Mv \in C \Rightarrow \|v(x)\| < \psi(x),\ v(x) \geqslant o\ (0 \leqslant x \leqslant 1)$. Now use $K_\lambda = \{u : \|u(x)\| \leqslant \psi(x), u(x) \geqslant o \text{ for } 0 \leqslant x \leqslant 1 - \lambda\}$. (See the remarks following (6.4).)

6.2 Estimates by Lyapunov functions

Here our object is to describe certain estimates which not only generalize the estimates by pointwise norm bounds in a theoretical sense, but also yield results of practical interest which cannot be obtained by norm estimates. These estimates have the form $V(x, v(x)) \leqslant \psi^2(x)$ with certain functions $V(x, y)$ called *Lyapunov functions*. The case of norm bounds can be incorporated by choosing $V(x, y) = \|y\|^2$. *We must point out, however, that the results of Section* 4 *cannot be obtained by simply inserting this special function V into the succeeding results.*

We restrict ourselves in the following to differential operators of the first order (for second order operators see Problem 6.6). Suppose that $M : R \to S$ is an operator as in (4.40). Moreover, let $v \in R$, $\psi \in C_1[0, l]$, $\psi \geqslant o$, and assume that $V(x, y)$ denotes a continuously differentiable real-valued function on $[0, l] \times \mathbb{R}^n$. We are interested in estimates of the form

$$V(x, v(x)) \leqslant \psi^2(x) \qquad (0 \leqslant x \leqslant l) \tag{6.5}$$

derived from properties of Mv. In particular, v may be a solution of the problem

$$u'(x) + f(x, u(x)) = o \qquad (0 < x \leqslant l), \qquad u(0) = r^0. \tag{6.6}$$

By applying the continuity principle in a special way, we shall derive the following results, where $z \in R_0$ denotes a function satisfying $z \succ o$, $\psi_\lambda = \psi + \lambda z$, and

$$\dot{V}(x, y) := V_x(x, y) - V_y(x, y) f(x, y).$$

Theorem 6.1. *If* $V(0, v(0)) < \psi_\lambda^2(0)$ *for* $\lambda > 0$ *and if*

$$[V(x, v(x))]^{1/2} V_y(x, v(x)) Mv(x) < 2V(x, v(x))\psi_\lambda'(x) - \dot{V}(x, v(x))\psi_\lambda(x) \tag{6.7}$$

for all $x \in (0, l]$ *and* $\lambda > 0$ *satisfying* $V(x, v(x)) = \psi_\lambda^2(x)$, *then* (6.5) *holds.*

Proof. Let $\lambda \geqslant 0$ denote the smallest number $\geqslant 0$ such that $V(x, v(x)) \leqslant \psi_\lambda^2(x)\ (0 \leqslant x \leqslant l)$, and suppose that $\lambda > 0$. Then

$$V(\xi, v(\xi)) = \psi_\lambda^2(\xi) \tag{6.8}$$

for some $\xi \in [0, l]$. This number ξ cannot equal zero. For $\xi > 0$, however, one obtains in addition

$$V_x(\xi, v(\xi)) + V_y(\xi, v(\xi)) v'(\xi) \geqslant 2\psi_\lambda'(\xi)\psi_\lambda(\xi).$$

When in this inequality $v'(\xi)$ is replaced by $Mv(\xi) - f(\xi, v(\xi))$ and, moreover, relation (6.8) is properly used, one obtains an inequality which contradicts (6.7). Hence $\lambda = 0$, so that (6.5) is true. □

As in Theorem 4.2, the inequalities required of ψ_λ may be replaced by separate inequalities for z and ψ. In this way the following statement for solutions of the initial value problem (6.6) is obtained.

Corollary 6.1a. *Suppose that v is a solution of problem* (6.6) *and that the function $z \succ o$ satisfies* $0 < 2V(x, v(x))z'(x) - \dot{V}(x, v(x))z(x)$ $(0 < x \leqslant l)$. *Then*

$$\left.\begin{aligned} 0 \leqslant 2V(x, v(x))\psi'(x) & \\ - \dot{V}(x, v(x))\psi(x) & \qquad (0 < x \leqslant l) \\ V(0, r^0) \leqslant \psi^2(0) & \end{aligned}\right\} \Rightarrow \quad V(x, v(x)) \leqslant \psi^2(x) \quad (0 \leqslant x \leqslant l).$$

For example, $z(x) = \exp(Nx)$ satisfies the above condition, if

$$0 < 2V(x, v(x))N - \dot{V}(x, v(x)) \qquad (0 < x \leqslant l). \tag{6.9}$$

In general, one may want to have an estimate of a norm $\|v\|$ of v, but still apply the above results instead of the theory in Section 4. Of course, such a norm estimate can be achieved, if $\|v(x)\|$ can be estimated by a function of $V(x, v(x))$.

The above theorem and corollary can be used to obtain results on the stability of solutions similarly as Theorem 4.8 was used in Section 4.5.2. As an example, we shall derive a simple statement which can be applied to prove *global* asymptotic stability, where the convergence to a stationary value is proved not only for solutions with initial values r^0 close to the stationary value, but for all $r^0 \in \mathbb{R}^n$, or at least all r^0 in which one is interested. (See also Problems 6.4 and 6.5.)

For simplicity, we consider only an initial value problem

$$u'(x) + f(u(x)) = o, \qquad u(0) = r^0 \tag{6.10}$$

with $f \in C_1^n(\mathbb{R}^n)$ independent of x and assume $f(o) = o$, so that $\bar{u}(x) \equiv o$ is a solution of the differential equation. We write $u(x, r^0)$ for the solution of (6.10) and denote by $d(r^0) \leqslant \infty$ the largest d such that $u(x, r^0)$ is defined on (can be extended to) $[0, d)$. Now let $V(x, y) = V(y)$, too, be independent of x, $V \in C_1(\mathbb{R}^n)$ and $\dot{V}(y) = -V'(y)f(y)$.

Theorem 6.2. *Suppose that for a given $r^0 \neq o$ in $\mathbb{R}^n$ a function $\varphi \in C_1[0, \infty)$ exists which satisfies $\varphi(x) \geqslant 0$ $(0 \leqslant x < \infty)$, $\varphi(0) \geqslant 1$, and*

$$0 \leqslant 2V(u(x, r^0))\varphi'(x) - \dot{V}(u(x, r^0))\varphi(x) \qquad \textit{for} \quad 0 \leqslant x < d \quad \textit{with} \quad d = d(r^0). \tag{6.11}$$

Moreover, let

$$V(y) > 0 \quad \textit{for} \quad y \neq o, \qquad V(y) \to \infty \quad \textit{for} \quad \|y\| \to \infty. \tag{6.12}$$

Then $u(x, r^0)$ *is defined on* $[0, \infty)$ *and*

$$V(u(x, r^0)) \leqslant V(r^0)\varphi^2(x) \quad \textit{for} \quad 0 \leqslant x < \infty. \tag{6.13}$$

Proof. We apply Corollary 6.1a to $v(x) = u(x, r^0)$ and an arbitrary $l \in (0, d)$. Since the solution does not vanish on $[0, l]$ and hence $V(u(x, r^0)) \geqslant c > 0$ $(0 \leqslant x \leqslant l)$, there exists a number N such that (6.9) is satisfied. Therefore, according to the corollary, the inequality in (6.13) holds for all $x \in [0, l]$ and hence for all $x \in [0, d)$.

If $d < \infty$, then $\|u(x, r^0)\|_2 \to \infty$ for $x \to d$. On the other hand, $V(u(x, r^0))$ is bounded on $[0, d)$, due to the estimate already proved. This contradicts the second condition in (6.12), so that necessarily $d = \infty$. □

Since the unknown quantities $u(x, r^0)$ and $d(r^0)$ occur in condition (6.11) for the bound function φ, one need know some further "global" properties of V in order to apply Theorem 6.2. Then this result can be used to improve the bound iteratively, as, for instance, in the following corollary.

Corollary 6.2a. *Suppose that assumption* (6.12) *is satisfied.*

(i) *If* $\dot{V}(y) \leqslant 0$ *for all* $y \in \mathbb{R}^n$, *then* $V(u(x, r^0)) \leqslant V(r^0)$ $(0 \leqslant x < \infty)$.

(ii) *If the assumption in* (i) *holds and a constant* κ_0 *exists such that*

$$2V(y)\kappa_0 + \dot{V}(y) \leqslant 0 \quad \textit{for all} \quad y \in \mathbb{R}^n \quad \textit{with} \quad V(y) \leqslant V(r^0), \tag{6.14}$$

then $V(u(x, r^0)) \leqslant V(r^0) \exp(-2\kappa_0 x)$.

Proof. (i) is verified using $\varphi(x) \equiv 1$. Then (ii) follows using the estimate in (i) and $\varphi(x) = \exp(-\kappa_0 x)$. □

Of course, the iterative process for improving the bound φ may be continued. For example, requiring

$$2V(y)\kappa_1(x) + \dot{V}(y) \leqslant 0 \quad \text{for} \quad y \quad \text{with}$$
$$V(y) \leqslant V(r^0) \exp(-2\kappa_0 x) \quad (0 \leqslant x < \infty)$$

one obtains $V(u(x, r^0)) \leqslant V(r^0) \exp(-2 \int_0^x \kappa_1(t)\, dt)$.

Remark. If it is known that for a given r^0 the values $u(x, r^0)$ $(0 \leqslant x < d(r^0))$ can lie only in a certain domain $\Omega \subset \mathbb{R}^n$, the function V, too, need only be defined on this domain. Then the theorem and corollary above remain true, if some obvious changes are made, such as replacing $y \in \mathbb{R}^n$ with $y \in \Omega$, etc.

Example 6.1. We consider again the differential equations (4.54) in Example 4.16 and assume that $A\bar{y} + b = o$ for some $\bar{y} \succ o$. As before we are interested only in initial values $r^0 \succ o$ (at $x = 0$), for which then $u(x, r^0) \succ o$, as far as this solution is defined. We shall apply Corollary 6.2a to the transformed problem (4.55) for $v(x) = u(x) - \bar{y}$. Here only initial values $s^0 = r^0 - \bar{y} \succ -\bar{y}$ are considered and $v(x, s^0) \succ -\bar{y}$ on the interval of definition. Consequently, $V(y)$ need be defined only for $y \succ -\bar{y}$.

We choose a positive definite diagonal $n \times n$ matrix $D = (d_i\,\delta_{ik})$ and

$$V(y) = \sum_{i=1}^{n} d_i[y_i - \bar{y}_i \log(1 + y_i\bar{y}_i^{-1})].$$

Then

$$-\dot{V}(y) = y^{\mathrm{T}}By \qquad \text{with} \quad B = -\tfrac{1}{2}(DA + A^{\mathrm{T}}D). \tag{6.15}$$

Clearly, $V(y)$ satisfies (6.12). If B is positive semidefinite, it follows from Corollary 6.2a(i) that $v(x, s^0)$ is defined for $0 \leqslant x < \infty$ and $V(v(x, s^0)) \leqslant V(s^0)$ $(0 \leqslant x < \infty)$. If B is even positive definite, the inequality in (6.14) holds for all $y \succ -\bar{y}$ with $V(y) \leqslant V(s^0)$ and a suitable number $\kappa_0 > 0$, which could be computed for a given s^0. Consequently,

$$V(u(x, r^0) - \bar{y}) \leqslant V(r^0 - \bar{y}) \exp(-2\kappa_0 x) \qquad (0 \leqslant x < \infty).$$

One can also determine a constant $\sigma > 0$ such that

$$\|y\|^2 \leqslant \sigma^2 V(y) \qquad \text{for} \quad y \succ -\bar{y} \quad \text{with} \quad V(y) \leqslant V(s^0);$$

and thus obtain estimates of $\|u(x, r^0) - \bar{y}\|$.

In the special case considered next, the occurring constants will be calculated for some values r^0. We state here only that $\lim_{x\to\infty} u(x, r^0) = \bar{y}$ *for all* $r^0 \succ o$, *if a positive definite diagonal matrix* D *exists such that* B *in* (6.15) *is also positive definite.* □

Example 6.2. For the differential equations (4.56) treated in Example 4.17 we calculated $\bar{y}^{\mathrm{T}} = (10, 10)$. Choosing here $D = I$ we obtain $B = \frac{1}{10}\begin{pmatrix} 5 & 0 \\ 0 & 1 \end{pmatrix}$ and $V(y) = k(y_1) + k(y_2)$ with $k(t) = t - 10 \log(1 - 10^{-1}t)$.

Let $s^0 = r^0 - \bar{y}$ be given and define $\alpha \in (0, 10)$ and $\beta > 0$ to be those constants for which $k(-\alpha) = V(s^0) = k(\beta)$. Since $k(0) = k'(0) = 0$ and $k''(t) = 10(10 + t)^{-2}$, one estimates

$$\sigma^{-2}t^2 \leqslant k(t) \leqslant \tau^{-2}t^2 \qquad \text{for} \quad -\alpha \leqslant t \leqslant \beta \quad \text{with}$$

$$\sqrt{5}\sigma = 10 + \beta, \quad \sqrt{5}\tau = 10 - \alpha.$$

Consequently, with $20\kappa_0 = \tau^2$

$$\sigma^{-2}\|y\|_2^2 \leqslant V(y) \leqslant \tfrac{1}{2}\kappa_0^{-1} y^{\mathrm{T}} B y \qquad \text{for} \quad y \succ -\bar{y}, \quad V(y) \leqslant V(s^0).$$

According to the discussion in the preceding example, these inequalities yield

$$\|u(x, r^0) - \bar{y}\|_2 \leqslant \gamma \exp(-\kappa_0 x) \qquad (0 \leqslant x < \infty)$$
$$\text{with} \quad \gamma = \sigma(V(r^0 - \bar{y}))^{1/2}.$$

Table 14 contains some corresponding numerical results for vectors s^0 with components $s_1^0 = s_2^0$. □

TABLE 14

Constants γ, κ_0 describing the error bound in Example 6.2

$s_1^0 = s_2^0$	γ	κ_0
0.001	$1.415E - 03$	$9.997E - 01$
0.01	$1.417E - 02$	$9.972E - 01$
0.1	$1.430E - 01$	$9.721E - 01$
1.0	$1.567E \quad 00$	$7.555E - 01$

Problems

6.3 Apply Corollary 6.1a to the equations in Problem 4.10, using $V(y) = \frac{1}{2}y_2^2 + \int_0^{y_1} g(t)\,dt$. In particular, prove the stability of $u(x) \equiv o$.

In the following two problems the stability theorems of Lyapunov will be incorporated in our estimation theory, at least for the special case where the functions f and V depend only on y and are defined for all $y \in \mathbb{R}^n$. The reader may also prove more general forms of these theorems in an analogous way.

6.4 Suppose that v denotes a nontrivial solution of the differential equation in (6.10). Use Corollary 6.1a to show that $V(v(x)) \leqslant V(r(0))$ $(0 \leqslant x \leqslant 1)$, if (*) $V(0) = 0$, $V(y) > 0$, and $\dot{V}(y) \leqslant 0$ for $o \neq y \in \mathbb{R}^n$. Then employ this estimate to prove the following statement (first stability theorem of Lyapunov): *If* (*) *holds, then the trivial solution is stable.* (Observe that to each $\varepsilon > 0$ there exists a $\mu > 0$ such that $V(y) < \mu \Rightarrow \|y\|_2 < \varepsilon$, and to each $\mu > 0$ there exists a $\delta > 0$ such that $\|y\|_2 < \delta \Rightarrow V(y) \leqslant \mu$.)

6.5 For the differential equation in (6.10), the second stability theorem of Lyapunov assumes the following form. *If* $V(0) = 0$, $\dot{V}(0) = 0$, $V(y) > 0$, $\dot{V}(y) < 0$ *for* $o \neq y \in \mathbb{R}^n$, *then the trivial solution is asymptotically stable.*

This statement can be derived from Corollary 6.1a. The antitone function $V(v(x))$ converges toward a limit $\gamma \geqslant 0$. If $\gamma > 0$, an estimate $V(v(x)) \leqslant V(v(0)) \exp(-2\kappa x)$ for some $\kappa > 0$ can be obtained. If $\gamma = 0$, no sequence $\{x^{(\nu)}\}$ exists such that $x^{(\nu)} \to \infty$ and $\|v(x^{(\nu)})\| \geqslant \varepsilon_0 > 0$. Carry out the details.

6.6 Estimates by means of Lyapunov functions can also be derived for second order differential operators. Suppose that M is defined by (4.6) and $V: \mathbb{R}^n \to \mathbb{R}$ is a twice continuously differentiable function with derivatives $V'(y) \in \mathbb{R}^{1,n}$, $V''(y) \in \mathbb{R}^{n,n}$. Then define

$$W(x) = V(v(x)), \quad \dot{W}(x) = V'(v(x))v'(x),$$

$$\ddot{W}(x) = (v'(x))^{\mathrm{T}} V''(v(x))v'(x) + V'(v(x))f(x, v(x), v'(x)).$$

Moreover, let ψ_λ have the properties described in the first paragraph of Theorem 4.1, with the following exception. Now we require that $W \leqslant \psi_\lambda^2$ for some λ instead of $\|v\|^2 \leqslant \psi_\lambda^2$. Under these conditions prove the following result.

If $W(0) < \psi_\lambda^2(0)$ and $W(1) < \psi_\lambda^2(1)$ for $\lambda > 0$ and if

$$[W(x)]^{1/2} V'(v(x))Mv(x) < -2W(x)\psi_\lambda''(x) - \dot{W}(x)\psi_\lambda'(x) + \ddot{W}(x)\psi_\lambda(x)$$

for all $x \in (0, 1)$ and $\lambda > 0$ satisfying $W(x) = \psi_\lambda^2(x)$ and $\dot{W}(x) = 2\psi_\lambda'(x)\psi_\lambda(x)$, then $W(x) \leqslant \psi^2(x)$ $(0 \leqslant x \leqslant 1)$.

NOTES

Section 2

Differential inequalities which involve order intervals have first been considered by Müller [1927a,b] while investigating initial value problems for first order differential equations (cf. Corollary 2.1b). For such problems the use of such inequalities and also the use of quasi-antitone majorants analogous to those in Problem 2.2 are now a common tool (see, for example, Walter [1970]). The presentation for second order operators essentially is taken from Schröder [1977a]. The results in Example 2.4 are taken from Marcowitz [1973].

Section 3

Most of the results are taken from Schröder [1975b]. A special case where in particular, $f(x, u, u')$ does not depend on u', was already treated by Schröder [1972a]. Bernfeld and Lakshmikantham [1974] considered another special case in which $f(x, y, p)$ is quasi-antitone with respect to y and each component f_i satisfies a Nagumo condition with respect to p_i.

For certain second order functional differential equations on the interval $[0, \infty)$ Lan [1971] proved the existence of a solution u^* with $0 \leqslant u^*(x) \leqslant R$ $(0 \leqslant x < \infty)$, using the modification method and Tychonov's fixed-point theorem. Lan applied his result to certain laminar boundary layer problems, of which the problem in Section 3.4.2 is a simple example. (For such problems see also McLeod and Serrin [1968], Hartman [1972], and the references given therein. Moreover, compare Problem 3.11.) Hübner [1979] carried out the existence proof for the singular problem in Section 3.4.3.

Results of the type considered here were also derived for periodic boundary conditions. Schmitt [1972] considered equations $-u'' + f(x, u) = 0$; Bebernes and Schmitt [1973] treated also equations $-u'' + f(x, u, u') = 0$ obtaining estimates $u^* \in [\varphi, \psi]$, $(u^*)' \in [\Phi, \Psi]$ for constants φ, ψ under certain special conditions on f; Bebernes [1974] investigated the same problem requiring certain a priori estimates for derivatives u'.

Several authors extended the theory of comparison functions to boundary value problems in Banach spaces, in part motivated by applications of the line method to parabolic differential equations. See, for example, Schmitt and Thompson [1975], Thompson [1975], Schmitt and Volkmann [1976], Chandra and Lakshmikantham [1978], and the references in these papers.

Further related results will be discussed in the Notes to Section 6.

Section 4

The theory of this section was developed by Schröder [1977c]. Concerning qualitative results on stability as discussed in Section 4.5.2 and further results in this area see Brauer and Nohel [1969], for example. For mathematical models (4.54) in biology consult Goel *et al.* [1971] and May [1974]. For an application of iterative procedures to such problems see Ames and Adams [1975]. Harrison [1977] applied the method of two-sided bounds to estimate the solution of problem (4.56).

Section 5

For a constant function ψ the results in Section 5.3 are due to Hartman [1960, 1964]. In particular, the proof of Theorem 5.6 and the inequalities in Problem 5.2 and (5.21) were used in Hartman's theory. Lasota and Yorke [1972] replaced the inequality in Problem 5.2 by the condition that $\langle p, p\rangle + \langle y, f(x, y, p)\rangle \geqslant -\gamma(1 + \|x\| + \langle y, y\rangle)$ for some constant γ. The results presented here were proved by Schröder [1977c]. For other generalizations of the results of Hartman [1960] see Hartman [1974], Lan [1975], and the references therein. Also, see the Notes to Section 6. Howes [1979] investigated singularly perturbed semilinear systems, applying a result on norm estimates by Kelley [1977].

Section 6

Lyapunov functions are a standard tool in the stability theory of differential equations (see Brauer and Nohel [1969], for example). Our approach emphasizes the possibility of quantitative estimates. The result at the end of Example 6.1 is due to Goh [1976, 1977]. This author used a Lyapunov function as occurring in this example.

Many authors have considered modifications and generalizations of the results on two-sided estimates in Section III.4 and the results of Hartman contained in Section 5. Most authors were mainly interested in providing a unified approach to obtain theorems from which one can derive the more special results mentioned above and also other special results such as estimates $u(x) \in \Omega$ $(0 \leqslant x \leqslant 1)$ for a suitable set $\Omega \subset \mathbb{R}^n$ (Bebernes and Schmitt [1972], Gustafson and Schmitt [1973]). In some of the papers only periodic boundary conditions are treated. The derivation of more special results from such general theorems is not always straightforward, but often requires to employ additional tools.

Work of this type is contained, for example, in papers by Knobloch [1971], Bebernes [1974], Mawhin [1974], Knobloch and Schmitt [1977], and the book by Gaines and Mawhin [1977]. These authors also employ Lyapunov-type functions; the degree of mapping method is generally used for the existence proof. In particular, Gaines and Mawhin [1977] emphasize the use of topological degree techniques, applying their *coincidence degree* theory. Lyapunov-type functions occur also in Bernfeld and Lakshmikantham [1974] and Bernfeld *et al.* [1975]. In the latter paper, the method of modification is used and its relation to the degree of mapping method is discussed. For applications of Lyapunov-type functions to second order problems, see Hartman [1974], Howes [1979], and Kelley [1977, 1979]. For further results in this area we refer the reader to the references in the papers and books mentioned.

The approach used in this book can also be adapted to problems not explicitly treated here. For example, the results of Section 4 can be generalized to problems for functions $u(x)$ with values in an inner product space. In a similar manner one can develop a still more general theory which covers both two-sided estimates and norm estimates (Schröder [1980]). Moreover, the techniques in Chapters IV and V may be applied to delay differential equations (in $\mathbb{R}$, $\mathbb{R}^n$, or abstract spaces). For references on results of this type see Chandra [1974] and Lakshmikantham *et al.* [1978].

References

Adams, E., and Ames, W. F. [1979]. On contracting interval iteration for nonlinear problems in R^N, *Nonlinear Analysis* **3**,773–794.

Adams, E., and Scheu, G. [1975]. Zur numerischen Konstruktion konvergenter Schrankenfolgen für Systeme nichtlinearer, gewöhnlicher Anfangswertaufgaben, *in* "Interval Mathematics" (K. Nickel, ed.), pp. 279–287. Springer-Verlag, Berlin and New York. See also *Z. Angew. Math. Mech.* **56** (1976), T 270–272.

Adams, E., and Spreuer, H. [1975]. Uniqueness and stability for boundary value problems with weakly coupled systems of nonlinear integro-differential equations and applications to chemical reactions, *J. Math. Anal. Appl.* **49**, 393–410.

Akô, Kiyoshi [1965]. Subfunctions for ordinary differential equations, *J. Fac. Sci. Univ. Tokyo Sect, I*, **12**, 17–43.

Akô, Kiyoshi [1967–1970]. Subfunctions for ordinary differential equations II–V. *Funkcial. Akvac.* **10** (1967), 145–162; **11** (1968), 111–129, 185–195; **12** (1969), 239–249.

Albrecht, J. [1961]. Monotone Iterationsfolgen und ihre Verwendung zur Lösung linearer Gleichungssysteme, *Numer. Math.* **3**, 345–358.

Albrecht, J. [1962]. Fehlerschranken und Konvergenzbeschleunigung bei einer monotonen oder alternierenden Iterationsfolge, *Numer. Math.* **4**, 196–208.

Alefeld, G., and Varga, R. S. [1976]. Zur Konvergenz des symmetrischen Relaxationsverfahrens, *Numer. Math.* **25**, 291–295.

Alexander, A. E., and Johnson, P. [1949]. "Colloid Science." Oxford Univ. Press (Clarendon), London and New York.

Amann, H. [1976]. Fixed point equations and nonlinear eigenvalue problems in ordered Banach spaces, *SIAM Rev.* **18**, 620–709.

Amann, H., and Crandall, N. [1978]. On some existence theorems for semi-linear elliptic equations, *Indiana Univ. Math. J.* **27**, 779–790.

Ames, W. F., and Adams, E. [1975]. Monotonically convergent numerical two-sided bounds for a differential birth and death process, *in* "Interval Mathematics" (K. Nickel, ed.), pp. 135–140. Springer-Verlag, Berlin and New York.

Ames, W. F., and Ginsberg, M. [1975]. Bilateral algorithms and their applications, *in* "Computational Mechanics" (J. T. Oden, ed.), pp. 1–31. Springer-Verlag, Berlin and New York.

Anderson, N., and Arthurs, A. M. [1968]. Variational solutions of nonlinear Poisson–Boltzmann boundary-value problems, *J. Math. Phys.* **9**, 2037–2038.

Ando, T. [1980]. Inequalities for *M*-matrices, *Linear and multilinear Algebra*, to appear.

Aronszajn, N., and Smith, K. T. [1957]. Characterization of positive reproducing kernels. Applications to Green's functions, *Amer. J. Math.* **79**, 611–622.

Babuska, I., Prager, M., and Vitasek, E. [1966]. "Numerical Processes in Differential Equations." Wiley, New York.

Barker, G. P. [1972]. On matrices having an invariant cone, *Czechoslovak Math. J.* **22**(97), 49–68.

Barker, G. P., and Schneider, H. [1975]. Algebraic Perron–Frobenius theory, *Linear Algebra Appl.* **11**, 219–233.

Barrett, J. H. [1962]. Two-point boundary problems for linear self-adjoint differential equations of the fourth order with middle term, *Duke Math. J.* **29**, 543–554.

Bauer, F. L. [1974]. Positivity and Norms I. Lecture Notes, Techn. Univ. München.

Bazley, N. W., and Fox, D. W. [1966]. Error bounds for approximations to expectation values of unbounded operators, *J. Math. Phys.* **7**, 413–416.

Beauwens, R. [1976]. Semistrict diagonal dominance, *SIAM J. Numer. Anal.* **13**, 109–112.

Beauwens, R. [1979]. Factorization iterative methods, *M*-operators and *H*-operators, *Numer. Math.* **31**, 335–357.

Bebernes, J. W. [1963]. A subfunction approach to a boundary value problem for ordinary differential equations, *Pacific J. Math.* **13**, 1053–1066.

Bebernes, J. W. [1974]. A simple alternative problem for finding periodic solutions of second order ordinary differential systems, *Proc. Amer. Math. Soc.* **42**, 121–127.

Bebernes, J. W., and Schmitt, K. [1972]. An existence theorem for periodic boundary value problems for systems of second order differential equations, *Arch. Mat. Brno* **8**, 173–176.

Bebernes, J. W., and Schmitt, K. [1973]. Periodic boundary value problems for systems of second order differential equations, *J. Differential Equations* **13**, 32–47.

Bebernes, J. W., and Wilhelmsen, R. [1971]. A remark concerning a boundary value problem, *J. Differential Equations* **10**, 389–391.

Beckenbach, E. F., and Bellman, R. [1961]. "Inequalities." Springer-Verlag, Berlin and New York.

Bellman, R. [1957]. On the non-negativity of Green's functions, *Boll. Un. Mat.* **12**, 411–413.

Bellmann, R. E., and Kalaba, R. E. [1965]. "Quasilinearization and Nonlinear Boundary-value Problems." American Elsevier, New York.

Ben-Israel, A. [1970]. On cone-monotonicity of complex matrices, *SIAM Rev.* **12**, 120–123.

Berman, A., and Plemmons, R. J. [1972]. Monotonicity and the generalized inverse, *SIAM J. Appl. Math.* **22**, 155–161.

Berman, A., and Plemmons, R. J. [1979]. "Nonnegative Matrices in the Mathematical Sciences." Academic Press, New York.

Bernfeld, S. R., and Chandra, J. [1977]. Minimal and maximal solutions of nonlinear boundary value problems, *Pacific J. Math.* **71**, 13–20.

Bernfeld, S. R., and Lakshmikantham, V. [1974]. "An Introduction to Nonlinear Boundary Value Problems." Academic Press, New York.

Bernfeld, S. R., Ladde, G. S., and Lakshmikantham, V. [1975]. Existence of solutions of two point boundary value problems for nonlinear systems, *J. Differential Equations* **18**, 103–110.

Beurling, A., and Deny, J. [1958]. Espaces de Dirichlet I. Le cas élémentaire, *Acta Math.* **99**, 203–224.

Beyn, W. J. [1978]. Schwach majorisierende Elemente und die besonderen Monotonie-Eigenschaften von Randwertaufgaben zweiter Ordnung, Preprint, Univ. of Münster.

Bodewig, E. [1959]. "Matrix Calculus." Wiley (Interscience), New York.

Bohl, E. [1968]. An iteration method and operators of monotone type, *Arch. Rational Mech. Anal.* **29**, 395–400.

Bohl, E. [1974]. "Monotonie: Lösbarkeit und Numerik bei Operatorgleichungen." Springer-Verlag, Berlin and New York.

Bohl, E., and Lorenz, J. [1979]. Inverse monotonicity and difference schemes of higher order. A summary for two-point boundary value problems, *Aequationes Math.* **19**, 1–36.

Bramble, J. H., and Hubbard, B. E. [1964a]. New monotone type approximations for elliptic problems, *Math. Comp.* **18**, 349–367.

Bramble, J. H., and Hubbard, B. E. [1964b]. On a finite difference analogue of an elliptic boundary problem which is neither diagonally dominant nor of non-negative type, *J. Math. Phys.* **43**, 117–132.

Brauer, F., and Nohel, J. A. [1969]. "The Qualitative Theory of Ordinary Differential Equations." Benjamin, New York.

Briš, N. I. [1954]. On boundary problems for the equation $\varepsilon y'' = f(x, y, y')$ for small ε's, *Dokl. Akad. Nauk SSSR* **95**, 429–432 [NASA translation NASA TT F-10, 839 (1967)].

Bromberg, E. [1956]. Nonlinear bending of a circular plate under normal pressure, *Commun. Pure Appl. Math.* **9**, 633–659.

Browder, F. E. [1963]. Non-linear elliptic boundary value problems, *Bull. Amer. Math. Soc.* **69**, 862–874.

Burger, E. [1957]. Eine Bemerkung über nicht-negative Matrizen, *Z. Angew. Math. Mech.* **37**, 227.

Čapligin, S. A. [1950]. "New Methods in the Approximate Integration of Differential Equations" (in Russian). Gosudarstv. Izdat. Tehn.-Teoret. Lit., Moscow.

Chandra, J. [1974]. A comparison result for a boundary value problem for a class of nonlinear differential equations with deviating argument, *J. Math. Anal. Appl.* **47**, 573–577.

Chandra, J., and Davis, P. Wm. [1974]. A monotone method for quasilinear boundary value problems, *Arch. Rational Mech. Anal.* **54**, 257–266.

Chandra, J., and Lakshmikantham, V. [1978]. Comparison Principle and Nonlinear Boundary Value Problems in Banach Spaces. Preprint.

Čičkin, E. S. [1960]. A non-oscillation theorem for a linear self-adjoint differential equation of fourth order (Russian), *Izv. Vysš. Učebn. Zaved. Matematika* No. 4 (27), 206–209, MR 25 #1348.

Čičkin, E. S. [1962]. A theorem on a differential inequality for multi-point boundary-value problems (Russian), *Izv. Vysš. Učebn. Zaved. Matematika* No. 2 (27), 170–179, MR 26 #1541.

Clément, Ph., and Peletier, L. A. [1979]. On positive, concave solutions of two point nonlinear eigenvalue problems, *J. Math. Anal. Appl.* **69**, 329–340.

Cohen, D. S. [1967]. Positive solutions of a class of nonlinear eigenvalue problems, *J. Math. Mech.* **17**, 209–215.

Cohen, D. S. [1971]. Multiple stable solutions of nonlinear boundary value problems arising in chemical reactor theory, *SIAM J. Appl. Math.* **20**, 1–13.

Cohen, D. S. [1972]. Multiple solutions of singular perturbation problems, *SIAM J. Math. Anal.* **3**, 72–82.

Cohen, D. S., and Laetsch, T. W. [1970]. Nonlinear boundary value problems suggested by chemical reactor theory, *J. Differential Equations* **7**, 217–226.

Collatz, L. [1942]. Einschließungssatz für die charakteristischen Zahlen von Matrizen, *Math. Z.* **48**, 221–226.

Collatz, L. [1950]. Über die Konvergenzkriterien bei Iterationsverfahren für lineare Gleichungssysteme, *Math. Z.* **53**, 149–161.

Collatz, L. [1952]. Aufgaben monotoner Art, *Arch. Math.* **3**, 366–376.

Collatz, L. [1958]. Fehlerabschätzungen bei Randwertaufgaben partieller Differentialgleichungen mit unendlichem Grundgebiet, *Z. Angew. Math. Phys.* **9**, 118–128.

Collatz, L. [1960]. "The Numerical Treatment of Differential Equations." Springer-Verlag, Berlin and New York.

Collatz, L. [1963]. "Eigenwertaufgaben mit technischen Anwendungen." Akademische Verlagsgesellschaft, Leipzig.

Collatz, L. [1964]. "Funktionalanalysis und Numerische Mathematik." Springer-Verlag, Berlin and New York.

Collatz, L., and Schröder, J. [1959]. Einschließen der Lösungen von Randwertaufgaben, *Numer. Math.* **1**, 61–72.

Coppel, W. A. [1971]. "Disconjugacy." Springer-Verlag, Berlin and New York.

Cottle, R. W. [1975]. On Minkowski matrices and the linear complementary problem, *in* "Optimization and Optimal Control" (R. Bulirsch, W. Oettli, and J. Stoer, eds.), pp. 18–26. Springer-Verlag, Berlin and New York.

Cottle, R. W., and Veinott, A. F. [1972]. Polyhedral sets having a least element, *Math. Programming* **3**, 238–249.

Courant, R., and Hilbert, D. [1962]. "Methods of Mathematical Physics," Vol. II. Wiley (Interscience), New York.

Deimling, K. [1977]. "Ordinary Differential Equations in Banach Spaces," Lecture Notes 596. Springer-Verlag, Berlin and New York.

Deimling, K., and Lakshmikantham, V. [1979a]. On existence of extremal solutions of differential equations in Banach spaces, *Nonlinear Analysis* **3**, 563–568.

Deimling, K., and Lakshmikantham, V. [1979b]. Existence and comparison theorems for differential equations in Banach spaces, *Nonlinear Analysis* **3**, 569–576.

Demetrius, L. [1974]. Invariant ideals of positive operators, *SIAM Rev.* **16**, 428–440.

de Rham, G. [1952]. Sur un théorème de Stieltjes relatif à certaines matrices, *Acad. Serbe Sci. Publ. Inst. Math.* **4**, 133–134.

Deuel, J., and Hess, P. [1975]. A criterion for the existence of solutions of non-linear elliptic boundary value problems, *Proc. Roy. Soc. Edinburgh* **74A** (1974/75), 49–54.

Douglas, J., Dupont, T., and Serrin, J. [1971]. Uniqueness and comparison theorems for nonlinear elliptic equations in divergence form, *Arch. Rational Mech. Anal.* **42**, 157–168.

Edwards, R. E. [1965]. "Functional Analysis." Holt, New York.

Egervàry, E. [1954]. On a lemma of Stieltjes on matrices, *Acta Sci. Math.* (*Szeged*) **15**, 99–103.

Elsner, L. [1970]. Monotonie und Randspektrum bei vollstetigen Operatoren, *Arch. Rational Mech. Anal.* **36**, 356–365.

Epheser, H. [1955]. Über die Existenz der Lösungen von Randwertaufgaben mit gewöhnlichen, nichtlinearen Differentialgleichungen zweiter Ordnung, *Math. Z.* **61**, 435–454.

Erbe, L. H. [1970]. Nonlinear boundary value problems for second order differential equations, *J. Differential Equations* **7**, 459–472.

Falkner, V. M., and Skan, S. W. [1931]. Solutions of the boundary layer equations, *Philos. Mag.* **12**, 865–896.

Fan, K. [1958]. Topological proofs for certain theorems on matrices with non-negative elements, *Monatsh. Math.* **62**, 219–237.

Feynman, R. P., Leighton, R. B., and Sands, M. [1966]. "The Feynman Lectures on Physics," Vol. 2. Addison-Wesley, Reading, Massachusetts.

Fiedler, M., and Pták, V. [1962]. On matrices with nonpositive off-diagonal elements and positive principal minors, *Czechoslovak Math. J.* **12** (87), 382–400.

Fiedler, M., and Pták, V. [1966]. Some generalizations of positive definiteness and monotonicity, *Numer. Math.* **9**, 163–172.

Fountain, L., and Jackson, L. [1962]. A generalized solution of the boundary value problem for $y'' = f(x, y, y')$, *Pacific J. Math.* **12**, 1251–1272.

Friedman, A. [1969]. "Partial Differential Equations." Holt, New York.

Frobenius, C. [1908]. Über Matrizen aus positiven Elementen, *Sitzungsber. Preuss. Akad. Wiss. Berlin* 471–476; (1909), 514–518.

Frobenius, C. [1912]. Über Matrizen aus nicht-negativen Elementen, *Sitzungsber. Preuss. Akad. Wiss. Berlin* 456–477.

Gaines, R. E. [1972]. A priori bounds and upper and lower solutions for nonlinear second-order boundary-value problems, *J. Differential Equations* **12**, 291–312.

Gaines, R. E., and Mawhin, J. L. [1977]. "Coincidence Degree, and Nonlinear Differential Equations." Springer-Verlag, Berlin and New York.

Gantmacher, F. R., and Krein, M. G. [1960]. "Oszillationsmatrizen, Oszillationskerne und kleine Schwingungen mechanischer Systeme." Akademie-Verlag, Berlin.

Gautschi, W. [1975]. Computational methods in special functions—a survey, *in* "Theory and Applications of Special Functions" (A. Askey, ed.), pp. 1–98. Academic Press, New York.

Geiringer, H. [1948]. "On the Solution of Systems of Linear Equations by Certain Iteration Methods," Reissner Anniversary Volume, pp. 365–393. Edwards, Ann Arbor, Michigan.

Gerschgorin, S. [1930]. Fehlerabschätzung für das Differenzenverfahren zur Lösung partieller Differentialgleichungen, *Z. Angew. Math. Mech.* **10**, 373–382.

Gerschgorin, S. [1931]. Über die Abgrenzung der Eigenwerte einer Matrix, *Izv. Akad. Nauk SSSR Ser. Mat.* **7**, 749–754.

Glasshoff, K., and Werner, B. [1979]. Inverse order-monotonicity of monotone L-operators with applications to quasilinear and free boundary value problems, *J. Math. Anal. Appl.*, to appear.

Goel, N. S., Maitra, C. S., and Montrall, E. W. [1971]. "Nonlinear Models of Interacting Populations." Academic Press, New York.

Goh, B. S. [1976]. Global stability in two species interactions, *J. Math. Biol.* **3**, 313–318.

Goh, B. S. [1977]. Global stability in many-species systems, *The American Naturalist* **III**, No. 977, 135–143.

Goheen, H. E. [1949]. On a lemma of Stieltjes on matrices, *Amer. Math. Monthly* **56**, 328–329.

Gorenflo, R. [1973]. Über S. Gerschgorins Methode der Fehlerabschätzung bei Differenzenverfahren, *in* "Numerische, insbesondere approximationstheoretische Behandlung von Funktionalgleichungen" (R. Ansorge and W. Törnig, eds.), pp. 128–143, Lecture Notes 333. Springer-Verlag, Berlin and New York.

Greenspan, D., and Parter, S. V. [1965]. Mildly nonlinear elliptic partial differential equations and their numerical solution II, *Numer. Math.* **7**, 129–146.

Gustafson, G. B., and Schmitt, K. [1973]. Periodic solutions of hereditary differential systems, *J. Differential Equations* **13**, 567–587.

Hadley, G. [1962]. "Linear Programming." Addison-Wesley, Reading, Massachusetts.

Harrison, G. W. [1977]. Dynamic Models with Uncertain Parameters, Preprint, Institute of Ecology, Univ. of Georgia, Athens, Georgia.

Hartman, P. [1960]. On boundary value problems for systems of ordinary, nonlinear, second order differential equations, *Trans. Amer. Math. Soc.* **96**, 493–509.

Hartman, P. [1964]. "Ordinary Differential Equations." Wiley, New York.

Hartman, P. [1972]. On the swirling flow problem, *Indiana Univ. Math. J.* **21**, 849–855.

Hartman, P. [1974]. On two-point boundary value problems for nonlinear second order systems, *SIAM J. Math. Anal.* **5**, 172–177.

Heidel, J. W. [1974]. A second-order nonlinear boundary value problem, *J. Math. Anal. Appl.* **48**, 493–503.

Hess, P. [1975]. Problèmes aux limites non linéaires dans les domaines non bornés, *C. R. Acad. Sci. Paris* **281A**, 555–557.

Hess, P. [1976]. On the solvability of nonlinear elliptic boundary value problems, *Indiana Univ. Math. J.* **25**, 461–466.

Hess, P. [1977]. Nonlinear elliptic problems in unbounded domains, *in* "Theory of Nonlinear Operators" (R. Kluge, ed.), pp. 105–110. Akademie-Verlag, Berlin.

Hess, P. [1978]. On a second-order nonlinear elliptic boundary value problem, *in* "Nonlinear Analysis" (L. Cesari, R. Kannan, and H. F. Weinberger, eds.), pp. 99–107, Academic Press, New York.

Hopf, E. [1927]. Elementare Bemerkungen über die Lösung partieller Differentialgleichungen zweiter Ordnung vom elliptischen Typus, *Sitzungsber. Preuss. Akad. Wiss.* **19**, 147–152.

Householder, A. S. [1958]. The approximate solution of matrix problems, *J. Assoc. Comput. Mech.* **5**, 205–243.

Householder, A. S. [1964]. "The Theory of Matrices in Numerical Analysis." Blaisdell, New York.

Howes, F. A. [1978]. Boundary-interior layer interactions in nonlinear singular perturbation theory, *Mem. Amer. Math. Soc.* **203**.

Howes, F. A. [1979]. Singularly perturbed semilinear systems, *Stud. Appl. Math.* **61**, 185–209.

Hübner, G. [1979]. Vektorielle Randwertaufgaben 2. Ordnung bei gewöhnlichen Differentialgleichungen, Diplomarbeit, Math. Institut, Univ. Köln.

Jackson, L. K. [1968]. Subfunctions and second-order ordinary differential inequalities, *Advances in Math.* **2**, 307–363.

Jackson, L. K., and Klaasen, G. [1971]. A variation of the topological method of Wazewski, *SIAM J. Appl. Math.* **20**, 124–130.

Jackson, L. K., and Schrader, K. W. [1967]. Comparison theorems for nonlinear differential equations, *J. Differential Equations* **3**, 248–255.

Jameson, G. [1970]. "Ordered Linear Spaces." Springer-Verlag, Berlin and New York.

Johnston, J. B., Price, G. B., and van Vleck, F. S. [1966]. "Linear Equations and Matrices." Addison-Wesley, Reading, Massachusetts.

Kahan, W. [1958]. Gauß–Seidel methods of solving large systems of linear equations, Doctoral Thesis, Univ. of Toronto.

Kamke, E. [1932]. Zur Theorie der Systeme gewöhnlicher Differentialgleichungen II, *Acta Math.* **58**, 57–85.

Kamke, E. [1951]. "Differentialgleichungen, Lösungsmethoden und Lösungen I." Akademische Verlagsgesellschaft, Leipzig.

Kaneko, I. [1978]. Linear complimentary problems and characterizations of Minkowski matrices, *Linear Algebra and Appl.* **20**, 111–129.

Kantorovič, L. [1939]. The method of successive approximations for functional equations, *Acta Math.* **71**, 63–97.

Karlin, S. [1959]. "Mathematical Methods and Theory in Games, Programming and Economics I, II." Addison-Wesley, Reading, Massachusetts.

Karlin, S. [1968]. "Total Positivity I." Stanford Univ. Press, Stanford, California.

Keller, H. B. [1969a]. Elliptic boundary value problems suggested by nonlinear diffusion processes, *Arch. Rational Mech. Anal.* **35**, 363–381.

Keller, H. B. [1969b]. Positive solutions of some nonlinear eigenvalue problems, *J. Math. Mech.* **19**, 279–295.

Keller, H. B. [1972]. Existence theory for multiple solutions of a singular perturbation problem, *SIAM J. Math. Anal.* **3**, 86–92.

Keller, H. B., and Cohen, D. S. [1967]. Some positone problems suggested by nonlinear heat generation, *J. Math. Mech.* **16**, 1361–1376.

Keller, H. B., and Reiss, E. L. [1958]. Iterative solutions for the non-linear bending of circular plates, *Commun. Pure. Appl. Math.* **11**, 273–292.

Kelley, J. L., and Namioka, I. [1963]. "Linear Topological Spaces." van Nostrand-Reinhold, Princeton, New Jersey.

Kelley, W. G. [1977]. A geometric method of studying two point boundary value problems for second order systems, *Rocky Mountain J. Math.* **7**, 251–263.

Kelley, W. G. [1979]. A nonlinear singular perturbation problem for second order systems, *SIAM J. Math. Anal.* **10**, 32–37.

Klee, V. [1968]. Maximal separation theorems for convex sets, *Trans. Amer. Math. Soc.* **134**, 133–147.

Klee, V. [1969]. Separation and support properties of convex sets—a survey, *in* "Control Theory and the Calculus of Variations" (A. V. Balakrishnan, ed.), pp. 235–303. Academic Press, New York.

Knobloch, H. W. [1963]. Eine neue Methode zur Approximation periodischer Lösungen nichtlinearer Differentialgleichungen zweiter Ordnung, *Math. Z.* **82**, 177–197.

Knobloch, H. W. [1969]. Second order differential inequalities and a nonlinear boundary value problem, *J. Differential Equations* **4**, 55–71.

Knobloch, H. W. [1971]. On the existence of periodic solutions for second order vector differential equations, *J. Differential Equations* **9**, 67–85.

Knobloch, H. W., and Schmitt, K. [1977]. Non-linear boundary value problems for systems of differential equations, *Proc. Roy. Soc. Edinburgh Sect. A* **78**, 139–159.

Koehler, G. J., Whinston, A. B., and Wright, G. P. [1975]. "Optimization Over Leontief Substitution Systems." American Elsevier, New York.

Köthe, G. [1966]. "Topologische Lineare Räume I." Springer-Verlag, Berlin and New York.

Kotelyanskii, D. M. [1952]. Some properties of matrices with positive elements, *Mat. Sb.* **31** (73), 497–506.

Krasnosel'skii, M. A. [1964a]. "Positive Solutions of Operator Equations." Noordhoff, Groningen.

Krasnosel'skii, M. A. [1964b]. "Topological Methods in the Theory of Nonlinear Integral Equations." Macmillan, New York.

Krasnosel'skii, M. A. [1968]. "Translation Along Trajectories of Differential Equations," Transl. of Russian Monographs Ser. Vol. **19**. American Mathematical Society, Providence, Rhode Island.

Krein, M. G. [1939]. Sur les fonctions de Green nonsymétriques oscillatoires des opérateurs différentiels ordinaires, *C. R. Dokl. Acad. Sci. URSS* **25**, 643–646.

Krein, M. G., and Rutman, M. A. [1948]. Linear operators leaving invariant a cone in a Banach space, *Usp. Mat. Nauk* **3**, No. 1 (23), 3–95; [*Amer. Math. Soc. Transl.* **26** (1950)].

Küpper, T. [1976a]. Einschließungsaussagen für gewöhnliche Differentialoperatoren, *Numer. Math.* **25**, 201–214.

Küpper, T. [1976b]. Pointwise lower and upper bounds for eigenfunctions of ordinary differential operators, *Numer. Math.* **27**, 111–119.

Küpper, T. [1978a]. Einschließungsaussagen bei Differentialoperatoren zweiter Ordnung durch punktweise Ungleichungen, *Numer. Math.* **30**, 93–101.

Küpper, T. [1978b]. Einschließungsaussagen für Differentialoperatoren vierter Ordnung und ein Verfahren zur Berechnung von Schrankenfunktionen, *Manuscripta Math.* **26**, 259–291.

Küpper, T. [1979]. On minimal nonlinearities which permit bifurcation from the continuous spectrum. *Math. Mech. in the Appl. Sci.* **1**, 572–580.

Lakshmikantham, V. [1979]. Comparison results for reaction-diffusion equations in a Banach space, *in* Proc. Conf. "A Survey on the Theoretical and Numerical Trends in Nonlinear Analysis." GIUS. LATERZA & FIGLI S.p.A., Bari, 121–156.

Lakshmikantham, V., and Leela, S. [1969]. "Differential and Integral Inequalities I." Academic Press, New York.

Lakshmikantham, V., Leela, S., and Moauro, V. [1978]. Existence and uniqueness of solutions of delay differential equations on a closed subset of a Banach space, *Nonlinear Analysis* **2**, 311–327.

Lan, C. C. [1971]. On functional-differential equations and some laminar boundary layer problems, *Arch. Rational Mech. Anal.* **42**, 24–39.

Lan, C. C. [1975]. Boundary value problems for second and third order differential equations. *J. Differential Equations* **18**, 258–274.

Lasota, A., and Yorke, J. A. [1972]. Existence of solutions of two-point boundary value problems for nonlinear systems, *J. Differential Equations* **11**, 509–518.

Lemmert, R. [1977]. Über die Invarianz einer konvexen Menge in Bezug auf Systeme von gewöhnlichen, parabolischen und elliptischen Differentialgleichungen. *Math. Ann.* **230**, 49–56.

Leontief, W. W. [1951]. "The Structure of American Economy 1919–1939," 2d ed. Oxford Univ. Press, London and New York.

Levin, A. Ju. [1963]. Some problems bearing on the oscillation of solutions of linear differential equations, *Sov. Math.* **4**, 121–124.

Levin, A. Ju. [1964]. Distribution of the zeros of solutions of a linear differential equation, *Sov. Math.* **5**, 818–821.

Levin, A. Ju. [1969]. Non-oscillation of solutions of the equation $x^{(n)} + p_1(t)x^{(n-1)} + \cdots + p_n(t)x = 0$, *Russ. Math. Surveys* **24**, 43–99.

Lions, J. [1969]. "Quelques Méthodes de Résolution des Problèmes aux Limites non Linéaires." Dunod, Gauthier-Villars, Paris.

Littman, W. [1959]. A strong maximum principle for weakly L-subharmonic functions, *J. Math. Mech.* **8**, 761–770.

Littman, W. [1963]. Generalized subharmonic functions: monotonic approximations and an improved maximum principle, *Ann. Scuola Norm. Sup. Pisa* (3) **17**, 207–222.

Lorenz, J. [1977]. Zur Inversmonotonie diskreter Probleme, *Numer. Math.* **27**, 227–238.

Luss, D. [1968]. Sufficient conditions for uniqueness of the steady state solutions in distributed parameter systems, *Chem. Eng. Sci.* **23**, 1249–1255.

McLeod, J. B. [1971]. The existence of axially symmetric flows above a rotating disc, *Proc. Roy. Soc. Ser. A* **324**, 391–414.

McLeod, J. B., and Serrin, J. [1968]. The existence of similar solutions for some laminar boundary layer problems, *Arch. Rational Mech. Anal.* **31**, 288–303.

McNabb, A. [1961]. Strong comparison theorems for elliptic equations of second order, *J. Math. Mech.* **10**, 431–440.

Mammana, G. [1931]. Decomposizione delle espressioni differenziali lineari omogenee in prodotti di fattori simbolici e applicazione relativa allo studio delle equazioni differenziali lineari, *Math. Z.* **33**, 186–231.

Mangasarian, O. L. [1968]. Characterizations of real matrices of monotone kind, *SIAM Rev.* **10**, 439–441.

Marcowitz, U. [1973]. Fehlerabschätzung bei Anfangswertaufgaben für Systeme von gewöhnlichen Differentialgleichungen mit Anwendung auf das REENTRY-Problem. Diss. Math. Institut, Univ. Köln; partly published in *Numer. Math.* **24** (1975), 249–275.

Marek, I. [1967]. u_0-positive operators and some of their applications, *SIAM J. Appl. Math.* **15**, 484–494.

Markus, L., and Amundson, N. R. [1968]. Nonlinear boundary-value problems arising in chemical reactor theory, *J. Differential Equations* **4**, 102–113.

Martin, R. H., Jr. [1975]. Remarks on differential inequalities in Banach spaces, *Proc. Amer. Math. Soc.* **53**, 65–71.

Mawhin, J. [1974]. Boundary value problems for nonlinear second-order vector differential equations, *J. Differential Equations* **16**, 257–269.

May, R. M. [1974]. "Stability and Complexity in Model Ecosystems," 2nd ed. Princeton Univ. Press, Princeton, New Jersey.

Mehmke, R. [1892]. Über das Seidelsche Verfahren, um lineare Gleichungen bei einer sehr großen Anzahl der Unbekannten durch sukzessive Annäherung aufzulösen, *Mosk. Math. Samml.* **16**, 342–345.

Metelmann, K. [1973]. Ein Kriterium für den Nachweis der Totalnichtnegativität von Bandmatrizen, *Linear Alg. Appl.* **7**, 163–171.

Meyer, Jr., C. D., and Stadelmaier, M. W. [1978]. Singular M-matrices and inverse positivity, *Linear Alg. Appl.* **22**, 139–156.

Meyn, K. H., and Werner, B. [1979]. Maximum and monotonicity principles for elliptic boundary value problems in partitioned domains, *Applicable Anal.*, to appear.

Minty, G. I. [1962]. Monotone (nonlinear) operators in Hilbert space, *Duke Math. J.* **29**, 341–346.

Moore, R. E. [1975]. Two-sided approximation to solutions of nonlinear operator equations—a comparison of methods from classical analysis, functional analysis, and interval analysis, *in* "Interval Mathematics" (K. Nickel, ed.), pp. 31–47. Springer-Verlag, Berlin and New York.

Moré, J. J. [1972]. Nonlinear generalizations of matrix diagonal dominance with application to Gauss–Seidel iterations, *SIAM J. Numer. Anal.* **9**, 357–378.

Moré, J., and Rheinboldt, W. C. [1973]. On P- and S-functions and related classes of n-dimensional nonlinear mappings, *Linear Alg. Appl.* **6**, 45–68.

Müller, K. H. [1971]. Zum schwachen Zeilensummenkriterium bei nichtlinearen Gleichungssystemen, *Computing* **7**, 153–171.

Müller, M. [1927a]. Über das Fundamentaltheorem in der Theorie der gewöhnlichen Differentialgleichungen, *Math. Z.* **26**, 619–645.

Müller, M. [1927b]. Über die Eindeutigkeit der Integrale eines Systems gewöhnlicher Differentialgleichungen und die Konvergenz einer Gattung von Verfahren zur Approximation dieser Integrale, *Sitzungsber. Heidelberger Akad. Wiss., Math.-Naturwiss. Kl.* **9**, 3–38.

Murray, J. D. [1968]. A simple method for obtaining approximate solutions for a class of diffusion-kinetics enzyme problems: I. General class and illustrative examples, *Math. Biosci.* **2**, 379–411.

Nagumo, M. [1937]. Über die Differentialgleichung $y'' = f(x, y, y')$, *Proc. Phys.-Math. Soc. Japan* **19**, 861–866.

Nagumo, M. [1942]. Über das Randwertproblem der nicht linearen gewöhnlichen Differentialgleichungen zweiter Ordnung, *Proc. Phys.-Math. Soc. Japan* (3) **24**, 845–851.

Nekrasov, P. A. [1892]. Zum Problem der Auflösung von linearen Gleichungssystemen mit einer großen Anzahl von Unbekannten durch sukzessive Approximation, *Ber. Petersburger Akad. Wiss.* **5**, 1–18.

Neumann, M., and Plemmons, R. J. [1978]. Convergent nonnegative matrices and iterative methods for consistent linear systems, *Numer. Math.* **31**, 265–279.

Nickel, K. L. [1976]. Stability and convergence of monotonic algorithms, *J. Math. Anal. Appl.* **54**, 157–172.

Nickel, K. [1979]. The construction of a priori bounds for the solution of a two point boundary value problem with finite elements I, *Computing* **23**, 247–265.

Ortéga, J. M., and Rheinboldt, W. C. [1970]. "Iterative Solution of Nonlinear Equations in Several Variables." Academic Press, New York.

Ostrowski, A. [1937]. Über die Determinanten mit überwiegender Hauptdiagonale, *Comment. Math. Helv.* **10**, 69–96.

Ostrowski, A. [1956]. Determinanten mit überwiegender Hauptdiagonale und die absolute Konvergenz von linearen Iterationsprozessen, *Comment. Math. Helv.* **30**, 175–210.

Parter, S. V. [1965]. Mildly nonlinear elliptic differential equations and their numerical solution I, *Numer. Math.* **7**, 113–128.

Payne, L. E. [1976]. Some remarks on maximum principles, *J. Analyse Math.* **30**, 421–433.

Peixoto, M. M. [1949]. Generalized convex functions and second order differential inequalities, *Bull. Amer. Math. Soc.* **55**, 563–572.

Perron, O. [1907]. Zur Theorie der Matrices, *Math. Ann.* **64**, 248–263.

Plemmons, R. J. [1972]. Monotonicity and iterative approximations involving rectangular matrices, *Math. Comp.* **26**, 853–858.

Plemmons, R. J. [1976]. M-matrices leading to semiconvergent splittings, *Linear Alg. Appl.* **15**, 243–252.

Plemmons, R. J. [1977]. M-matrix characterizations I—nonsingular M-matrices, *Linear Alg. Appl.* **18**, 175–188.

Pólya, G. [1922]. On the mean-value theorem corresponding to a given linear homogeneous differential equation, *Trans. Amer. Math. Soc.* **24**, 312–324.

Pólya, G., and Szegö, G. [1925]. "Aufgaben und Lehrsätze aus der Analysis I." Springer-Verlag, Berlin and New York.

Poole, G., and Boullion, T. [1974]. A survey on M-matrices, *SIAM Rev.* **16**, 419–427.

Price, H. S. [1968]. Monotone and oscillation matrices applied to finite difference approximations, *Math. Comp.* **22**, 489–516.

Protter, M. H., and Weinberger, H. F. [1967]. "Maximum Principles in Differential Equations." Prentice-Hall, Englewood Cliffs, New Jersey.

Redheffer, R. M. [1962a]. An extension of certain maximum principles, *Monatsh. Math.* **66**, 32–42.

Redheffer, R. M. [1962b]. Eindeutigkeitssätze bei nichtlinearen Differentialgleichungen, *J. Reine Angew. Math.* **211**, 70–77.

Redheffer, R. M. [1973]. Gewöhnliche Differentialungleichungen mit quasimonotonen Funktionen in normierten linearen Räumen, *Arch. Rational Mech. Anal.* **52**, 121–133.

Redheffer, R. M. [1977]. Fehlerabschätzung bei nichtlinearen Differentialungleichungen mit Hilfe linearer Differentialungleichungen, *Numer. Math.* **28**, 393–405.

Redheffer, R. M., and Walter, W. [1975]. Flow-invariant sets and differential inequalities in normed spaces, *Appl. Anal.* **5**, 149–161.

Reed, M., and Simon, B. [1978]. "Methods of Modern Mathematical Physics IV, Analysis of Operators." Academic Press, New York.

Rheinboldt, W. C. [1970]. On M-functions and their application to nonlinear Gauss–Seidel iterations and to network flows, *J. Math. Anal. Appl.* **32**, 274–307.

Rheinboldt, W. C., and Vandergraft, J. S. [1973]. A simple approach to the Perron–Frobenius theory for positive operators on general partially-ordered finite-dimensional linear spaces, *Math. Comp.* **27**, 139–145.

Robert, F. [1976]. Autour du théorème de Stein–Rosenberg, *Numer. Math.* **27**, 133–141.

Rockafellar, R. T. [1968]. "Convex Analysis." Princeton Univ. Press, Princeton, New Jersey.

Russell, R. D., and Shampine, L. F. [1972]. A collocation method for boundary value problems, *Numer. Math.* **19**, 1–28.

Sassenfeld, H. [1951]. Ein hinreichendes Konvergenzkriterium und eine Fehlerabschätzung für die Iteration in Einzelschritten bei linearen Gleichungen, *Z. Angew. Math. Mech.* **31**, 92–94.

Sattinger, D. H. [1972]. Monotone methods in nonlinear elliptic and parabolic boundary value problems, *Indiana Univ. Math. J.* **21**, 979–1000.

Sattinger, D. H. [1973]. "Topics in Stability and Bifurcation Theory," Lecture Notes 309. Springer-Verlag, Berlin and New York.

Sawashima, I. [1964]. On spectral properties of some positive operators, *Nat. Sci. Rep. Ochanomizu Univ.* **15**, 53–64.

Schaefer, H. H. [1955]. Neue Existenzsätze in der Theorie nichtlinearer Integralgleichungen, *Berichte Verhandl. Sächsische Akad. Wiss. Leipzig Math.-Naturwiss. Kl.* **101**, 7.

Schaefer, H. H. [1960]. Some spectral properties of positive linear operators, *Pacific J. Math.* **10**, 1009–1019.

Schaefer, H. H. [1966]. "Topological Vector Spaces." Macmillan, New York.

Schaefer, H. H. [1974]. "Banach Lattices and Positive Operators." Springer-Verlag, Berlin and New York.

Schäfke, F. W. [1968]. Zum Zeilensummenkriterium, *Numer. Math.* **12**, 448–453.

Schmitt, K. [1968]. Boundary value problems for nonlinear second order differential equations, *Monatsh. Math.* **72**, 347–354.

Schmitt, K. [1970]. A nonlinear boundary value problem, *J. Differential Equations* **7**, 527–537.

Schmitt, K. [1972]. Periodic solutions of systems of second-order differential equations, *J. Differential Equations* **11**, 180–192.

Schmitt, K., and Thompson, R. [1975]. Boundary value problems for infinite systems of second-order differential equations, *J. Differential Equations* **18**, 277–295.

Schmitt, K., and Volkmann, P. [1976]. Boundary value problems for second order differential equations in convex subsets of a Banach space, *Trans. Amer. Math. Soc.* **218**, 397–405.

Schneider, H. [1953]. An inequality for latent roots applied to determinants with dominant principal diagonal, *J. London. Math. Soc.* **28**, 8–20.

Schneider, H. [1954]. The elementary divisors, associated with 0, of a singular *M*-matrix, *Proc. Edinburgh Math. Soc.* **10**, 108–122.

Schneider, H. [1965]. Positive operators and an inertia theorem, *Numer. Math.* **7**, 11–17.

Schneider, H., and Turner, R. E. L. [1972]. Positive eigenvectors of order-preserving maps, *J. Math. Anal. Appl.* **37**, 506–515.

Schrader, K. W. [1967]. Boundary-value problems for second-order ordinary differential equations, *J. Differential Equations* **3**, 403–413.

Schrader, K. W. [1968]. Solutions of second order ordinary differential equations, *J. Differential Equations* **4**, 510–518.

Schröder, J. [1956a]. Anwendung funktionalanalytischer Methoden zur numerischen Behandlung von Gleichungen, *Z. Angew. Math. Mech.* **36**, 260–261.

Schröder, J. [1956b]. Nichtlineare Majoranten beim Verfahren der schrittweisen Näherung, *Arch. Math.* **7**, 471–484.

Schröder, J. [1959]. Anwendung von Fixpunktsätzen bei der numerischen Behandlung nichtlinearer Gleichungen in halbgeordneten Räumen, *Arch. Rational Mech. Anal.* **4**, 177–192.

Schröder, J. [1961a]. Lineare Operatoren mit positiver Inversen, *Arch. Rational Mech. Anal.* **8**, 408–434.

Schröder, J. [1961b]. Fehlerabschätzung mit Rechenanlagen bei gewöhnlichen Differentialgleichungen erster Ordnung, *Numer. Math.* **3**, 39–61, 125–130.

Schröder, J. [1961c]. A unified theory for estimation and iteration in metric spaces and partially ordered spaces, Rep. 237, Mathematics Research Center, Univ. Wisconsin, Madison, Wisconsin.

Schröder, J. [1962]. Invers-monotone Operatoren, *Arch. Rational Mech. Anal.* **10**, 276–295.

Schröder, J. [1963]. Monotonie-Eigenschaften bei Differentialgleichungen, *Arch. Rational Mech. Anal.* **14**, 38–60.

Schröder, J. [1965a]. Randwertaufgaben vierter Ordnung mit positiver Greenscher Funktion, *Math. Z.* **90**, 429–440; continued *ibid.* **92** (1966), 75–94; **96** (1967), 89–110.

Schröder, J. [1965b]. Differential inequalities and error bounds, *in* "Error in Digital Computation 2" (L. B. Rall, ed.), pp. 141–179. Wiley, New York.

Schröder, J. [1966]. Operator-Ungleichungen und ihre numerische Anwendung bei Randwertaufgaben, *Numer. Math.* **9**, 149–162.

Schröder, J. [1968a]. On linear differential inequalities, *J. Math. Anal. Appl.* **22**, 188–216.

Schröder, J. [1968b]. Monotonie-Aussagen bei quasilinearen elliptischen Differentialgleichungen und anderen Problemen, *in* "Numerische Mathematik, Differentialgleichungen, Approximationstheorie" (L. Collatz, G. Meinardus, and H. Unger, eds.), pp. 341–361, Birkhäuser Verlag, Basel and Stuttgart.

Schröder, J. [1970a]. Range–domain implications for linear operators, *SIAM J. Appl. Math.* **19**, 235–242.

Schröder, J. [1970b]. Proving inverse-positivity of linear operators by reduction, *Numer. Math.* **15**, 100–108.

Schröder, J. [1971]. Range–domain implications for concave operators, *J. Math. Anal. Appl.* **33**, 1–15.

Schröder, J. [1972a]. Einschließungsaussagen bei Differentialgleichungen, *in* "Numerische Lösung nichtlinearer partieller Differential- und Integrodifferentialgleichungen" (R. Ansorge and W. Törnig, eds.), pp. 23–49. Springer-Verlag, Berlin and New York.

Schröder, J. [1972b]. Duality in linear range–domain implications, *in* "Inequalities III" (O. Sisha, ed.), pp. 321–332, Academic Press, New York.

Schröder, J. [1975a]. Inverse-positive linear operators, Rep. 75-4 and 75-5, Math. Inst., Univ. Köln.

Schröder, J. [1975b]. Upper and lower bounds for solutions of generalized two-point boundary value problems, *Numer. Math.* **23**, 433–457.

Schröder, J. [1977a]. Inclusion statements for operator equations by a continuity principle, *Manuscripta Math.* **21**, 135–171.

Schröder, J. [1977b]. Inverse-monotone nonlinear differential operators of the second order, *Proc. Roy. Soc. Edinburgh* **79A**, 193–226.

Schröder, J. [1977c]. Pointwise norm bounds for systems of ordinary differential equations, Report 77-14, Math. Inst., Univ. of Köln; partly published in *J. Math. Anal. Appl.* **70** (1979), 10–32.

Schröder, J. [1978]. *M*-matrices and generalizations using an operator theory approach, *SIAM Rev.* **20**, 213–244.

Schröder, J. [1980]. Estimates for solutions of boundary value problems by means of positive homogeneous functions, Rep. 80-01, Math. Inst., Univ. of Köln.

Schülgen, C. [1972]. Irreduzible und primitive Operatoren in geordneten Banachräumen, Diplomarbeit, Math. Inst., Univ. of Köln.

Serrin, J. [1970]. On the strong maximum principle for quasilinear second order differential inequalities, *J. Functional Anal.* **5**, 184–193.

Shampine, L. F. [1966]. Monotone iterations and two-sided convergence, *SIAM J. Numer. Anal.* **3**, 607–615.

Shampine, L. F. [1968]. Boundary value problems for ordinary differential equations, *SIAM J. Numer. Anal.* **5**, 219–242.

Simader, C. G. [1972]. "On Dirichlet's Boundary Value Problem." Springer-Verlag, Berlin and New York.

Söllner, K. [1975]. Konvergenzaussagen bei Differenzenverfahren für elliptische Differentialgleichungen, Diplomarbeit, Math. Inst., Univ. of Köln.

Spreuer, H., Adams, E., and Srivastava, U.N. [1975]. Monotone Schrankenfolgen für gewöhnliche Randwertaufgaben bei schwach gekoppelten nichtlinearen Systemen, *Z. Angew. Math. Mech.* **55**, 211–218.

Stakgold, I. [1974]. Global estimates for nonlinear reaction and diffusion, *in* "Ordinary and Partial Differential Equations" (B. D. Sleeman and I. M. Michael, eds.), pp. 252–266. Springer-Verlag, Berlin and New York.

Stampacchia, G. [1966]. "Equations Elliptiques du Second ordre à Coefficients Discontinus." Les presses de l'Univ. de Montréal.

Stečenko, V. Ja. [1966]. Criteria of irreducibility of linear operators (Russian), *Usp. Mat. Nauk.* **21**, No. 5, 265–267.

Stein, P., and Rosenberg, R. L. [1948]. On the solution of linear simultaneous equations by iteration, *J. London Math. Soc.* **23**, 111–118.

Steinmetz, W. J. [1974]. On a nonlinear singular perturbation boundary value problem in gas lubrication theory, *SIAM J. Appl. Math.* **26**, 816–827.

Stieltjes, T. J. [1887]. Sur les racines de l'équation $X_n = 0$, *Acta Math.* **9**, 385–400.

Stoer, J., and Bulirsch, R. [1973]. "Einführung in die Numerische Mathematik II." Springer-Verlag, Berlin and New York (2nd ed., 1978).

Stoss, H. J. [1970]. Monotonie-Eigenschaften bei Differentialungleichungen mit nichtkompaktem Grundbereich, *Numer. Math.* **15**, 61–73.

Stüben, K. [1975]. Einschließungsaussagen bei Differentialgleichungen mit periodischen Randbedingungen, Rep. 75–6, Math. Inst., Univ. of Köln.

Swanson, C. A. [1968]. "Comparison and Oscillation Theory of Linear Differential Equations." Academic Press, New York.

Szarski, J. [1947]. Sur un système d'inégalités differentielles, *Ann. Soc. Polon. Math.* **20**, 126–134.

Szarski, J. [1965]. "Differential Inequalities." Monografic Matematyczne, Tom 43, Warszawa.

Temple, G. [1929]. The computation of characteristic numbers and characteristic functions, *Proc. London Math. Soc.* (2) **29**, 257–280.

Thompson, R. C. [1975]. Differential inequalities for infinite second order systems and an application to the method of lines, *J. Differential Equations* **17**, 421–434.

Thompson, R. C. [1977]. An invariance property of solutions to second order differential inequalities in ordered Banach spaces, *SIAM J. Math. Anal.* **8**, 592–603.

Trottenberg, U. [1974]. An elementary calculus for Green's functions associated with singular boundary value problems, *in* "Ordinary and Partial Differential Equations," *Proc. Conf., Dundee*, pp. 434–438. Springer-Verlag, Berlin and New York.

Trottenberg, U. [1975]. Aufspaltungen linearer gewöhnlicher Differentialoperatoren und der zugehörigen Greenschen Funktionen, *Manuscripta Math.* **15**, 289–308.

Trottenberg, U., and Winter, G. [1979]. On faces and W-irreducible operators in finite-dimensional linear spaces, *Linear Alg. Appl.*, to appear.

Trudinger, N. S. [1967]. On Harnack type inequalities and their application to quasilinear elliptic equations, *Commun. Pure Appl. Math.* **20**, 721–747.

Vainberg, M. M. [1964]. "Variational Methods for the Study of Nonlinear Operators." Holden-Day, San Francisco, California.

Vandergraft, J. S. [1968]. Spectral properties of matrices which have invariant cones, *SIAM J. Appl. Math.* **16**, 1208–1222.

Vandergraft, J. S. [1976]. A note on irreducibility for linear operators on partially ordered finite dimensional vector spaces, *Linear Alg. Appl.* **13**, 139–146.

Varga, R. S. [1960]. Factorization and normalized iterative methods, *in* "Boundary Problems in Differential Equations" (R. E. Langer, ed.), pp. 121–142. Univ. Wisconsin Press, Madison, Wisconsin.

Varga, R. S. [1962]. "Matrix Iterative Analysis." Prentice Hall, Englewood Cliffs, New Jersey.

Varga, R. S. [1976]. On recurring theorems on diagonal dominance, *Linear Alg. Appl.* **13**, 1–9.

Volkmann, P. [1972]. Gewöhnliche Differentialungleichungen mit quasimonoton wachsenden Funktionen in topologischen Vektorräumen, *Math. Z.* **127**, 157–164.

Volkmann, P. [1973]. Über die Invarianz konvexer Mengen und Differentialungleichungen in einem normierten Raume, *Math. Ann.* **203**, 201–210.

Volkmann, P. [1974]. Gewöhnliche Differentialgleichungen mit quasimonoton wachsenden Funktionen in Banachräumen, *in* "Ordinary and Partial Differential Equations," *Proc. Conf., Dundee* pp. 439–443. Springer-Verlag, Berlin and New York.

Volkmann, P. [1976]. Über die positive Invarianz einer abgeschlossenen Teilmenge eines Banachschen Raumes bezüglich der Differentialgleichung $u' = f(t, u)$, *J. Reine Angew. Math.* **285**, 59–65.

von Mises, R., and Pollaczek-Geiringer, H. [1929]. Praktische Verfahren der Gleichungsauflösung. *Z. Angew. Math. Mech.* **9**, 58–77, 152–164.

Voß, H. [1977]. Existenz und Einschließung positiver Lösungen von superlinearen Integralgleichungen und Randwertaufgaben. Preprint 77/2, Inst. Angew. Math., Univ. of Hamburg.

Wake, G. C. [1976]. A functional boundary value problem, *Math. Cronicle* **4**, 163–170.

Walter, W. [1967]. Bemerkungen zu Iterationsverfahren bei linearen Gleichungssystemen, *Numer. Math.* **10**, 80–85.

Walter, W. [1969]. Gewöhnliche Differential-Ungleichungen im Banachraum, *Arch. Math.* **20**, 36–47.

Walter, W. [1970]. "Differential and Integral Inequalities." Springer-Verlag, Berlin and New York.

Walter, W. [1971]. Ordinary differential inequalities in ordered Banach spaces, *J. Differential Equations* **9**, 253–261.

Wazewski, T. [1950]. Systèmes des équations et des inégalités différentielles ordinaires aux deuxièmes membres monotones et leurs applications, *Ann. Soc. Polon. Math.* **23**, 112–166.

Werner, B. [1975]. Monotonie und finite Elemente bei elliptischen Differentialgleichungen, *in* "Numerische Behandlung von Differentialgleichungen" (A. Ansorge, L. Collatz, G. Hämmerlin, and W. Törnig, eds.), pp. 309–329, Birkhäuser, Basel.

Werner, J. [1969]. Einschließungssätze bei nichtlinearen gewöhnlichen Randwertaufgaben und erzwungenen Schwingungen, *Numer. Math.* **13**, 24–38.

Werner, J. [1970]. Einschließungssätze für periodische Lösungen der Liénardschen Differentialgleichung, *Computing* **5**, 246–252.

Wielandt, H. [1950]. Unzerlegbare, nicht negative Matrizen, *Math. Z.* **52**, 642–648.

Willson, A. N. Jr [1971]. A useful generalization of the P_0 matrix concept, *Numer. Math.* **17**, 62–70.

Willson, A. N. Jr. [1973]. A note on pairs of matrices and matrices of monotone kind, *SIAM J. Numer. Anal.* **10**, 618–622.

Witsch, K. [1978]. Projektive Newton-Verfahren und Anwendungen auf nichtlineare Randwertaufgaben, *Numer. Math.* **31**, 209–230.

Wong, Y. K. [1955]. Some properties of the proper values of a matrix, *Proc. Amer. Math. Soc.* **6**, 891–899.

Young, D. M. [1971]. "Iterative Solution of Large Linear Systems," Academic Press, New York.

Zarantonello, E. H. [1960]. Solving functional equations by contractive averaging, MRC Rep. 160, Madison, Wisconsin.

Zurawski, W. [1975]. Beweis der Inverspositivität von gewöhnlichen Differentialgleichungen vierter Ordnung mit Hilfe der Theorie der reproduzierenden Kerne, Diplomarbeit, Math. Inst., Univ. Köln.

Notation index

$\leqslant, <, \geqslant, >$ 2,3

$\dot{\leqslant}, \dot{<}, \dot{\geqslant}, \dot{>}$ 2,7,238

$\preccurlyeq, \prec, \succcurlyeq, \succ$ 5

$\succ_p$ 21

$\not\leqslant, \not<$, etc., negation of signs above

$(R, \leqslant)$ 2

$(R, \leqslant) \subset (S, \leqslant)$ 2

$(R, K), (R, K, \leqslant)$ 4

$\| \|_z$ 11,12,13,38

$\| \|_0$ 182

$\| \|_1$ 39,182

$\| \|_2$ 299

$\| \|_\infty$ xvi ,39

$[u, v], (u, v]$, etc. 8

$[u, v]_R$ 8

$[u, v]_m$ xvi

$[u : v], (u : v]$, etc. 8

$\langle y, \eta \rangle$ 299

$\langle u, v \rangle_0, \langle u, v \rangle_1$ 182

$\langle u, v, w \rangle$ 199,288

$u^\sharp$ 200,281

u^+, u^- 186

$|A|, A^+, A^-$, for matrix A 38

A^T, for matrix A 39

C^*, C_H, for matrix C 299

$A^c, \partial A, \overline{A}$, for set A 14

$A_X^c, \partial_X A, \overline{A}_X$, for set A 14

$A^i, \delta A, \bar{A}$, for set A 14

K_s, for cone K 5

R^*, for space R 9

$\kappa(M, N)$ 106

$\rho(T)$ 23

$\sigma(A, B)$ 49

$\sigma(C)$ 299

$\sigma(T)$ 113

$\tau(C)$ 299

ω 199

B_p, set 21

$C_m, C'_m, C_{j,m}, C'_{j,m}, C_m^n$, etc., function spaces xvi

D_m, function space xvi

D^+, D^-, derivatives 165

e, function, vector xvi

H_1 181
I_n 38
$L[R, S]$ 10
L_s 63
M 39
M_s 63
o 3
O 6,10
P, set 302
$P(v)$, set 304
Q, set 302
$\mathbb{R}[0, 1]$, $\mathbb{R}_b[0, 1]$, $\mathbb{R}^n[0, 1]$, etc. xvi
$\mathbb{R}^{n,m}$ 6
R_M 62
Z 39,105
Z(W) 105

Subject index

Definitions are indicated by italic page numbers.

A

Absorption, method of
  abstract formulation, 237
  applications, 66, 178–180, 213–216, 253
Admissible family
  of elements, *238*
  of functions, *166*
  of sets, *236*
Admissible starting vectors for iteration, *60*, 152
Algebraically closed set, *14*, 26
  equivalent properties, 15
Algebraic boundary of set, *14*, 26
  equivalent definition for convex set, *15*
Algebraic closure of set, *14*
Algebraic interior of set, *14*, *see also* Core of set
Antitone operator, *2*
Archimedian ordered linear space, *6*
Assumptions (A_1), (A_2), *30*

B

Bifurcation problems, *see* Eigenvalue problems, nonlinear
Boundary maximum principle, 67, 81, 84, 173
Boundary value problems, existence and estimation of solutions; *see also* Generalized boundary value problems; Range–domain implications for differential operators
  norm bounds, pointwise, for second order problems, 323–332, 341
  two-sided bounds for second order problems
    scalar-valued, 198–220, 226–228, 230
    vector-valued, 287–295, 297–299, 340–341
    various other problems
      of mixed order, 298
      of third order, 296–297
    various other estimates, 341, 342

C

Caratheodory condition, *181*
Collocation Newton method, *226*, 232
Comparison functions, *199*
  weak, *220*
Conditions, S_l, S_m, S_r, *165*, *166*
Cone, *4*, *see also* Wedge
  described by functionals, *9*
    completely described, *9*
  dual, *10*
  positive, *4*
Continuation theorem, 35
Convex set
  described by functionals, *17*
    completely described, *17*
  extremal subset of, *16*
    relation to strong order, 20
  nonsupport point of, *18*
  support point of, *18*
Core of set, *14*
  equivalent definition for convex set, 15
  relation to topological interior, 16, 20
Coupled, *see* Differential operator; Function

D

Degree of mapping method for existence proofs, 217–219, 278, 341
Difference methods, convergence proof by majorant method
  for abstract equations, 122–125, 129–130
  for boundary value problems, 125–129, 130–132
    by method of z-splitting, 130–131
    by perturbation method, 131–132
Differential operator
  coupled
    strongly, *277*
    weakly, *248*, *277*
  of divergence form, *174*
  formally self-adjoint, *63*
  inverse-monotone, *see* Operator, abstract
  inverse-positive, *63*
  positive definite, *74*
  pre-inverse-monotone, *see* Operator, abstract
  pre-inverse-positive, *64*
  monotone definite, *188*, 229
  in normed linear space, 250–251, 254–257, 341, 342
  quasi-antitone, *277*, *see also* Operator, abstract
  quasi-monotone, *see* Operator, abstract
  regular, *63*
Disconjugate, *see* Nonoscillation
Distance condition in proving flow invariance, *273*
Dual conditions in inverse-positivity, 148–149

E

Eigenelements, eigenfunctions, eigenvectors
  positive, existence of, 23, 106, 154, 157
  strictly positive, existence of, 36–37, 41, 45, 52–53, 83, 89, 108, 117, 154
Eigenvalue problems, nonlinear, 174, 179, 201–202, 206, 215–216, 231, 283, 289, 292–293, 296–297, 298–299, 314–316
Eigenvalues
  comparison of, 52–53, 155
  estimation of, 51–52, 90, 107–108, 113
  positive, existence of, 53, 106, 117, 154
Error estimation, numerical
  for boundary value problems, 138–144, 226–228, 314–316
  for initial value problems, 134–138, 224–226, 285–286
  principle of, 132, 224
Extremal, *see* Convex set

F

Face of cone, 155
Fixed-point theorem of Schauder, 23
Flow-invariance, *273*
Fourth order differential operators
  inverse-positivity, 99–104, 144–148
  norm-estimate for special case, 311
Function
  coupled, *277*
  M-function, *247*
  not-coupled, *277*
  off-diagonally antitone, *244*, *247*
  quasi-antitone, *247*, *277*
  Z-function, *244*
Functional, *see also* Functionals, set of
  concave, *17*
  extended, *17*
  point, *9*
  positive, *9*

continuity of, 13
supporting, *18*
Functional-differential operators, 248–251, 279–281, 287–295, 298–299, 305
Functionals, set of
describing cone, *9*, 22
completely, *9*
describing convex set, *17*
completely, *17*

G

Generalized boundary operator, 91–92
Generalized boundary value problems
linear, existence of solution, 183–184
nonlinear, existence and estimation of solution, 220–223
Generalized inverse-monotonicity of second order differential operators, *181*
basic results, 186, 189
for monotone definite operators, 188
for noncompact intervals, 188
Generalized inverse-positivity of second order differential operators, *182*
stronger type of, 195
sufficient conditions,
by majorizing functions, 190, 195
by positive definiteness, 187
Green function
estimation of, 191–192, 193
generalized, formal, 156
positivity of, 86, 100, 153, 156, 157, 185, 198
total positivity of, 156

H

H-matrix, 152

I

Infinite matrices, examples, 114, 119
Initial value problems, vector-valued, estimations for, *see also* Inverse-monotonicity of first order differential operators
Lyapunov functions, estimation by, 335–340
stability, results on, 336–339, 339–340
norm bounds, pointwise, 316–323
stability, results on, 319–323
two-sided bounds, 285, 295
various other estimates, 333–335
Integral operators, inverse-positivity, 114–115, 120
Internal set, *see* Core of set
Interval arithmetic, description, 8
Inverse-monotone abstract operator, *160*, *see also* inverse-monotonicity of specific operators
general description, 237
main results, 239
for order-quasilinear operators, 242
Inverse-monotonicity of first order differential operators
in normed linear spaces, 254–256
scalar-valued operators, 67, 171, 173, 251–254, 257, 259
sequence-valued operators, 256–257, 259
Inverse-monotonicity of nonlinear functions, 245–248
Inverse-monotonicity of second order differential operators, *see also* Generalized inverse-monotonicity of second order differential operators
basic results, 163, 167
under constraints, 174
divergence form, for operators of, 174–176
for noncompact intervals, 178–180, 229
sufficient conditions
by absorption method, 178–180
by generalized inverse-monotonicity, 194
by transformation, 161, 229
for vector-valued operators, 248–249
Inverse-positive abstract operator, *30*, *see also* inverse-positivity of specific operators
basic results, 31–32
for cones with empty interior, 118–121
dual conditions, 148–149
$M(W)$-operator, *105*
noninvertible operators, relation to, 36–37
sets of operators, 34–36
Inverse-positivity of matrices, *39*
basic results, 39–40
for M-matrices, 42
conditions for
by eigenvalue theory, 50
by positive definiteness, 45
by reduction method, 97, 129, 130–131

determinant theory, 47, 151
strict, *39*
Inverse-positivity of second order differential operators; *see also* Generalized inverse-positivity of second order differential operators; Inverse-monotonicity of second order differential operators
basic theorem, 64
conditions for
by absorption method, 66, 93–94
by construction of majorizing functions, 68–71
by eigenvalue theory, 88, 157
by oscillation theory, 87
by positive definiteness, 75, 157, 187
under constraints, 76–78
local, 70–71
for noncompact intervals, 90–94
noninvertible operators, relation to, 83
for sets of operators, 71–74
strict, 81–85
stronger types of, 78–85
strongest result on, for regular operators, 83
for vector-valued operators, 109–112
strict, 110
Irreducibility of matrices and operators, *39*, *116*
conditions for, 39, 49, 117–118
W-irreducibility, *116*, 117, 155
Iterative procedures
abstract equations, 24, 26, 27, 115–116, 231
boundary value problems, 219–220
matrix equations, 54–61, 152

L

Leontief matrix, *153*
Leray–Schauder theorem, 23, *see also* Degree of mapping method for existence proofs
Limit functions for fourth order differential operators, *101*, *146*
Line method for parabolic problems, 259, 273, 341
Lower solution, *199*
Lyapunov functions, 335–340, 341–342
Lyapunov theorem on matrices, 149

M

Majorant, quasi-antitone, 286–287
Majorizing element, *31*
strictly, *32*
weakly, *32*
Majorizing family of elements, *238*
of functions, *166*
of sets, defined by intervals, 251–253, 334
Majorizing function, *64*
construction of, general results, 68–71
Matrix
diagonally dominant, *39*
weakly, *39*
diagonally positive, *39*
strictly, *39*
inverse-positive, *39*
strictly, *39*
irreducible, *39*, *see also* Irreducibility of matrices and operators
M-matrix, *39*
off-diagonally negative, *39*
positive, *38*
strictly, *38*
of positive types, 151
pre-inverse-positive, *31*, 40
strictly, *32*, 42
weakly, *32*, 42
reducible, *39*
Z-matrix, *39*
Maximal solution, *219*, *258*
Maximum principle, *see* Boundary maximum principle
M-function, *247*
Minimal solution, *219*, *258*
Minkowski determinant, 151
M-matrix, *39*
M_0-matrix, *45*
Modification method for existence proofs
application, 200–216, 230, 286–299, 324–332, 342
description, 159, 235
Monotone definite operator, *188*, 229
Monotone operator, *2*, 229
Monotonicity theorem on inverse-positive linear operators, 31
M-operator, *105*
M(*W*)-operator, *105*

N

Nagumo condition, *211*, 216–217, 230
  one-sided, 209, *211*, 212, 230, 293–295
Natural order, 3, 4
Noninvertible operators, *see* Eigenelements, eigenfunctions, eigenvectors; Eigenvalues
Nonoscillation, interval of, *87*, 88, 156
Normal form, real, of a matrix, 300–301

O

o-continuous, *12*
o-convergence, *12*
o-limit, *12*
o-monotone, 229
Operator, abstract
  antitone, *2*
  (C, K)-inverse, *30*, *234*
  compact, results on, 22–23
  inverse, *23*
  inverse-monotone, *160*
  inverse-positive, *30*
    strictly, *30*
  invertible, *23*
  irreducible, *116*, 117–118
  $\mathscr{L}$-quasi-antitone, *109*, 243
  $\mathscr{L}$-quasilinear, *242*
  monotone, *2*, 229
  monotone definite, *188*, 299
  monotonically decomposable, *26*
    linearly, *10*
  of monotonic kind, 150
  M-operator, *105*
  M(W)-operator, *105*
  order monotone, 229
  order-quasilinear, *242*
  positive, *10*
    boundedness of, 13
  pre-inverse-monotone, *242*
  pre-inverse-positive, *31*
    strictly, *32*
    weakly, *32*
    with respect to a strong order, *32*
  quasi-antitone, *243*
  quasi-interior, 154
  quasi-monotone, *243*
  seminonsupport, 154
  strongly positive, 154
  u_0-positive, 154
  W-irreducible, *116*, 117–118
  W-pre-inverse-positive, *32*
  (W_1, W_2)-connected, *155*
  Z-operator, *105*
  Z(W)-operator, *105*
Order, *2*, *see also* ordered spaces
  Archimedian, *6*
    equivalent conditions, 6, 19
  bounded set, *25*
  cone, *4*
  continuous, 12
  convergence, *12*
  convex, *24*
  interval, *8*
  lexicographic, *3*
  limit, *12*
  linear, *3*
  natural, *3*, *4*
    strict, *3*
  relation, *2*
  strict, *5*
  strong, *7*
    relation to extremal set, 20
  stronger than, *2*, *7*
  topology induced by $\| \ \|_z$, 12
  weaker than, *2*, *7*
Ordered Banach space, *23*
  normal, *25*
Ordered limit space, *24*
Ordered linear space, *3*
  Archimedian, *6*
Ordered space, *2*
Oscillation, 86–88

P

Partial differential equations, 153, 229, 231–232, 259
p-bounded below set, *21*
Periodic boundary conditions, 77–78, 177–178, 230, 316, 341
Perron–Frobenius theory, 53, 154; *see also* Eigenelements, eigenfunctions, eigenvectors; Eigenvalues
Positive, *see* specific terms
Pre-inverse-monotone; *see* Operator, abstract

Pre-inverse-positive; *see* Operator, abstract
Preorder, *8*
Property IM*, IP*, *182*
p-strictly positive element, *21*

Q

Quasi-antitone, *see* Operator, abstract; Function; Differential operator
Quasi-monotone operator, *243*

R

Range–domain implications
  for abstract operators, *233*
    continuity principle, 235–237
    for linear operators, 262–264
    two-sided bounds, 259–261, 264
Range–domain implications for differential operators
  norm bounds, pointwise
    for Dirichlet boundary terms, 301–312
    for first order operators, 316–317
    for periodic boundary conditions, 316
    for Sturm–Liouville boundary terms, 312–316
  two-sided bounds
    comparison with norm bounds, 282–283, 304–305
    for non-inverse-positive operators, 264–272, 274
    for vector-valued operators
      of first order, 285–286, 287
      of second order, 279–285, 286, 287
  various other estimates
    by combined bounds, 333–335
    by Lyapunov functions, 335–340, 341, 342
Reduction method for proving inverse-positivity
  fourth order differential operators, application to, 101–104
  general description, 95–96
  matrices, application to, 97–98, 129, 131
Regular splitting of matrix, *55*
Reproducing kernel, 208
Residual set of operator, *23*

S

Segment, *8*
Segmentally closed set, *14*
  equivalent properties for convex set, 15
  equivalent property for cone, 19
Separation theorem, 18
Singular differential operators, singular boundary value problems,
  on noncompact intervals, 90–94, 178–179, 180, 188, 213–216, 244–245, 296–297, 299, 340
  with singular coefficients, 212–213, 283–285, 297–298, 314–316
Singular perturbation problems, 164, 168, 201, 211, 227–228, 231
Smoothness conditions, relaxing of, 161, 165, 166, 200, 202, 206, 254, 278, *see also* Conditions S_l, S_m, S_r
Spectral radius of operator, *23*
Spectrum of operator, *23*
Splitting of differential operator, 156
Stability of difference method, 125
Stability for initial value problems, 319–323, 336–339, 339–340
Stein–Rosenberg theorem, 152, 154
Stieltjes matrix, 152
Strictly positive element, *5*
  equivalent properties, 19, 20
Strictly positive set, *119*
Sturmian oscillation theorem, 87

U

Uniqueness function, *254*
Upper solution, *199*

V

Variational problems, estimation of solution, 185

W

Wedge, *8*
  dual, *10*

W-pre-inverse-positive, *32*

X

X-closed, *14*
X-closure, *14*

Z

Z-function, *244*
Z-matrix, *39*
Z-operator, *105*
Z-pair of operators, *105*
Z(*W*)-operator, *105*
Z(*W*)-pair of operators, *105*